ROBUST SYSTEMS
THEORY AND APPLICATIONS

Adaptive and Learning Systems for Signal Processing, Communications, and Control

Editor: Simon Haykin

Werbos / THE ROOTS OF BACKPROPAGATION: From Ordered Derivatives to Neural Networks and Political Forecasting

Krstić, Kanellakopoulos, and Kokotović / NONLINEAR AND ADAPTIVE CONTROL DESIGN

Nikias and Shao / SIGNAL PROCESSING WITH ALPHA-STABLE DISTRIBUTIONS AND APPLICATIONS

Diamantaras and Kung / PRINCIPAL COMPONENT NEURAL NETWORKS: THEORY AND APPLICATIONS

Tao and Kokotović / ADAPTIVE CONTROL OF SYSTEMS WITH ACTUATOR AND SENSOR NONLINEARITIES

Tsoukalas / FUZZY AND NEURAL APPROACHES IN ENGINEERING

Hrycej / NEUROCONTROL: TOWARDS AN INDUSTRIAL CONTROL METHODOLOGY

Beckerman / ADAPTIVE COOPERATIVE SYSTEMS

Cherkassky and Mulier / LEARNING FROM DATA: CONCEPTS, THEORY, AND METHODS

Passino and Burgess / STABILITY OF DISCRETE EVENT SYSTEMS

Sánchez-Peña and Sznaier / ROBUST SYSTEMS THEORY AND APPLICATIONS

Vapnik / STATISTICAL LEARNING THEORY

ROBUST SYSTEMS THEORY AND APPLICATIONS

Ricardo S. Sánchez-Peña

Mario Sznaier

A Wiley-Interscience Publication
JOHN WILEY & SONS, INC.
New York / Chichester / Weinheim / Brisbane / Singapore / Toronto

This book is printed on acid-free paper. ⊚

Copyright © 1998 by John Wiley & Sons, Inc. All rights reserved.

Published simultaneously in Canada.

No part of this publication may be reproduced, stored in a retrieval system or transmitted in any form or by any means, electronic, mechanical, photocopying, recording, scanning or otherwise, except as permitted under Section 107 or 108 of the 1976 United States Copyright Act, without either the prior written permission of the Publisher, or authorization through payment of the appropriate per-copy fee to the Copyright Clearance Center, 222 Rosewood Drive, Danvers, MA 01923, (978) 750-8400, fax (978) 750-4744. Requests to the Publisher for permission should be addressed to the Permissions Department, John Wiley & Sons, Inc., 605 Third Avenue, New York, NY 10158-0012, (212) 850-6011, fax (212) 850-6008, E-Mail: PERMREQ@WILEY.COM.

Library of Congress Cataloging-in-Publication Data:
Sánchez-Peña, Ricardo
   Robust systems theory and applications / Ricardo Sánchez-Peña,
   Mario Sznaier.
      p.   cm.
   Includes bibliographical references and index.
   ISBN 0-471-17627-3
   1. Automatic control. 2. Control theory. 3. System
identification.   I. Sznaier, Mario. II. Title.
TJ213.S11544 1998
629.8--dc21   98-6420
                                                               CIP

Printed in the United States of America
10 9 8 7 6 5 4 3 2 1

To our parents
Lolita and Miguel, Rebeca and Marcos

CONTENTS

1 Introduction   1
- 1.1 General Control Problem / 1
  - 1.1.1 Experimental Phase / 2
  - 1.1.2 Simulation Phase / 3
  - 1.1.3 Theoretical Phase / 5
- 1.2 Why Feedback? / 11
- 1.3 Feedback Loop Trade-off / 14
- 1.4 Objectives of an Applied Theory / 17
- 1.5 Objectives of This Book / 19
  - 1.5.1 General Remarks / 19
  - 1.5.2 Scope / 19
  - 1.5.3 How to Use This Book / 21

2 SISO Systems   23
- 2.1 Introduction / 23
- 2.2 Well Posedness / 24
- 2.3 Nominal Internal Stability / 26
- 2.4 Robust Stability / 29
  - 2.4.1 Phase and Gain Margins / 29
  - 2.4.2 Global Dynamic Uncertainty / 32
- 2.5 Nominal Performance / 42
  - 2.5.1 Known Disturbance/Noise/Reference / 42
  - 2.5.2 Bounded Disturbances at the Output / 44
  - 2.5.3 Other Performance Criteria / 47
- 2.6 Robust Performance / 50
- 2.7 Extension to MIMO Systems / 55
- 2.8 Problems / 57

3 Stabilization — 61

3.1 Introduction / 61
3.2 Well Posedness and Internal Stability / 63
    3.2.1 Well Posedness / 63
    3.2.2 Internal Stability / 64
3.3 Open-Loop Stable Plants / 65
3.4 The General Case / 68
    3.4.1 Special Problems / 69
    3.4.2 The Output Feedback Case / 75
3.5 Controller Structure and Separation Principle / 78
3.6 Closed-Loop Mappings / 79
3.7 A Coprime Factorization Approach / 80
    3.7.1 Coprime Factorizations / 80
3.8 LFTs and Stability / 88
3.9 Problems / 90

4 Loop Shaping — 93

4.1 Introduction / 93
4.2 Nominal Performance / 97
4.3 Robust Stability / 98
4.4 Nominal Performance and Robust Stability / 99
4.5 Robust Performance / 100
    4.5.1 Sensor Uncertainty / 101
    4.5.2 Actuator Uncertainty / 102
4.6 Design Procedure / 105
4.7 Examples / 109
    4.7.1 Permanent Magnet Stepper Motor / 109
    4.7.2 Loop Shaping $Q(s)$ / 119
4.8 Related Design Procedures / 123
4.9 Problems / 125

5 $\mathcal{H}_2$ Optimal Control — 127

5.1 Introduction / 127
5.2 The Classical Linear Quadratic Regulator Problem / 128
5.3 The Standard $\mathcal{H}_2$ Problem / 133
5.4 Relaxing Some of the Assumptions / 141
5.5 Closed-Loop Properties / 142
    5.5.1 The LQR Case: Kalman's Inequality / 143

          5.5.2 Some Consequences of Kalman's Inequality / 145
          5.5.3 Stability Margins of Optimal $\mathcal{H}_2$ Controllers / 146
    5.6 Related Problems / 149
    5.7 Problems / 152

6 $\mathcal{H}_\infty$ Control   157

    6.1 Introduction / 157
    6.2 The Standard $\mathcal{H}_\infty$ Problem / 158
          6.2.1 Background: Hankel and Mixed Hankel–Toeplitz Operators / 161
          6.2.2 Proof of Theorem 6.1 / 176
    6.3 Relaxing Some of the Assumptions / 179
    6.4 LMI Approach to $\mathcal{H}_\infty$ Control / 180
          6.4.1 Characterization of All Output Feedback $\mathcal{H}_\infty$ Controllers / 183
          6.4.2 Connections with the DGKF Results / 188
    6.5 Limiting Behavior / 192
    6.6 The Youla Parametrization Approach / 193
    6.7 Problems / 203

7 Structured Uncertainty   207

    7.1 Introduction / 207
          7.1.1 Stability Margin / 210
    7.2 Structured Dynamic Uncertainty / 215
          7.2.1 Computation / 215
          7.2.2 Analysis and Design / 219
    7.3 Parametric Uncertainty / 223
          7.3.1 Introduction / 223
          7.3.2 Research Directions / 225
          7.3.3 Kharitonov's Theorem / 230
          7.3.4 Mapping Theorem / 233
    7.4 Mixed Type Uncertainty / 237
          7.4.1 Introduction / 237
          7.4.2 Mixed μ / 239
    7.5 Problems / 242

8 $\mathcal{L}^1$ Control   245

    8.1 Introduction / 245

8.2 Robust Stability Revisited / 247
    8.2.1 Robust Stability Under LTV Perturbations / 250
    8.2.2 Stability Under Time Invariant Perturbations / 253
8.3 A Solution to the SISO ℓ^1 Control Problem / 254
    8.3.1 Properties of the Solution / 262
    8.3.2 The MIMO Case / 266
8.4 Approximate Solutions / 267
    8.4.1 An Upper Bound of the ℓ^1 Norm / 268
    8.4.2 The $\star$ Norm / 269
    8.4.3 Full State Feedback / 273
    8.4.4 All Output Feedback Controllers for Optimal $\star$-Norm Problems / 276
8.5 The Continuous Time Case / 282
    8.5.1 Solution Via Duality / 282
    8.5.2 Rational Approximations to the Optimal $\mathcal{L}_1$ Controller / 285
8.6 Problems / 288

9 Model Order Reduction   293

9.1 Introduction / 293
9.2 Geometry of State-Space Realizations / 295
    9.2.1 Controllable/Unobservable Spaces / 295
    9.2.2 Principal Components / 297
9.3 Hankel Singular Values / 304
    9.3.1 Continuous Systems / 304
    9.3.2 Discrete Systems / 306
9.4 Model Reduction / 310
    9.4.1 Introduction / 310
    9.4.2 Hankel Operator Reduction / 310
    9.4.3 Balanced Realizations / 312
    9.4.4 Balanced Truncation / 314
9.5 Algorithms / 318
    9.5.1 Approximation Error / 319
9.6 Problems / 321

10 Robust Identification   323

10.1 Introduction / 323
10.2 General Setup / 324

- 10.2.1 Input Data / 325
- 10.2.2 Consistency / 326
- 10.2.3 Identification Error / 328
- 10.2.4 Convergence / 334
- 10.2.5 Validation / 338
- 10.3 Frequency-Domain Identification / 339
  - 10.3.1 Preliminaries / 340
  - 10.3.2 Sampling Procedure / 342
  - 10.3.3 Consistency / 344
  - 10.3.4 Identification Procedures / 346
- 10.4 Time-Domain Identification / 361
  - 10.4.1 Preliminaries / 362
  - 10.4.2 Identification Procedures / 366
- 10.5 Further Research Topics / 372
  - 10.5.1 Unstable Systems / 372
  - 10.5.2 Nonuniformly Spaced Experimental Points / 372
  - 10.5.3 Model Reduction / 373
  - 10.5.4 Continuous Time Plants / 373
  - 10.5.5 Sample Complexity / 373
  - 10.5.6 Mixed Time/Frequency Experiments / 374
  - 10.5.7 Mixed Parametric/Nonparametric Models / 374

11 Application Examples   377

- 11.1 SAC-C Attitude Control Analysis / 377
  - 11.1.1 Introduction / 377
  - 11.1.2 Linear Model / 378
  - 11.1.3 Design Constraints / 381
  - 11.1.4 Robustness Analysis / 383
  - 11.1.5 Simulations / 385
- 11.2 Controller Design for a D_2O Plant / 389
  - 11.2.1 Model of the Plant / 389
  - 11.2.2 Robustness Analysis / 391
  - 11.2.3 Controller Design / 397
- 11.3 X-29 Parametric Analysis / 400
  - 11.3.1 Linear Model / 400
  - 11.3.2 Results / 405
- 11.4 Control of a DC-to-DC Resonant Converter / 407
  - 11.4.1 Introduction / 407

xii CONTENTS

        11.4.2 The Conventional Parallel Resonant Converter / 407
        11.4.3 Small Signal Model / 409
        11.4.4 Control Objectives / 410
        11.4.5 Analysis of the Plant / 411
        11.4.6 Control Design / 414
        11.4.7 Controller Synthesis / 420
        11.4.8 Simulation Results / 421

Bibliography   427

A Mathematical Background   445

    A.1 Algebraic Structures / 445
        A.1.1 Field / 445
        A.1.2 Linear Vector Space / 446
        A.1.3 Metric, Norm, and Inner Products / 447
    A.2 Function Spaces / 449
        A.2.1 Introduction / 449
        A.2.2 Banach and Hilbert Spaces / 449
        A.2.3 Operator and Signal Spaces / 450
        A.2.4 Isomorphism / 452
        A.2.5 Induced Norms / 453
        A.2.6 Some Important Induced System Norms / 455
    A.3 Duality and Dual Spaces / 457
        A.3.1 The Dual Space / 457
        A.3.2 Minimum Norm Problems / 458
    A.4 Singular Values / 460
        A.4.1 Definition / 460
        A.4.2 Properties and Applications / 461

B System Computations   465

    B.1 Series / 465
    B.2 Change of Variables / 466
    B.3 State Feedback / 467
    B.4 State Estimation / 467
    B.5 Transpose System / 467
    B.6 Conjugate System / 467
    B.7 Addition / 468
    B.8 Output Feedback / 468

B.9   Inverse / 469
B.10 Linear Fractional Transformations / 470
B.11 Norm Computations / 475
     B.11.1 $\mathcal{H}_2$ Norm Computation / 475
     B.11.2 $\mathcal{H}_\infty$ Norm Computation / 476
     B.11.3 ℓ^1 Norm Computation / 476
B.12 Problems / 479

C Riccati Equations   481

Index   487

PREFACE

Robustness against disturbances and model uncertainty is at the heart of control practice. Indeed, in the (completely unrealistic) case where both all external disturbances and a model of the system to be controlled are exactly known, there is no need for feedback: Optimal performance can be achieved with an open loop controller.

The main ingredients of present day robust control theory were already present in the classical work of Bode and in many popular frequency domain-based design techniques. With the advent of state–space methods in the mid 1960s the issue of robustness took a backseat to other topics, but was never completely abandoned by the control community, especially control practitioners. Interest in robust control rose again in the late 1970s where it was shown that many popular control methods (including optimal LQR control and controller design based on the cascade of an observer and state feedback) led to closed-loop systems very sensitive to model perturbations. Moreover, widely accepted "ad-hoc" recipes for "improving robustness," such as using artificially large noise levels in the design, had precisely the opposite effect.

Robust control has undergone extensive developments in the past two decades, leading to powerful formalisms, such as $\mathcal{H}_\infty$, μ-synthesis/analysis and, more recently, ℓ^1 optimal control, that, coupled with newly developed control-oriented identification techniques, have been successfully applied to challenging practical problems. A salient feature of the framework is that it is oriented towards applications and thus is based on "practical," realistic assumptions.

There are many excellent books that cover specialized topics ($\mathcal{H}_2$, $\mathcal{H}_\infty$, ℓ^1, parametric uncertainty, linear matrix inequalities) with others scattered in the technical journals. Our intention in writing this book is to provide a self-contained overview of robust control that illustrates all the issues involved, ranging from the transformation of experimental signals from the physical plant to a set of models (robust identification), to the synthesis of a controller for that set of models (robust control). The purpose of the book is twofold: to serve as a textbook for courses at the Master's/beginning Ph.D. level and as a reference for control practitioners. It assumes that the reader has a background in classical and state–space control methods. In order to keep the text size at a manageable level, in some cases only basic results are covered, and the reader is referred to more specialized literature for further coverage.

In all cases we have strived to link the theory with practical applications. To this end, in addition to the examples covered throughout the book, the last chapter contains several worked out application problems that stress specific practical issues: nonlinearities, unknown time delays, infinite dimensional plants, actuator and sensor limitations. They are all extracted from our practical experience in different engineering fields. Furthermore, due to the fact that most of the problems to be solved by the theory presented here are computer-intensive, we stress the algorithmic and computational aspect of the solution along with the mathematical theory.

We (and our graduate students) have tried to eliminate obvious mistakes. However, as anyone who has tried knows, it is virtually impossible to make a "perfect" book. We encourage readers to send corrections, comments and general feedback to either one of the authors. Finally, as attested by the list of more than 300 references, we have tried to give credit where it is due. However, owing to the sheer volume of literature published on robust control, we may have inadvertently failed to do so on occasion. We apologize in advance to readers or authors who may feel that this is the case, and we encourage them to send us comments.

ACKNOWLEDGMENTS

We owe a special debt to Professor Thanasis Sideris from whom both of us learned a great deal (not only about robust control) and who was very influential in shaping our subsequent research. We would also like to acknowledge the important contribution that Professor Manfred Morari had in the development of the concepts presented in this book, and in advising Ricardo S. Sánchez Peña to make the right decisions during hard times in his career. Mario Sznaier is specially indebted to Professor Mark Damborg for his mentoring. Among Professor Damborg's many contributions, the two that stand out are attracting him to the field of control and making sure that he got a Ph.D. degree when he showed signs of enjoying life too much as a graduate student.

The reviews by Professors Peter Dorato and Roy Smith at initial stages of this book have been instrumental in encouraging the authors to carry out the project and in shaping its final form.

Preliminary versions of the book were tested in courses offered by the authors at Penn State University and the University of Buenos Aires. Feedback from students enrolled in these courses was fundamental to its fine tuning. We are also indebted to our graduate and post-doctoral students, Takeshi Amishima, Pablo Anigstein, Juanyu Bu, Tamer Inanc, Cecilia Mazzaro, Pablo Parrilo and Zi-Qin Wang, for reviewing the book in great detail and providing some of the material. In particular, Cecilia developed very efficiently the examples in Chapter 10 and Takeshi, Juanyu, and Zi-Qin contributed some of the material in Chapters 8 and 11.

The book was shaped to a large extent by interactions with our colleagues. In particular, many of the ideas and developments in Chapter 8 arose from research carried out jointly with Professor Franco Blanchini. Dr. Hector Rotstein influenced the presentation of the $\mathcal{H}_\infty$ material in Chapter 6. Professors S. P. Bhattacharyya, D. Bernstein, N. K. Bose, O. Crisalle, M. Dahleh, M. A. Dahleh, J. C. Doyle, M. Fan, C. V. Hollot, M. Khammash, A. Megretski, A. Packard, J. Boyd Pearson, A. Ray, A. Saberi, A. Stoorvogel, R. Suarez, R. Tempo and J. Vagners contributed (directly or indirectly) to this book through many discussions and comments during the years.

The first author would like to thank his colleagues at LAE, School of Engineering of the University of Buenos Aires and at the Argentine Space Agency (CONAE), especially Professors C. Godfrid and A. Perez, and Ing. Beto Alonso for important support. The financial support of CONAE, the University of Buenos Aires and Fundación Antorchas during his stays at Penn State in 1994 and 1996 are also gratefully acknowledged.

Mario Sznaier would like to thank the National Science Foundation, and in particular Dr. Kishan Baheti, Director of the Engineering Systems Program, for supporting his research program. This book would have not been possible without this support.

We are also indebted to Ms. Lisa Van Horn and Mr. George Telecki at John Wiley and Sons, Inc. for their assistance throughout this project.

Finally we would like to thank our children, Gadiel, Lucila and Pablo, and spouses, Mónica and Octavia, for their infinite patience, support, and understanding for the time this book has stolen from them.

RICARDO S. SÁNCHEZ-PEÑA
MARIO SZNAIER

Buenos Aires, Argentina
University Park, Pennsylvania

ROBUST SYSTEMS
THEORY AND APPLICATIONS

1

INTRODUCTION

1.1 GENERAL CONTROL PROBLEM

The purpose of this book is to introduce the reader to the theory of control systems, with particular emphasis on the applicability of the results to practical problems. Therefore it considers not only the *control* viewpoint but also the *identification* of systems as part of the problem. Also, as for any theory of systems oriented toward practical applications, robustness is essential and will be the underlying concept throughout the book. For these reasons the title of the book includes the following words: Theory, Applications, Robust, and Systems.

It is important to start with a *wide angle* view of the general control picture, so that we can locate ourselves in the particular approach focused by this book. For this reason, the general problem of implementing a controller, which actuates over a physical process in a prescribed way, will be divided into three well defined phases. These phases are all equally important but involve very different *technical tools*, which go from pure mathematics to microprocessor technology. In today's practical control problems, all phases should be considered and it is very difficult to have a clear knowledge of all the technical fields involved. Although in this book we concentrate on the theoretical and computational aspects of the whole problem, we first make a general overview of the disciplines involved at each stage and the interrelation among them. This is the basic objective of robust identification and control theory: to produce theoretical results that are closely related to the computational and experimental aspects of the control problem. In the next subsection we will describe each phase in detail.

1.1.1 Experimental Phase

This phase is where the control problem is generated and where the last test on the controller is performed. This means that the actual control problem to be solved comes always from a physical setup: an aircraft, a distillation plant, a robotic manipulator. The performance specifications are also formulated in terms of variables involved in this physical experience: stable dynamics at high angle of attack, constant concentration of the distilled product, fast response to commanded trajectories. Also, in many cases, once the practical problem has been formulated, an experiment should be performed to obtain a mathematical model as "close" as possible to the real physical system to be controlled. Finally, the last test to prove the effectiveness of the controller is performed on the physical system.

The experimental phase is basically technological. It involves engineering disciplines (electrical, mechanical, chemical, aeronautical, etc.) or even biology or economics, depending on what we define as the *system to be controlled*. It also involves knowledge on the sensors, actuators, and technical aspects of microprocessors and I/O devices, for computer control. We may as well include in this phase the low- and high-level programming, as in the case of communication programs in Assembler and control algorithm coding, respectively.

The interaction of the physical plant with its environment is performed by "real world" signals provided by the plant to the sensors and application of physical stimulations to the plant through the actuators, which modify its behavior. These input and output signals, as well as the physical laws that describe the system behavior, provide the only information we have of the physical plant. In many cases there are no physical laws that explain the plant behavior or they are extremely complex. A *trial and error* control, or direct experimentation using the input and output signals (without any analysis), is unrealizable, at least for the class of problems in which we are interested. A more elaborate and "scientific" procedure should (and can) be performed to control a given plant. Furthermore, in many cases trial and error would not even be practical because of costs (chemical plants) or simply to prevent accidents (aircraft, nuclear power plant).

The information on the system—input–output signals and physical laws—is never complete (to complicate things). The reason is that the physical laws, which may be applied to the (*sensor, actuator, plant*) behavior describe only a certain aspect of the problem, but not all. To make this point, consider the following example. To describe the output voltage on a given resistor through which a given current is circulating, we may use Ohm's law. Nevertheless, the temperature, gravitational field, and magnetic field could also influence the measurement, as well as the physics of the voltage sensor and current supply. A description of a particular physical process using all the many aspects involved could, to say the least, be impractical.[1] Sometimes, also for practical

---

[1] It is an epistemological question if a complete mathematical picture of physical *reality* is possible.

reasons, a simplification of the physical laws should be applied to describe the system. Actually, the laws are not modified, but instead new simplifying assumptions are added.

The above facts determine that all practical control problems have a certain amount of *uncertainty* in its more general meaning. It should be noted that this uncertainty is a consequence of the observation of the system and its environment, not due to the system itself. In particular, in the control jargon, this lack of knowledge is called *noise* when applied to the sensing procedure, *disturbance* for the external stimulations to the plant, and *uncertainty* when it applies to the description of the dynamical behavior of the plant by a mathematical model. These three elements—in particular, the latter—play a fundamental role in the orientation of control theory since the 1980s and in the motivation for Robust Identification and Control.

1.1.2 Simulation Phase

For many different reasons it is highly convenient to have a computational model of the plant. By this we mean a representation, in a computer, of the relevant inputs and outputs of the physical system and their relationship. This representation can be very elaborate as in a *hardware in the loop* scheme, where parts of the system are represented computationally and others are connected physically to the computer I/O ports.[2] Otherwise it can be a complete computational model of the plant obtained from a mathematical model that includes all relevant physical phenomena involved.

To illustrate some of the many reasons justifying a simulation of the plant, we cite the following "real world" examples, which will be described in greater detail in Chapter 11.

1. In many cases direct experimentation with the hardware is impossible, as in the case of satellites. In these situations, it is usual to start with a high-fidelity simulation, which includes models of the system (satellite), actuators (inertial wheels, gas jets, magnetic torque coils), sensors (solar and Earth magnetic field sensors), and environment (Sun Ephemeris and Earth magnetic field model along the orbit). These models include the nonlinear dynamics, flexible phenomena, and all the necessary elements to represent the system. According to required tests to be performed along the project, pieces of real hardware may be connected to this *hi-fi simulator* replacing their simulations, that is, sensors or actuators.

2. The cost of experimentation could be a strong reason to simulate a plant, like the case of many chemical processes, where the startup of

---

[2] It could be argued that in this case we have a combination of experimental and simulation phases. What we mean by simulation in this book is in general an object that is not exactly the real plant or environment. In any case, the important point is to realize that there are different phases in the control problem.

the process could involve many millions of dollars. The high costs could be a consequence of very expensive materials or long experimentation. Testing over a computer model provides a very fast and inexpensive way of verifying a controller (see D_2O plant in Chapter 11 and [228]).
3. A very important reason for simulation is when human lives are involved, as in the case of an experimental aircraft. Therefore a computer simulation gives a safe means of testing high-risk maneuvers (see NASA's X-29 experimental aircraft in Chapter 11).

The disciplines involved are related to the areas of computer science, optimization, and numerical methods. The mathematical models to be implemented by a computer code could be obtained from physical laws or from direct experimentation on the plant. In this last case, the classical Parameter Identification procedures ([184, 293]) or the very recent Robust Identification ([146]) techniques (see Chapter 10) could be used.

The main goal of the simulation phase is to have a copy as "close" as possible to the *experimental* plant. The word experimental has been emphasized to distinguish it from the *true* plant, which is not the same concept, as mentioned in the previous subsection and detailed next. For simulation, we are interested only in certain aspects of the physical system but not in all of them. Take, for example, the dynamics of the longitudinal axis of a rigid aircraft, which is a particular aspect to be considered not at all related to the thermal gradients of the wings of the same aircraft. Therefore, for simulation purposes, we only consider certain input and output signals that are relevant to our problem. Furthermore, even for a particular aspect of the problem, we may consider different levels of detail. Take the same example, where in one case we may consider the aircraft as a rigid body and in a more detailed analysis we may add the flexible modes. Therefore, for simulation purposes, we only consider the *experimental* plant, which is a particular aspect of the *true* plant, that arises in the experimental setup.

The trade-off in this phase of the problem is defined in terms of how "close" we want the computational model to represent the plant versus how "big" a computer will be needed to perform this task. In the latter, the main limitations are computation time and available memory. As the computational model describes the plant better and better, the computation time will grow, which could pose a serious limitation in many real-time applications. For example, even for reasonable size minicomputers, a complete model of the orbit, Earth magnetic field, Sun Ephemeris, gravitational and aerodynamic disturbance models, and satellite dynamics could take much more time for simulation than for real time. This is the case of low-orbit satellites with short orbital periods (SAC-C in Chapter 11 has a 90 minute period). Also, in many cases, a "good" computational model could involve a great amount of memory, like in many finite element representations of physical systems.

In any case, this trade-off determines that certain simplifying assumptions should be made when representing a physical system by a computer code.

Typical assumptions for the nonlinear model of an aircraft are rigid body dynamics and symmetry. The simplifying assumptions mentioned above, in the simulation phase, are only related to the computational constraints. Instead, for the theoretical analysis and synthesis phase, new hypothesis should be added to simplify the problem even more. This is related to the available mathematical tools to design and analyze a feedback loop, as will be seen next.

1.1.3 Theoretical Phase

Introduction In history and, in particular, in the history of science, humans have tried to solve first the simpler problems and continue with the more complicated ones. This, which seems almost a natural law of human behavior, also applies to the theory of control systems. For this reason in classical control theory, systems were first described as single-input single-output (SISO), while later on, the more general multiple-input multiple-output (MIMO) description was introduced by modern control theory through the state-space description. In the same line of thought, the optimization of a quadratic functional (LQG optimal control) appeared earlier than the $\mathcal{H}_\infty$ optimal control problem. In this latter case, the linear $\mathcal{H}_\infty$ control problem was stated ([329]) almost ten years before the first solutions ([22, 157, 309, 310]) to the $\mathcal{H}_\infty$ nonlinear control problem were produced.

At this point, there are not yet fully developed analysis and design tools to deal with control systems described by coupled nonlinear partial differential equation descriptions with (possibly) uncertain parameters. Therefore the theoretical phase of a control problem, which will be the main concern of this book, deals with a much more simplified mathematical model of the plant. Historically, the most usual system description for design and analysis of feedback systems has been finite-dimensional (lumped), linear, time invariant (FDLTI) mathematical models. For this class of models there is a well established theory, which also covers many of the most usual applications.[3]

A visual idea of all the simplifying assumptions the designer needs to make to obtain a FDLTI mathematical model from the physical system may be seen in Figure 1.1. The usual steps that should be taken to compute a FDLTI model, from a very complex mathematical representation of a particular plant, are presented there. Some applications, which will be detailed in Chapter 11, illustrate a few of the steps.

System Model At this point we will distinguish two different approaches that compute a FDLTI model from a physical system. The first one obtains a simplified version of a very complex mathematical model. This could be a set of partial differential equations or a time delayed or a high order set of ordinary differential equations. This complex mathematical model is usually

---

[3] Not because physical systems are themselves FDLTI, but because for most practical purposes they can be considered to be so.

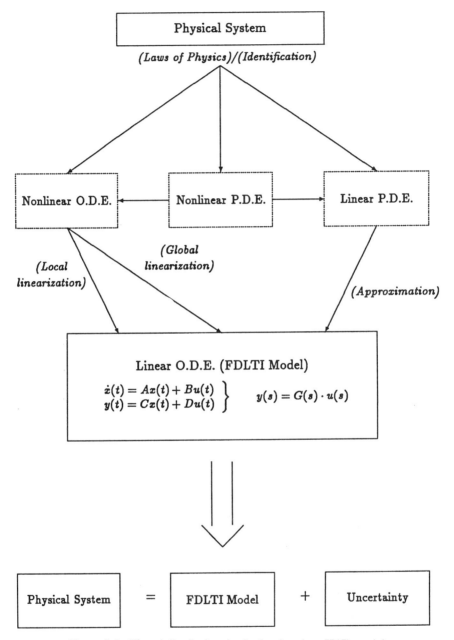

Figure 1.1. "Translation" of a physical system to a FDLTI model.

obtained from physical laws describing the system. This model simplification procedure is called *approximation* and is purely mathematical, once the model of the system, obtained from physical laws, is known.

There are a variety of methods that approximate systems described by partial differential equations (infinite-dimensional systems) by ordinary differential equations. Among others, we can mention finite elements, finite differences, averaging techniques, splines, and FFT methods ([24, 25, 139, 169]). More recently, the balanced realization and truncation of infinite-dimensional models provides simplified finite-dimensional models ([79, 134]). In the case of systems with time delays, which are also infinite dimensional, although the above methods can be used, the most usual and practical approach is to use Padé approximants.

The second approach is based on direct experimentation over the physical system and is called in general *identification*. This approach consists in computing a mathematical model, usually FDLTI, from the data measured at the system's output when a particular input signal is applied. The classical *parameter identification* procedures ([184, 293]) assume a particular mathematical structure for the model and, through an optimization method, obtain a set of parameters that minimize the "difference" between the model and the system's output, for the same applied input. More recent techniques called *robust identification* obtain directly a FDLTI model, without assuming any parameter structure, from the output data of the system when a particular set of signals are applied to the input. These latter techniques are related directly to robust control and will be described in greater detail in Chapter 10.

From Figure 1.1 we observe that sometimes we should linearize a nonlinear model of the system. The classical linearization technique consists in computing a linear version of the nonlinear model for a particular set of parameters (operating point) of the latter. This is called a local model ([189]), which is approximately valid only in a small neighborhood of the selected values of these parameters. In some applications, the parameters have fixed values and, for that reason, there is only one working point and one linear model ([268]). Otherwise, if the parameters can take different values, the designer should synthesize a linear controller for each linear model. Next, a test should be made (analytical, by simulation or experimentation) to guarantee a smooth transition from one controller to another (see Example below). This is called *gain scheduling* and is a well known procedure to control nonlinear plants. Recent results in this area attack this problem through the use of linear parameter varying (LPV) controllers ([222]).

Also, for certain classes of nonlinear models (linear analytic models), a state transformation and a nonlinear state feedback can be computed, which produce a (global) linear model ([89, 153, 156]). Although this is not a robust result,[4] it is very useful in many practical applications ([1, 2, 44, 106, 107]), for example, robotic manipulators, aircraft/spacecraft control, and motor control.

---

[4] The state-feedback law, which linearizes the model, depends on *exact* knowledge of the nonlinear model; therefore feedback linearization is not robust against uncertainty in the nonlinear model.

Disturbances The designer not only assumes certain properties of the system (e.g., linearity, time invariance) to obtain a simpler model but also needs to make certain hypotheses on the disturbances applied to the system by its environment. In classical and modern control, the disturbances were usually assumed to be certain classes of signals: steps, ramps, impulses, sinusoids. In a stochastic approach, the covariance matrix of the stochastic process representing the disturbance was assumed to be known exactly.

Very seldom does the designer have a clear knowledge of the disturbing signals acting on the system. Therefore the less one assumes about their nature, the higher the possibility of producing an effective control, despite these disturbances. As a matter of fact, a control design procedure that optimally rejects a step disturbing signal will no longer be optimal if the actual disturbance is a square wave. For this reason, in robust control, the only assumption on the disturbances is the fact that they are bounded (usually in energy), which is a very realistic hypothesis. If the designer has some extra information on the frequency distribution of the disturbances, that information could be incorporated into the design procedure and produce a less conservative result. Note that even in this case, there is no assumption on the *form* of the disturbance.

Uncertainty Due to the assumptions on the system and the disturbances, a fair amount of uncertainty between the mathematical model and the *experimental* results will appear. In the simulation phase, the simplifying hypotheses are made due to computational restrictions (time, memory). Instead, in the theoretical phase, the assumptions that generate the uncertainty are a consequence of the mathematical tools that the analyst/designer has decided to use.

We can evaluate separately the effect on the feedback loop of the uncertainty due to the modeling of the system and due to the external disturbances. From basic control courses we know that the latter has no effect on the stability of the loop, at least in linear systems,[5] although it can degrade the performance. On the other hand, the uncertainty in the model can have a critical effect on the stability of the closed loop. Specifically, we know that if the design of a stabilizing controller has been based on an erroneous plant model, the loop can be unstable. This is clear from the basics of control systems, as in the case of models with uncertain gains or phases. Take, for example, a controller designed for a model of the plant with a gain that exactly doubles the gain of the actual plant. If this model has a gain margin smaller than 2 (and the designer doesn't know it), the controller could make the loop unstable.

From the above arguments, we can conclude that the uncertainty in the model of the plant is more critical than the lack of knowledge of the disturbances, because it affects the stability of the closed-loop feedback system.

---

[5] In nonlinear systems, both the uncertainty in the model and the unknown disturbances can destabilize the closed-loop system.

Surprisingly, model uncertainty has only started to be considered quantitatively in the assumptions of control theory since the 1980s. This recent theory, which considers uncertainty as part of the problem, is called robust control and, together with robust identification, is the main subject of this book. Throughout its chapters, the following fact will be stressed: *the design of a controller should be based not only on a particular model, but also on the uncertainty this model has with respect to the actual physical plant.* Uncertainty is related directly to feedback, as will be seen in Section 1.2.

Example The three phases described above are closely related and equally important. The final solution of the control problem, from the purely speculative theoretical step to the final physical implementation of the controller of the plant, depends on all of them. Each phase has its own characteristics. Although in the experimental phase there are no approximations, it is not possible to elaborate any control strategy directly from the input and output signals. The computational model used in simulations is always an intermediate step, because it already contains simplifying assumptions, which limit its applicability. Finally, no matter how strong (necessary and sufficient) the conditions for stability and performance are, they will always have to be verified in the other two stages.

A typical example where it is possible to see clearly these three steps in the control problem is the case of aircraft control design. A very complete model of the aircraft can be obtained from classical mechanics. The set of equations of the aircraft's attitude are highly nonlinear in the states, even when assuming symmetry and considering the system as rigid. These equations will serve as the mathematical structure for the model used in simulations. The values of the parameters of this model are computed through a parameter identification process ([189]), which is described in Section 1.1.3. In this procedure, a set of inputs (physical and computational, respectively) are applied to both the "real" aircraft and its "computational" model. The output data are compared and the simulation model parameters are tuned until the difference between both the actual plant and its computational model is as small as possible.

In the theoretical phase, a set of linear models are obtained by linearizing, at different velocities, angles of attack, and altitudes, the general nonlinear model. We obtain a grid of working points according to the above parameters and a linear model for each point in the grid. A linear controller is designed for each of the linear models in this set. When the flight conditions (altitude, velocity, angle of attack) change, the general control strategy should determine the point on the grid to which these new conditions (approximately) correspond. The control action is performed by the linear controller, which corresponds to this point. As mentioned previously, this is called gain scheduling.

Next in the simulation phase, the designer should verify, using the computational nonlinear model, that when the flight conditions and the linear controllers are changed, the effect on the overall system does not affect the

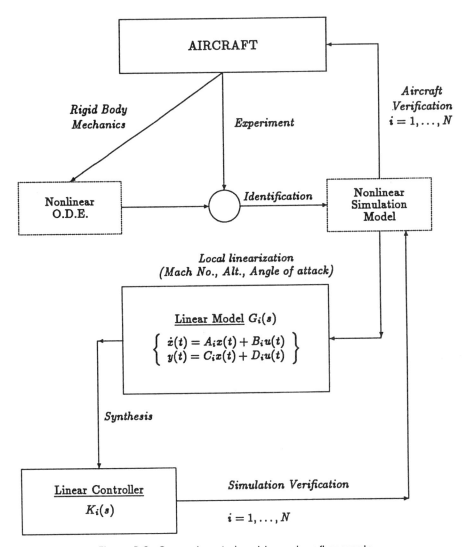

Figure 1.2. General control problem: aircraft example.

stability and performance. Each linear controller is effective in a small neighborhood of its corresponding flight condition. Therefore we need to verify that the change from one controller to another is smooth enough. Although a nonlinear analysis could be made in certain applications, the most common approach is to leave this evaluation for the simulation stage.[6]

---

[6] Recent methods produce a linear parameter varying (LPV) controller that performs the classical gain scheduling but takes into account the complete nonlinear model in terms of stability, thus avoiding in many cases the simulation stage ([222]).

Finally, in the experimental phase, once the design has been tested extensively over many different simulated conditions, the pilot performs a test on the real aircraft[7] of the control strategy for different maneuvers. The set of maneuvers can gradually be expanded until the pilot feels confident with the control design. Otherwise, from this test, the designer can obtain useful "first hand" information that can be used to review both the theoretical and simulation stages. As a consequence, in many cases, a new mathematical or computational model may be necessary. This new model could have less restrictive assumptions; that is, instead of a rigid body model it can incorporate vibrational effects.

Schematically, the three phases are illustrated in Figure 1.2. The use of robust identification and control in the theoretical phase can give a clear idea of the whole picture. This is because relevant conclusions can be drawn at this early stage of the problem by using the knowledge on the bounds of model uncertainty as well as realistic assumptions on the disturbances (bounded sets). This information can be obtained experimentally from the plant, by robust identification procedures. In many cases, the theoretical results can reduce the amount of experimental and simulation work and decrease the number of iterations among the three phases.

1.2 WHY FEEDBACK?

The concept of feedback is related directly to control theory to a point that we could consider the latter as the theory of feedback systems. This is why we should understand the theoretical reasons why a feedback loop should be used to control, in practice, a system. Although this could seem a trivial question for an advanced control course, the fact that uncertainty has been included quantitatively in control theory only very recently indicates the opposite.

Let us take, for example, the feedback system of Figure 1.3, where S represents the physical system to be controlled, and $K(s)$ and $d(s)$ the mathematical

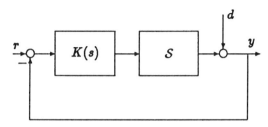

Figure 1.3. Feedback loop.

---

[7] In many cases an intermediate test can be performed by the pilot on a *flight simulator*, which contains a computational model of the aircraft dynamics combined with real hardware of sensors and actuators.

12 INTRODUCTION

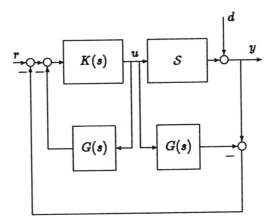

Figure 1.4. Physical system and mathematical model.

models of the linear controller and the external disturbance at the output of the plant, respectively. In Figure 1.4, we add and subtract inside the loop a linear mathematical model of the plant $G(s)$, so that the feedback loop remains unchanged. Finally, in Figure 1.5 we redefine the connection between the models of the controller and the plant as

$$C(s) \triangleq K(s)\left[I + G(s)K(s)\right]^{-1} \tag{1.1}$$

This last definition can be interpreted as another way of connecting the controller, used in *internal model control* (see [211]). The objective of these transformations inside the feedback loop is to leave the feedback signal $f(s)$

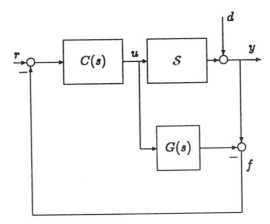

Figure 1.5. Feedback and uncertainty.

expressed only in terms of its necessary components; that is,

$$f(s) = d(s) + \underbrace{[S - G(s)]}_{\Delta} u(s) \qquad (1.2)$$

From the above we see that the need for a feedback signal is due **exclusively** to the uncertain elements in the loop: disturbance $d(s)$ and model uncertainty Δ.

The disturbance is considered as unknown because otherwise, if we knew exactly the *type* of signal and the *time* at which it disturbs the loop, another signal could be injected in the loop that counteracts the effect of $d(s)$. In classical and modern control, which generally assume knowledge of the *type* of signal (step, ramp, sinusoid), there is no certainty in the moment the disturbance will appear. In robust control, the hypotheses are relaxed and the disturbances are assumed to have bounded energy.[8]

Model uncertainty Δ represents the fact that a mathematical model does not copy *exactly* the relevant physical phenomena taking place in the system. In the jargon, it is directly called *uncertainty*. The main difference between classical/modern control theories and robust control is the fact that, in the latter, uncertainty is incorporated explicitly in the hypothesis of the problem. Therefore, in robust control, the word **model** is not equivalent to **system**, the latter meaning physical system or plant. Specifically, the system is treated mathematically as a *family of models* or set, represented by a nominal model $G(s)$ (the same one used in classical/modern control) and *bounded* uncertainty Δ.

Model uncertainty should *always* be bounded, otherwise with absolutely no knowledge of the system, the problem becomes ill-posed, the reason being the following. Given any controller connected to a completely uncertain system, there is always the *possibility* of making the closed loop unstable. A control designer should state conditions that do not allow *any* possibility of instability. Thus there is no way to design a controller for a completely uncertain system, which could provide a minimum certainty on the loop stability. Instead, it is possible to guarantee stability when working on a model with bounded uncertainty.

The goal of robust control is to compute the least conservative conditions providing *certainty* on loop stability and performance of an *uncertain model* (bounded family of models) that represents a physical system. When these properties, stability and performance, refer to the nominal model they are called *nominal*. When they refer to the complete family of models or uncertain model they are called *robust*.

Next, let us imagine ideally that there is exact knowledge of $d(s)$ and an exact mathematical representation of the system $\mathcal{S}$, that is, $\mathcal{S} \equiv G(s)$. By the

---

[8] In this book this will be the case, although, in general, bounds on power or absolute value can also be assumed.

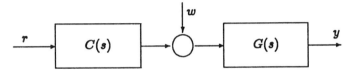

Figure 1.6. Physical connection among components.

arguments in the above paragraphs, without loss of generality we can assume $d(s) \equiv 0$ and therefore $f(s) \equiv 0$. In this case there is no need for feedback, which is reasonable under these utopian conditions. This is so because these conditions establish that the *physical* and *mathematical* manipulation of the system and its disturbance are equivalent. Any desired output could be obtained or stabilization of $G(s)$ could be achieved by conveniently designing an open-loop controller $C(s)$.

Nevertheless, these assumptions do not include the physical connection between the controller and the system. Through any physical connection (the electrical signals from the D/A of a computer controller to the actuator) there is a possibility of having external disturbances (electrical noise, quantization) as seen in Figure 1.6. If the system is open loop unstable (e.g., inverted pendulum), there exist input disturbances w that could produce an undesirable diverging output. Again, the lack of knowledge of possible disturbances entering the loop at different points makes open-loop control a useless choice.

There is no way to avoid feedback unless we have:

- Complete and exact knowledge of the disturbances entering the system at all possible points.
- The exact mathematical representation of the complete system (plant, actuators, sensors) and the controller.

In other words, *if we already have complete knowledge of the situation, there is no need for feedback*. This is almost a tautology, because if we have *complete certainty* already, what extra benefit would we obtain from control theory being the situation already *under control*?

Finally, we can extend what was said at the beginning of this section. Control theory is related to uncertainty and, as a consequence, to feedback. Therefore we could also call it the theory for uncertain systems, which is the basic objective of robust control and robust identification as well.

1.3 FEEDBACK LOOP TRADE-OFF

In the previous section we have established the connection between uncertainty and feedback. The presence of uncertainty, both in the modeling and

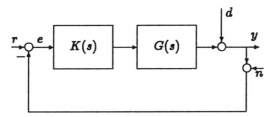

Figure 1.7. Feedback loop trade-off.

on the external disturbances, will be unavoidable in any practical problem. Therefore we need to understand the algebraic restrictions posed by the feedback loop on the design problem. This trade-off derived from the feedback loop makes the problem nontrivial. Let us see this through an example.

Consider the feedback loop of Figure 1.7, where $G(s)$ represents the nominal model of the plant, $K(s)$ a model of the controller, $y(s)$ the output, $r(s)$ a reference signal, $e(s) \triangleq r - y$ the tracking error, $d(s)$ a disturbance, and $n(s)$ the measurement noise. The relations among them are the following:

$$y(s) = \underbrace{G(s)K(s)\left[I + G(s)K(s)\right]^{-1}}_{T(s)} \cdot [r(s) - n(s)]$$

$$+ \underbrace{\left[I + G(s)K(s)\right]^{-1}}_{S(s)} \cdot d(s) \quad (1.3)$$

$$e(s) = r(s) - y(s) = S(s) \cdot [r(s) - d(s)] + T(s) \cdot n(s) \quad (1.4)$$

We call $S(s)$ the *sensitivity function* and $T(s)$ the *complementary sensitivity function*. The limit of the relative (infinitesimal) variations in the closed loop versus the (infinitesimal) variation in the open loop $L(s) = G(s)K(s)$ provides an alternative definition for the sensitivity (take for simplicity a SISO model) as follows:

$$\frac{\delta T(s)}{T(s)} \rightarrow S(s) \cdot \frac{\delta L(s)}{L(s)} \quad (1.5)$$

This variation may represent uncertainty in the modeling. If the variation is not infinitesimal, we can compute the relation between open- and closed-loop models considering the influence of a signal $u_\Delta(s)$ on the output of the plant model $y_\Delta(s)$. These signals are related to the relative uncertainty of the loop as follows:

$$u_\Delta(s) = \frac{\Delta L(s)}{L(s)} y_\Delta(s)$$

where the uncertain loop is $[1 + \Delta L(s)/L(s)]\, L(s)$. Inside the feedback loop both signals are related by

$$y_\Delta(s) = T(s) \cdot u_\Delta \left[s, \frac{\Delta L(s)}{L(s)} \right] \qquad (1.6)$$

This interpretation of model uncertainty will be explained in greater detail in Chapters 2 and 4.

From their definitions, it is easy to see that $S(s)$ and $T(s)$ satisfy the following equation:

$$\boxed{T(s) + S(s) = I} \qquad (1.7)$$

where I is the identity matrix. This equation places a serious constraint when designing a controller that should guarantee stability and performance, as well as robustness to model uncertainty.

According to equations (1.3) and (1.4), for disturbance rejection at the output, to minimize the influence of reference and disturbance in the tracking error, or the (infinitesimal) robustness problem of equation (1.5), the designer should minimize the "size" of $S(s)$. Instead, if the objective is rejection of measurement noise at the output, at the tracking error signal e, or the robustness problem of equation (1.6), the "size" of $T(s)$ should be minimized. From the restriction in (1.7) this is not possible simultaneously.

To solve this problem, first we should define an adequate "size" for a MIMO transfer matrix, particularly the sensitivity matrices, and by means of this measure we should try to solve the trade-off in (1.7). In later chapters we will define this measure and present different methodologies to solve the trade-off. Some of them are based on a more *intuitive* approach, as is the case of loop shaping (Chapter 4). The others—$\mathcal{H}_2$ (Chapter 5) and $\mathcal{H}_\infty$ optimal control (Chapter 6), μ-synthesis (Chapter 7), and ℓ^1 (Chapter 8) optimal control—solve the problem by means of an algorithmic procedure.

The important point of this section is to realize that any well defined control problem should include implicitly in its requirements this trade-off. Otherwise the problem may be trivial, as in the following "textbook" problem:

Problem 1.1 *Design a stabilizing controller for the system* $g(s) = (s+1)/(s-2)$, *which also rejects disturbing steps at the output.*

Clearly a controller $k(s) = q/s$ with $q > 2$ solves the problem. This means that an infinite control action may be proposed being that q has no upper bound. From the practical point of view, this is obviously unrealizable because of actuator bounds and high-frequency model uncertainty. The point is that even in the formulation of a "theoretical" problem, the trade-off should be stated implicitly.

1.4 OBJECTIVES OF AN APPLIED THEORY

In the preceding sections we presented some ideas concerning model uncertainty in "real world" applications, the use of feedback, and the connections between the three basic phases of a control problem. From them we can extract minimal requirements for a systems theory to be applicable in practical problems. In particular, we will concentrate on how Robust System Theory satisfies these requirements.

- **Practical Assumptions.** This requirement reflects the need to have simplifying assumptions that are related not only to the mathematical simplicity but also to the applicability of the theory. In general, it means that the standing assumptions should be as unrestrictive as possible, in the sense of being more "realistic."

    In the case of robust control it pertains to the connection between the three phases described in Section 1.1. As an example, assuming disturbances as bounded energy signals is always more realistic than assuming they are steps. Also, from the practical point of view, instead of having a particular model as a description of the system, we can represent it by a family of models. This family is described by a nominal model and uses the information on the uncertainty bound.

- **Strong Results.** From the mathematical point of view, by *strong* we mean necessary and sufficient conditions. These conditions are therefore *equivalent* to a desired result, which can be verified (analysis) or implemented (synthesis). From a more practical point of view, we seek conditions that are neither unrealistic (too optimistic) nor conservative.

    In particular, the conditions we are interested in are on stability and performance of a certain control loop. As an example, the verification of a necessary condition for stability is not enough to guarantee it; therefore it is optimistic. On the other hand, there could be no controller that satisfies a sufficient condition for achieving performance. That does not mean there is no controller achieving that performance, therefore a conservative result. In both cases the results are inconclusive. Instead, we look for a *strong* mathematical condition (so it can be tested), which is equivalent to a particular desired result (stability and/or performance).

- **Computability.** The results obtained from the analysis and synthesis procedures should have not only a mathematical representation but also a computer representation. This means that the analysis conditions to be tested should be computable and the synthesis procedures implementable by means of an algorithmic code.

    In particular, at this point we may discuss what "computable" means exactly. We could argue that, although a code that implements an algorithm can be written, it may happen that either the code or the algorithm are very inefficient. This could lead to a more general discussion between tractable and untractable computational problems ([226]) and efficiency

18 INTRODUCTION

of algorithms and codes. Therefore the main point could be clarified as follows. *The computational implementation of a control problem (analysis or synthesis) should not be the* **bottleneck** *of the solution to the problem.*

In the following example we will see how robust system theory meets the above requirements.

Example Consider the closed-loop system of Figure 1.8. The system S is represented by a family of models of the form

$$S \equiv \{G_o(s)[I + W(s)\Delta] \mid \|\Delta\| < 1\} \tag{1.8}$$

This description considers specifically the uncertainty between the system and the set of models by means of Δ and the weight $W(s)$. The output data from an experiment performed on the actual system have been processed by a *robust identification* algorithm (Chapter 10). From this procedure we compute the nominal model $G_o(s)$ and $W(s)$. Therefore we have quantified the relation between the actual system, the nominal model, and the uncertainty.

The objective is to achieve closed-loop stability for all members of the family, defined as *robust stability*. If from the experimental data we obtain a family of models that "covers" the actual plant, robust stability will guarantee stability of the actual closed-loop system with the controller connected to the real plant. As will be explained in Chapters 2 and 4, an equivalent condition for robust stability is the following:

$$\left\| K(s)G_o(s)[I + K(s)G_o(s)]^{-1} W(s) \right\|_\infty \leq 1 \tag{1.9}$$

By means of the $\mathcal{H}_\infty$ optimal procedure (Chapter 6), a controller that achieves the minimum of the above analysis condition can be calculated. Both the analysis condition and the synthesis algorithm can be computed efficiently by means of commercial software ([20, 126, 262]).

From the point of view of the three phases described in Section 1.1, both the analysis and synthesis theoretical results provide implementable conditions to guarantee closed-loop stability in the final experimental verification phase.

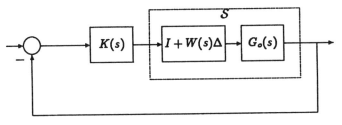

Figure 1.8. Robust system theory example.

1.5 OBJECTIVES OF THIS BOOK

1.5.1 General Remarks

The main objective of this book is to introduce the techniques used in the theoretical phase of the control problem from the point of view of Robust System Theory. Therefore, according to the discussion in Section 1.1, the theoretical results will take into account, as well, the problems that arise in the other two phases.

The title of the book includes *Robust System Theory* because it not only considers robust control but also has a chapter devoted to robust identification. The words *Theory* and *Applications* appear in the title because not only theory is presented but also practical "real world" problems. This is consistent with the objective of bridging the gap between both.

The book is oriented to students of engineering departments with a background in classical and modern control.

The main analysis and design methods are complemented by an elaborated example, either in the same chapter or in Chapter 11. The latter has a group of worked out applications that stress particular practical issues: nonlinearities, unknown time delays, infinite-dimensional plants, and actuator and sensor limitations. The three appendices—Mathematical Background, System Computations, and Riccati Equations—make the book as self-contained as possible.

1.5.2 Scope

Chapter 2 gives the first robustness concepts applied to SISO control systems described by Laplace transforms. It relates these new ideas to the classical control stability, performance, and robustness concepts. It is a simple introduction to the main issues of robust control, for students or engineers with a classical control background.

Chapter 3 presents the internal stabilization problem from two different points of view. The first is a state-space approach based on solving four special problems first introduced in [104]. This also serves as an introduction to the tools used in Chapters 5 and 6. The second is a more classical approach, based on the use of coprime factorizations. As a by-product we obtain a parametrization of all stabilizing controllers and achievable closed-loop maps. These results are instrumental in solving the ℓ^1 control problem addressed in Chapter 8.

Chapter 4 introduces the four basic problems in robust control: nominal and robust stability and nominal and robust performance for MIMO systems in a linear fractional transformation setup. The objective is to present a practical synthesis methodology that is not mathematically involved and could be helpful to the student/engineer who developed an "intuition" designing with classical control tools (Bode plots). The fact that this methodology for multivariable systems is not a simple generalization of the procedure for SISO systems (Chapter 2) is also explained. This is the case of combined plant input uncertainty (uncertain actuator models) and high condition number plants, presented in its last section.

Chapters 5 and 6 present the optimal $\mathcal{H}_2$ and $\mathcal{H}_\infty$ control problems. These problems are solved first using a state-space approach using the tools introduced in Chapter 3. Alternatively, by using the Youla parametrization the problems are recast into an approximation form and solved using either projections (in the $\mathcal{H}_2$ case) or Nehari's theorem ($\mathcal{H}_\infty$). Finally, in the $\mathcal{H}_\infty$ case we present a third approach based on the bounded real lemma and the use of linear matrix inequalities. This approach allows us to easily address general (even singular) output feedback problems.

Chapter 7 discusses more general types of uncertainties. In the case of structured dynamics, it defines the structured singular value (μ) and presents its computation and use in both analysis and design. In the case of parametric uncertainty or mixed type uncertainties, it covers the analysis, the synthesis results being under development. This chapter also gives a good idea of the open problems in this area, which are still undergoing intense research.

Chapter 8 presents an overview of the relatively new subject of optimal ℓ^1 control theory. While the problem was formulated in 1986 [315], with the first solution methods for the simpler SISO case appearing around that time [80, 81, 83], this is presently a very active research area, where a complete theory is just starting to emerge and where most of the results appear only in the specialized literature. The chapter starts by discussing the ℓ^1 framework following the same induced-norm paradigm used in Chapter 2 to address stability and performance in the $\mathcal{H}_\infty$ context. The ℓ^1 optimal problem is motivated as the problem of rejecting bounded persistent disturbances (as opposed to bounded energy signals in the $\mathcal{H}_\infty$ framework) or the problem of achieving stability against linear time varying model uncertainty. For simplicity most of the chapter is devoted to SISO systems, showing that in this simpler case the problem can be recast, through the use of duality as a finite-dimensional linear programming problem. Since this theory may result in very large order controllers, the chapter contains some very recent results on synthesizing suboptimal controllers using an LMI-based approach as an alternative to the usually high-order optimal controllers. Finally, it is shown that in the continuous time case $\mathcal{L}^1$ optimal controllers are infinite-dimensional and the issue of finite-dimensional approximations is addressed.

Chapter 9 has a threefold objective. First, it gives a nice geometrical interpretation of the structural properties of the state-space models in terms of principal components and directions, as related to controllability and observability. Second, this same interpretation is useful when evaluating numerical issues in state-space descriptions: For example, how "near" is the model from uncontrollability? This is useful for the student to distinguish between the "theoretical" model and the "numerical" model implemented in a computer. This is also related to the "measure" given by the Hankel singular

OBJECTIVES OF THIS BOOK 21

Table 1.1. Road map for self-study, by topic

Stabilization	Robust Analysis	Robust Synthesis	$\mathcal{L}^1$	Robust Indentification
2	3	3	3	9
3	6	6	8	10
4	11.1–11.2	7		
		11		

values. Finally, the third objective is to provide an introduction to balanced realizations as a part of model order reduction. A classical model reduction algorithm is presented. This material is closely related to the one presented in the next chapter.

Chapter 10 is a tutorial on a very recent (*circa* 1990) research subject that connects the robust control methodology with the experimental modeling of systems. It presents the conceptual problem of worst-case deterministic identification but also details some of the main algorithmic procedures. Both time- and frequency-based experiments are considered, and a flexible beam example illustrates the use of some of these procedures along the chapter. It also provides a wide range of bibliographic references for the researcher.

Chapter 11 gives a good idea of the issues that appear in practical problems. It presents the implications of applying robust identification and control procedures to several practical application problems.

1.5.3 How to Use This Book

This book is intended to be used either as a reference for control engineers or as a graduate level textbook. Table 1.1 gives an outline by topic of how to use the book for self-study. If used as a texbook, there are several alternatives depending on the background of the students and the amount of

Table 1.2. Using this book for a one- or two-semester course

First Course Prereq: Linear Systems	Second Course Prereq: First Course
2	7
3	8[a]
4	9
5[a]	10
6.1, 6.2, 6.4	11
7.1–7.3	
11.1	

[a] Optional chapters.

time available. At the simplest level it could be used as a one-semester, beginning graduate level course serving as an introduction to robust control and requiring as a prerequisite a standard course on linear systems theory. This one-semester course would cover Chapters 2–5, portions of Chapters 6 and 7, and some of the examples in Chapter 11. Since Chapter 8 is largely independent, it could replace Chapter 5 if the students have already taken an optimal control course. As a follow-up to this course, a second advanced graduate level course would cover Chapters 7–11. These options are outlined in Table 1.2.

2

SISO SYSTEMS

2.1 INTRODUCTION

In this chapter we address the issues of nominal and robust stability and performance problems in single-input single-output (SISO) systems, represented either by a transfer function or by state-space equations. The analysis is performed proceeding from stability of the nominal model of the plant to the final objective of robust control: robust performance.

As mentioned in the introductory chapter, robustness measures are related to the type of uncertainty considered. The classical uncertainty description for SISO models, gain and phase margins, are presented first, along with an analysis of their limitations. To overcome these limitations, a more general robustness analysis based on more realistic descriptions is presented. Performance is based on rejection or tracking of *sets* of disturbances, noises, or reference signals, which lead to conditions on the loop sensitivity functions. The above conditions for both performance and stability will affect the selection of the loop transfer function as mentioned in Chapter 1.

For the sake of simplicity and clarity, in this chapter the definitions of uncertainty and performance are restricted to SISO systems. Generalization to MIMO systems is postponed until Chapter 4, where we will also introduce additional issues that do not arise in the simpler SISO case. Following the standard notation, we will use lowercase to denote SISO transfer functions and uppercase for MIMO systems. Throughout the chapter we will use packed notation to represent state-space realizations, that is,

$$g(s) = C(sI - A)^{-1}B + D \equiv \left[\begin{array}{c|c} A & B \\ \hline C & D \end{array}\right]$$

2.2 WELL POSEDNESS

Consider the feedback loop of Figure 2.1, where the input signals $[u_1(s), u_2(s)]$, the outputs $[y_1(s), y_2(s)]$, and the errors $[e_1(s), e_2(s)]$ satisfy the following equations:

$$y_1(s) = g(s)e_1(s); \quad e_1(s) = u_1(s) - y_2(s) \tag{2.1}$$

$$y_2(s) = k(s)e_2(s); \quad e_2(s) = u_2(s) + y_1(s) \tag{2.2}$$

Next, we define well posedness of a feedback loop, a standing assumption that should be made in any control problem. This avoids trivial situations in the loop, as in the case where $g(s) = -k^{-1}(s)$.

Definition 2.1 *The feedback loop of Figure 2.1 is said to be well posed if and only if all possible transfer functions between the inputs $u_1(s)$ and $u_2(s)$ and all outputs $e_1(s)$, $e_2(s)$, $y_1(s)$, and $y_2(s)$ exist and are proper.*

In other words, we can say that a feedback system is well posed when for *well defined* inputs, we obtain *well defined* outputs. In fact, it can be proved (Problem 1) that it is sufficient to verify well defined outputs $[e_1(s), e_2(s)]$ or $[y_1(s), y_2(s)]$. It should be clear that well posedness of the feedback loop is a mathematical prerequisite for any control problem in which stability and performance are to be defined. It is also clear that in practical situations, feedback loops are always well posed.

Before proving an equivalent condition for well posedness, we state in compact form the relationship between the inputs u_i and errors e_i for the feedback system of Figure 2.1:

$$\begin{bmatrix} u_1(s) \\ u_2(s) \end{bmatrix} = \begin{bmatrix} 1 & k(s) \\ -g(s) & 1 \end{bmatrix} \begin{bmatrix} e_1(s) \\ e_2(s) \end{bmatrix}$$

$$\triangleq P(s) \begin{bmatrix} e_1(s) \\ e_2(s) \end{bmatrix} \tag{2.3}$$

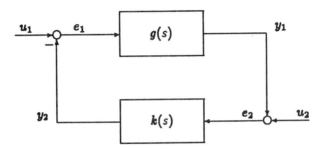

Figure 2.1. Feedback loop model to evaluate well posedness and (nominal) internal stability.

Therefore to have *well defined* outputs $e_1(s)$ and $e_2(s)$, we need the inverse of $P(s)$ to exist and to be proper. Otherwise, either the outputs will not be well defined or could be nonproper. We will illustrate this by means of a simple example.

Example 2.1 *Consider the following model of a plant and a controller, connected as in the feedback loop of Figure 2.1:*

$$g(s) = \frac{(1-s)}{(2+s)}, \quad k(s) = \frac{(s+7)}{(s+5)} \tag{2.4}$$

The transfer function between u_2 and e_2 is

$$[1+g(s)k(s)]^{-1} = \frac{(s+2)(s+5)}{(s+17)} \tag{2.5}$$

which is improper. Therefore the loop is not well posed. Note that, in practice, the model of the plant is strictly proper. In that case, it is easy to see that the above transfer function would have been proper.

Next, we state necessary and sufficient conditions for well posedness. Assume that the plant and controller have the following state-space realizations:

$$g(s) \equiv \left[\begin{array}{c|c} A_g & B_g \\ \hline C_g & D_g \end{array}\right] \tag{2.6}$$

$$k(s) \equiv \left[\begin{array}{c|c} A_k & B_k \\ \hline C_k & D_k \end{array}\right] \tag{2.7}$$

Then we have the following result.

Lemma 2.1 *The following conditions are all equivalent:*

1. *The system in Figure 2.1 is well posed.*
2. *$P(s)$ has a proper inverse.*
3. *$1 + g(s)k(s)$ has a proper inverse.*
4. *$1 + D_g D_k \neq 0$.* ← because of the steady state?

Proof. All transfer functions between $u_i(s)$ and $e_j(s)$, $i, j = 1, 2$ in (2.3) are well defined if and only if the inverse of $P(s)$ is proper and exists for *almost all s*. The latter and the fact that to guarantee well posedness it is enough that these four transfer functions are well defined (Problem 1) prove the equivalence between items 1 and 2. It is clear that if both the plant and the controller are proper, $P(s)$ will have a proper inverse if and only if its determinant, $[1 + g(s)k(s)]$, has a proper inverse; which proves 2 and 3 are equivalent. For properness, $[1 + g(s)k(s)]^{-1}$ should exist for almost all s and, in particular, for $s \to \infty$. Equivalently,

$$\lim_{s \to \infty} 1 + g(s)k(s) \neq 0 \qquad (2.8)$$

The state-space representation of this limit is $1 + D_g D_k \neq 0$, which proves 3 and 4 are equivalent. The above arguments still hold in the case where we consider y_1 and y_2 as the outputs of the closed loop (Problem 1). □

2.3 NOMINAL INTERNAL STABILITY

The classical concept of stability of a linear time invariant system, described either as a transfer function or by state-space equations, between a given input–output pair of signals, is well known. Therefore we will not repeat here these definitions already stated by classical and modern control theories. Nevertheless, we will point out the idea of *internal stability* of a feedback loop, which was not stated explicitly. To this end we will clearly distinguish the stability from an input–output point of view from the stability of a feedback loop.

Consider again the feedback loop of Figure 2.1. It is not difficult to see that the stability of a certain input–output pair (all the poles of the corresponding transfer function in the open left-half complex plane $\mathbb{C}_-$) does not guarantee that all input–output pairs will be stable (in the same sense). This is illustrated by the following simple example.

Example 2.2 *Consider the following system and controller:*

$$g(s) = \frac{(s+1)}{(s-1)(s+3)}, \qquad k(s) = \frac{(s-1)}{(s+1)}$$

connected as in Figure 2.1. The transfer function from the input u_2 to the output y_1 is given by $T_{y_1 u_2} = 1/(s+4)$, which is stable in the usual sense. However, the transfer function between the input u_1 and the output y_1 is $T_{y_1 u_1} = (s+1)/(s-1)(s+4)$, which is obviously unstable. As we will see next, this is caused by the cancellation of the unstable plant pole at $s = 1$ by a zero of the controller.

This example shows that there is a difference between the stability of a certain system, considered as a mapping between its input and output,[1] which we define as input–output stability, and stability of a feedback loop, which will be defined next. In the latter, we must guarantee that all possible input–output pairs be stable, which leads to the concept of internal stability.

---
[1] Even in the MIMO case with several inputs and outputs.

Definition 2.2 *The feedback loop of Figure 2.1 is internally stable if and only if all transfer functions obtained from all input–output pairs have their poles in $\mathbb{C}_-$ (input–output stable).*

It is easy to prove (Problem 2) that to verify internal stability it is sufficient to check the input–output stability of the four transfer functions between the inputs $[u_1(s), u_2(s)]$ and the outputs $[e_1(s), e_2(s)]$. This is summarized in the following lemma.

Lemma 2.2 *Let*

$$C_\ell(s) \equiv \left[\begin{array}{c|c} A_c & B_c \\ \hline C_c & D_c \end{array} \right] \qquad (2.9)$$

be a minimal state-space realization of the closed-loop system of Figure 2.1, between the inputs $[u_1(s), u_2(s)]$ and outputs $[e_1(s), e_2(s)]$ (or $[y_1(s), y_2(s)]$). Then the following conditions are equivalent:

1. *The feedback loop in Figure 2.1 is internally stable.*
2. *The eigenvalues of A_c are in the open left-half complex plane $\mathbb{C}_-$.*
3. *The four transfer functions obtained between inputs $[u_1(s), u_2(s)]$ and outputs $[e_1(s), e_2(s)]$ (or $[y_1(s), y_2(s)]$) have their poles in $\mathbb{C}_-$; that is, $C_\ell(s)$ is input–output stable.*

Proof. $1 \iff 3$ is left as an exercise (Problem 2). $2 \Rightarrow 3$ follows immediately from Definition 2.2. Finally, if the four transfer functions are stable, then all the eigenvalues of A_c are in $\mathbb{C}_-$, since minimality of (2.9) precludes pole–zero cancellations. This shows that $3 \Rightarrow 2$. □

It is important to note that the concept of internal stability, although not explicitly stated, has existed among practicing engineers for many years. A proof of this is the basic controller design rule of *no unstable pole/non-minimum phase zero cancellation between plant and controller* should be allowed. A misleading explanation for this rule is that, numerically, it is impossible to cancel **exactly** a zero and a pole (why is it misleading?). According to the definition of internal stability, we now know that the argument behind this "practical" rule is the following: with unstable pole–zero cancellations, there will always be an unstable input–output pair in the loop (and hence an internal signal could grow unbounded, damaging the system). This will be proved next.

Lemma 2.3 *The feedback loop in Figure 2.1 is internally stable if and only if $[1 + g(s)k(s)]^{-1}$ is stable and there are no right-half plane (RHP) pole–zero cancellations between the plant and the controller.*

Proof. First factorize both plant and controller into *coprime* numerator and denominator polynomials (which do not share common roots), as follows:

$$g(s) = \frac{n_g(s)}{d_g(s)}, \quad k(s) = \frac{n_k(s)}{d_k(s)} \tag{2.10}$$

We assume that

$$[1 + g(s)k(s)]^{-1} = \frac{d_g(s)d_k(s)}{n_g(s)n_k(s) + d_k(s)d_g(s)} \tag{2.11}$$

is stable. If $n_g(s)$ and $d_k(s)$ share common roots in the RHP, these will also be roots of $n_g(s)n_k(s) + d_k(s)d_g(s)$. The same occurs if $n_k(s)$ and $d_g(s)$ share common roots in the RHP. Therefore in the first case, the transfer function

$$k(s)[1 + g(s)k(s)]^{-1} = \frac{d_g(s)n_k(s)}{n_g(s)n_k(s) + d_k(s)d_g(s)} \tag{2.12}$$

is unstable, because the RHP roots of the denominator will not be canceled by the numerator [this is precluded by the coprimeness of the pairs $(n_g(s), d_g(s))$ and $(n_k(s), d_k(s))$]. In the second case, the transfer function $[1 + g(s)k(s)]^{-1} g(s)$ will be unstable for the same reason. Note that the complementary sensitivity $[1 + g(s)k(s)]^{-1} g(s)k(s)$ remains always stable, with or without RHP cancellations, if the sensitivity is stable. On the other hand, if there are no RHP pole–zero cancellations, all these transfer functions are stable. As these are all possible transfer functions, the loop is internally stable, which concludes the proof. □

Note also that, according to the definition, any stable pole–zero cancellation in the loop will not modify the internal stability. However, from the point of view of performance, the cancellation could be performed over a stable although undesirable mode of the system. This means that a certain input–output pair of Figure 2.1 will have "poor" performance. This can be understood better by means of an example.

Example 2.3 *Consider the following system and controller:*

$$g(s) = \frac{(1-s)(s^2 + 0.2s + 1)}{(s+1)(s+3)(s+5)}, \quad k(s) = \frac{(s+1)(s+5)}{s(s^2 + 0.2s + 1)}$$

connected as in Figure 2.1. The design objectives are to internally stabilize the loop and to track with y_1 a step input injected at u_2 with not more than 25% overshoot. The controller selected meets the performance specifications, since the closed loop is internally stable and $y_1(s)/u_2(s) = (1-s)/(s+1)^2$, which has an adequate time response. However, the transfer function $y_2(s)/u_2(s) =$

$(s+3)(s+5)/(s+1)(s^2+0.2s+1)$ *presents undesirable oscillations due to the lightly damped poles at* $s \approx -0.1 \pm j$.

This situation can be avoided by simply extending the definition of internal stability to encompass more general regions, rather than just the (open) left-half complex plane. Therefore, according to the desired location of the closed-loop poles, an adequate region in $\mathbb{C}_-$ can be defined, and no pole–zero cancellation between plant and controller should be performed outside this region.

2.4 ROBUST STABILITY

As mentioned in Chapter 1, a given robustness margin is related to a specific type of model uncertainty. Therefore by first evaluating the stability robustness margins used in classical control theory we can analize their limitations and propose more realistic and general uncertainty descriptions.

2.4.1 Phase and Gain Margins

The phase ϕ_m and gain margins g_m are used as *robustness* measures to evaluate a classical controller design. Here, loosely speaking, "robustness" means the ability of a system to respond adequately, in terms of performance and stability, even when the open-loop model $\ell(s) = g(s)k(s)$ used in the design does not exactly match the physical system, due to the existence of uncertainty.

Both measures can be clearly interpreted in terms of the Nyquist plot and the Nyquist stability criteria. In this plot, ϕ_m and g_m represent the "distance" in angle and absolute value, respectively, to the critical point $z = -1$, as shown in Figure 2.2.

The usual practice of evaluating both margins separately assumes implicitly that both types of uncertainty (phase and gain) act on the loop one at a time. As a consequence, these margins are effective as analysis tools **only** when the model of the plant has either phase **or** gain uncertainty. Care must be taken when using ϕ_m and g_m for stability robustness analysis of SISO closed-loop systems, because they do not guarantee *robust stability* for the more realistic situation where both phase and gain are **simultaneously** affected by uncertainty. Here *robust stability* means the stability of *all* possible models that are described by the combination of the nominal model $g_o(s)$ and the uncertainty $\delta = c \cdot e^{j\phi}$, where c and ϕ are *uncertain* values contained inside the intervals $I_c = [c_{min}, c_{max}]$ and $I_\phi = [\phi_{min}, \phi_{max}]$, respectively (Figure 2.3). This set of plant models can be defined as the *family of models* $\mathcal{F} \stackrel{\triangle}{=} \{g(s): g(s) = g_o(s)[ce^{j\phi}], \ c \in I_c, \ \phi \in I_\phi\}$. We will clarify these concepts by means of the following examples.

30 SISO SYSTEMS

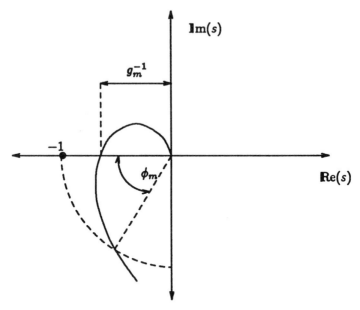

Figure 2.2. Phase and gain margins.

Example 2.4 *Consider the Nyquist plot of Figure 2.4, in which both ϕ_m and g_m have adequate values. Nevertheless, with small simultaneous perturbations in the phase and gain of the loop, the plot will encircle the critical point $z = -1$. Therefore, in this case, the minimum distance from the Nyquist plot to the critical point (the minimum return difference) gives a much better assessment of the robust stability properties of the system than ϕ_m or g_m.*

Example 2.5 *Consider the Nyquist plot of Figure 2.5. In this case, the gain margin $g_m \to \infty$, although the phase margin is very small. If we know that the only source of uncertainty in the loop is in the gain, a small ϕ_m poses no threat to stability. Therefore, in this case, the controller designed guarantees robust stability.*

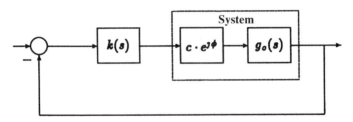

Figure 2.3. Phase and gain margins related to the uncertainty in the model.

ROBUST STABILITY

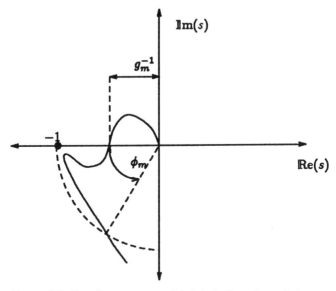

Figure 2.4. Simultaneous uncertainty in both gain and phase.

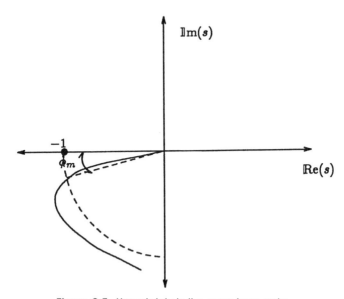

Figure 2.5. Uncertainty in the open-loop gain.

From both examples we conclude the following:

- A given stability margin is a good indicator of the robust stability properties of the system only when it is associated with a specific type of model uncertainty.
- The use of gain and phase margins in the analysis of control systems is limited to the cases where there is *a priori* knowledge that the uncertainty is either on the phase or in the gain, but not on both simultaneously. If that is not the case, these margins provide only a necessary (but not sufficient) condition to test robust stability.
- The uncertain system can be analyzed as a *family of models* that can be described mathematically by a nominal model and a bounded uncertainty set. In the Nyquist plot this corresponds to a nominal plot and an uncertainty region surrounding it.
- If there is simultaneous uncertainty in phase and gain, it is important to analyze the *type* of uncertainty considered, because it will determine the *form* of the set of Nyquist plots in the complex plane.
- The stability margin can be computed as the minimum distance between the family of models and the critical point in the Nyquist plot. Therefore it is directly related with the region describing the model set in $\mathbb{C}$ and, as a consequence, with the type of uncertainty. This can be expressed as follows:

$$\boxed{\text{Type of uncertainty}} \longleftrightarrow \boxed{\text{Stability margin}}$$

Phase and gain margins are special measures that are useful only when either the uncertainty is "angular" or "modular," respectively. They can be considered as special cases of a more general (and realistic) uncertainty description that will be presented next.

2.4.2 Global Dynamic Uncertainty

More realistic descriptions of model uncertainty have been developed, leading to controller designs that perform better in practice. One such description is the *global dynamic uncertainty*. The name arises since this uncertainty is related to the uncertainty in the system dynamics and covers globally the complete model of the plant. This type of uncertainty description can be used when the order of the differential equations describing the plant is unknown.[2] This situation arises in practice when certain physical phenomena (elasticity, flexibility) are not included in the plant description, or when the physical description of the system is too complicated to lead to tractable problems.

---

[2] A different branch of control theory also dealing with uncertain systems is adaptive control. It has as one of its basic assumptions that the order of the equations describing the system is known. Today, there is an important effort geared towards relaxing this assumption.

In many cases it describes situations where a constraint in the computation time (real-time control of a fast process) imposes a limit on the order of the mathematical models that can be handled. As a consequence, these models are approximated by a set of equations of lower order, and the neglected dynamics are lumped into the uncertainty.

Global dynamic uncertainty can also be used to describe linearization errors, where a nonlinear system is linearized around a nominal operating point. Although the resulting model is valid only in a neighborhood of the nominal operating point, the nonlinear effects outside this neighborhood can be bounded and interpreted as dynamic uncertainty. Typical examples of this situation appear in the modeling of aircrafts, which are described by a general nonlinear mathematical model. The latter is linearized at different altitudes, velocities, and angles of attack, yielding a set of linear models, each one valid at a particular operating point. An uncertainty bound "covering" the linearization errors in a neighborhood of each linearization point can be interpreted as dynamic uncertainty.

Finally, many systems may be described by partial differential equations or time delayed equations, both called *infinite-dimensional* systems. In these cases, an approximation ([24, 25, 79, 134, 169]) of the general equation should be made so that a finite-dimensional model is obtained. The approximation error can be interpreted as dynamic uncertainty because the approximation process eliminates all the high-order dynamics of the general mathematical model. In particular, uncertainty in the time delay can be modeled as dynamic uncertainty ([268]) in a very simple way. Examples of the latter and of the dynamic uncertainty of a flexible structure due to its approximation error are presented in Chapters 10 and 11.

The approach adopted by robust control theory is to describe a physical system by means of an uncertain model. The latter is defined as a set of models and described in terms of a nominal plant together with bounded uncertainty. In classical and modern control theories the controller is designed using the nominal model of the plant. Instead, robust control theory attempts to design a single controller guaranteeing that certain properties such as stability and a given level of performance are achieved for *all* members of the set of models. In this case, these properties are said to be *robust*. In the following subsections, different sets or families of models will be described.

Multiplicative Dynamic Uncertainty An uncertain model with multiplicative dynamic uncertainty is shown in Figure 2.6. It describes a physical system as a set $\mathcal{G}$ of mathematical models, as follows:

$$\mathcal{G} = \{g(s): g(s) = g_o(s)\left[1 + \delta W(s)\right], \ \delta \in \mathbb{C}, \ |\delta| < 1\} \qquad (2.13)$$

The set $\mathcal{G}$ is the family of models and is characterized by a nominal plant $g_o(s)$, a fixed weighting function $W(s)$, and a class of bounded uncertainty δ. The nominal model $g_o(s)$ corresponds to the case where there is no uncertainty, that is, $\delta = 0$. Without loss of generality, the bound on the uncertainty

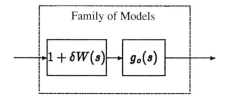

Figure 2.6. Dynamic multiplicative uncertainty.

δ can be taken to be one, because any other bound can be absorbed into the weight $W(s)$.

The weighting function $W(s)$ represents the "dynamics" of the uncertainty, or in other words its "frequency distribution." A graphical interpretation of this uncertainty weight can be seen in Figure 2.7. In this plot we can observe a typical weighting function, with 20% uncertainty at low frequencies and 100% at higher ones. A complete lack of knowledge of the system (100% error) means in practice that there is no phase information (i.e., the plant transfer function $g(s)$ can have either a positive or negative real part). Thus, in order to guarantee stability of the closed-loop system, the controller must render the nominal loop function $g_o(j\omega)k(j\omega)$ small enough at frequencies above ω_o to guarantee that the Nyquist plot does not encircle the critical point, regardless of the phase of $g(j\omega)k(j\omega)$. This point illustrates an important limitation for robust stabilization that should be taken into account in any controller design.

The situation where the uncertainty is small at low frequencies and increases at higher frequencies (although not nearly as dramatically as in the example discussed above) is the usual case arising in practice. For instance, for mechanical systems a better model can be obtained to describe slow

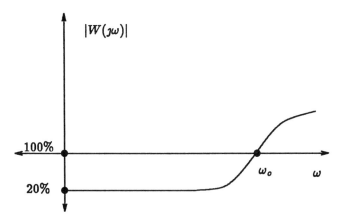

Figure 2.7. Frequency distribution of uncertainty.

mechanical resonances than faster ones (involving mechanical flexibility). This situation can also be observed in models of electrical circuits. For applications in lower frequency bands, the model is simpler and usually provides a good description of the physical reality. To describe the response of the circuit at higher frequencies, other physical phenomena should be included in the model (parasitic capacitances and inductances) increasing its complexity. Alternatively, these additional dynamics can be treated as dynamic uncertainty. The weighting function $W(s)$ can be obtained explicitly in many applications (see Chapter 11) or by using the techniques known as robust identification, which will be described in Chapter 10.

An important restriction in the problem of robust stabilization is that each member of the family of models is required to have the same number of unstable poles. This requirement can be explained using a simple argument based on the Nyquist graphical criteria (which is a necessary and sufficient condition for stability of a single model) as follows. Suppose that any two members of a family have a different number of RHP poles, then their Nyquist plots will encircle the critical point a different number of times. As a consequence, the region between both plots will include the critical point $z = -1$. For the family of models considered, the set of Nyquist plots of this family "covers" this region and also includes $z = -1$. On the other hand, a model that goes from (closed-loop) stability to instability should have its Nyquist plot go across the critical point, because it changes the number of encirclements. Thus it will be difficult to detect if an element of the family is (closed-loop) unstable since the whole set of models already includes the point $z = -1$. Hence we cannot derive equivalent conditions for (closed-loop) stability of all members, based only on their frequency response (Nyquist plot). For this reason, to guarantee that the number of RHP poles of all members of $\mathcal{G}$ remains constant, we will assume, without loss of generality, that $W(s)$ is stable (why?) and that no unstable pole–zero cancellations take place when forming $g_o(s)[1 + \delta W(s)]$. Perturbations δ satisfying this latter condition will be termed allowable.

A more general uncertainty description can be obtained by replacing $\delta \in \mathbb{C}$ by a real rational transfer function $\Delta(s)$, analytic in the RHP and such that $\|\Delta(s)\|_\infty \stackrel{\triangle}{=} \sup_{j\omega} |\Delta(j\omega)| < 1$. As before, we require that no unstable pole–zero cancellations occur when forming $g_o(s)[1 + \Delta(s)W(s)]$. As will be seen next, robust stability of this general model set just described and the one in (2.13) yield the same condition. Thus, for simplicity, in most of this chapter we adopt the description in (2.13), which assigns the knowledge of the "frequency" distribution of uncertainty to $W(s)$, and the bound to $\delta \in \mathbb{C}$. Finally, note that, in order for the problem to make sense, we should always assume bounded uncertainty, otherwise there is no possible design that can stabilize the whole set simultaneously (why?). A condition guaranteeing stability of all elements of the family $\mathcal{G}$, that is, *robust stability* of $g_o(s)$, is derived next.

36 SISO SYSTEMS

Theorem 2.1 *Assume the nominal model $g_o(s)$ is (internally) stabilized by a controller $k(s)$ (Figure 2.8). Then all members of the family $\mathcal{G}$ will be (internally) stabilized by the same controller if and only if the following condition is satisfied:*

$$\|T(s)W(s)\|_\infty \stackrel{\Delta}{=} \sup_{j\omega} |T(j\omega)W(j\omega)| \leq 1 \qquad (2.14)$$

with $T(s) \stackrel{\Delta}{=} g_o(s)k(s)[1+g_o(s)k(s)]^{-1}$ the complementary sensitivity function of the closed-loop system.

Proof. From Lemma 2.3, it follows that:

1. $[1+g(s)k(s)]^{-1}$ stable for all members $g(s) \in \mathcal{G}$.
2. No RHP pole–zero cancellation between $k(s)$ and any member $g(s) \in \mathcal{G}$.

are equivalent to robust (internal) stability.

First we prove that (2.14) is equivalent to condition 1 above, when the nominal loop is internally stable.

$$[1+g(s)k(s)]^{-1} \text{ stable} \qquad \forall g(s) \in \mathcal{G} \qquad (2.15)$$

$$\iff 1+g_o(s)k(s)[1+W(s)\delta] \neq 0 \qquad \forall |\delta| < 1, s \in \mathbb{C}_+ \qquad (2.16)$$

$$\iff 1+g_o(s)k(s) \neq g_o(s)k(s)W(s)\delta \qquad \forall |\delta| < 1, s \in \mathbb{C}_+ \qquad (2.17)$$

$$\iff \delta^{-1} \neq \frac{g_o(s)k(s)W(s)}{1+g_o(s)k(s)} \qquad \forall |\delta| < 1, s \in \mathbb{C}_+ \qquad (2.18)$$

$$\iff \left|\frac{g_o(s)k(s)W(s)}{1+g_o(s)k(s)}\right| \leq 1 \qquad \forall s \in \mathbb{C}_+ \qquad (2.19)$$

$$\iff \|T(s)W(s)\|_\infty \leq 1 \qquad (2.20)$$

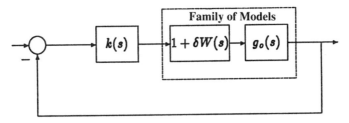

Figure 2.8. Family of models with dynamic multiplicative uncertainty.

The "if" portion of equation (2.19) has been obtained using the fact that, when

$$\left|\frac{g_o(s)k(s)W(s)}{1+g_o(s)k(s)}\right| > |\delta^{-1}|$$

a $\delta \in \mathbb{C}$, $|\delta| < 1$ can always be found such that equality holds. The "only if" part is trivial. For equation (2.20), we have used the maximum modulus theorem, which states that since $T(s)W(s)$ is analytical in $\mathbb{C}_+$ (because of stability of the nominal closed loop and $W(s)$), $|T(s)W(s)|$ achieves its maximum on the boundary of $\mathbb{C}_+$, that is, the $\jmath\omega$ axis. Next, we complete the proof of the theorem.

($\Longrightarrow$) It is clear that robust (internal) stability implies nominal (internal) stability and equation (2.14) (equivalent to condition 1).

($\Longleftarrow$) On the other hand, it is clear that nominal (internal) stability and equation (2.20) imply condition 1. To complete the proof we need to show that condition 2 also holds. From Lemma 2.3 we have that nominal internal stability implies no RHP pole–zero cancellation between $g_o(s)$ and $k(s)$. Hence for any RHP pole of the open-loop $p_\star \in \mathbb{C}_+$, $T(p_\star) = 1$ (why?). Using (2.14), which is equivalent to condition 1, we have $|W(p_\star)| \le 1$ and hence $1 + \delta W(p_\star) \neq 0$, $\forall |\delta| < 1$. Because $p_\star$ is arbitrary, this implies no RHP pole–zero cancellations between $k(s)$ and any member $g(s) \in \mathcal{G}$ (condition 2). □

Using Figure 2.9, we can interpret condition (2.20) graphically, in terms of the family of Nyquist plots corresponding to the set of loops. First observe that (2.20) is equivalent to

$$|1 + g_o(s)k(s)| \ge |g_o(s)k(s)W(s)|, \quad \forall s = \jmath\omega \qquad (2.21)$$

For a given frequency ω, the locus of all points $z(\jmath\omega) = g_o(\jmath\omega)k(\jmath\omega) + g_o(\jmath\omega)k(\jmath\omega)W(\jmath\omega)\delta$, $\delta \in \mathbb{C}$, $|\delta| < 1$ is a disk $\mathcal{D}(\omega)$, centered at $g_o(\jmath\omega)k(\jmath\omega)$ with radius $r = |g_o(\jmath\omega)k(\jmath\omega)W(\jmath\omega)|$. Since $|1 + g_o(\jmath\omega)k(\jmath\omega)|$ is the distance between the critical point and the point of the *nominal* Nyquist plot corresponding to the frequency $\jmath\omega$, it follows that condition (2.20) is equivalent to requiring that, for each frequency ω, the uncertainty disk $\mathcal{D}(\omega)$ excludes the critical point $z = -1$. Therefore robust stability for SISO systems can be checked graphically by drawing the envelope of all Nyquist plots formed by the set of circles centered at the nominal plot, with radii $|g_o(\jmath\omega)k(\jmath\omega)W(\jmath\omega)|$, and checking whether or not this envelope encloses the critical point $z = -1$. In the MIMO case, although an equivalent condition can also be obtained (see Chapter 4), there is no such graphical interpretation.

The frequency at which $|W(\jmath\omega_\star)| = 1$ corresponds to the limit for which there is 100% uncertainty in the model of the plant. Therefore another interpretation of condition (2.14) is that frequency $\omega_\star$ is the upper limit for

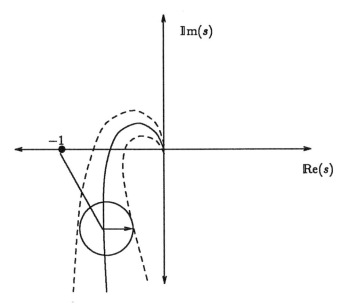

Figure 2.9. Set of Nyquist plots of the family of models.

the bandwidth of the complementary sensitivity function $T(s)$. At frequencies above ω_*, the gain $|T(j\omega)|$ should be less than one, to achieve robust stability.

Gain and phase uncertainty in the loop can be considered as special cases of the dynamic multiplicative description. $W(s) = 1$ and $\delta \in \mathbb{R}, |\delta| < 1$ correspond to gain uncertainty. On the other hand $W(s) = -2$ and $\delta = \frac{1}{2}(1 - e^{j\phi}) \in \mathbb{C}, \phi \in [0, 2\pi) \Rightarrow |\delta| < 1$ corresponds to phase uncertainty. In both cases there is an extra *structure* on the uncertainty δ, which should be a consequence of the extra *a priori* information on model uncertainty, that is, only phase or gain is uncertain. For this reason, the condition of Theorem 2.1 could be conservative if applied to this case and if the *a priori* information is correct. Otherwise it is more prudent to consider the unstructured type of uncertainty described by the set $\mathcal{G}$.

Several examples will serve to illustrate the above concepts.

Example 2.6 *Consider the following nominal model $g_o(s)$ and a second possible plant $g_1(s)$:*

$$g_1(s) = \frac{300}{(s+1)(s+3)(s+100)} \quad (2.22)$$

$$g_o(s) = \frac{3}{(s+1)(s+3)} \quad (2.23)$$

These two plants can be described using the following family, characterized by multiplicative uncertainty represented by a weight $W(s)$ and a bound on δ:

$$\mathcal{G} = \{g(s) : g(s) = g_o(s)[1 + \delta W(s)], \ |\delta| \leq 1\} \qquad (2.24)$$

$$W(s) = \frac{s}{s + 100} \qquad (2.25)$$

It is easily verified that g_1 corresponds to $\delta = -1$. Note that the set $\mathcal{G}$ also includes many other plant models. For instance, for $\delta = 1$ we obtain

$$g_2(s) = \frac{6(s + 50)}{(s + 1)(s + 3)(s + 100)} \qquad (2.26)$$

Therefore if we only need to include the models $g_1(s)$ and $g_o(s)$ or even a finite set, the description could be unnecessarily conservative. As a consequence, any design that applies to all members of $\mathcal{G}$ could be conservative as well (what will be sacrificed?).

In the next example we illustrate the use of multiplicative uncertainty to describe uncertainty in the location of the zeros of a transfer function.

Example 2.7 *Consider a set of models with uncertainty in the high-frequency dynamics ($\omega > 120$ rad/s) of the numerator polynomial.*

$$\mathcal{G} = \left\{ \frac{3[1 + \delta/5 + s(\delta/100)]}{(s + 1)(s + 3)}, \ |\delta| < 1 \right\} \qquad (2.27)$$

$$g_o(s) = \frac{3}{(s + 1)(s + 3)} \quad \text{for } \delta = 0 \qquad (2.28)$$

This uncertainty can be represented as multiplicative dynamic uncertainty using the weighting function $W(s) = 0.2[1 + (s/20)]$. Its graphical representation is similar to the one of Figure 2.7. The system is known with a 20% relative error up to 10 rad/s. Above 100 rad/s the model has no information on the system that may be useful for control design ($|W(j\omega)| > 1$). According to condition (2.20), this frequency is the upper limit for the bandwidth of the complementary sensitivity function $T(s)$, to achieve robust stability.

Other Uncertainty Descriptions Additional types of dynamic uncertainty descriptions, similar to (2.13), can be formulated as follows:

$$\mathcal{G}_a = \{g_o(s) + \delta W_a(s), \ |\delta| < 1\} \qquad (2.29)$$

$$\mathcal{G}_q = \left\{ g_o(s)[1 + \delta W_q(s)]^{-1}, \ |\delta| < 1 \right\} \qquad (2.30)$$

$$\mathcal{G}_i = \left\{ g_o(s)[1 + \delta W_i(s) g_o(s)]^{-1}, \ |\delta| < 1 \right\} \qquad (2.31)$$

40 SISO SYSTEMS

These descriptions are defined as additive, quotient, and inverse dynamic uncertainties, respectively. The difference between the description of uncertainty by any of the above depends on the specific application under consideration. Take, for example, the case where a high-order model (even infinite-dimensional) $g(s)$ must be approximated by a lower-order one $g_r(s)$. The approximation error can be considered as *additive dynamic* uncertainty. The set of models "centered" at the nominal $g_r(s)$, which includes the high-order one $g(s)$, can be defined as follows:

$$\mathcal{G}_a \triangleq \{g_r(s) + \delta W_a(s), \ |\delta| < 1\} \tag{2.32}$$

The weight $W_a(s)$ can be obtained from the frequency responses of the approximation error (or its upper bound) and $g(s)$. A systematic way to obtain this uncertainty description and its corresponding uncertainty weight will be presented in Chapter 10.

In the same way that multiplicative uncertainty seems to be the "natural" structure to describe uncertainty in the zero locations, *quotient dynamic* uncertainty is the least conservative way to describe open-loop pole uncertainty. This will be illustrated in the following example.

Example 2.8 *Consider the following set of mathematical models, which can be described as a family with* quotient dynamic *uncertainty:*

$$\mathcal{G}_q = \left\{ \frac{1}{[s + 3(2 + \delta)]}, \ |\delta| < 1 \right\} \tag{2.33}$$

$$\triangleq \left\{ g_o(s)\left[1 + \delta_q W_q(s)\right]^{-1}, \ |\delta_q| < 1 \right\} \tag{2.34}$$

which leads to the following nominal model and uncertainty weight:

$$g_o(s) = \frac{1}{s+6}, \quad W_q(s) = \frac{3}{s+6} \tag{2.35}$$

If, instead, the same set would have been described by a family of multiplicative dynamic uncertain models, there would have been an unnecessarily large number of models, and as a consequence the description would have been conservative. If we take the same nominal model, the new uncertainty weight $W_m(s)$ and bound γ_m should be defined such that

$$\{\delta_m W_m(s), \ \delta_m \in \mathbb{C}, \ |\delta_m| < \gamma_m\} \tag{2.36}$$

"covers" the following set:

$$\left\{ \frac{3\delta}{s + 3(2+\delta)}, \ \delta \in \mathbb{C}, \ |\delta| < 1 \right\} \tag{2.37}$$

obtained by considering $(g(s) - g_o(s))/g_o(s)$, $g(s) \in \mathcal{G}_q$. The weight $W_m(s) = 3/(s+3)$ and bound $\gamma_m = 1$ achieve the above condition but include in the new set of models $\mathcal{G}_m$ second-order systems, which were not included in $\mathcal{G}_q$.

We conclude that, depending on the specific application, the description of model uncertainty should be selected appropriately among the above descriptions. If *all* the uncertainty is included in one particular description, it is called global dynamic uncertainty. Otherwise, in many cases, we may adopt two or more uncertainty descriptions to better "fit" the family of models. By this we mean that, taking advantage of the information on the "structure" of the uncertainty in the model, we select the smallest possible set. This is important because, in general, a robust controller should stabilize and provide performance for all possible plants in the set of models; therefore the larger this set, the more conservative the design.

The type of uncertainty description that combines information of different portions of the plant and adopts different uncertainties for each part, that is, different δ's, is called a *structured dynamic* uncertainty description. For example, consider the following set of models with both uncertain zero and uncertain pole locations:

$$\mathcal{G} = \left\{ \frac{s + 2(1 + 0.3\delta_z)}{s + 3(1 + 0.5\delta_p)}, \ |\delta_i| < 1, \ i \in \{z, p\} \right\} \qquad (2.38)$$

$$\stackrel{\triangle}{=} \left\{ g_o(s) \frac{[1 + \delta_1 W_z(s)]}{[1 + \delta_2 W_p(s)]}, \ |\delta_i| < 1, \ i = 1, 2 \right\} \qquad (2.39)$$

In Chapter 7, robustness analysis and design tools for this more general type of uncertainty description will be presented.

Following the proof of Theorem 2.1, necessary and sufficient conditions for the robust stability of the model sets described in Table 2.1 can be obtained (Problem 3), where $S(s) \stackrel{\triangle}{=} [1 + g_o(s)k(s)]^{-1}$ is the sensitivity function of the nominal closed-loop system. We can conclude from this table that different stability conditions correspond to different uncertainty descriptions. This is coherent with the fact stated before, that stability margins are related to a

Table 2.1. Global dynamic uncertainty descriptions and the corresponding robust stability conditions ($|\delta| < 1$)

$\mathcal{G}$	Robust Stability
$g_o(s)[1 + \delta W_m(s)]$	$\|W_m(s)T(s)\|_\infty \leq 1$
$g_o(s) + \delta W_a(s)$	$\|W_a(s)k(s)S(s)\|_\infty \leq 1$
$g_o(s)[1 + \delta W_q(s)]^{-1}$	$\|W_q(s)S(s)\|_\infty \leq 1$
$g_o(s)[1 + \delta W_i(s)g_o(s)]^{-1}$	$\|W_i(s)g(s)S(s)\|_\infty \leq 1$

specific type of uncertainty description. This was the case with the classical phase and gain margins.

The robust stability conditions in Table 2.1 can be interpreted also as stability margins. Take the case of multiplicative uncertainty and suppose we bound it as $|\delta_m| < \gamma_m$, with γ_m a positive real number. Then it is easy to prove that the equivalent robust stability condition for this new set of models is the following:

$$\|T(s)W(s)\|_\infty \leq \frac{1}{\gamma_m} \qquad (2.40)$$

The above means that $\|T(s)W(s)\|_\infty$ can be interpreted as a stability margin for the set of models with multiplicative dynamic uncertainty in the same way g_m is the stability margin for a set of models with gain uncertainty. This is so because the infinity norm of this particular function gives the measure of how much dynamic multiplicative uncertainty γ_m can be tolerated before there exists at least one model in the set that is closed-loop unstable.

It follows that the controller that yields the largest stability margin (with respect to multiplicative uncertainty) can be found by solving the following optimization problem:

$$\inf_{\text{stabilizing } k(s)} \|T(s)W(s)\|_\infty \qquad (2.41)$$

The above is an $\mathcal{H}_\infty$ optimal control, which will be analyzed in detail in Chapter 6.

2.5 NOMINAL PERFORMANCE

Performance will be defined in the frequency domain in terms of the sensitivity functions, according to the compromises stated in Chapter 1. First, the classical and modern control concept of performance as rejection (tracking) of known disturbances and/or noise (references) is stated. Based on the practical limitations of this approach, we state a more realistic definition of performance as rejection (tracking) of sets of disturbances and/or noise (references).

2.5.1 Known Disturbance/Noise/Reference

In classical control theory, a measure of performance of a closed-loop system is based on its ability to reject (in steady state) *known* disturbances, noise, or measurement errors, which may appear at different parts of the loop, that is, sensors, actuators, or outputs. In this context, we define a *known* signal as one having a particular form (sine, step, impulse, etc.), which is known beforehand by the designer, although it is not known at what particular time it will disturb the system. Similarly, a measure of performance may be posed as the ability of the loop to follow *known* reference signals with zero steady-state error. It is a well known fact that rejection of exogenous disturbances and tracking references are equivalent problems. If we consider input signals that are polynomial in time, then we can classify stable systems according to the highest degree of the input that can be tracked with finite steady-state error. An internally stable system is said to be of *type n* if it can track (or reject) a polynomial input of degree up to n with zero steady-state error. It is easy to show that a type n FDLTI system must satisfy the condition $\lim_{s\to 0} s^n g(s)k(s) = K \neq 0$. For instance, in order to track a step input (or reject a step disturbance), the system must be at least of type 1 (i.e., $\lim_{s\to 0} sg(s)k(s) \triangleq K_v \neq 0$). Similarly, to track a ramp reference (or reject a ramp disturbance), the system must be at least of type 2 ($\lim_{s\to 0} s^2 g(s)k(s) \triangleq K_a \neq 0$).

In modern control theory, the same concept of performance has been used: rejection of *known* disturbances in the state or output of the system or tracking a *known* reference signal. In addition, the concept of optimal control is introduced. In this framework the controller seeks to minimize a certain functional, which quantifies the compromise between tracking (rejection, stabilization) speed and the control signal energy. It is possible to state this optimization problem in both a deterministic and a stochastic context. In the latter, the perturbing signal is considered as a stochastic process of *known* covariance. In either case, the optimal controller that minimizes the functional will remain optimal as long as the disturbance or the reference signal matches exactly the assumptions made in the design process. If this hypothesis is not met, there is no guarantee of optimal, or even "good," disturbance rejection or tracking [100].

We may conclude that, in both cases, the fact that the designer should have a clear knowledge of the disturbance, noise, or reference to be tracked poses a serious constraint. If he/she designs for the rejection of a particular signal, but it turns out that the disturbing signal is different, performance could seriously be degraded. In the same case, if the design has been performed by an optimization method, the performance no longer will be optimal. Conceptually, this is the same limitation we found for closed-loop stability. If we stabilize a nominal model, there is no guarantee that the closed loop will remain stable if the model is changed. For these reasons, classical and modern control techniques require a clear knowledge of the nominal plant model and the external signals for correct (or optimal) stabilization and performance.

In practice, the designer does not have such detailed information. For this reason, the assumptions on the system and external signals should be relaxed. For the case of uncertainty in the system description, the approach pursued in the last section was to consider a family of models, rather than a single nominal model. Similarly, in the case of performance, a *family* of disturbances, noise, or references will be considered. An example of a practical situation

44 SISO SYSTEMS

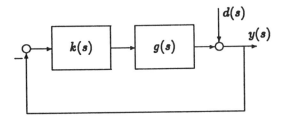

Figure 2.10. Feedback loop with output disturbance.

where this relaxed assumption is necessary is the case of robotic manipulators, when different trajectories need to be tracked in situations where the workspace is changing with time. Also, we could have sinusoidal disturbances with frequencies contained in a certain uncertainty *band*, or even less information, as the case of energy-bounded signals. In the next section we present the performance analysis when defined with respect to a family of external signals, acting on the *nominal* model of the system.

2.5.2 Bounded Disturbances at the Output

Consider the block diagram of Figure 2.10, with output $y(s) = S(s)d(s) = [1 + g(s)k(s)]^{-1}d(s)$. If the performance objective is to minimize the effect of the disturbance $d(s)$ at the output $y(s)$, a stabilizing controller that makes $S(s)$ as "small" as possible needs to be designed. The trivial solution to this problem is $k(s) \to \infty$, so that $S(s) \to 0$, but there is no guarantee of nominal stability. Furthermore, as mentioned in the introductory chapter, a fundamental constraint of any feedback loop is the equation $S(s) + T(s) = 1$. Therefore in this case we would have $T(s) \to 1$ at all frequencies. According to condition (2.14) of the last section, this implies that robust stability can be guaranteed *only* if the uncertainty in the plant model is less than 100% error *at all frequencies*.

As a consequence, we should seek a controller that makes $S(s)$ "small" in a certain frequency range of interest for the particular application on hand (see Figure 2.11). This could be the case when output disturbance rejection needs to be achieved at low frequencies, as in the position control of large mechanical systems with very low natural resonance frequencies. Further-

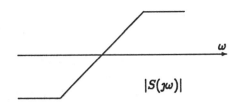

Figure 2.11. Desired frequency distribution for $S(s)$.

NOMINAL PERFORMANCE 45

more, there could be situations where there is a certain *a priori* knowledge of the frequency contents of the disturbances. The frequency range where $S(s)$ needs to be "small" or the frequency content of the disturbances can be represented by the weights $W_y(s)$ and $W_d(s)$, respectively. These weights are dynamic systems, which represent the knowledge of the frequency bands of interest for performance and the disturbance frequency content. Both can be incorporated into the block diagram of the system as in Figure 2.12.

Our next step is to define precisely the meaning of "small" $S(s)$. The size of a transfer function, in particular the sensitivity function, will be measured by its induced norm. For a general LTI operator $A : x \to A \star x$ the induced norm is defined as follows:

$$\|A\|_{\alpha \to \beta} \triangleq \sup_{\|x\|_\alpha \leq 1} \|A \star x\|_\beta \qquad (2.42)$$

In particular, the operation $\star$ will be equivalent to convolution, if both input and output are time signals, or product, in the case where they are represented by their Laplace transforms. Therefore the induced norm will depend entirely on the norms we adopt to measure both the input and the output. For convenience, in most of this text, we will adopt the energy of the signals as a measure of their size. In this case, according to Parseval's theorem, for both representations of a signal, time or frequency, their norm will be the same, that is, $\|x(t)\|_2 = \|x(s)\|_2$. With this in mind, we define nominal performance as follows:

Definition 2.3 *Nominal performance of the feedback loop of Figure 2.12 is achieved if and only if the weighted output remains bounded by unity, that is, $\|W_y(s)y(s)\|_2 \leq 1$, for all disturbances in the set $\{d \in \mathcal{L}_2, \ \|d\|_2 \leq 1\}$, and for all other external inputs to the system equal to zero.*

Without loss of generality, we have considered unitary bounds in both cases, because any other bound can be absorbed into the weights $W_y(s)$ and $W_d(s)$, by linearity.

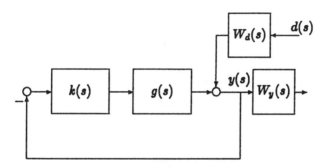

Figure 2.12. Augmented feedback loop with performance weights.

Checking nominal performance by using Definition 2.3 directly requires a search over all bounded $\mathcal{L}_2$ disturbances, which is clearly not possible. Hence, as in the case of robust stability, it is necessary to find a computationally verifiably equivalent condition. This condition will be obtained by exploiting the relationship between the $\mathcal{L}_2$ to $\mathcal{L}_2$ induced norm of a LTI system g and its frequency response.

Theorem 2.2 *The feedback system of Figure 2.12 achieves nominal performance, as defined in Definition 2.3, if and only if*

$$\|W_y(s)S(s)W_d(s)\|_\infty \triangleq \sup_{j\omega} |W_y(j\omega)S(j\omega)W_d(j\omega)| \leq 1 \qquad (2.43)$$

Proof. The operator that maps the input signal $d(s)$ to the weighted output $W_y(s)y(s)$ is $W_y(s)S(s)W_d(s)$. The induced norm from $\mathcal{L}_2$ to $\mathcal{L}_2$ is the infinity norm (see Appendix A). Therefore

$$\sup_{\|d\|_2 \leq 1} \|W_y y\|_2 = \|W_y y\|_{\mathcal{L}_2 \to \mathcal{L}_2} \leq 1 \qquad (2.44)$$

$$\iff \|W_y(s)S(s)W_d(s)\|_\infty \leq 1 \qquad (2.45)$$

Hence nominal performance is equivalent to equation (2.43). $\square$

Note that this condition is similar to the one for robust stability (2.14), although applied to a different transfer function. Nevertheless, there is an important difference between both conditions. The one for robust stability can be interpreted *qualitatively*. In other words, either all members of $\mathcal{G}$ are closed-loop stable, or there is at least one model that is not. There is no "gradual" boundary separation between the set that satisfies (2.14) and its complement. Instead, nominal performance can gradually be relaxed, since it is a *quantitative* property of the closed loop. Specifically, this means that the upper bound of the weighted output to be bounded (or minimized) can be changed to a number $\gamma \geq 1$. This new bound will "gradually" degrade the performance.

This separation between *qualitative* and *quantitative* properties of systems is well known and has nothing to do with robustness but with the definitions of stability and performance. It has been pointed out, because it will be important in the next section, when considering robust performance.

As before, the optimal controller, in the sense of providing optimal distubance rejection, can be found by solving an $\mathcal{H}_\infty$ optimal control problem, in this case of the form

$$\min_{\text{stabilizing } k(s)} \|W_y(s)S(s)W_d(s)\|_\infty \qquad (2.46)$$

The solution to this type of problem will be discussed in detail in Chapter 6.

Note that the only assumption on the disturbances is that they have bounded energy. There is no hypothesis whatsoever on the *shape* of these signals, as in the usual setting of classical and modern control. The weighting functions $W_d(s)$ and $W_y(s)$ only provide an *a priori* knowledge of the frequency content of these signals but do not determine their shape.

A graphical interpretation of the nominal performance condition can be obtained by means of a Nyquist plot (see Figure 2.13). To this end, define $W(s) \triangleq W_y(s) \cdot W_d(s)$ and note that (2.43) is equivalent to

$$|W(j\omega)| \leq |1 + g_o(j\omega)k(j\omega)|, \quad \forall \omega \qquad (2.47)$$

Consider, for each frequency $j\omega$, a disk $\mathcal{D}(j\omega)$ centered at $z = -1$, with radius $r = |W(j\omega)|$. Then (2.47) can be interpreted graphically as nominal performance being achieved if and only if, for every frequency $j\omega$, the disk $\mathcal{D}(j\omega)$ does not include the Nyquist plot of $g_o(j\omega)k(j\omega)$, the nominal loop.

2.5.3 Other Performance Criteria

There can be as many performance criteria as input and output signal pairs may be obtained from the feedback loop of Figure 2.14. The ones that make sense, from a practical point of view, are presented in Table 2.2. In all cases the inputs are $\mathcal{L}_2$ signals (i.e., bounded energy signals) and a generic weight has been considered, which should be interpreted accordingly. In each case, the necessary and sufficient conditions for nominal performance were

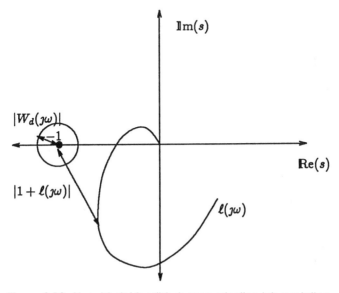

Figure 2.13. Nyquist plot for disturbance rejection interpretation.

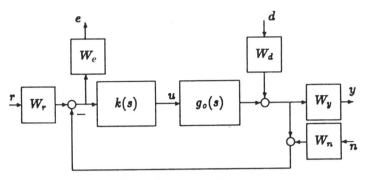

Figure 2.14. General feedback loop.

obtained by considering the $\mathcal{L}_2$ to $\mathcal{L}_2$ norm induced by the corresponding input–output pairs.

Take as an example a tracking problem, whose performance can be defined as follows:

Definition 2.4 *Nominal performance for the tracking problem is achieved if and only if the weighted tracking error remains bounded by unity, that is, $\|W_e(s)e(s)\|_2 \leq 1$, for all reference signals in the set $\{r \in \mathcal{L}_2, \; \|r\|_2 \leq 1\}$, and for all other inputs to the system equal to zero.*

In this case, the tracking error is $e(s) = W_r(s)r(s) - y(s)$ and the weighted error is $W_e(s)e(s) = W_e(s)S(s)W_r(s)r(s)$. We have defined $W_r(s)$ and $W_e(s)$ as the weights representing the frequency content of the reference signals and the frequency band where the tracking error should be bounded, respectively. It follows that in this case nominal performance is achieved if and only if

$$\|W_e(s)S(s)W_r(s)\|_\infty \leq 1 \qquad (2.48)$$

Table 2.2. Nominal performance and equivalent conditions

Input–Output Pair	Nominal Performance
$d \to y$ $r \to e$ $d \to e$ $n \to e$	$\|W(s)S(s)\|_\infty \leq 1$
$r \to u$ $d \to u$ $n \to u$	$\|W(s)k(s)S(s)\|_\infty \leq 1$
$n \to y$	$\|W(s)T(s)\|_\infty \leq 1$

NOMINAL PERFORMANCE

Hence the controller that achieves optimal nominal performance can be obtained by solving the following optimization problem:

$$\min_{\text{stabilizing } k(s)} \|W_e(s)S(s)W_r(s)\|_\infty \qquad (2.49)$$

Nominal performance cannot always be cast in terms of a weighted sensitivity norm, as illustrated by the following example.

Definition 2.5 *Nominal performance for the measurement noise attenuation is achieved if and only if the weighted output noise error remains bounded by unity, that is, $\|W_y(s)y(s)\|_2 \leq 1$, for all possible measurement noise in the set $\{n \in \mathcal{L}_2 : \|n\|_2 \leq 1\}$, and for all other inputs to the system equal to zero.*

In this case, the equivalent condition is a function of the complementary sensitivity function $T(s)$, since $y(s) = T(s)W_n(s)n(s)$. Hence the condition for nominal performance becomes

$$\|W_y(s)T(s)W_n(s)\|_\infty \leq 1 \qquad (2.50)$$

where $W_n(s)$ and $W_y(s)$ are the weights representing the frequency content of the measurement noise and the frequency band where the noisy output should be bounded, respectively.

In this case, the condition for nominal performance can be interpreted graphically by means of the inverse Nyquist plot of $g_o(j\omega)k(j\omega)$ (see

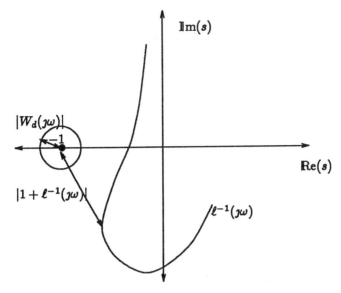

Figure 2.15. Inverse Nyquist plot noise attenuation interpretation.

Figure 2.15). Note that (2.50) is equivalent to

$$\left|1 + [g(j\omega)k(j\omega)]^{-1}\right| \geq |W(j\omega)|, \quad \forall \omega \tag{2.51}$$

where $W(s) \triangleq W_y(s) \cdot W_n(s)$. Hence, in this case, nominal performance is achieved if and only if, at each frequency, the disk centered at $z = -1$ with radius $r = |W(j\omega)|$ does not include the inverse Nyquist plot of $g_o(j\omega)k(j\omega)$.

Finally, we like to remark that, due to the fact that the bounds for input and output signals have been stated in terms of their energy, all the conditions for nominal performance are expressed in terms of the infinity norm. An equivalent formulation can be obtained when the input signals are taken as bounded persistant signals. In this case, the natural norm to measure the "size" of the inputs and the outputs is the time-domain $\mathcal{L}_\infty$ norm (i.e., the $\sup_t$). Since the $\mathcal{L}_\infty$ to $\mathcal{L}_\infty$ induced norm is the $\mathcal{L}_1$ operator norm (see Appendix A), in this case the conditions for nominal performance will have the form of a weighted $\mathcal{L}_1$ norm. The problem of optimizing this norm ($\mathcal{L}_1$ optimal control theory) will be addressed in Chapter 8.

2.6 ROBUST PERFORMANCE

The final goal of robust control is to achieve the performance requirement on all members of the family of models (i.e., robust performance), with a single controller. Clearly, robust stability and nominal performance are prerequisites for achieving robust performance (why?). Next, we will establish a necesary and sufficient condition for robust performance, by making use of the conditions for nominal performance and robust stability. To this end we consider the case of performance defined as disturbance rejection at the output for a family of models described by dynamic multiplicative uncertainty as represented in Figure 2.16.

Definition 2.6 *The feedback loop of Figure 2.16 achieves robust performance if and only if $\|W_y(s)y(s)\|_2 \leq 1$, for all possible disturbances in the set $\{d \in \mathcal{L}_2 \mid \|d\|_2 \leq 1\}$, for all inputs to the system equal to zero, and for all models in the set $\mathcal{G} = \{g : g(s) = [1 + W_\delta(s)\delta]g_o(s), \ \delta \in \mathbb{C}, \ |\delta| < 1\}$.*

Recall that the necessary and sufficient conditions for nominal performance (2.43) and (2.14) and robust stability are

$$\boxed{\text{Nominal performance}} \iff \|W_d(s)S(s)\|_\infty \leq 1 \tag{2.52}$$

$$\boxed{\text{Robust stability}} \iff \|W_\delta(s)T(s)\|_\infty \leq 1 \tag{2.53}$$

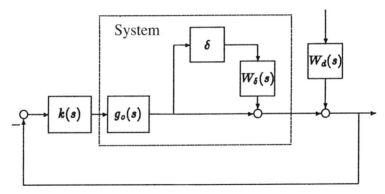

Figure 2.16. Disturbance rejection at the output for a family of model with multiplicative uncertainty.

Robust performance is equivalent to condition (2.43), but with the nominal model $g_o(s)$ replaced by the family of models $g = g_o \left[1 + W_\delta(s)\delta\right]$, $\delta \in \mathbb{C}$, $|\delta| < 1$. We use this fact to obtain a necessary and sufficient condition for robust performance.

Theorem 2.3 *A necessary and sufficient condition for robust performance of the family of models in Figure 2.16 is*

$$\left\| |W_d(j\omega)S(j\omega)| + |T(j\omega)W_\delta(j\omega)| \right\|_\infty \leq 1 \quad (2.54)$$

Proof. Replacing the nominal model $g_o(s)$ in (2.43) by the family of models $g(s)$ yields the following condition for robust performance:

$$\left\| W_d(s) \left[1 + \ell(s)(1 + W_\delta(s)\delta)\right]^{-1} \right\|_\infty \leq 1, \quad |\delta| < 1$$

$$\iff \left| W_d(j\omega)S(j\omega) \left[1 + T(j\omega)W_\delta(j\omega)\delta\right]^{-1} \right| \leq 1, \quad |\delta| < 1$$

$$\iff |W_d(j\omega)S(j\omega)| \leq |1 + T(j\omega)W_\delta(j\omega)\delta|, \quad |\delta| < 1, \forall \omega$$

$$\iff |W_d(j\omega)S(j\omega)| \leq 1 - |T(j\omega)W_\delta(j\omega)|, \quad \forall \omega \quad (2.55)$$

$$\iff |W_d(j\omega)S(j\omega)| + |T(j\omega)W_\delta(j\omega)| \leq 1, \quad \forall \omega \quad (2.56)$$

$$\iff \left\| |W_d(j\omega)S(j\omega)| + |T(j\omega)W_\delta(j\omega)| \right\|_\infty \leq 1 \quad (2.57)$$

where $\ell(j\omega) = g_o(j\omega)k(j\omega)$. To prove the "if" part of (2.55), we consider $\delta = \delta_\star$ such that $T(j\omega)W_\delta(j\omega)\delta_\star$ is real and negative. The "only if" part is immediate. □

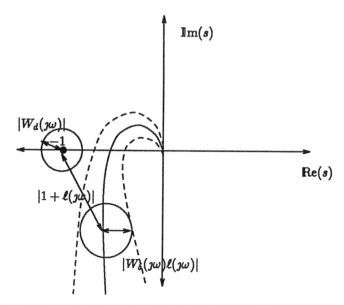

Figure 2.17. Nyquist plot for robust performance interpretation.

As before, a graphical interpretation of the robust performance condition can be obtained by means of the Nyquist plot of Figure 2.17. Note that (2.54) is equivalent to

$$|1 + \ell(j\omega)| - (|W_\delta(j\omega)\ell(j\omega)| + |W_d(j\omega)|) \geq 0 \qquad (2.58)$$

From the figure we see the robust performance requirement combines both the graphical conditions for robust stability of Figure 2.9 and nominal performance of Figure 2.13. Robust performance is equivalent to the disk centered at $z_1 = -1$ with radius $r_1 = |W_d(j\omega)|$ and the disk centered at $z_2 = \ell(j\omega)$ with radius $r_2 = |W_\delta(jw)\ell(j\omega)|$ being disjoint. Clearly, this is more restrictive than achieving the robust stability and nominal performance conditions separately.

Nevertheless, in the SISO case, by scaling the nominal performance and/or robust stability conditions conveniently, we can guarantee robust performance. From the practical point of view, there is an important difference when scaling one or the other. In the first case, as we mentioned before, the condition can be relaxed by changing the upper bound of the output to be bounded. In the second case, condition (2.14) is *exclusive*, therefore either we guarantee internal closed-loop stability for all members of the set or not. Scaling can be accomplished by means of a new "control-oriented" identification of the plant, yielding a new nominal model and a "smaller" uncertainty region. This is not a trivial task, as will be seen in Chapter 10.

Therefore the usual practice is to scale performance, once condition (2.14) is achieved. This is performed by means of a parameter γ, as follows:

$$\left\| \gamma |W_d(j\omega)S(j\omega)| + |W_\delta(j\omega)T(j\omega)| \right\|_\infty \leq 1 \quad \gamma \in [0,1] \quad (2.59)$$

As we have seen before, nominal performance and robust stability can be verified by computing the infinity norm of an appropriately weighted closed-loop transfer function. Synthesis of an optimal controller, which maximizes *either* nominal performance *or* robust stability, is a well established procedure, known as $\mathcal{H}_\infty$ optimal control (see Chapter 6).

However, the condition for robust performance (2.54) cannot be expressed in terms of a closed-loop weighted infinity norm. Rather, a new measure μ, the *structured singular value*, must be used. This measure will be defined in Chapter 7, where we will indicate how to compute it, a procedure known as *μ-analysis*. The synthesis of controllers that optimize this measure, *μ-synthesis*, will also be explained in the same chapter.

In the next paragraph we briefly describe the relation between robust performance and *structured uncertainty*.

Consider the block diagram in Figure 2.18, where two different types of uncertainty have been considered, $W_\delta \delta_1$ and $W_d \delta_2$. The set of models involving both types of uncertainties is the following:

$$\mathcal{G}_s = \left\{ g_o(s) \left[1 + W_d(s)\delta_2\right]^{-1} \left[1 + W_\delta(s)\delta_1\right], \ |\delta_i| < 1, \ i = 1, 2 \right\} \quad (2.60)$$

Next, we obtain a condition for the robust stability of this family of models. The proof follows the one in Theorem 2.1, replacing the multiplicative uncertainty set by (2.60).

Lemma 2.4 *Robust performance of the family of models in Figure 2.16 is*

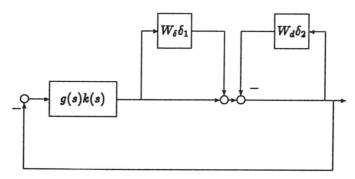

Figure 2.18. Robust performance interpreted as robust stability of sets with structured uncertainty.

equivalent to the robust stability of the family of models described by equation (2.60), shown in Figure 2.18.

Proof. Robust stability of all members of the set $\mathcal{G}_s$ is equivalent to

$$1 + \ell(j\omega)[1 + W_\delta(j\omega)\delta_1][1 + W_d(j\omega)\delta_2]^{-1} \neq 0,$$
$$\forall \omega, \ \forall |\delta_1| < 1 \ |\delta_2| < 1$$
$$\iff 1 + \ell(j\omega) + \ell(j\omega)W_\delta(j\omega)\delta_1 + W_d(j\omega)\delta_2 \neq 0,$$
$$\forall \omega, \ \forall |\delta_1| < 1 \ |\delta_2| < 1$$
$$\iff |1 + \ell(j\omega)| - |\ell(j\omega)W_\delta(j\omega)| - |W_d(j\omega)| \geq 0, \quad \forall \omega$$
$$\iff |T(j\omega)W_\delta(j\omega)| + |S(j\omega)W_d(j\omega)| \leq 1, \quad \forall \omega$$
$$\iff \big\||T(s)W_\delta(s)| + |S(s)W_d(s)|\big\|_\infty \leq 1$$

where $\ell(s) = g(s)k(s)$ is the feedback loop. The above equation is identical to (2.54). □

It follows that robust performance *can always* be cast into robust stability of a set of models with structured uncertainty ([102]). Another interpretation is the following. The problem of rejecting a family of disturbances weighted by $W_d(s)$ can be incorporated as a new type of model uncertainty, as in the block diagram of Figure 2.18.

As mentioned before, the measure over the closed-loop system corresponding to this problem is not a usual system norm. The computation of this new measure should take into account the structural information on the uncertainty of the system (see Chapter 7). The transformation from robust performance to structured robust stability can be represented by means of linear fractional transformations (LFT) as in Figure 2.19 (see Appendix B). In this case δ_2 is called the *performance block*.

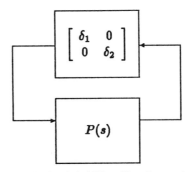

Figure 2.19. Structured robust stability with extra performance block δ_2.

2.7 EXTENSION TO MIMO SYSTEMS

Concerning the extension of the above concepts to MIMO systems, we can state the following:

- There is an equivalent concept for pole–zero cancellations, which is not straightforward as in SISO systems.
- The infinity norm is also used to evaluate nominal performance and robust stability, by means of the singular values.
- The robust stability equivalent condition changes when the uncertainty in the model appears at the input (e.g., actuator uncertainty) or at the output (e.g., sensor uncertainty). The condition number of the nominal model plays an important role in the former case.
- In general, scaling by γ the nominal performance condition, as a method to achieve robust performance, is not as effective as in the SISO case.
- The robust performance condition cannot be computed directly as in equation (2.54). Instead, a minimization problem should be solved.

We conclude that the tools for the analysis of stability and performance robustness in MIMO systems are not a simple extension of the ones used in SISO systems, although we can find many similarities.

A usual procedure to operate with multivariable systems is to apply the tools for SISO models to each input–output pair of the MIMO system. We will show, by means of a simple example, that, in general, this is not correct.

Example 2.9 *In this example we show how the analysis by "loop at a time" may lead to wrong results in the case of a MIMO control problem. Consider the following nominal plant and controller (see Figure 2.20):*

$$G(s) = \begin{bmatrix} \dfrac{s+2}{s} & -\dfrac{1}{2(s+1)} \\ \dfrac{(s+1)(s+2)}{s^2} & \dfrac{1}{2s} \end{bmatrix} \quad (2.61)$$

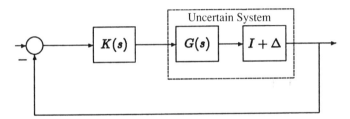

Figure 2.20. Feedback loop of uncertain system.

56 SISO SYSTEMS

$$K(s) = \begin{bmatrix} \dfrac{1}{s+2} & \dfrac{s}{2(s+1)(s+2)} \\ 0 & 1 \end{bmatrix} \quad (2.62)$$

For the above design, the nominal loop $L(s) = G(s)K(s)$ and the complementary sensitivity functions are

$$L(s) = \dfrac{1}{s}\begin{bmatrix} 1 & 0 \\ \dfrac{s+1}{s} & 1 \end{bmatrix} \quad (2.63)$$

$$T(s) = \dfrac{1}{s+1}\begin{bmatrix} 1 & 0 \\ 1 & 1 \end{bmatrix} \quad (2.64)$$

The "loop at a time" analysis is based on using the tools for SISO systems on the problem. To this end, each loop is analyzed by means of the gain and phase margins, while the other loop is closed. This can be seen in Figure 2.21, where the first loop has been opened, while the second one remains closed. The equations in this case are the following:

$$u_2 = -y_2 \implies y_1 = -\dfrac{u_1}{s} \quad (2.65)$$

Similarly, closing the first loop and evaluating the second one, we obtain $y_2 = -u_2/s$. As a consequence of this analysis, there is an infinite gain margin and $90°$ phase margin in both loops. This analysis is based on the fact that the uncertainties in the system do not appear simultaneously in both channels. Next, we show that with small uncertain gains in both channels, the closed-loop system can be unstable.

To this end, take the multiplicative uncertainty of Figure 2.20. If we consider a perturbed model $(I + \Delta)G(s)$ instead of the nominal, the characteristic polynomial will be modified. Take, for example, an uncertainty of the form

$$\Delta = \begin{bmatrix} \delta_1 & \delta_2 \\ 0 & 0 \end{bmatrix} \quad (2.66)$$

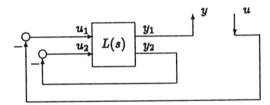

Figure 2.21. "Loop at a time" analysis.

The new characteristic polynomial of the closed-loop system is the following:

$$s^2 + s(2 + \delta_1 + \delta_2) + (1 + \delta_1 + \delta_2) = 0 \tag{2.67}$$

It is easy to verify that the following uncertainty destabilizes the closed loop:

$$\Delta_o = \begin{bmatrix} -\frac{1}{2} & -\frac{1}{2} \\ 0 & 0 \end{bmatrix} \tag{2.68}$$

The perturbed model produces a closed-loop pole at $s = 0$. The "size" of the perturbation can be measured by means of a matrix norm. In particular, we may use the maximum singular value, which results in $\bar{\sigma}(\Delta_o) = \sqrt{2}/2$.

We have considered multiplicative uncertainty, therefore we might as well try to use the tools described in this chapter for SISO systems. To this end we measure the stability margin using the equivalent condition (2.14) (the weight in this case is $W(s) = I$). If we use as the spatial norm $|\cdot|$ for the complementary sensitivity the maximum singular value, we obtain $\|T(s)\|_\infty = 1.618$. Therefore the "smallest" (measured with the same norm) perturbation that destabilizes the loop is its inverse, that is, $\bar{\sigma}(\Delta_*) \approx 0.618$. This minimum perturbation can be found if we allow Δ to be a full complex matrix. Nevertheless, by using the special structure of the uncertainty, it can be shown that the smallest destabilizing perturbation is indeed (2.68), which has norm $\bar{\sigma}(\Delta_o) = \sqrt{2}/2$. This result is not as optimistic as we could have expected from the "loop at a time" analysis, nor is it as pessimistic as the analysis neglecting the special structure of Δ. This is reasonable because we are now allowing a much richer (and more practical) set of uncertainties, which affect the plant model.

We will see in Chapter 4 that the equivalent condition (2.14) for robust stability still holds for MIMO systems, by replacing the magnitude by a matrix norm.

2.8 PROBLEMS

1. Prove that well posedness can be verified by considering only outputs e_1 and e_2 of Figure 2.1.

2. Prove that to verify internal stability, it is enough to check the input-output stability of the four transfer functions $F(s): u_i \to e_j$ with $i, j = 1, 2$.

3. Prove all entries of Table 2.1.

4. Represent as a linear fractional transformation diagram ($[M(s), \Delta]$ structure, see Appendix B), specifying $M(s)$ and Δ, the following set of models:

    (a) A model $g(s)$ with a positive feedback of δ.

(b) A model $g(s)$ multiplied at the input and output by blocks $(1 + \delta_1)$ and $(1 + \delta_2)$, respectively.

(c) A model $g(s)$ multiplied at the input and output by blocks $(1 + \delta_1)$ and $(1 + \delta_2)^{-1}$, respectively.

In all cases there is a controller $k(s)$ that stabilizes the nominal model. Is it possible to find necessary and sufficient conditions for robust stability in all the cases?

5. Consider the following family of models:

$$\mathcal{G} = \left\{ \frac{1}{(s + 2 + \delta)}, \; |\delta| < 1, \; \delta \in \mathbb{C} \right\}$$

and a constant controller $k \in \mathbb{R}$. Find the conditions on the latter to achieve nominal and robust stability. (*Hint:* Use the $[M(s), \Delta]$ structure.) Which is the "worst" model in the family in terms of stability?

6. Solve algebraically or by means of a root locus the last problem for $\delta \in \mathbb{R}$. Compare both results in the case $\delta \in \mathbb{C}$ and explain differences and similarities.

7. Design the simplest controller $k(s)$ that achieves nominal and robust stability and performance for the following set of multiplicative dynamic uncertainty models:

$$\mathcal{G} = \left\{ [1 + W_\delta(s)\delta] \frac{1}{(s+5)}, \; |\delta| < 1, \; \delta \in \mathbb{C} \right\}$$

$$W_\delta(s) = \tfrac{1}{2}(1 + s)$$

Performance is measured by the energy bound of the weighted tracking error for all energy-bounded reference signals.

$$\|W_e(s)e(s)\|_2 \leq 1, \quad \forall \|r(s)\|_2 < 1$$

$$W_e(s) = \frac{1}{2}\left(1 + \frac{1}{s + \epsilon}\right)$$

where ϵ is a positive "small" number.

8. Consider the following family of models:

$$\mathcal{G} = \left\{ \frac{e^{-s\tau}}{(sp+1)}, \quad \tau = \tau_o + \delta, \quad \delta \in [-\bar{\delta}, \bar{\delta}] \right\}$$

where τ_o, $\bar{\delta}$, and p are known. This uncertain model represents in a simple way the thermal dynamics of a person in the shower. The dynamics of the pole $1/p$ is the body thermal model. The time τ_o is the nominal delay for the arrival of the hot water through the pipeline and $\bar{\delta}$ is its uncertainty. Let us consider this everyday practical problem, for which we have developed certain intuition, from the point of view of robust control.

(a) Represent the set $\mathcal{G}$ in terms of a nominal model $g(s)$ with multiplicative uncertainty $[1 + \Delta(s, \delta)]$. Draw a plot of the uncertainty magnitude $|\Delta(\jmath\omega, \delta)|$ as a function of frequency.
(b) Assuming a controller $k(s)$ stabilizes the nominal model, give sufficient conditions to guarantee robust stability. Are these conditions conservative or tight?
(c) Based on the above conditions, design a constant controller $k(s) = k \in \mathbb{R}$ that achieves them. Assume $p = \tau_o = 2\bar{\delta} = 1$.
(d) Which is the "worst" value for δ, from the point of view of robust stability?

9. Prove that no RHP pole–zero cancellation between $g(s) = n_g(s)/d_g(s)$ and $k(s) = n_k(s)/d_k(s)$ (assuming no trivial cancellation between $n_{g(k)}(s)$ and $d_{g(k)}(s)$) is equivalent to $f(s) \stackrel{\Delta}{=} n_g(s)n_k(s) + d_g(s)d_k(s)$ having roots in $\mathbb{C}_+$ that are not shared by any of the numerator or denominator polynomials.

10. Prove that stability of $[1 + g(s)k(s)]^{-1}$ is equivalent to $f(s)$ having all RHP roots included in the RHP roots of $d_g(s)d_k(s)$.

3

STABILIZATION

3.1 INTRODUCTION

Most control problems can be summarized as follows:

Design a controller $K(s)$ that internally stabilizes a given plant $G(s)$ and such that the closed-loop system satisfies some performance specifications, *or show that no controller meeting these specifications exists.*

For instance, consider the SISO system shown in Figure 3.1. As we saw in the previous chapter, the problem of achieving optimal nominal performance can be stated as finding a controller $K(s)$ that minimizes $\|W(s)S(s)\|_\infty$, that is,

$$\min_{K(s) \text{ int. stab.}} \|W(s)S(s)\|_\infty = \min_{K(s)} \left\| \frac{W(s)}{1 + K(s)G(s)} \right\|_\infty \qquad (3.1)$$

where $W(s)$ is a suitable weighting function. However, at this point it is not clear how to proceed to solve this (infinite-dimensional) optimization problem. Following a *classical control* approach, we could attempt to optimize $\|W(s)S(s)\|_\infty$ working directly with $K(s)$. However, the following difficulties are immediately apparent:

1. The set of internally stabilizing controllers is hard to characterize in terms of $K(s)$, as the following example shows.

   Example 3.1 *Consider a SISO plant with the pole–zero configuration shown in Figure 3.2 and assume for simplicity that we limit our search to*

62 STABILIZATION

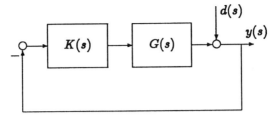

Figure 3.1. A disturbance rejection problem.

static controllers, that is, $K(s) = K > 0$. Then a simple root-locus argument shows that in this case the set of stabilizing controllers of this form is not even connected.

2. Even in cases where the set of stabilizing controllers can be characterized, the resulting optimization problem is hard to solve since the objective function (3.1) is a *nonconvex* function of the optimization variable [in this case $K(s)$].

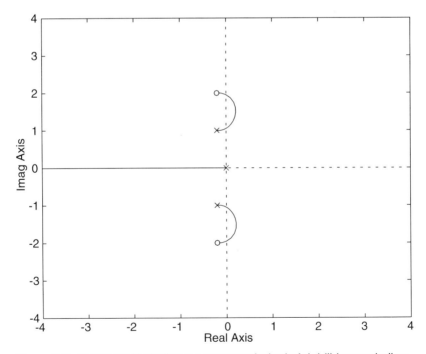

Figure 3.2. Root locus illustrating a nonconnected set of stabilizing controllers.

Alternatively, we could attempt to use tools from *optimal control theory* such as LQR. However, this will require recasting the performance specifications in terms of Q and R, the state and control weighting matrices used in the LQR performance index. Since there are no rules that will allow for exactly expressing specifications of the form (3.1) in terms of Q and R, in practice this leads to a "box-with-a-crank" approach, where the weighting matrices are adjusted through multiple trial-and-error type iterations.

In order to avoid these difficulties, we will parametrize all the controllers that stabilize a given plant in terms of a free parameter $Q(s)$ (a stable proper transfer matrix). This parametrization, originally developed by Youla and co-workers in the continuous time case [323, 324] (see also [94] for an alternative derivation) and, independently, by Kucera [178] for discrete time systems, is usually known as the Youla or YBJK parametrization, although sometimes it is also denoted as the Q-parametrization. A similar parametrization for multidimensional systems was developed by Guiver and Bose [140].

The main result of the chapter shows that all the closed-loop mappings are affine functions of the parameter $Q(s)$. Thus any convex specification given in terms of a closed-loop transfer function translates into a convex specification in $Q(s)$, and the resulting synthesis problem reduces to a convex (albeit infinite-dimensional) optimization problem [59]. We will use these results in later chapters to solve the $\mathcal{H}_2$, $\mathcal{H}_\infty$, and ℓ^1 optimal control problems.

The first step in achieving the parametrization is to extend the concepts of well posedness and internal stability introduced in Chapter 2 to the general case of MIMO plants.

3.2 WELL POSEDNESS AND INTERNAL STABILITY

While the concepts of well posedness and internal stability of MIMO systems are very similar to those introduced in Chapter 2 for the SISO case, there are nevertheless some differences, mostly related to the tools used to assess whether or not a given loop exhibits these properties. In this section we briefly reexamine these concepts and introduce the appropriate changes.

3.2.1 Well Posedness

For multivariable systems, the definition of well posedness is the same as in the scalar case (see Definition 2.1). Thus Lemma 2.1 can still be used to assess whether or not the feedback loop is well posed, provided that the scalar 1 there is replaced by the identity matrix I and in item 4, different from zero ($\neq 0$) is interpreted as *nonsingular*. Nevertheless, in spite the similarity with the SISO case, the lack of well posedness in MIMO systems may not be detected easily, as illustrated by the following example.

Example 3.2 *Consider the following plant model and controller:*

$$G(s) = \begin{bmatrix} \dfrac{1}{s+1} & \dfrac{s+2}{s+1} \\ 1 & \dfrac{1}{s+1} \end{bmatrix}, \quad K(s) = I \qquad (3.2)$$

The elements of the corresponding output sensitivity function $S_o(s)$ are all improper:

$$[I + G(s)K(s)]^{-1} = \begin{bmatrix} (s+1) & -(s+1) \\ -\dfrac{(s+1)^2}{(s+2)} & (s+1) \end{bmatrix} \qquad (3.3)$$

If instead the controller is

$$K(s) = \begin{bmatrix} 1 & 0 \\ -1 & 0 \end{bmatrix} \qquad (3.4)$$

the sensitivity function is not well defined, because $I + G(s)K(s)$ is singular for all s.

3.2.2 Internal Stability

The same concept of internal stability presented in Definition 2.2, as well as Lemma 2.2, can be applied to the multivariable case. The loop interconnection shown in Figure 3.3 is internally stable if and only if all possible input–output transfer functions are stable. On the other hand, the generalization of Lemma 2.3 is not straightforward. The difficulty here is that the concept of RHP pole–zero cancellation (as interpreted in SISO systems) in

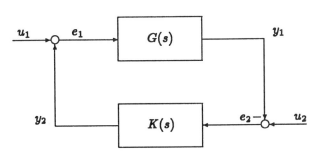

Figure 3.3. Loop interconnection to assess well posedness and internal stability.

transfer matrices does not always compromise internal stability. In fact, it is entirely possible for a transfer matrix to have poles and zeros at the same location of $\mathbb{C}$ (but not in the same input–output "channel"), as we illustrate with the following example.

Example 3.3 *Consider the plant*

$$G(s) = \begin{bmatrix} \dfrac{s+1}{s+2} & 0 \\ 0 & \dfrac{s+2}{s+1} \end{bmatrix} \quad (3.5)$$

From the definitions of multivariable poles and zeros, we have that both are "simultaneously" located at $z_1 = p_1 = -1$ and $z_2 = p_2 = -2$. However, it can easily be verified that a realization containing the modes $\lambda_1 = -1$ and $\lambda_2 = -2$ is both controllable and observable and therefore minimal. Moreover, a loop having $G(s)$ as the closed-loop transfer function from $[u_1 \ u_2]$ to $[e_1 \ e_2]$ is certainly internally stable.

A generalization of Lemma 2.3 will be deferred until Section 3.7, where we introduce the concept of coprime factorizations. In the sequel, to develop a (state-space based) parametrization of all stabilizing controllers we only need the following MIMO version of Lemma 2.2.

Lemma 3.1 *Let*

$$G_{cl}(s) \equiv \left[\begin{array}{c|c} A_c & B_c \\ \hline C_c & D_c \end{array} \right] \quad (3.6)$$

be a minimal state-space realization of the closed-loop system of Figure 3.3, between the inputs $[u_1(s), u_2(s)]$ and outputs $[e_1(s), e_2(s)]$ (or $[y_1(s), y_2(s)]$). Then the following conditions are equivalent:

1. *The feedback loop is internally stable.*
2. *The eigenvalues of A_c are in the open left-half complex plane $\mathbb{C}_-$.*
3. *The four transfer functions obtained between the inputs $[u_1(s), u_2(s)]$ and outputs $[e_1(s), e_2(s)]$ (or $[y_1(s), y_2(s)]$) have their poles in $\mathbb{C}_-$.*
4. *The transfer matrix $\begin{bmatrix} I & -K \\ G & I \end{bmatrix}$ is invertible in $\mathcal{RH}_\infty$.*

Proof. The proof is virtually identical to the proof of Lemma 2.2. □

3.3 OPEN-LOOP STABLE PLANTS

In the case of open-loop stable plants, the set of all linear time invariant, finite-dimensional controllers that internally stabilize the plant has a simple expression, easily interpreted.

Theorem 3.1 *Assume that the interconnection shown in Figure 3.3 is well posed and $G(s) \in \mathcal{RH}_\infty$. Then the set of all FDLTI controllers that internally stabilize the loop is given by*

$$K(s) = Q(s)[I - G(s)Q(s)]^{-1},$$
$$Q(s) \in \mathcal{RH}_\infty, \ \det[I - G(\infty)Q(\infty)] \neq 0$$
(3.7)

Proof. ($\Rightarrow$) Assume that the interconnection is well posed and that $K(s)$ internally stabilizes $G(s)$. We need to show that there exists $Q(s) \in \mathcal{RH}_\infty$ such that $K(s) = Q(s)[I - G(s)Q(s)]^{-1}$. To this effect define

$$Q(s) \triangleq T_{y_2 u_2}(s) = (I + KG)^{-1}K = K(I + GK)^{-1}$$
(3.8)

By assumption, $Q(s)$ is well defined and in $\mathcal{RH}_\infty$. Solving equation (3.8) for K yields

$$K(s) = Q(s)[I - G(s)Q(s)]^{-1}$$
(3.9)

($\Leftarrow$) To complete the proof we need to show that any controller of the form (3.7) internally stabilizes the loop. Recall that checking internal stability is equivalent to checking that the four transfer matrices obtained between the inputs $[u_1(s), u_2(s)]$ and the outputs $[y_1(s), y_2(s)]$ in Figure 3.3 are in $\mathcal{RH}_\infty$. Straightforward calculations using (3.7) yield

$$\begin{bmatrix} y_1 \\ y_2 \end{bmatrix} = \begin{bmatrix} (I - GQ)G & GQ \\ -QG & Q \end{bmatrix} \begin{bmatrix} u_1 \\ u_2 \end{bmatrix}$$
(3.10)

The proof now follows immediately from the fact that by assumption both $G(s)$ and $Q(s)$ are in $\mathcal{RH}_\infty$. $\square$

A simple physical interpretation of the parametrization (3.7) can be obtained by elementary block manipulation techniques as follows. First note

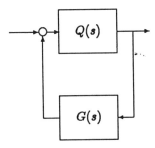

Figure 3.4. Realizing $Q(s)[I - G(s)Q(s)]^{-1}$

OPEN-LOOP STABLE PLANTS   67

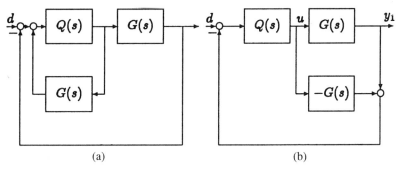

(a)   (b)

Figure 3.5. Physical interpretation of the parametrization for stable plants.

that $Q(s)\,[I - G(s)Q(s)]^{-1}$ can be realized with the block diagram shown in Figure 3.4. Replacing the block $K(s)$ in Figure 3.1 with this realization yields the block diagram in Figure 3.5a. Finally, elementary loop manipulations lead to the block diagram in Figure 3.5b. From this last block diagram we can see that, in the case of a stable plant, the parametrization (3.7) amounts to *cancelling out* the plant (by subtracting in the feedback signal) and replacing it with any *arbitrary* stable dynamics $Q(s)$.

This interpretation also highlights the reason why this approach will fail for the case of open-loop *unstable* plants. While $-G(s)$ cancels the unstable

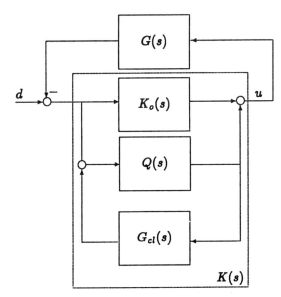

Figure 3.6. Prestabilizing an open-loop unstable plant. G_{cl} denotes the prestabilized plant $(I+GK_o)^{-1}G$.

dynamics $G(s)$ in *some* of the paths (such as T_{ud}), there are some others (e.g., T_{y_1d}) where this cancellation does not take place, yielding an unstable closed-loop transfer function. Thus a different approach is needed to deal with the general case where the plant is not necessarily open-loop stable. Note in passing that one may attempt to parametrize all the controllers that stabilize an open-loop unstable plant by first prestabilizing the plant and then applying the parametrization (3.7) to the prestabilized plant, as illustrated in Figure 3.6. However, it can be shown that while this approach generates stabilizing controllers, it does not generate *all* the controllers that stabilize the plant, *unless* the prestabilizing controller $K_o(s)$ is itself open-loop stable.

3.4 THE GENERAL CASE

In this section we develop a parametrization of all stabilizing controllers for the general case where the plant is not necessarily open-loop stable. The main result shows the existence of a separation-like structure, where all the stabilizing controllers have a familiar observer-based structure. The state-space approach that we use follows after the recent paper [185], although the observer-based structure of all stabilizing controllers was already pointed out as early as 1984 [103].

In order to use the results of this section in later chapters, we consider the slightly more general structure shown in Figure 3.7. Here the inputs to the plant have been partitioned into two groups: w, containing exogenous disturbances and reference signals, and u, the control inputs. Likewise, the outputs have been partitioned into z, containing outputs subject to performance specifications, and y, containing the outputs available to the controller. Throughout this section we assume that the corresponding state-space real-

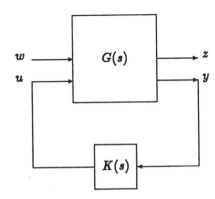

Figure 3.7. The generalized plant.

THE GENERAL CASE

ization of $G(s)$ is given by

$$\left[\begin{array}{c|cc} A & B_1 & B_2 \\ \hline C_1 & D_{11} & D_{12} \\ C_2 & D_{21} & D_{22} \end{array}\right] \qquad (3.11)$$

where the pairs (A, B_2) and (C_2, A) are stabilizable and detectable, respectively. Clearly, this assumption is required for the stabilization problem to be solvable.

3.4.1 Special Problems

In the sequel we consider four special cases of the general structure (3.11), commonly referred to as *full information* (FI), *disturbance feedforward* (DF), *full control* (FC), and *output estimation* (OE). The desired parametrization for the general *output feedback* (OF) case will be obtained by combining these cases.

Full Information This problem corresponds to the case where the plant has the following form:

$$G_{FI} = \left[\begin{array}{c|cc} A & B_1 & B_2 \\ \hline C_1 & D_{11} & D_{12} \\ I & 0 & 0 \\ 0 & I & 0 \end{array}\right] \qquad (3.12)$$

Note that the outputs available to the controller include both the states of the plant *and* the exogenous perturbation, that is,

$$y(t) = \begin{bmatrix} x(t) \\ w(t) \end{bmatrix}$$

hence the name.

Rather than attempting to parametrize all the controllers that stabilize a given FI plant, in this subsection we obtain a weaker result, namely, a parametrization of all the possible control actions. Specifically, we define two controllers K and K' to be equivalent (denoted $K \sim K'$) if the corresponding closed-loop transfer matrices are identical, that is,

$$K \sim K' \iff F_\ell(G, K) = F_\ell(G, K') \qquad (3.13)$$

This equivalence relation induces a partition of all the stabilizing controllers into equivalence classes (an equivalence class is formed by all the controllers that generate the same control action). As we will see in the sequel, in order to obtain the set of all output feedback stabilizing controllers it suffices to obtain the set of all equivalence classes.

Lemma 3.2 *Let F be a constant feedback matrix such that $A + B_2F$ is stable. Then the set of all stabilizing FI controllers[1] can be parametrized as*

$$K_{FI}(s) = [F \quad Q(s)], \quad Q \in \mathcal{RH}_\infty \tag{3.14}$$

Proof. ($\Rightarrow$) Simple algebra shows that closing the loop with the controller (3.14) yields

$$T_{zw}(s) = T_1(s) + T_2(s)Q(s)$$

$$T_1 = \left[\begin{array}{c|c} A + B_2F & B_1 \\ \hline C_1 + D_{12}F & D_{11} \end{array}\right] \tag{3.15}$$

$$T_2 = \left[\begin{array}{c|c} A + B_2F & B_2 \\ \hline C_1 + D_{12}F & D_{12} \end{array}\right]$$

which is stable since $A + B_2F$ and Q are stable.

($\Leftarrow$) Assume now that a FI stabilizing controller $K_{FI} = [K_1 \quad K_2]$ is given. To complete the proof we need to show that there exists $Q \in \mathcal{RH}_\infty$ such that $K_{FI} \sim [F \quad Q]$. To this effect, make the change of variable $v = u - Fx$, and let Q denote the *closed-loop* transfer function from w to v obtained when closing the loop with K_{FI}. Since K_{FI} stabilizes G_{FI}, $Q \in \mathcal{RH}_\infty$. Moreover, we have that

$$u = Fx + v = K_{FI}\begin{bmatrix} x \\ w \end{bmatrix}$$

$$= [F \quad Q]\begin{bmatrix} x \\ w \end{bmatrix} \tag{3.16}$$

Hence $K_{FI} \sim [F \quad Q]$. □

Disturbance Feedforward In this case, the plant has the following state-space realization:

$$G_{DF}(s) = \left[\begin{array}{c|cc} A & B_1 & B_2 \\ \hline C_1 & D_{11} & D_{12} \\ C_2 & I & 0 \end{array}\right] \tag{3.17}$$

In addition, we assume that $A - B_1C_2$ is stable. As we show next, with this additional hypothesis the FI and DF problems are equivalent.

Lemma 3.3 *Consider G_{DF} and assume that $A - B_1C_2$ is stable. Then the following results hold:*

1. *The feedback matrix $K_{DF}(s)$ internally stabilizes G_{DF} if and only if $K_{DF}\begin{bmatrix} C_2 & I \end{bmatrix}$ internally stabilizes G_{FI}. In addition, $F_\ell(G_{DF}, K_{DF}) = F_\ell(G_{FI}, K_{DF}\begin{bmatrix} C_2 & I \end{bmatrix})$.*

---

[1] In the sense of generating all possible control actions, as discussed before.

THE GENERAL CASE   71

2. Given any feedback system $K_{FI}(s)$ that stabilizes G_{FI}, then controller $K_{DF} = F_\ell(P_{DF}, K_{FI})$ internally stabilizes G_{DF}, where

$$P_{DF} = \left[\begin{array}{c|cc} A - B_1 C_2 & B_1 & B_2 \\ 0 & 0 & I \\ \hline \begin{bmatrix} I \\ -C_2 \end{bmatrix} & \begin{bmatrix} 0 \\ I \end{bmatrix} & \begin{bmatrix} 0 \\ 0 \end{bmatrix} \end{array}\right] \quad (3.18)$$

Moreover, $F_\ell(G_{FI}, K_{FI}) = F_\ell [G_{DF}, F_\ell(P_{DF}, K_{FI})]$.

Proof. Result 1 follows immediately from the relationship between the FI and DF problems, illustrated in Figure 3.8.

For result 2, note that in this case the state and output equations can be solved to obtain $\begin{bmatrix} x \\ w \end{bmatrix}$ in terms of $\begin{bmatrix} y \\ u \end{bmatrix}$, yielding

$$\begin{bmatrix} u \\ x \\ w \end{bmatrix} = \left[\begin{array}{c|cc} A - B_1 C_2 & B_1 & B_2 \\ 0 & 0 & I \\ I & 0 & 0 \\ -C_2 & I & 0 \end{array}\right] \begin{bmatrix} y \\ u \end{bmatrix} \quad (3.19)$$

Denote by $\hat{x}$ the states of the "observer" P_{DF} and by $\hat{w}$ the estimated value of the disturbance. From Figure 3.9 we have that a state-space realization of

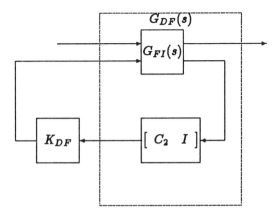

Figure 3.8. Relationship between the FI and DF problems.

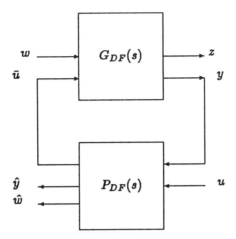

Figure 3.9. Interconnecting $G_{DF}(s)$ and $P_{DF}(s)$.

the interconnection G_{DF}–P_{DF} is given by

$$\begin{aligned}
\dot{x}(t) &= Ax(t) + B_1 w(t) + B_2 u(t) \\
\dot{e}(t) &= (A - B_1 C_2) e(t) \\
z(t) &= C_1 x(t) + D_{11} w(t) + D_{12} u(t) \\
\hat{x}(t) &= x(t) - e(t) \\
\hat{w}(t) &= C_2 e(t) + w(t)
\end{aligned} \quad (3.20)$$

where $e \triangleq x - \hat{x}$. Finally, eliminating the (stable) noncontrollable mode $e(t)$ yields

$$F_\ell(G_{DF}, P_{DF}) = \left[\begin{array}{c|cc} A & B_1 & B_2 \\ \hline C_1 & D_{11} & D_{12} \\ I & 0 & 0 \\ 0 & I & 0 \end{array}\right] = G_{FI} \quad (3.21)$$

Thus $F_\ell(G_{FI}, K_{FI}) = F_\ell [F_\ell(G_{DF}, P_{DF}), K_{FI}]$. The desired result follows now from the fact that $F_\ell [F_\ell(G_{DF}, P_{DF}), K_{FI}] = F_\ell [G_{DF}, F_\ell(P_{DF}, K_{FI})]$ (see Appendix B). □

This lemma shows that as long as the additional assumption concerning the stability of $A - B_1 C_2$ holds, the FI and DF problems are equivalent. This equivalence is exploited next to parametrize all the stabilizing controllers for the DF case. Note in passing that the proof of the lemma clearly shows the necessity of the additional assumption, since the matrix $A - B_1 C_2$ is precisely the error dynamics in (3.20).

Combining Lemmas 3.2 and 3.3, we have that a class of stabilizing controllers for the DF problem is given by

$$K_{DF}(s) = F_\ell(P_{DF}, [F \quad Q]), \quad Q \in \mathcal{RH}_\infty \qquad (3.22)$$

where F is any matrix such that $A + B_2 F$ is stable. Moreover, simple algebra shows that (3.22) can be rewritten as

$$K_{DF}(s) = F_\ell(J_{DF}, Q), \quad Q \in \mathcal{RH}_\infty \qquad (3.23)$$

where

$$J_{DF} = \left[\begin{array}{c|cc} A + B_2 F - B_1 C_2 & B_1 & B_2 \\ F & 0 & I \\ -C_2 & I & 0 \end{array} \right] \qquad (3.24)$$

Indeed a stronger result holds: (3.23) parametrizes *all* the stabilizing DF controllers. This result is summarized in the following lemma.

Lemma 3.4 *Assume that $A - B_1 C_2$ is stable. Then (3.23) parametrizes all the controllers that internally stabilize G_{DF}.*

Proof. ($\Rightarrow$) The fact that $K_{DF}(s)$ internally stabilizes G_{DF} follows immediately from Lemmas 3.2 and 3.3.
($\Leftarrow$) Consider now a DF stabilizing controller K_{DF}. We need to show that K_{DF} can be written in the form (3.23) for some $Q \in \mathcal{RH}_\infty$. Let

$$\hat{J}_{DF} = \left[\begin{array}{c|cc} A & B_1 & B_2 \\ -F & 0 & I \\ C_2 & I & 0 \end{array} \right] \qquad (3.25)$$

and define $Q_o = F_\ell(\hat{J}_{DF}, K_{DF})$. Since $\hat{J}_{DF}$ and G_{DF} share the same triple (A, B_2, C_2) and since K_{DF} stabilizes G_{DF}, it follows that $Q_o \in \mathcal{RH}_\infty$. Moreover, by using the formulas for composition of LFTs given in Appendix B we have that $F_\ell(J_{DF}, Q_o) = F_\ell\left[J_{DF}, F_\ell(\hat{J}_{DF}, K_{DF})\right] = F_\ell(J_{tmp}, K_{DF})$, where

$$J_{tmp} = \left[\begin{array}{cc|cc} A + B_2 F - B_1 C_2 & -B_2 F & B_1 & B_2 \\ -B_1 C_2 & A & B_1 & B_2 \\ \hline F & -F & 0 & I \\ -C_2 & C_2 & I & 0 \end{array} \right]$$

$$= \left[\begin{array}{cc|cc} A - B_1 C_2 & -B_2 F & B_1 & B_2 \\ 0 & A + B_2 F & 0 & 0 \\ \hline 0 & -F & 0 & I \\ 0 & C_2 & I & 0 \end{array} \right] = \left[\begin{array}{cc} 0 & I \\ I & 0 \end{array} \right] \qquad (3.26)$$

Thus $F_\ell(J_{DF}, Q_o) = F_\ell(J_{tmp}, K_{DF}) = K_{DF}$. $\square$

74 STABILIZATION

Full Control In this case the plant has the following realization:

$$G_{FC} = \left[\begin{array}{c|cc} A & B_1 & [I\ 0] \\ \hline C_1 & D_{11} & [0\ I] \\ C_2 & D_{21} & [0\ 0] \end{array}\right] \quad (3.27)$$

The name stems from the fact that the control input

$$u(t) = \begin{bmatrix} u_1(t) \\ u_2(t) \end{bmatrix}$$

affects independently the states x and the performance outputs z.

Rather than obtaining directly a parametrization of all stabilizing controllers for this case, which would essentially follow the same steps of the previous two cases, we will exploit these results by using the concept of *duality*: given a plant $G(s)$, its algebraic dual is obtained by simply transposing it. Consider now the interconnection of Figure 3.7 and the duals of both the controller and the plant. It can easily be checked that

$$T_{zw}^T = [F_\ell(G, K)]^T = F_\ell(G^T, K^T) \quad (3.28)$$

It follows then that K internally stabilizes a plant G if and only if its dual K^T stabilizes G^T. Thus a parametrization of all controllers that stabilize a given plant G can be obtained by simply transposing the corresponding parametrization for its dual G^T. In particular, comparing the realizations (3.12) and (3.27) it is immediate that the FI and FC problems are algebraic duals. It follows that a parametrization of all FC stabilizing controllers can be obtained by simply transposing (3.14). This result is summarized in the following lemma.

Lemma 3.5 *Let L be a constant feedback matrix such that $A + LC_2$ is stable. Then the set of all stabilizing FC controllers[2] can be parametrized as*

$$K_{FC} = \begin{bmatrix} L \\ Q(s) \end{bmatrix}, \quad Q \in \mathcal{RH}_\infty \quad (3.29)$$

Output Estimation Finally, we consider the case where the plant has the realization

$$G_{OE} = \left[\begin{array}{c|cc} A & B_1 & B_2 \\ \hline C_1 & D_{11} & I \\ C_2 & D_{21} & 0 \end{array}\right] \quad (3.30)$$

Note that if $B_2 \equiv 0$, then this reduces to the problem of estimating the output z from the noisy measurements y, hence the name.

---
[2] In the same sense as in the FI case.

It is easily seen that the DF and OE estimation problems are algebraic duals. Thus the relationship between OE and FC is similar to that of the DF and FI problems. In particular, the following properties hold.

Lemma 3.6 *Assume that $A - B_2C_1$ is stable. Then:*

1. *The feedback matrix $K_{OE}(s)$ internally stabilizes $G_{OE}(s)$ if and only if $\begin{bmatrix} B_2 \\ I \end{bmatrix} K_{OE}(s)$ internally stabilizes $G_{FC}(s)$. Moreover, in this case we have that*

$$F_\ell(G_{OE}, K_{OE}) = F_\ell\left(G_{FC}, \begin{bmatrix} B_2 \\ I \end{bmatrix} K_{OE}\right)$$

2. *Given any feedback matrix $K_{FC}(s)$ that stabilizes $G_{FC}(s)$, then controller $K_{OE}(s) = F_\ell(P_{OE}, K_{FC})$ internally stabilizes $G_{OE}(s)$, where*

$$P_{OE} = \left[\begin{array}{c|c} A - B_2C_1 & 0 \quad [I \;\; -B_2] \\ \hline C_1 & 0 \quad [0 \;\; I] \\ C_2 & I \quad [0 \;\; 0] \end{array}\right] \quad (3.31)$$

Moreover, $F_\ell(G_{FC}, K_{FC}) = F_\ell[G_{OE}, F_\ell(P_{OE}, K_{FC})]$

3. *The set of all internally stabilizing OE controllers can be parametrized as $K_{OE} = F_\ell(J_{OE}, Q)$, $Q \in \mathcal{RH}_\infty$, where*

$$J_{OE} = \left[\begin{array}{c|cc} A + LC_2 - B_2C_1 & L & -B_2 \\ \hline C_1 & 0 & I \\ C_2 & I & 0 \end{array}\right] \quad (3.32)$$

3.4.2 The Output Feedback Case

We return now to the problem of parametrizing all the controllers that internally stabilize the plant (3.11). This parametrization will be obtained by combining some of the four special cases discussed before.

Theorem 3.2 *Consider the general output feedback problem for the plant (3.11). Let F and L be constant matrices such that $A + LC_2$ and $A + B_2F$ are stable. Then all the controllers that internally stabilize G are given by*

$$K(s) = F_\ell(J, Q), \quad Q \in \mathcal{RH}_\infty, \; \det[I + D_{22}Q(\infty)] \neq 0 \quad (3.33)$$

where

$$J = \left[\begin{array}{c|cc} A + B_2F + LC_2 + LD_{22}F & -L & B_2 + LD_{22} \\ \hline F & 0 & I \\ -(C_2 + D_{22}F) & I & -D_{22} \end{array}\right] \quad (3.34)$$

76 STABILIZATION

Proof. For simplicity, we will assume for the time being that $D_{22} = 0$. This assumption will be removed at the end of the proof. Consider the state-space realization of $G(s)$, which under this assumption will be denoted $\hat{G}(s)$:

$$\begin{aligned} \dot{x}(t) &= Ax(t) + B_1 w(t) + B_2 u(t) \\ z(t) &= C_1 x(t) + D_{11} w(t) + D_{12} u(t) \\ y(t) &= C_2 x(t) + D_{21} w(t) \end{aligned} \quad (3.35)$$

Using the change of variable $v = u - Fx$, we have that (3.35) can be partitioned into two subsystems as illustrated in Figure 3.10, where

$$G_1 = \left[\begin{array}{c|cc} A + B_2 F & B_1 & B_2 \\ \hline C_1 + D_{12} F & D_{11} & D_{12} \end{array} \right] \quad (3.36)$$

and

$$G_{tmp} = \left[\begin{array}{c|cc} A & B_1 & B_2 \\ -F & 0 & I \\ \hline C_2 & D_{21} & 0 \end{array} \right] \quad (3.37)$$

Since G_1 is stable by choice of F, it follows that a controller $K(s)$ internally stabilizes $\hat{G}(s)$ if and only if it stabilizes G_{tmp}. Moreover, G_{tmp} has an output estimation structure, since $A + B_2 F$ is stable by construction. Hence, from Lemma 3.6, we have that all the controllers that stabilize G_{tmp} (and therefore

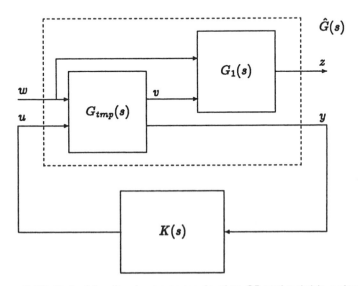

Figure 3.10. Output feedback as a cascade of an OE and a stable system.

$\hat{G}(s))$ are given by $K(s) = F_\ell(\hat{J}, Q)$, where

$$\hat{J} = \left[\begin{array}{c|cc} A + B_2F + LC_2 & -L & B_2 \\ \hline F & 0 & I \\ -C_2 & I & 0 \end{array}\right] \quad (3.38)$$

To complete the proof we need to remove the assumption $D_{22} = 0$. This can be achieved through a standard *loop shifting technique*, consisting in using the change of variable

$$\hat{y} = y - D_{22}u$$

This change of variable subtracts $D_{22}u$ from the output of the plant and wraps a feedback D_{22} around the controller, resulting in the structure shown in Figure 3.11, where $\hat{G}$ has $D_{22} = 0$. Hence the controller $\hat{K}$ is given by $\hat{K} = F_\ell(\hat{J}, Q)$. The original controller K can be recovered by wrapping the feedback $-D_{22}$ around $\hat{K}$ as shown in Figure 3.12. Thus, in the general OF case, we have that $K = F_\ell(J, Q)$, where simple algebra shows that

$$J = \left[\begin{array}{c|cc} A + B_2F + LC_2 + LD_{22}F & -L & B_2 + LD_{22} \\ \hline F & 0 & I \\ -(C_2 + D_{22}F) & I & -D_{22} \end{array}\right] \quad (3.39)$$

Finally, $\det(I + D_{22}Q(\infty)) \neq 0$ is a necessary and sufficient condition for the interconnection of Figure 3.12 to be well posed. □

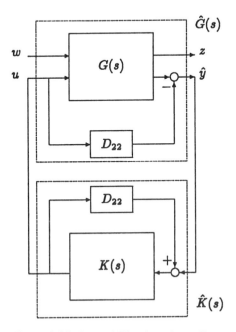

Figure 3.11. Loop shifting transformation.

78 STABILIZATION

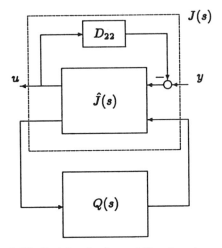

Figure 3.12. Undoing the loop shifting transformation.

3.5 CONTROLLER STRUCTURE AND SEPARATION PRINCIPLE

Consider again the parametrization of all OF stabilizing controllers developed in last section, $K(s) = F_\ell(J, Q)$. A block diagram of this parametrization is shown in Figure 3.13. Note that J has a very familiar structure. Indeed, if we neglect for a moment the additional output and input associated with Q (e and v), then J is precisely a classical observer-based output feedback controller, having F and L as the controller and observer gains, respectively.

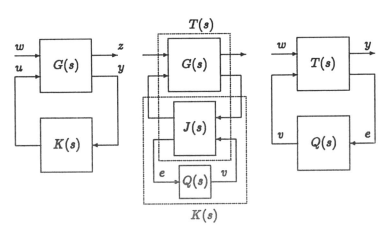

Figure 3.13. Parameterization of all stabilizing controllers.

Thus the parametrization (3.33) can be interpreted as an observer-based controller, where a nominal classical observer is modified so that it produces an additional error signal $e \triangleq y - y_{estim}$. All stabilizing controllers are obtained by filtering this error with arbitrary *stable* dynamics Q and injecting the resulting signal $v = Qe$ back into the observer, where it gets added to the control signal $u = F\hat{x}$. Recall from standard linear systems theory that the estimation error e is uncontrollable from the input u. Thus the additional dynamics Q sees no feedback. Since the nominal controller stabilizes the plant and since Q is stable, in hindsight it is not surprising that the family of augmented controllers also internally stabilizes the plant. The surprising result is the fact that this modified observer paradigm generates *all* stabilizing controllers.

The existence of a separation property is now apparent. All output feedback controllers can be synthesized by solving two decoupled problems: a state-feedback (or FI) problem to obtain a matrix F such that $A + B_2 F$ is stable and an output estimation (or FC) problem to obtain L such that the filtering dynamics $A + LC_2$ are stable. All controllers are obtained by combining the solution to these problems to form the modified observer J and connecting arbitrary stable dynamics Q between e and v. Alternatively, this result can be established from the partition shown in Figure 3.10: here G_1 was obtained essentially using state-feedback and G_{tmp} corresponds to an output estimation problem.

3.6 CLOSED-LOOP MAPPINGS

Recall from the example in the introduction that one of the reasons for seeking a parametrization of all stabilizing controllers is to recast performance specifications given in terms of closed-loop transfer functions into a form having a simpler dependence on the free parameter (in this case Q). As we show next, an additional advantage of the parametrization is that the resulting closed-loop mapping are affine in Q. To see this, consider again the block diagram of Figure 3.13. We have that $T_{zw} = F_\ell[G, F_\ell(J, Q)] = F_\ell(T, Q)$, where

$$T = F_\ell(G, J) \triangleq \begin{bmatrix} T_{11} & T_{12} \\ T_{21} & T_{22} \end{bmatrix}$$

Since, as pointed out before, e is uncontrollable from v, $T_{22} = 0$. Thus T_{zw} reduces to

$$T_{zw} = T_{11} + T_{12}Q(I - T_{22}Q)^{-1}T_{21} = T_{11} + T_{12}QT_{21} \quad (3.40)$$

an affine function of Q. Simple algebra using the explicit expressions given in Appendix B yields the following state-space realization for T:

80 STABILIZATION

$$T = \begin{bmatrix} T_{11} & T_{12} \\ T_{21} & T_{22} \end{bmatrix}$$

$$= \left[\begin{array}{cc|cc} A + B_2 F & -B_2 F & B_1 & B_2 \\ 0 & A + LC_2 & B_1 + LD_{21} & 0 \\ \hline C_1 + D_{12}F & -D_{12}F & D_{11} & D_{12} \\ 0 & C_2 & D_{21} & 0 \end{array} \right] \quad (3.41)$$

3.7 A COPRIME FACTORIZATION APPROACH

In this section we provide an alternative derivation of the Youla parametrization based on the use of coprime factorizations [94, 324, 314]. Besides its historical importance, this approach allows for a simple derivation of some well known results on the simultaneous stabilization of multiple plants and the related problem of strong stabilization, that is, stabilization with stable controllers.

3.7.1 Coprime Factorizations

Definition 3.1 *Two transfer matrices M_r and N_r in $\mathcal{RH}_\infty$ are right coprime over $\mathcal{RH}_\infty$ if there exist two matrices $X_r \in \mathcal{RH}_\infty$ and $Y_r \in \mathcal{RH}_\infty$ such that the following Bezout identity holds:*

$$X_r M_r + Y_r N_r = I \quad (3.42)$$

Similarly, $M_l, N_l \in \mathcal{RH}_\infty$ are said to be left coprime over $\mathcal{RH}_\infty$ if there exist matrices $X_l \in \mathcal{RH}_\infty$ and $Y_l \in \mathcal{RH}_\infty$ such that

$$M_l X_l + N_l Y_l = I \quad (3.43)$$

Remark From these definitions it can easily be shown that if M_r and N_r have a common right factor $U_r \in \mathcal{RH}_\infty$ (i.e., $M_r = \tilde{M} U_r$ and $N_r = \tilde{N} U_r$), then $U_r^{-1} \in \mathcal{RH}_\infty$. A similar property holds for left-coprime matrices. Thus coprimeness among elements of the ring of stable, proper transfer matrices is a generalization of the SISO concept of *no common RHP zeros between numerator and denominator* ([72, 314]).

Every proper, real rational transfer matrix $G(s)$ can be factored as

$$G(s) = N(s)M(s)^{-1} = \tilde{M}(s)^{-1}\tilde{N}(s) \quad (3.44)$$

where $N, M, \tilde{N}$, and $\tilde{M}$ are in $\mathcal{RH}_\infty$ and where N and M are right coprime and $\tilde{N}$ and $\tilde{M}$ are left coprime. These factorizations are known as a *right-coprime* and a *left-coprime* factorization of $G(s)$, respectively. In the next lemma we provide a state-space based algorithm to obtain such factorizations starting from a stabilizable and detectable realization of $G(s)$.

Lemma 3.7 *Let*

$$G(s) \equiv \left[\begin{array}{c|c} A & B \\ \hline C & D \end{array}\right] \qquad (3.45)$$

be a stabilizable and detectable realization of $G(s)$ and F and L any matrices such that $A + BF$ and $A + LC$ are stable. Define

$$\left[\begin{array}{c} M(s) \\ N(s) \end{array}\right] \equiv \left[\begin{array}{c|c} A + BF & B \\ F & I \\ \hline C + DF & D \end{array}\right] \qquad (3.46)$$

and

$$[\tilde{N}(s) \quad \tilde{M}(s)] \equiv \left[\begin{array}{c|cc} A + LC & B + LD & L \\ \hline C & D & I \end{array}\right] \qquad (3.47)$$

Then (3.46) and (3.47) are right- and left-coprime factorizations of $G(s)$, respectively.

Proof. Since $A + BF$ and $A + LC$ are Hurwitz by construction, it follows that all the transfer matrices defined in (3.46) and (3.47) are in $\mathcal{RH}_\infty$. Moreover, it is easily seen that M and $\tilde{M}$ are not singular. Using the formulas for transfer matrix inversion and multiplication given in Appendix B, we have that

$$M^{-1}(s) \equiv \left[\begin{array}{c|c} A & -B \\ \hline F & I \end{array}\right] \qquad (3.48)$$

and

$$N(s)M^{-1}(s) \equiv \left[\begin{array}{cc|c} A + BF & BF & B \\ 0 & A & -B \\ \hline C + DF & DF & D \end{array}\right]$$

$$\equiv \left[\begin{array}{cc|c} A + BF & 0 & 0 \\ 0 & A & -B \\ \hline C + DF & -C & D \end{array}\right] = G(s) \qquad (3.49)$$

where the last equality was obtained by using the similarity transformation

$$T = \left[\begin{array}{cc} I & -I \\ 0 & I \end{array}\right]$$

and elimination of the (stable) uncontrollable modes of the realization. Similarly, it can easily be shown that $\tilde{M}^{-1}(s)\tilde{N}(s) = G(s)$. To show that the pairs M, N and $\tilde{M}, \tilde{N}$ are right and left coprime, respectively, define

$$[X_r(s) \quad Y_r(s)] \equiv \left[\begin{array}{c|cc} A + LC & -(B + LD) & L \\ \hline F & I & 0 \end{array}\right] \qquad (3.50)$$

and
$$\begin{bmatrix} Y_l(s) \\ X_l(s) \end{bmatrix} \equiv \left[\begin{array}{c|c} A+BF & L \\ F & 0 \\ -(C+DF) & I \end{array} \right] \qquad (3.51)$$

Clearly, these four matrices are in $\mathcal{RH}_\infty$. Using the formulas for transfer matrix multiplication yields (after some tedious algebra and eliminating unobservable and uncontrollable modes)

$$\begin{bmatrix} X_r(s) & Y_r(s) \\ -\tilde{N}(s) & \tilde{M}(s) \end{bmatrix} \begin{bmatrix} M(s) & -Y_l(s) \\ N(s) & X_l(s) \end{bmatrix} = \begin{bmatrix} I & 0 \\ 0 & I \end{bmatrix} \qquad (3.52)$$

$\square$

Remark A factorization of the form (3.52) is called a *double-coprime factorization* of the transfer matrix $G(s)$. The state-space formulas presented here for computing double-coprime factorizations are due to Nett et al. [215].

Using these concepts we are ready now to present a MIMO extension of Lemma 2.3.

Theorem 3.3 *Assume that the plant and controller in the feedback loop shown in Figure 3.3 admit the following coprime factorizations:* $G(s) = N(s)M^{-1}(s) = \tilde{M}^{-1}(s)\tilde{N}(s)$, $K(s) = N_k(s)M_k^{-1}(s) = \tilde{M}_k^{-1}(s)\tilde{N}_k(s)$. *Assuming well posedness, the following items are all equivalent:*

1. *The feedback loop is internally stable.*
2. $\begin{bmatrix} M & -N_k \\ N & M_k \end{bmatrix}$ *is invertible in* $\mathcal{RH}_\infty$.
3. $\begin{bmatrix} \tilde{M}_k & \tilde{N}_k \\ -\tilde{N} & \tilde{M} \end{bmatrix}$ *is invertible in* $\mathcal{RH}_\infty$.

Proof. (1 $\iff$ 2) Recall from Lemma 3.1 that internal stability of the loop is equivalent to the transfer matrix $\begin{bmatrix} I & -K \\ G & I \end{bmatrix}$ being invertible in $\mathcal{RH}_\infty$. Explicit calculations using the (right) coprime factorizations of G and K yield

$$\begin{bmatrix} I & -K \\ G & I \end{bmatrix}^{-1} = \begin{bmatrix} M & 0 \\ 0 & M_k \end{bmatrix} \begin{bmatrix} M & -N_k \\ N & M_k \end{bmatrix}^{-1} \qquad (3.53)$$

Using the (right) coprimeness of M, N and M_k, N_k, it can be shown that the matrices

$$\begin{bmatrix} M & 0 \\ 0 & M_k \end{bmatrix} \quad \text{and} \quad \begin{bmatrix} M & -N_k \\ N & M_k \end{bmatrix}$$

are right coprime.

A COPRIME FACTORIZATION APPROACH

Since $M, M_k \in \mathcal{RH}_\infty$ and the matrices in the right-hand side of (3.53) only share common elements invertible in $\mathcal{RH}_\infty$, it follows that:

$$\begin{bmatrix} I & -K \\ G & I \end{bmatrix}^{-1} \in \mathcal{RH}_\infty \iff \begin{bmatrix} M & -N_k \\ N & M_k \end{bmatrix}^{-1} \in \mathcal{RH}_\infty \quad (3.54)$$

The equivalence of items 1 and 3 can be established proceeding in the same way using the left-coprime factorizations of G and K by noting that

$$\begin{bmatrix} I & -K \\ G & I \end{bmatrix}^{-1} \in \mathcal{RH}_\infty \iff \begin{bmatrix} \tilde{M}_k & \tilde{N}_k \\ -\tilde{N} & \tilde{M} \end{bmatrix}^{-1} \in \mathcal{RH}_\infty \quad (3.55)$$

□

Next, we show that given a stabilizable plant $G(s)$ there exists an internally stabilizing controller $K_o(s) = U_o V_o^{-1} = \tilde{V}_o^{-1} \tilde{U}_o$ such that the factors $U_o, V_o, \tilde{U}_o,$ and $\tilde{V}_o$ are precisely the components of a double-coprime factorization of G. This result constitutes the first step in the alternative derivation of the Youla parametrization.

Lemma 3.8 *If $G(s)$ is stabilizable, and detectable then there exist right-coprime factors U_o, V_o and left-coprime factors $\tilde{U}_o, \tilde{V}_o$ in $\mathcal{RH}_\infty$ such that*

$$\begin{bmatrix} \tilde{V}_o & \tilde{U}_o \\ -\tilde{N} & \tilde{M} \end{bmatrix} \begin{bmatrix} M & -U_o \\ N & V_o \end{bmatrix} = \begin{bmatrix} I & 0 \\ 0 & I \end{bmatrix} \quad (3.56)$$

Moreover, the controller $K_o(s) = U_o(s)V_o^{-1}(s) = \tilde{V}_o^{-1}(s)\tilde{U}_o(s)$ internally stabilizes the loop.

Proof. The first part of the proof follows immediately from Lemma 3.7 by selecting $\tilde{V}_o = X_r, \tilde{U}_o = Y_r, V_o = X_l,$ and $U_o = Y_l$. To show that the controller K internally stabilizes the loop, note that with this choice of factors we have that

$$\begin{bmatrix} M & -U_o \\ N & V_o \end{bmatrix}^{-1} = \begin{bmatrix} \tilde{V}_o & \tilde{U}_o \\ -\tilde{N} & \tilde{M} \end{bmatrix} \in \mathcal{RH}_\infty \quad (3.57)$$

Internal stability of the loop follows now from Theorem 3.3. □

Lemma 3.8 provides a simple way of finding a stabilizing controller if a double-coprime factorization of the plant is known. As we show next, this controller is *exactly* the familiar observer-based controller obtained by cascading an observer with gain L with a state-feedback gain F. A state-space realization of such an observer-based controller is given by (see Figure 3.14)

$$\begin{aligned} \dot{\hat{x}} &= A\hat{x} + Bu + L[\hat{y} - (-y)] \\ \hat{y} &= C\hat{x} + Du \\ u &= F\hat{x} \end{aligned} \quad (3.58)$$

84 STABILIZATION

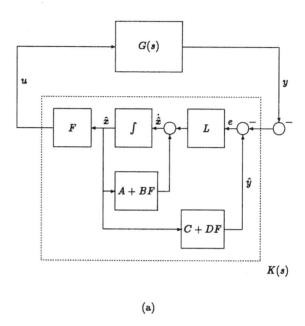

(a)

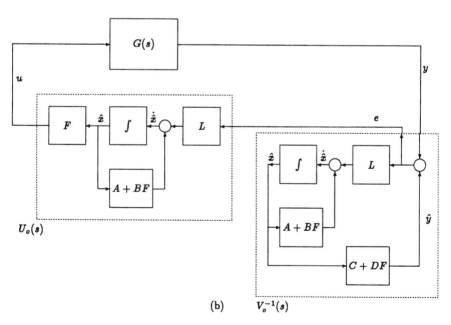

Figure 3.14. The controller of Lemma 3.8 as an observer-based controller: (a) standard realization and (b) factored as $U_o V_o^{-1}$.

Define $e \triangleq \hat{y} + y$ and replace u by the last equation:

$$u = \left[\begin{array}{c|c} A+BF & L \\ \hline F & 0 \end{array}\right] e$$
$$= U_o(s)e$$

$$y = e - \hat{y} = \left[\begin{array}{c|c} A+BF & L \\ \hline -(C+DF) & I \end{array}\right] e$$
$$= V_o(s)e$$
(3.59)

Thus

$$u = U_o(s)e = U_o(s)V_o(s)^{-1}y \qquad (3.60)$$

which coincides with the controller obtained in Lemma 3.8.

Next, we use the results of Theorem 3.3 and Lemma 3.8 to obtain an alternative derivation of the Youla parametrization.

Theorem 3.4 *The set of all controllers that internally stabilize the plant $G(s)$ can be parametrized as*

$$K(s) = U(s)V(s)^{-1} = \tilde{V}(s)^{-1}\tilde{U}(s) \qquad (3.61)$$

where

$$U = U_o - MQ, \qquad V = V_o + NQ$$
$$\tilde{U} = \tilde{U}_o - Q\tilde{M}, \qquad \tilde{V} = \tilde{V}_o + Q\tilde{N}$$
(3.62)

$Q \in \mathcal{RH}_\infty$ *is a free parameter, and where $N, M, \tilde{N}, \tilde{M}, U_o, V_o, \tilde{U}_o,$ and $\tilde{V}_o$ are selected as in Lemma 3.8.*

Proof. First, we establish that for any $Q \in \mathcal{RH}_\infty$ the controller K internally stabilizes the loop. To this effect consider the product

$$\begin{bmatrix} \tilde{V} & \tilde{U} \\ -\tilde{N} & \tilde{M} \end{bmatrix} \begin{bmatrix} M & -U \\ N & V \end{bmatrix}$$
$$= \begin{bmatrix} \tilde{V}_o + Q\tilde{N} & \tilde{U}_o - Q\tilde{M} \\ -\tilde{N} & \tilde{M} \end{bmatrix} \begin{bmatrix} M & -U_o + MQ \\ N & V_o + NQ \end{bmatrix} = \begin{bmatrix} I & 0 \\ 0 & I \end{bmatrix} \qquad (3.63)$$

where the last equality follows from equation (3.56) and the facts that $\tilde{N}M = \tilde{M}N$ and $\tilde{U}_oV_o = \tilde{V}_oU_o$. Stability of the loop follows now from Theorem 3.3.

To complete the proof we need to show that any stabilizing controller K can be written in the form (3.61) for some $Q \in \mathcal{RH}_\infty$. Start by considering a coprime factorization of K, $K = UV^{-1} = \tilde{V}^{-1}\tilde{U}$. Define

$$Z = \tilde{M}V + \tilde{N}U$$
$$\tilde{Z} = \tilde{V}M + \tilde{U}N$$
(3.64)

86 STABILIZATION

Since K stabilizes the loop, from Theorem 3.3 it follows that

$$\begin{bmatrix} \tilde{V} & \tilde{U} \\ -\tilde{N} & \tilde{M} \end{bmatrix}^{-1} \in \mathcal{RH}_\infty \tag{3.65}$$

and

$$\begin{bmatrix} M & -U \\ N & V \end{bmatrix}^{-1} \in \mathcal{RH}_\infty \tag{3.66}$$

Thus

$$\begin{bmatrix} M & -U \\ N & V \end{bmatrix}^{-1} \begin{bmatrix} \tilde{V} & \tilde{U} \\ -\tilde{N} & \tilde{M} \end{bmatrix}^{-1} = \begin{bmatrix} \tilde{Z} & 0 \\ 0 & Z \end{bmatrix}^{-1} \in \mathcal{RH}_\infty \tag{3.67}$$

Hence $Z^{-1} \in \mathcal{RH}_\infty$. Finally, define

$$Q = M^{-1}\left(U_o - UZ^{-1}\right) \tag{3.68}$$

Note that since $Z^{-1} \in \mathcal{RH}_\infty$, then $MQ \in \mathcal{RH}_\infty$. With this definition of Q we have that

$$\begin{aligned}
V_o + NQ &= V_o + NM^{-1}\left(U_o - UZ^{-1}\right) \\
&= V_o + \tilde{M}^{-1}\tilde{N}\left(U_o - UZ^{-1}\right) \\
&= \tilde{M}^{-1}\left(\tilde{M}V_o + \tilde{N}U_o - \tilde{N}UZ^{-1}\right) \\
&= \tilde{M}^{-1}\left(I - \tilde{N}UZ^{-1}\right) \\
&= \tilde{M}^{-1}\left(Z - \tilde{N}U\right)Z^{-1} = VZ^{-1}
\end{aligned} \tag{3.69}$$

Combining this result with equation (3.68) we have that

$$(U_o - MQ)(V_o + NQ)^{-1} = UZ^{-1}ZV^{-1} = UV^{-1} = K \tag{3.70}$$

To complete the proof we need to show that the transfer matrix Q defined in (3.68) is in $\mathcal{RH}_\infty$. This follows from the fact that both MQ and NQ are in $\mathcal{RH}_\infty$ and that M and N are coprime. The formula for the left-coprime factorization of K can be proved proceeding in a similar fashion. $\square$

It is interesting to examine the parametrization (3.61) in the case where the plant $G(s)$ is open-loop stable. In this case we can take $M = \tilde{M} = -I$, $N = \tilde{N} = -G$, $U_o = \tilde{U}_o = 0$, and $V_o = \tilde{V}_o = I$ (this amounts to taking $K_o = 0$). With this choice (3.61) reduces to

$$K(s) = Q(I - GQ)^{-1} = (I - QG)^{-1}Q \tag{3.71}$$

A COPRIME FACTORIZATION APPROACH

which coincides precisely with the parametrization (3.7) derived in Section 3.3.

Consider now the general parametrization (3.61). By using the matrix inversion lemma, this expression can be rewritten as

$$\begin{aligned}
K &= (U_o - MQ)(V_o + NQ)^{-1} \\
&= (U_o - MQ)V_o^{-1}(I + NQV_o^{-1})^{-1} \\
&= K_o \left[I - NQ(I + V_o^{-1}NQ)^{-1}V_o^{-1} \right] - MQV_o^{-1}(I + NQV_o^{-1})^{-1} \\
&= K_o - (K_oN + M)Q(I + V_o^{-1}NQ)^{-1}V_o^{-1} \\
&= K_o - \tilde{V}_o^{-1}Q(I + V_o^{-1}NQ)^{-1}V_o^{-1}
\end{aligned} \quad (3.72)$$

Assume now that $G(s)$ is not necessarily stable, but such that the controller $K_o = U_o V_o^{-1}$ is by itself open-loop stable, and denote by $G_{cl} = G(I + K_o G)^{-1} = N\tilde{V}_o$, i.e. the plant *prestabilized* using the controller K_o. In this case (3.72) can be rewritten as

$$\begin{aligned}
K &= K_o - \tilde{V}_o^{-1} Q V_o^{-1}(I + V_o^{-1} NQV_o^{-1})^{-1} \\
&= K_o + Q_r(I - N\tilde{V}_o Q_r)^{-1} \\
&= K_o + Q_r(I - G_{cl} Q_r)^{-1}
\end{aligned} \quad (3.73)$$

where $Q_r \triangleq -\tilde{V}_o^{-1} Q V_o \in \mathcal{RH}_\infty$ since by assumption $K_o \in \mathcal{RH}_\infty$. The alternative expression (3.73) can be interpreted as the controller K being obtained through the two-stage process mentioned in Section 3.3 and illustrated in Figure 3.6. First the plant is prestabilized using the controller K_o yielding G_{cl}. A family of stabilizing controllers is then obtained by applying the Youla parametrization for stable plants to G_{cl}. From the discussion above it should be clear that, as long as K_o is stable, this family contains all stabilizing controllers. On the other hand, if K_o is unstable, this derivation fails, since $Q_r \notin \mathcal{RH}_\infty$. Indeed, it can be shown that this two-stage procedure will recover all stabilizing controllers *only if* the plant can be stabilized using a stable controller (i.e., when it is strongly stabilizable).

We end this section by exploring the connections between the parametrization (3.33) obtained using a state-state approach and the parametrization (3.61) obtained in this section. From (3.72) it follows that (3.61) is equivalent to the following LFT:

$$K = F_\ell(\tilde{J}, Q), \qquad \tilde{J} = \begin{bmatrix} U_o V_o^{-1} & \tilde{V}_o^{-1} \\ -V_o^{-1} & -V_o^{-1} N \end{bmatrix} \quad (3.74)$$

Straightforward calculations using the state-space realizations (3.46) and (3.50) and (3.51) yield

$$\tilde{J} = \left[\begin{array}{c|cc} A + BF + LC + LDF & L & B + LD \\ \hline F & 0 & I \\ -(C + DF) & -I & -D \end{array} \right] \quad (3.75)$$

which coicides precisely with (3.34) up to an additional "−" sign in $\tilde{J}_{11}$ and $\tilde{J}_{21}$, due to the "−" sign in front of y_1 in the block diagram of Figure 3.3.

3.8 LFTs AND STABILITY

Before closing this chapter we want to briefly address the issue of stability of a special LFT that will arise later in the book in the context of $\mathcal{H}_\infty$ control. To this effect we begin by introducing the concepts of inner and co-inner transfer matrices.

Definition 3.2 *A LTI system with transfer matrix $T(s) \in \mathcal{RH}_\infty$ is called inner if $T\tilde{\ }(s)T(s) = I$, where $T\tilde{\ }(s) \triangleq T^T(-s)$ (see Appendix B). A square inner transfer function is said to be allpass. Similarly, $T(s) \in \mathcal{RH}_\infty$ is said to be co-inner if $T(s)T\tilde{\ }(s) = I$.*

As we show next, inner systems are norm preserving. This result will become important in the context of both $\mathcal{H}_2$ and $\mathcal{H}_\infty$ control.

Lemma 3.9 *Let U be a LTI inner system and consider a signal $z \in \mathcal{L}_2[0, \infty)$. Then $\|Uz\|_2 = \|z\|_2$.*

Proof. Since $U(s)$ is real rational, we have that $U\tilde{\ }(j\omega) = U^T(-j\omega) = U^*(j\omega)$. The proof follows now from Parseval's theorem, since

$$\begin{aligned}
\|Uz(t)\|_2^2 &= \int_0^\infty [Uz(t)]^T [Uz(t)]\, dt \\
&= \frac{1}{2\pi} \int_{-\infty}^\infty z^*(j\omega) U^*(j\omega) U(j\omega) z(j\omega)\, d\omega \\
&= \frac{1}{2\pi} \int_{-\infty}^\infty z^*(j\omega) z(j\omega)\, d\omega \\
&= \int_0^\infty z^T(t) z(t)\, dt = \|z(t)\|_2^2
\end{aligned} \quad (3.76)$$

□

Given an inner system U and a rational transfer function, we are interested in characterizing the properties of the interconnection shown in Figure 3.15.

Lemma 3.10 *Consider the interconnection shown in Figure 3.15, where $U \in$*

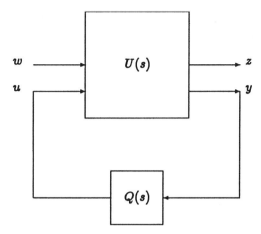

Figure 3.15. Interconnection of an inner system and a rational transfer matrix.

$\mathcal{RH}_\infty$ is inner. Partition U conformally to the inputs and outputs as

$$U = \begin{pmatrix} U_{11} & U_{12} \\ U_{21} & U_{22} \end{pmatrix}$$

and assume that $U_{21}^{-1} \in \mathcal{RH}_\infty$. Then the following statements are equivalent:

1. The interconnection is well posed and internally stable, with $\|T_{zw}\|_\infty < 1$.
2. $Q \in \mathcal{RH}_\infty$ and $\|Q\|_\infty < 1$.

Proof. (2 $\Rightarrow$ 1) The fact that $U^\sim U = I$ implies that $\|U_{22}\|_\infty \leq 1$ (since U_{22} is a compression of U and matrix compressions do not increase the norm). Since $\|Q\|_\infty < 1$ then a simple small gain argument shows that $(I - U_{22}Q)^{-1} \in \mathcal{RH}_\infty$. Since both $U, Q \in \mathcal{RH}_\infty$, well posedness and internal stability follow immediately. To show that $\|T_{zw}\|_\infty < 1$, note that, since U is inner, then

$$\left\| \begin{bmatrix} w \\ u \end{bmatrix} \right\|_2^2 = \left\| \begin{bmatrix} z \\ y \end{bmatrix} \right\|_2^2 \tag{3.77}$$

which implies

$$\|z\|_2^2 \leq \|w\|_2^2 - (1 - \|Q\|_\infty^2)\|y\|_2^2 \tag{3.78}$$

Since $\|Q\|_\infty < 1$, it follows that $\|z\|_2^2 < \|w\|_2^2$, or equivalently $\|T_{zw}\|_\infty < 1$.

(1 $\Rightarrow$ 2) Assume now that the interconnection is internally stable and such that $\|T_{zw}\|_\infty < 1$. To show that $\|Q\|_\infty < 1$ assume to the contrary that there exists a frequency ω_o and a constant vector $\hat{y}$ such that

$$\|Q(j\omega_o)\hat{y}\| \geq \|\hat{y}\| \tag{3.79}$$

(here we use the frequency-domain characterization of the $\mathcal{H}_\infty$ norm, i.e., $\|Q\|_\infty = \sup_\omega \bar{\sigma}[Q(j\omega)]$). Define the input $\hat{w} \triangleq U_{21}^{-1}(I - U_{22}Q)\hat{y}$. Then the control action and outputs corresponding to this input are precisely $\hat{u} = Q\hat{y}$ and $\hat{z} = U_{11}\hat{w} + U_{12}Q\hat{y}$. Since U is inner, we have that

$$\|\hat{z}\|_2^2 + \|\hat{y}\|_2^2 = \|\hat{w}\|_2^2 + \|\hat{u}\|_2^2 \tag{3.80}$$

which combined with (3.79) yields

$$\|\hat{z}\|_2^2 \geq \|\hat{w}\|_2^2 \tag{3.81}$$

However, this last equation contradicts the assumption that $\|T_{zw}\|_\infty < 1$. Thus $\bar{\sigma}[Q(j\omega)] < 1$, $\forall \omega \iff \|Q\|_\infty < 1$. Finally, to show that $Q \in \mathcal{RH}_\infty$, start by considering a (right) coprime factorization of Q, $Q = NM^{-1}$, $N, M \in \mathcal{RH}_\infty$. Since by assumption the interconnection is internally stable, we have that

$$Q(I - U_{22}Q)^{-1} = N(M - U_{22}N)^{-1} \in \mathcal{RH}_\infty \tag{3.82}$$

It follows then that the winding number of $\det(M - U_{22}N)$ as s traverses the Nyquist contour is 0. From internal stability, combined with the facts that $\|U_{22}\|_\infty \leq 1$ and $\|Q\|_\infty < 1$, we have that $\det[I - \alpha U_{22}(j\omega)Q(j\omega)] \neq 0$ for all ω and all $\alpha \in [0, 1]$. Since $\det(I - \alpha U_{22}Q) = \det(M - \alpha U_{22}N) \det(M^{-1})$, it follows that the winding number of $\det[M(j\omega) - \alpha U_{22}(j\omega)N(j\omega)]$ is also 0 for all ω, $\alpha \in [0, 1]$. In particular, for $\alpha = 0$ we get that the winding number of $\det[M(j\omega)]$ is zero and thus $Q \in \mathcal{RH}_\infty$. $\square$

3.9 PROBLEMS

1. Consider the nonminimum phase, open loop unstable plant

$$P(s) = \frac{s-2}{s-1}$$

   (a) Parametrize all the stabilizing controllers.
   (b) Parametrize all the stabilizing controllers that guarantee zero steady-state tracking error to a step input.

2. Parametrize all the controllers that stabilize the plant

$$P(s) = \frac{(s+1)(s-2)}{s(s-1)}$$

   Compare the parametrization against the one obtained in Problem 3.1

3. The purpose of this problem is to explore the effect of nonminimum phase zeros on performance. Consider the feedback system shown in Figure 3.16, where $P(s) = (s - 10)/(s + 2)$. After analyzing the plant it was determined that a desirable closed-loop transfer function $T(s)$ is:

$$T(s) = \frac{P(s)C(s)}{1 + PC} = \frac{0.1s + 1}{s + 1}$$

Can this transfer function be achieved by an internally stabilizing controller? If yes, design the controller. If no, fully justify your answer and then design a controller that achieves a closed-loop function such that:

(a) It has robustness properties as close as possible to those of the target system.

(b) It has the same steady-state step-tracking characteristics as the target system.

4. A system is said to be *strongly stabilizable* if it is stabilizable using an open-loop stable controller. Show that a SISO system is strongly stabilizable if and only if there are an even number of real poles in between every pair of RHP real zeros (including zeros at $s = \infty$, if the plant is strictly proper). This condition is known as the *parity interlacing property*. Hint: Use the coprime factorization form of the Youla parametrization and show that if this condition fails then $Y - NQ$ must necessarily have a RHP zero.

5. Consider now the simultaneous stabilization problem of finding a single controller $K(s)$ that simultaneously stabilizes two SISO plants $G_1(s)$ and $G_2(s)$. Show that such a controller exists, if and only if, $G(s) \doteq G_1(s) - G_2(s)$ is strongly stabilizable.[3]

6. Consider the feedback system shown in Figure 3.16, where $P(s) = (s^2 - 1)/(s^2 - 4)$, and the following performance specifications:

(a) Zero steady-state to a step input.

(b) Perfect rejection of a sinusoidal disturbance with frequency $\omega_o = 1$ rad/sec.

(c) (Open-loop) stable controller.

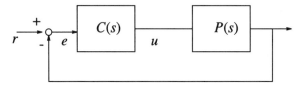

Figure 3.16. Feedback Loop for Problem 3.3.

---

[3] It has been recently shown [51] that the problem of determining whether or not there exists a controller that simultaneously stabilizes three or more plants is undecidable.

Indicate which of these specifications can be achieved with an internally stabilizing controller. Design an internally stabilizing controller that simultaneously achieves as many of these specifications as possible. [*Note:* These specifications are ordered by their desirability, with the most desirable listed first. If you find out that all of these are *not* achievable simultaneously, then when designing your controller you should drop the least important specifications (i.e., higher numbers) first.]

7. Prove, for the following system, that a realization containing poles at $s = 0$ and $s = -2$ is minimal (controllable and observable):

$$G(s) = \begin{bmatrix} \dfrac{s}{s+2} & 0 \\ 0 & \dfrac{s+2}{s} \end{bmatrix} \qquad (3.83)$$

4

LOOP SHAPING

4.1 INTRODUCTION

This chapter extends the analysis results of Chapter 2 to MIMO systems. Both the similarities with the scalar results and the main differences will be stressed. Notable among the latter are the appearance of the condition number of the plant as a new analysis parameter, and the different effects of actuator and sensor uncertainty. Based on these analysis results, the frequency-domain design methodology known as *loop shaping* ([101, 258]) is introduced. This is a generalization of the Bode method used in classical control ([90, 99, 120]).

The problem will be stated in terms of the interconnection of linear fractional transformations (LFT, see Appendix B). This formulation allows for treating, in a unified framework, the models of the plant, controller, and uncertainty and performance weights. As will be shown in the sequel, any linear robust control problem can be posed in this way.

Example 4.1 *Consider the robust tracking problem shown in Figure 4.1. The objective is to synthesize a controller such that, for all elements of the family of plants described by the model*

$$\mathcal{G} = \{[I + \Delta(s)W_\Delta(s)]\, G_o(s), \quad \Delta \in \mathcal{H}_\infty, \ \|\Delta(s)\|_\infty < 1\}$$

the resulting closed-loop system is internally stable and tracks a reference signal of the form

$$\{r(s), \ \|r(s)\|_2 \leq 1\}$$

with tracking error bounded by 1, that is, $\|\tilde{e}(s)\|_2 \leq 1$.

94 LOOP SHAPING

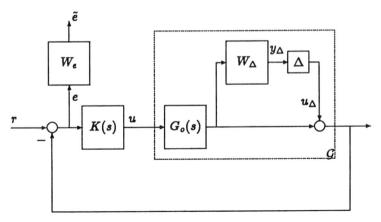

Figure 4.1. Robust tracking problem with sensor uncertainty.

In Figure 4.2 this problem is recast into a LFT form, with the performance input $w = r$ and output $z = \tilde{e}$ connected to the nominal block $M(s)$. The latter includes the nominal model, the controller, and the uncertainty and performance weights, that is, all the input data of the problem. The upper block represents bounded model uncertainty affecting the nominal plant.

Partitioning the nominal system $M(s)$ into four blocks conformally to the input and output vectors, $[u_\Delta, r]$ and $[y_\Delta, \tilde{e}]$, respectively, yields[1]

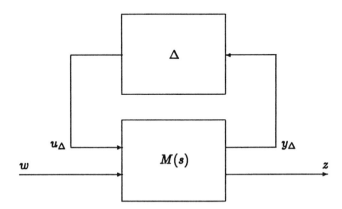

Figure 4.2. Statement of the problem as a LFT.

---

[1] This can be obtained either proceeding directly from the block diagram of Figure 4.1 or by computing the linear fractional transformation $M(s) = F_\ell [P(s), K(s)]$, where $P(s)$ denotes the transfer matrix from the inputs $[u_\Delta, r, u]$ to the outputs $[y_\Delta, \tilde{e}, e]$.

INTRODUCTION

$$M(s) = \begin{bmatrix} M_{11}(s) & M_{12}(s) \\ M_{21}(s) & M_{22}(s) \end{bmatrix}$$
$$= \begin{bmatrix} -W_\Delta(s)T_o(s) & W_\Delta(s)T_o(s) \\ -W_e(s)S_o(s) & W_e(s)S_o(s) \end{bmatrix} \quad (4.1)$$

where $S_o(s) = [I + G_o(s)K(s)]^{-1}$ and $T_o(s) = G_o(s)K(s)[I + G_o(s)K(s)]^{-1}$ denote the output sensitivity and its complement, respectively. Note that according to the definition of internal stability introduced in Chapter 3, in this case the input–output stability of all elements of $M(s)$ is not enough to guarantee that the closed-loop system is internally stable (why?).

Example 4.2 Consider the control system shown in Figure 4.3, where the actual plant has been modeled as a nominal plant $G_o(s)$ subject to inverse dynamic uncertainty, yielding the following family:

$$\mathcal{G} = \left\{ [I - G_o(s)W_\Delta(s)\Delta(s)]^{-1} G_o(s), \ \Delta \in \mathcal{H}_\infty, \ \|\Delta(s)\|_\infty < 1 \right\}$$

The objective is to design a controller that internally stabilizes this family and such that, in all cases, the control action u, weighted by $W_u(s)$, remains bounded, in the presence of measurement noise $n(s)$ described by

$$\{n(s), \ \|n(s)\|_2 \leq 1\}$$

For analysis purposes, we need to recast this problem into the standard form of Figure 4.2, now with input vectors $[u_\Delta, n]$ and outputs $[y_\Delta, \tilde{u}]$. To this end, we open the loop and compute the matrix $P(s)$, which takes as inputs $[u_\Delta, n, u]$ and generates the outputs $[y_\Delta, \tilde{u}, e]$.

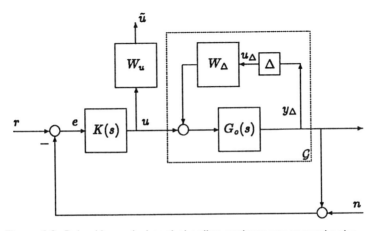

Figure 4.3. Robust bounded control action under measurement noise.

96 LOOP SHAPING

To compute $P(s)$, by linearity, we obtain all transfer functions between each input and output, setting all other inputs equal to zero. In this case, the resulting nominal system is given by

$$P(s) = \begin{bmatrix} \begin{bmatrix} G_o(s)W_\Delta(s) & 0 \\ 0 & 0 \end{bmatrix} & \begin{bmatrix} G_o(s) \\ W_u(s) \end{bmatrix} \\ \begin{bmatrix} -G_o(s)W_\Delta(s) & -I \end{bmatrix} & \begin{bmatrix} -G_o(s) \end{bmatrix} \end{bmatrix}$$
$$= \begin{bmatrix} P_{11}(s) & P_{12}(s) \\ P_{21}(s) & P_{22}(s) \end{bmatrix} \tag{4.2}$$

The above matrix has been partitioned into four blocks, according to the dimensions of the inputs $[u_\Delta, n]$ and u, and the outputs $[y_\Delta, \tilde{u}]$ and e. Once the controller $K(s)$ is designed, a robustness analysis can be performed by interconnecting it to $P(s)$ via a LFT, as follows:

$$F_\ell[P(s), K(s)] = P_{11}(s) + P_{12}(s)K(s)[I - P_{22}(s)K(s)]^{-1} P_{21}(s)$$
$$= M(s) = \begin{bmatrix} S_o G_o W_\Delta & -T_o \\ -W_u T_i W_\Delta & -W_u K S_o \end{bmatrix} \tag{4.3}$$

where $T_i(s) = K(s)G_o(s)[I + K(s)G_o(s)]^{-1}$ denotes the input *complementary sensitivity matrix. The block separation in the matrix $M(s)$ corresponds to the inputs u_Δ and n and the outputs y_Δ and $\tilde{u}$.*

As in Chapter 2, in the sequel we assume that the model uncertainty Δ is linear, time invariant, stable, and bounded. Moreover, without loss of generality, we take the bound to be one,[2] that is,

$$\|\Delta(s)\|_\infty \overset{\triangle}{=} \sup_\omega \bar{\sigma}[\Delta(j\omega)] < 1 \tag{4.4}$$

since any other bound can be absorbed into the nominal block $M(s)$ of Figure 4.2. Thus, for the remainder of this chapter, Δ will be assumed to be an element of $\mathcal{BH}_\infty$, the (open) unit ball in $\mathcal{H}_\infty$. In the case of MIMO systems, it is clear why the name of *global* or *unstructured* uncertainty is adopted. Even in the case where the uncertainty matrix is a constant, that is, $\Delta \in \mathbb{C}^{r \times r}$, it contains no additional structural information we could take advantage of, as could be the case when Δ is diagonal, unitary, or block diagonal. In many practical problems, this structural information is available and, as a consequence, describing the family of plants by means of global uncertainty may be unnecessarily conservative. On the other hand, the tools to analyze sets of models with structured uncertainty are computationally more demanding, as will be seen in further chapters. Therefore a compromise

---

[2] Since $\Delta(s) \in \mathcal{H}_\infty$, the supremum of $\bar{\sigma}[\Delta(s)]$ is achieved at the boundary of its analytic region $\mathbb{C}_+$ ([170]), that is, $s = j\omega$.

between conservatism and computational complexity should be adopted for each specific application.

As in Chapter 2, here the uncertainty and performance weights, $W_\Delta(s)$, $W_e(s)$, and $W_u(s)$, respectively, represent the frequency information available on the model uncertainty and input or output signals. In the multivariable case, in addition to the frequency content information, they contain structural information as well, since they are, in general, block-diagonal matrices.

Continuing with the analysis initiated in Chapter 3, in the sequel we evaluate the stability, performance, and robustness of the standard interconnection shown in Figure 4.2. This analysis entails the following steps:

Nominal Stability. Guarantee the internal stability of the nominal closed-loop system represented by $M(s)$ (Chapter 3).

Nominal Performance. Guarantee that for the nominal plant (i.e. $\Delta \equiv 0$) the energy of the output z due to exogenous disturbances $\{w \in \mathcal{L}_2, \|w(s)\|_2 \leq 1\}$ is bounded by 1, that is, $\|z(s)\|_2 \leq 1$.

Robust Stability. Guarantee the internal stability of the set of closed-loop systems represented by the LFT interconnection of $M(s)$ and $\Delta(s)$.

Robust Performance. Guarantee that, for all possible plants, the energy of the output z due to exogenous disturbances $\{w \in \mathcal{L}_2, \|w(s)\|_2 \leq 1\}$ is bounded by 1, that is, $\|z(s)\|_2 \leq 1$ for all $\{\Delta(s) ; \Delta(s) \in \mathcal{BH}_\infty\}$.

4.2 NOMINAL PERFORMANCE

This second analysis step assumes that nominal stability has already been established, for instance, using the tools covered in Chapter 3. Therefore in the remainder of this section, we will assume that $M(s)$ is stable. The problem statement of Figure 4.2 includes as special cases all possible performance objectives mentioned in Chapter 2. In reference to this figure, we define nominal performance as follows:

Definition 4.1 *The feedback loop shown in Figure 4.2 achieves nominal performance if and only if for the nominal model $M(s)$ ($\Delta(s) \equiv 0$), the energy of the output z due to disturbances with energy bounded by 1 is also bounded by 1. Without loss of generality, this can be stated as follows:*

$$\|z(s)\|_2 \leq 1, \quad \forall \{w \in \mathcal{L}_2, \|w(s)\|_2 \leq 1\} \tag{4.5}$$

This definition represents a worst-case analysis of the effect of exogenous disturbances on the closed-loop system. This means that the bound on the output z should be achieved for all elements in the set of disturbances and therefore for the *worst disturbance*. Since the transfer matrix relating w with z for the nominal case ($\Delta(s) \equiv 0$) is $M_{22}(s)$, it is natural to state an equivalent nominal performance condition in terms of this matrix.

Lemma 4.1 *Assume that the feedback interconnection of Figure 4.2 is internally stable. Then it achieves nominal performance if and only if*

$$\|M_{22}(s)\|_\infty \leq 1$$

Proof. The proof follows from the definition of nominal performance and the fact that (see Appendix A):

$$\sup_{\|w\|_2 \leq 1} \|z\|_2 = \|M_{22}(s)\|_\infty \qquad (4.6)$$

□

Thus the problem of establishing whether or not a system achieves nominal performance reduces to the problem of computing the $\mathcal{H}_\infty$ norm of a transfer matrix. This can be accomplished by using the algorithms proposed in [57, 104, 247] and briefly reviewed in Appendix B. Similarly, the problem of designing a controller that optimizes the nominal performance of the closed-loop system can be reduced to the problem of minimizing the $\mathcal{H}_\infty$ norm of a suitable transfer matrix. This leads to the well known $\mathcal{H}_\infty$ control problem addressed in Chapter 6.

4.3 ROBUST STABILITY

Consider again the interconnection shown in Figure 4.2. The transfer matrix relating the input and output signals is given by the LFT between the nominal and the uncertainty blocks, that is,

$$\begin{aligned} z(s) &= F_u [M(s), \Delta(s)] w(s) \qquad (4.7) \\ &= \left\{ M_{22}(s) + M_{21}(s)\Delta(s) [I - M_{11}(s)\Delta(s)]^{-1} M_{12}(s) \right\} w(s) \end{aligned}$$

where $F_u(\cdot, \cdot)$ denotes the upper LFT defined in Appendix B. If we assume nominal internal stability, all matrices $M_{ij}(s)$, $i, j = 1, 2$ in equation (4.7) are stable. Thus the only way to destabilize the interconnection is through the uncertainty $\Delta(s)$ appearing in the element $[I - M_{11}(s)\Delta(s)]^{-1}$. This observation leads to the following robust stability condition:

Lemma 4.2 *The interconnection of Figure 4.2 is robustly stable ($\forall \Delta(s) \in \mathcal{BH}_\infty$) if and only if the system is nominally internally stable and $\|M_{11}\|_\infty \leq 1$.*

Proof. We will prove that the condition $\|M_{11}\|_\infty \leq 1$ is equivalent to $[I - M_{11}(s)\Delta(s)]^{-1}$ input–output stable, for all $\Delta(s) \in \mathcal{BH}_\infty$.
($\Longrightarrow$) From standard singular value properties we have

$$\underline{\sigma}[I - M_{11}(s)\Delta(s)] \geq 1 - \bar{\sigma}[M_{11}(s)\Delta(s)] > 1 - \bar{\sigma}[M_{11}(s)] \qquad (4.8)$$

Since $\|M_{11}\|_\infty \leq 1$ and M_{11} is stable, from the maximum modulus theorem we have that $\bar{\sigma}[M_{11}(s)] \leq 1$, $\forall s \in \mathbb{C}_+$. Hence

$$\underline{\sigma}[I - M_{11}(s)\Delta(s)] > 0 \tag{4.9}$$

$$\iff \det[I - M_{11}(s)\Delta(s)] \neq 0, \quad \forall s \in \mathbb{C}_+ \tag{4.10}$$

($\impliedby$) Assume there exists $s_\star \in \mathbb{C}_+$ such that $\bar{\sigma}[M_{11}(s_\star)] > 1$. Consider the singular value decomposition $M_{11}(s_\star) = U\Sigma V^*$ and define $\Delta = \alpha V U^*$ with $\alpha = 1/\bar{\sigma}[M_{11}(s_\star)] < 1$. Then $\Delta \in \mathcal{BH}_\infty$ and

$$\det[I - M_{11}(s_\star)\Delta] = \det[I - \alpha U\Sigma U^*] \tag{4.11}$$

$$= \det(I - \alpha\Sigma) \tag{4.12}$$

$$= \prod_{i=1}^n (1 - \alpha\sigma_i) = 0 \tag{4.13}$$

since $(1 - \alpha\sigma_1) = 0$. To complete the proof we need to show that internal nominal stability of the closed-loop and input–output stability of $[I - M_{11}(s)\Delta(s)]^{-1}$, for all $\Delta(s) \in \mathcal{BH}_\infty$, are equivalent to robust internal stability.

($\implies$) All transfer matrices that map inputs to outputs in Figure 4.2 contain the product of stable matrices (because of nominal internal stability) and $[I - M_{11}(s)\Delta(s)]^{-1}$, which is also stable.

($\impliedby$) Consider $u_\Delta = \Delta y_\Delta(s) + n(s)$ in Figure 4.2. The transfer function from $n(s)$ to $u_\Delta(s)$ is given by $T_{u_\Delta,n} = [I - \Delta(s)M_{11}(s)]^{-1}$. Stability of this transfer matrix follows from the assumption of robust internal stability of the interconnection. □

4.4 NOMINAL PERFORMANCE AND ROBUST STABILITY

Clearly, in any practical situation, we should require robust stability of the closed-loop system. However, since there is usually more than one controller achieving robust stability, it is natural to attempt to use these additional degrees of freedom to optimize performance. Toward this goal one could attempt to optimize performance only for the nominal model while stabilizing all members of the family of plants. From the discussion in the two previous sections, it follows that a necessary and sufficient condition for simultaneously achieving nominal performance and robust stability can be obtained by combining Lemmas 4.1 and 4.2, that is,

$$\boxed{\text{Nominal performance}} + \boxed{\text{Robust stability}}$$
$$\iff$$
$$\max\{\|M_{11}\|_\infty, \|M_{22}\|_\infty\} \leq 1 \tag{4.14}$$

while this condition is useful for *analysis* purposes (since it can easily be checked), attempting to use it for synthesis leads to a *multidisk* problem (a problem that entails synthesizing a controller that satisfies simultaneously two or more $\mathcal{H}_\infty$ conditions). At the present time there is no known analytical solution to these problems, although in some cases these are amenable to efficient numerical solutions [217]. An alternative approach is to obtain a *sufficient* condition exploiting the properties of singular values. Since matrix dilations do not decrease norms it follows that

$$\max\{\|M_{11}\|_\infty, \|M_{22}\|_\infty\} \leq 1 \Longleftarrow \left\|\begin{matrix} M_{11} \\ M_{22} \end{matrix}\right\|_\infty \leq 1 \qquad (4.15)$$

The problem of synthesizing a controller such that the closed-loop system satisfies this sufficient condition is known as a *mixed sensitivity minimization* problem and can be solved using standard $\mathcal{H}_\infty$ tools covered in Chapter 6. Note in passing that a controller achieving nominal performance and robust stability can lead to severe performance degradation when confronted with off-nominal conditions. This situation can be avoided by synthesizing a controller achieving robust performance, that is, a guaranteed performance level for all members of the family. As we will see next, it turns out that this can be accomplished by solving a scaled mixed sensitivity problem. Thus, given the difficulties involved in solving exactly a nominal performance/robust stability problem and the fact that a tractable sufficient condition leads to a problem having similar complexity to that of (approximately) solving a robust performance problem, we will devote most of the remainder of the chapter to the solution of the latter.[3]

4.5 ROBUST PERFORMANCE

Definition 4.2 *The feedback loop shown in Figure 4.2 achieves robust performance if and only if all members of the family of plants described by the interconnection of $M(s)$ and $\Delta(s)$ achieve nominal performance. Equivalently, the performance output z should have bounded energy for all bounded energy disturbances and all models in the set. As we have seen before, without loss of generality, this can be restated as follows:*

$$\|z(s)\|_2 \leq 1, \quad \forall \{w \in \mathcal{L}_2, \ \|w(s)\|_2 \leq 1\}, \quad \text{and} \quad \forall \Delta(s) \in \mathcal{BH}_\infty \qquad (4.16)$$

Using Lemma 4.1 to impose that *all* members of the family of plants

---

[3] The situation is different in the case of multiobjective problems, where assessing performance and stability requires using *different* induced norms. Most of the results currently available for these problems are limited to nominal performance/robust stability [301].

achieve nominal performance yields the following necccsary and sufficient condition for robust performance:

$$\|F_u[M(s),\Delta(s)]\|_\infty \leq 1, \quad \forall \Delta(s) \in \mathcal{BH}_\infty \qquad (4.17)$$

This condition cannot be checked easily, since it depends on the uncertainty $\Delta(s)$. Therefore it would be desirable to obtain a simpler condition depending only on the nominal system $M(s)$. However, it turns out that condition (4.17) cannot be stated any longer in terms of the $\mathcal{H}_\infty$ norm of a suitable transfer matrix involving only the nominal plant. Thus the *exact* assessment of robust performance will be postponed until Chapter 7, where new tools will be developed. Instead, at this point we will exploit the conditions for nominal performance and robust stability to obtain a *sufficient* robust stability condition for two different cases: (i) sensor uncertainty and (ii) actuator uncertainty.

4.5.1 Sensor Uncertainty

Consider the robust tracking problem of Figure 4.1. Applying Lemmas 4.1 and 4.2 to this case yields the following nominal performance and robust stability conditions:

$$\boxed{\text{Nominal performance}} \iff \|W_e(s)S_o(s)\|_\infty \leq 1 \qquad (4.18)$$

$$\boxed{\text{Robust stability}} \iff \|W_\Delta(s)T_o(s)\|_\infty \leq 1 \qquad (4.19)$$

Imposing nominal performance on all members of the family of models leads to

$$\boxed{\text{Robust performance}}$$

$$\iff \bar{\sigma}\left\{W_e(j\omega)\left[I+(I+\Delta(j\omega)W_\Delta(j\omega))G_o(j\omega)K(j\omega)\right]^{-1}\right\} \leq 1 \qquad (4.20)$$

$$\iff \bar{\sigma}\left\{W_e(j\omega)\left[(I+G_o(j\omega)K(j\omega))+\Delta(j\omega)W_\Delta(j\omega)G_o(j\omega)K(j\omega)\right]^{-1}\right\} \leq 1$$

$$\iff \bar{\sigma}\left\{W_e(j\omega)S_o(j\omega)\left[I+\Delta(j\omega)W_\Delta(j\omega)T_o(j\omega)\right]^{-1}\right\} \leq 1 \qquad (4.21)$$

$$\Longleftarrow \bar{\sigma}\left[W_e(j\omega)S_o(j\omega)\right]\bar{\sigma}\left\{\left[I+\Delta(j\omega)W_\Delta(j\omega)T_o(j\omega)\right]^{-1}\right\}$$

$$= \frac{\bar{\sigma}\left[W_e(j\omega)S_o(j\omega)\right]}{\underline{\sigma}\left[I+\Delta(j\omega)W_\Delta(j\omega)T_o(j\omega)\right]} \leq 1 \qquad (4.22)$$

102 LOOP SHAPING

$$\Longleftarrow \frac{\bar{\sigma}\,[W_e(j\omega)S_o(j\omega)]}{1 - \bar{\sigma}\,[W_\Delta(j\omega)T_o(j\omega)]} \leq 1 \qquad (4.23)$$

$$\Longleftrightarrow \bar{\sigma}\,[W_e(j\omega)S_o(j\omega)] + \bar{\sigma}\,[W_\Delta(j\omega)T_o(j\omega)] \leq 1 \quad \text{for all } \omega \qquad (4.24)$$

where we used the fact that $\|\Delta(s)\|_\infty < 1$. At this point the following comments are in order:

- While condition (4.24) is similar to the one obtained for SISO systems, in this case it is only *sufficient*.
- From equation (4.23) it is clear that the nominal performance margin we had for the nominal plant $G_o(s)$ decreases as the "size" of the uncertainty $\Delta(s)$ increases.
- As in the univariable case, a method to design for robust performance is to synthesize a controller that minimizes the robust stability condition while scaling the performance requirement (using αW_e with $\alpha < 1$) until (4.24) is satisfied.
- If the system is recast into the standard LFT form, condition (4.24) becomes

$$\bar{\sigma}\,[M_{22}(j\omega)] + \bar{\sigma}\,[M_{11}(j\omega)] \leq 1, \quad \forall \omega \qquad (4.25)$$

This condition is no longer valid in the case of actuator uncertainty. Rather, as we will see next, in this case a different, more restrictive, condition should be used.

4.5.2 Actuator Uncertainty

The case of uncertainty in the input to the nominal model changes the sufficient condition for robust performance, due to the nonconmutative nature of the systems involved.[4] Consider the control problem of Figure 4.4 subject to the same performance requirements as before, but where the uncertainty appears now in the actuator. For simplicity we have assumed that the plant G_o is invertible. It can be shown that in this case the nominal performance condition (4.18) remains unchanged, while the only change in the robust stability condition (4.19) is that the input complementary sensitivity T_i should be used instead of $T_o(s)$; that is,

$$\boxed{\begin{array}{c}\text{Nominal}\\\text{performance}\end{array}} \Longleftrightarrow \|W_e(s)S_o(s)\|_\infty \leq 1 \qquad (4.26)$$

$$\boxed{\begin{array}{c}\text{Robust}\\\text{stability}\end{array}} \Longleftrightarrow \|W_\Delta(s)T_i(s)\|_\infty \leq 1 \qquad (4.27)$$

Imposing nominal performance over all members of the family yields the following robust performance condition:

---
[4] Of course, this situation does not arise in the SISO case.

ROBUST PERFORMANCE

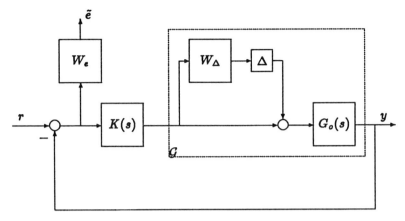

Figure 4.4. Robust tracking problem with actuator uncertainty.

Robust performance

$$\iff \bar{\sigma}\left\{W_e\left[I + G_o(I + \Delta W_\Delta)K\right]^{-1}\right\} \leq 1 \quad (4.28)$$

$$\iff \bar{\sigma}\left\{W_e\left[(I + G_oK) + G_o\Delta W_\Delta K\right]^{-1}\right\} \leq 1 \quad (4.29)$$

$$\iff \bar{\sigma}\left\{W_eS_o\left[I + G_o\Delta W_\Delta KS_o\right]^{-1}\right\} \leq 1 \quad (4.30)$$

$$\iff \bar{\sigma}\left\{W_eS_oG_o\left[I + \Delta W_\Delta K(I + G_oK)^{-1}G_o\right]^{-1}G_o^{-1}\right\} \leq 1 \quad (4.31)$$

$$\Longleftarrow \bar{\sigma}[W_eS_o]\,\bar{\sigma}[G_o]\,\bar{\sigma}\left[G_o^{-1}\right]\bar{\sigma}\left[(I + \Delta W_\Delta T_i)^{-1}\right] \leq 1 \quad (4.32)$$

$$\Longleftarrow \frac{\bar{\sigma}[W_eS_o]\,\kappa[G_o]}{1 - \bar{\sigma}[W_\Delta T_i]} \leq 1 \quad (4.33)$$

$$\iff \bar{\sigma}[W_eS_o]\,\kappa[G_o] + \bar{\sigma}[W_\Delta T_i] \leq 1 \quad \text{for all } \omega \quad (4.34)$$

where $\kappa[G_o(j\omega)]$ denotes the condition number of the nominal model. Note that (4.34) is more restrictive than the (sufficient) robust performance condition corresponding to sensor uncertainty. This is due to the fact that the condition number of the nominal model has a minimum value of 1. Therefore, for the same uncertainty level (in the sense of $\|W_\Delta T_i\|_\infty \sim \|W_\Delta T_o\|_\infty$) the value of $\bar{\sigma}[W_e(j\omega)S_o(j\omega)]$ required in order to satisfy equation (4.34) is smaller than the value required to meet (4.24), resulting in a lower nominal performance level (since a smaller W_e must be used in the first case).

The condition number $\kappa[G_o(j\omega)]$ represents, as a function of frequency, the relation between the maximum and minimum gain of the nominal model. This characteristic is exclusive of MIMO systems and denotes the "balance" between the input–output channels of $G_o(s)$. The ideal situation from this point of view is when the nominal model has the same gain for any input vector direction. This is the case, for instance, where the system matrix is unitary at each frequency. Note that a high condition number by itself is not a problem, except when combined with actuator uncertainty. This situation arises in practice in the case of high-purity distillation plants ([287]), leading to a challenging control problem.

A conceptual explanation of the difference between this case and the sensor uncertainty one is the following. When the uncertainty appears at the output of the nominal model, the interconnection between the output of the controller and the model input is not subject to uncertainty. Therefore we may consider the series connection of $G_o(s)$ and $K(s)$ as a single transfer matrix $L_o(s) = G_o(s)K(s)$. Instead, in the input uncertainty case, the uncertainty $\Delta(s)$ "separates" the connection between $K(s)$ and $G_o(s)$. This means that the controller output channels and model input ones are "mixed up." This does not pose any problems for robust stability, because the controller and the nominal plant can be connected through the output of the latter as in $L_i(s) = K(s)G_o(s)$. But for certain performance requirements this is not possible. Therefore if the channels are "mixed," an output of the controller that should go through a small gain direction of $G_o(s)$ could conceivably enter the plant through a high-gain direction and vice versa. This effect becomes worse as the relation between the higher and lower gains of the model (condition number) becomes higher.

In many cases, if $\kappa[G_o(\jmath\omega)]$ is very high in the frequency region of interest, condition (4.34) can guarantee very little performance for all the members of the family of models. In those cases, a new identification of the set of models should be made, attempting to reduce the size of the uncertainty $\Delta(s)$. Alternatively, we could seek a less conservative robust performance condition. The latter approach is covered in Chapter 7, while the former is addressed in Chapter 10.

In order to avoid these difficulties, one could be tempted to try to model a plant subject to actuator uncertainty as a set of models subject to output multiplicative uncertainty (possibly having different bounds). In this case, we should change the input uncertainty $\Delta_i(s)$ to an "equivalent" output uncertainty matrix $\Delta_o(s)$ so that both descriptions cover all elements of the family of plants. This can be accomplished as follows:

$$\mathcal{G}_i \stackrel{\triangle}{=} \{G_o(s)[I + \Delta_i(s)], \quad \bar{\sigma}[\Delta_i(s)] < 1, \quad \forall s = \jmath\omega\} \tag{4.35}$$

$$\mathcal{G}_o \stackrel{\triangle}{=} \{[I + \Delta_o(s)]G_o(s), \quad \bar{\sigma}[\Delta_o(s)] < \gamma, \quad \forall s = \jmath\omega\} \tag{4.36}$$

$$\mathcal{G}_i \equiv \mathcal{G}_o \iff \Delta_o = G_o(s)\Delta_i G_o^{-1}(s) \tag{4.37}$$

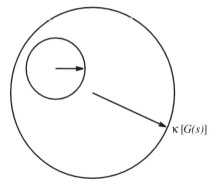

Figure 4.5. Input to output uncertainty transformation.

where again, for the sake of simplicity, we have assumed that $G_o(s)$ is invertible. To keep the description as "tight" as possible, the upper bound of the output uncertainty size, γ, should be as small as possible. While γ cannot be computed exactly, an upper bound can be obtained as follows:

$$\bar{\sigma}\left[\Delta_o(j\omega)\right] \leq \bar{\sigma}[G(j\omega)]\bar{\sigma}\left[\Delta_i(j\omega)\right] \bar{\sigma}\left[G^{-1}(j\omega)\right] \leq \kappa\left[G(j\omega)\right] \stackrel{\triangle}{=} \gamma_u, \quad \forall \omega \tag{4.38}$$

Note that, since $\gamma \leq \gamma_u$, the new uncertainty description might generate plants that were not included in the original family $\mathcal{G}_i$, leading to potentially conservative designs. This situation is illustrated in Figure 4.5.

4.6 DESIGN PROCEDURE

The simplest synthesis method derived from the analysis conditions described in previous subsections is known as *loop shaping*. The singular values of the (input or output) loop $L(s)$ are "shaped" so that the resulting system satisfies the sufficient condition for robust performance (and hence robust stability and nominal performance). This process is a natural generalization of the procedure used in the SISO case, where the magnitude Bode plot of $\ell(s) = g(s)k(s)$ is shaped.

Consider, without loss of generality, the tracking problem of Figure 4.1. The goal is to achieve robust performance in the presence of sensor multiplicative uncertainty. Assume, for simplicity, that the weights for both performance and stability are diagonal: $W_\Delta(s) = w_\Delta(s)I$ and $W_e(s) = w_e(s)I$. Define the output and input loop transfer matrices as $L_o(s) = G_o(s)K(s)$ and $L_i(s) = K(s)G_o(s)$. Restating conditions (4.18) and (4.19) in terms of these matrices and using standard singular value properties yields the following sufficient conditions for nominal performance and robust stability:

Nominal performance

$$\iff \bar{\sigma}\left\{W_e(j\omega)\left[I + L_o(j\omega)\right]^{-1}\right\} \leq 1, \quad \forall \omega \tag{4.39}$$

$$\iff \underline{\sigma}\left[I + L_o(j\omega)\right] \geq |w_e(j\omega)|, \quad \forall \omega \tag{4.40}$$

$$\impliedby \underline{\sigma}\left[L_o(j\omega)\right] \geq 1 + |w_e(j\omega)|, \quad \forall \omega \tag{4.41}$$

and

Robust stability

$$\iff \bar{\sigma}\left\{W_\Delta(j\omega)L_o(j\omega)\left[I + L_o(j\omega)\right]^{-1}\right\} \leq 1, \quad \forall \omega \tag{4.42}$$

$$\iff \bar{\sigma}\left\{\left[I + L_o^{-1}(j\omega)\right]^{-1}\right\} \leq 1/|w_\Delta(j\omega)|, \quad \forall \omega \tag{4.43}$$

$$\iff \underline{\sigma}\left[I + L_o^{-1}(j\omega)\right] \geq |w_\Delta(j\omega)|, \quad \forall \omega \tag{4.44}$$

$$\impliedby \underline{\sigma}\left[L_o^{-1}(j\omega)\right] - 1 \geq |w_\Delta(j\omega)|, \quad \forall \omega \tag{4.45}$$

$$\iff \bar{\sigma}\left[L_o(j\omega)\right] \leq 1/\left[1 + |w_\Delta(j\omega)|\right], \quad \forall \omega \tag{4.46}$$

A similar procedure (see Problem 4) leads to the following necessary conditions:

Nominal performance $\implies \underline{\sigma}\left[L_o(j\omega)\right] \geq |w_e(j\omega)| - 1, \quad \forall \omega \tag{4.47}$

Robust stability $\implies \bar{\sigma}\left[L_o(j\omega)\right] \leq 1/\left[|w_\Delta(j\omega)| - 1\right], \quad \forall \omega \tag{4.48}$

Note that the sufficient conditions (4.41) and (4.46) conflict since

$$\bar{\sigma}\left[L_o(j\omega)\right] \leq \left[1 + |w_\Delta(j\omega)|\right]^{-1} \leq 1 \leq \left[1 + |w_e(j\omega)|\right] \leq \underline{\sigma}\left[L_o(j\omega)\right] \tag{4.49}$$

for all ω. This conflict can be avoided by using the fact that performance and stability requirements are usually separated in frequency. Performance requirements are often more stringent at low frequencies. Thus $|w_e|$ is large at low frequencies and ~ 0 at high frequencies. On the other hand, uncertainty is usually higher at high frequencies so that $|w_\Delta(j\omega)| \gg 1$ for large ω and ~ 0

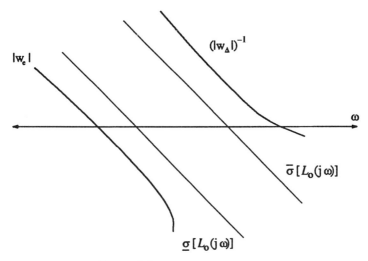

Figure 4.6. Loop shaping design.

for low frequencies. The design procedure exploits this separation to define admissible regions for L at low and high frequencies (see Figure 4.6). Specifically, the minimum (maximum) singular value of $L_o(s)$ should be shaped so that it falls above (below) the lower (upper) bound at small ω (high ω), that is,

$$|w_e(j\omega)| \leq \underline{\sigma}[L_o(j\omega)] \quad (\text{low } \omega) \tag{4.50}$$

$$\bar{\sigma}[L_o(j\omega)] \leq |w_\Delta(j\omega)|^{-1} \quad (\text{high } \omega) \tag{4.51}$$

where we used the facts that $[1 + |w_e(j\omega)|] \approx |w_e(j\omega)|$ at low frequencies and $[1 + |w_\Delta(j\omega)|]^{-1} \approx |w_\Delta(j\omega)|^{-1}$ at high frequencies. Proceeding in a similar fashion, the necessary conditions (4.47) and (4.48) can also be used to determine bounds for the extreme singular values of the output loop matrix $L_o(s)$.

Once an admissible band for the singular values of $L_o(s)$ has been established, the second step is to synthesize a controller so that the resulting $L_o(s) = G_o(s)K(s)$ satisfies these bounds. In this step, care must be taken so that the controller guarantees internal stability of the loop. In fact, in this design procedure, only sufficient (and/or necessary) conditions for nominal performance and robust stability are used, which do not guarantee nominal (internal) stability, and not even well posedness. As a consequence, both should be checked after designing $K(s)$.[5]

---

[5] This can be avoided by exploiting the Youla parametrization to perform the loop shaping over the set of stabilizing controllers. This technique is explained in the next section.

108 LOOP SHAPING

In order to achieve robust performance, the resulting closed-loop system should satisfy condition (4.24). This can be accomplished using the technique described above, coupled with scaling of the performance weight by $\alpha \in (0, 1]$ and searching for the highest α that satisfies

$$\bar{\sigma}\left[\alpha W_e(j\omega)S_o(j\omega)\right] + \bar{\sigma}\left[W_\Delta(j\omega)T_o(j\omega)\right] \leq 1, \quad \forall \omega \tag{4.52}$$

This method is simple and provides the designer with an intuitive tool to address the performance/robustness trade-off. In addition, it furnishes a clear idea of which constraint, robustness or performance, is more restrictive and some insight on the limiting factors. Nevertheless, there are several points that should be taken into consideration when using it.

1. Once the loop transfer matrix has been designed to satisfy the singular value constraints, the next step is to synthesize an appropriate controller. This controller is given by $K(s) = G_o^{-1}(s)L_o(s)$ only in cases where the nominal model is invertible, stable, and minimum phase. In the general case, a suitable controller can be synthesized by first separating the invertible (square, stable, minimum phase) from the non-invertible part of the plant as follows: $G_o(s) = G_{\overline{\text{inv}}}(s)G_{\text{inv}}(s)$. The controller is then given by $K(s) = G_{\text{inv}}^{-1}(s)G_+(s)L_o(s)$, where $G_+(s)$ is selected so that $G_{\overline{\text{inv}}}(s)G_+(s)$ is *allpass*. Since multiplication by an allpass matrix preserves the singular values (i.e., $\sigma_i[L_o(j\omega)] = \sigma_i[G_{\overline{\text{inv}}}(s)G_+(s)L_o(j\omega)]$), the resulting loop function satisfies the constraints. This is illustrated by the following simple SISO example:

$$g_o(s) = \frac{(s+1)(s-5)}{(s+4)(s-3)} \stackrel{\triangle}{=} g_{\text{inv}}(s)g_{\overline{\text{inv}}}(s) \tag{4.53}$$

From the loop shaping design, it is desirable to have $|\ell(j\omega)| = 1/|j\omega|$. The controller that satisfies the above condition for the loop transfer function $\ell(s)$ is

$$k(s) = \frac{(s+4)(s+3)}{(s+1)(s+5)}\frac{1}{s} = \frac{1}{s}g_{\text{inv}}^{-1}(s)g_+(s) \tag{4.54}$$

since $|(s-5)(s+3)/(s+5)(s-3)| = 1$ for all $s = j\omega$.

2. The sufficient conditions (4.50) and (4.51) are valid only for low and high frequencies, respectively. Thus as a final step in the synthesis process, the resulting loop function should be examined over the complete frequency range.

3. For the robust stability and nominal performance bounds, in the case of actuator uncertainty, there is no need to include the plant condition number. Nevertheless, it must be included with the scaling when we proceed to verify the robust performance sufficient condition.

4. Since the method is based on the use of sufficient conditions, it may be too conservative. In cases combining high-performance requirements with high levels of uncertainty, the bounds obtained from these conditions may overlap, so that the admissible region for L becomes empty. One solution is to carry out the design using the *necessary* rather than the sufficient conditions. In this case the synthesis must be followed by a verification of robust performance (which includes nominal performance and robust stability). Alternatively, a less conservative design procedure should be used (such as μ-synthesis convered in Chapter 7), or the uncertainty in the model should be decreased through an additional identification effort (see Chapter 10).

4.7 EXAMPLES

4.7.1 Permanent Magnet Stepper Motor

In this section we illustrate the use of the loop shaping technique with an application example. The system under consideration is a permanent magnet step motor driving a precision linear positioning table [53]. The objective is to track prespecified angular and velocity trajectories, with minimum tracking error and without excessive control action. Robustness against model uncertainty is also considered.

Description of the Plant Although the stepper motor is a nonlinear system, a linear model of the tracking error dynamics can be obtained through the use of feedback linearization ([156]). The state-space and transfer matrix representation of this linear model are given by

$$G(s) \equiv \left[\begin{array}{cccc|cc} -R/L & 0 & 0 & 0 & 1 & 0 \\ 0 & -R/L & -K_m/L & 0 & 0 & 1 \\ 0 & K_m/J & -B/J & 0 & 0 & 0 \\ 0 & 0 & 1 & 0 & 0 & 0 \\ \hline \multicolumn{4}{c|}{I_{4\times 4}} & \multicolumn{2}{c}{0_{4\times 2}} \end{array}\right] \qquad (4.55)$$

$$= \begin{bmatrix} \dfrac{1}{s+R/L} & 0 \\ 0 & \dfrac{s+B/J}{p(s)} \\ 0 & \dfrac{K_m}{Jp(s)} \\ 0 & \dfrac{K_m}{Jsp(s)} \end{bmatrix} \qquad (4.56)$$

110 LOOP SHAPING

$$p(s) = s^2 + s\left(\frac{R}{L} + \frac{B}{J}\right) + \left(\frac{RB + K_m^2}{JL}\right) \quad (4.57)$$

where R and L represent the resistance and inductance of the phase winding, J represents the inertia of the motor and translation table, B represents the viscous friction coefficient, and K_m represents the motor torque constant. The values of these constants have been obtained through identification procedures and can be found in [53]. The states of the plant are $x(t) = [\,e_{i_d}\ e_{i_q}\ e_\omega\ e_\theta\,]$, the tracking errors in direct, quadrature currents, angular velocity, and angular displacement, respectively. The inputs are the direct and quadrature voltages u_d and u_q and the outputs are all the states.

Nominal Stability and Performance First we will consider as performance objective tracking, with zero steady-state error, fast commands to the table. As an example we consider a 1.8 mm move of the table in 30 ms, following a desired angular and velocity trajectory. The velocity trajectory starts from zero and is linear during the first 5 ms. up to 1350 rpm, then it remains constant during 20 ms, and finally decreases linearly to zero in 5 ms. This represents a final angular rotation of slightly more than 200°.

Additionally, we will impose a bound on the control action. Both time response and control bound objectives can be represented by weights on the closed-loop matrices. Proceeding as in previous sections, bounds on the open-loop matrix $L(s)$ can be computed as functions of these weights. For simplicity, we consider as the frequency-domain bounds for nominal performance the $\sigma_1[L_\star(j\omega)]$ and $\sigma_2[L_\star(j\omega)]$, which were designed in [53] for the same performance objectives[6] (curves c and d in Figure 4.7). These will serve as upper and lower bounds on the open-loop transfer matrix to be designed.

Although we have considered in this chapter only lower bounds for performance, in general also upper bounds can be desirable. For example, when the performance objective is to bound the control action, as in this case, we define nominal performance as $\|W_u(s)u(s)\|_2 < 1$ for all bounded references $\|r(s)\|_2 \leq 1$. Here $W_u(s)$ is the control action weighting function. Proceeding as in previous sections we obtain the following sufficient condition over the loop, which guarantees this performance objective.

$$\text{NP} \iff \left\|W_u(s)K(s)[I + G(s)K(s)]^{-1}\right\|_\infty < 1 \quad (4.58)$$

$$\iff \left\|W_u(s)G^\dagger(s)L(s)[I + L(s)]^{-1}\right\|_\infty < 1 \quad (4.59)$$

$$\impliedby \frac{\bar{\sigma}\left[W_u(s)G^\dagger(s)\right]}{\underline{\sigma}\left[I + L^{-1}(s)\right]} < 1, \quad \forall s = j\omega \quad (4.60)$$

---

[6] Since $L_\star(s) = G(s)K_\star(s)$ has at most rank 2, we consider only the first and second singular values.

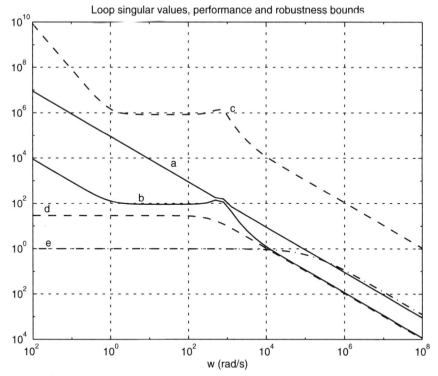

Figure 4.7. (a) $\bar{\sigma}[L(j\omega)]$, (b) $\sigma_2[L(j\omega)]$, (c) upper performance bound, (d) lower performance bound, and (e) upper robustness bound.

$$\Longleftarrow \bar{\sigma}\left[W_u(s)G^\dagger(s)\right] < \frac{1}{\bar{\sigma}\left[L(s)\right]} - 1, \quad \forall s = j\omega \quad (4.61)$$

$$\Longleftrightarrow \bar{\sigma}\left[L(s)\right] < \frac{1}{\bar{\sigma}\left[W_u(s)G^\dagger(s)\right] + 1}, \quad \forall s = j\omega \quad (4.62)$$

where $G^\dagger(s)$ is the left inverse (if it exists) of the model $G(s)$.

Considering the bounds c and d in Figure 4.7, we would ideally want an open-loop transfer matrix $L(s)$ with $\alpha\bar{\sigma}[L(j\omega)] = \sigma_2[L(j\omega)] \approx \ell/s$, $\alpha \in (0,1]$, that is, $L(s) = \text{diag}[\ell_i/s]$. However, it is clear from the structure of the plant $G(s)$ that a diagonal open-loop matrix cannot be achieved (since the plant has no right inverse). Thus we will try instead to obtain a *block diagonal* transfer matrix, of the form

$$L(s) = G(s)K(s) = \begin{bmatrix} * & 0 & 0 & 0 \\ 0 & * & * & * \\ 0 & * & * & * \\ 0 & * & * & * \end{bmatrix} \quad (4.63)$$

112 LOOP SHAPING

where $*$ represents nonzero elements. The simplest controller achieving this open-loop structure has the form

$$K(s) = \begin{bmatrix} k_{11}(s) & 0 & 0 & 0 \\ 0 & k_{22}(s) & k_{23}(s) & k_{24}(s) \end{bmatrix} \quad (4.64)$$

Furthermore, since the the open-loop plant is also block diagonal, it follows that we can design each block separately. For the first scalar block we can achieve the loop ℓ_1/s and compute a value of ℓ_1 such that $|\ell_1/j\omega|$ fits between the upper and lower bounds in Figure 4.7.

Before we start designing a controller that fits $L(s)$ between the nominal performance bounds, we should consider internal stability, which places additional constraints on the design. Although $G(s)$ has no right inverse, we can design $K(s) = D(s)G_s^\dagger(s)$, where $G_s^\dagger(s)$ is the left inverse of the stable part of the nominal model and $D(s) = \text{diag}[d_1(s), d_2(s)]$ is a diagonal system. Here we consider the following stable–unstable factorization of the model:

$$G(s) = \underbrace{\begin{bmatrix} \dfrac{1}{s+R/L} & 0 \\ 0 & \dfrac{s(s+B/J)}{p(s)} \\ 0 & \dfrac{sK_m}{Jp(s)} \\ 0 & \dfrac{K_m}{Jp(s)} \end{bmatrix}}_{G_s(s)} \underbrace{\begin{bmatrix} 1 & 0 \\ 0 & \dfrac{1}{s} \end{bmatrix}}_{G_u(s)} \quad (4.65)$$

such that $G_s^\dagger(s)G_s(s) = I$ with the stable left inverse being

$$G_s^\dagger(s) = \begin{bmatrix} (s+R/L) & 0 & 0 & 0 \\ 0 & 1 & \dfrac{JR}{K_mL} & \dfrac{RB+K_m^2}{K_mL} \end{bmatrix} \quad (4.66)$$

As a consequence,

$$\begin{aligned} T(s) &= G(s)K(s)[I+G(s)K(s)]^{-1} = G(s)[I+K(s)G(s)]^{-1}K(s) \\ &= G_s(s)[I+G_u(s)D(s)]^{-1}G_u(s)D(s)G_s^\dagger(s) \\ &= G_s(s)\begin{bmatrix} t_1 & 0 \\ 0 & t_2 \end{bmatrix}G_s^\dagger(s) = \begin{bmatrix} t_1 & 0 \\ 0 & t_2 I_{3\times 3} \end{bmatrix}G_s(s)G_s^\dagger(s) \end{aligned} \quad (4.67)$$

where we used the fact that $K(s)G(s) = D(s)G_u(s)$ and we have replaced the input complementary sensitivity $T_I(s) = [I+K(s)G(s)]^{-1}K(s)G(s)$ by diag$[t_1, t_2]$, where

$$t_1(s) \triangleq d_1(s)[1 + d_1(s)]^{-1} \quad \text{and} \quad t_2(s) \triangleq \frac{d_2(s)}{s}\left[1 + \frac{d_2(s)}{s}\right]^{-1}$$

Note that the input sensitivity $S_I(s) = [I + K(s)G(s)]^{-1}$, its complement $T_I(s)$ and $T(s)$ are all stable for an appropriate choice of $K(s)G(s) =$ diag$[d_1(s), d_2(s)/s]$. The poles at the origin of $G(s)$ and $K(s)$ will be cancelled by the zeros at the origin of $S_I(s)$. As a consequence $G(s)S_I(s)$ and $S_I(s)K(s)$ are also stable and internal stability is achieved.

It follows that the design specifications can be met with a controller of the form

$$K(s) = \begin{bmatrix} d_1(s)(s + R/L) & 0 & 0 & 0 \\ 0 & d_2(s) & \frac{JR}{K_m L}d_2(s) & \frac{RB + K_m^2}{K_m L}d_2(s) \end{bmatrix} \quad (4.68)$$

where the design transfer functions $d_1(s)$ and $d_2(s)$ should be selected to meet the performance and robustness specifications. Since in this case the performance bounds have been imposed directly over $L(s)$ (and not over $T(s)$ or $S(s)$), there is no need to verify the sensitivity or its complement. Nevertheless, the issue of integral action will be considered next.

Note that the input loop $L_I(s) = K(s)G(s)$ is diagonal and depends on $d_1(s)$ and $d_2(s)/s$. Therefore to achieve a loop of the form $L_I(s) = $ diag(ℓ_i/s), we only need to have integral action in $d_1(s)$. This may seem clear from the structure of model $G(s)$, which already has a pole at $s = 0$ in its element $g_{42}(s)$, but none in $g_{11}(s)$. As a consequence the block-diagonal structure of loop $L(s)$ in (4.63) might seem to have already integral action in the second block, but none in the first one; which should be provided by the controller element $d_1(s)$.

This is not at all true, since a pole at $s = 0$ in $L(s)$ or $L_I(s)$ only guarantees a multivariable zero in $S(s)$ at $s = 0$, which is not the same as $S(0) = 0_{4 \times 4}$. For example, take a simple controller $d_1(s) = (9 \times 10^4)/s$ and $d_2(s) = 90$, which meets the performance limits c and d of Figure 4.7. A second controller can be designed with some additional dynamics in the upper loop to provide a better fit of the upper bound at low frequencies (see Figure 4.8), as follows:

$$d_1(s) = 10^6 \frac{(10^{-4}s + 1)}{s(10^{-3}s + 1)}, \quad d_2 = 90 \quad (4.69)$$

In both cases, the sensitivity function at $s = 0$ can be obtained from equation (4.67):

114 LOOP SHAPING

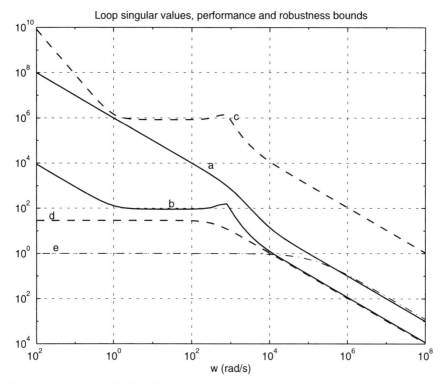

Figure 4.8. (a) $\bar{\sigma}[L(j\omega)]$, (b) $\sigma_2[L(j\omega)]$, (c) upper performance bound, (d) lower performance bound, and (e) upper robustness bound.

$$S(0) = I - T(0) = \begin{bmatrix} 0 & 0 & 0 & 0 \\ 0 & 1 & 0 & 0 \\ 0 & 0 & 1 & 0 \\ 0 & -\dfrac{K_m L}{RB + K_m^2} & -\dfrac{JR}{RB + K_m^2} & 0 \end{bmatrix} \quad (4.70)$$

where $t_1(0) = t_2(0) = 1$. The fact that the last row, which represents the effect of disturbances on the angular error e_θ, is nonzero means that constant disturbances are not completely eliminated at steady state, that is, no integral action.

To achieve integral action on the angular output, we could add an integral state to the nominal model. This is equivalent to having an integrator in the last element of controller $K(s)$ described before, that is, $k_{24}(s)$. If this is the case, the controller can no longer have the structure $K(s) = D(s)G_s^\dagger(s)$. Instead, we may use a more general controller, which also inverts the stable portion of the plant but has different elements in its last row,

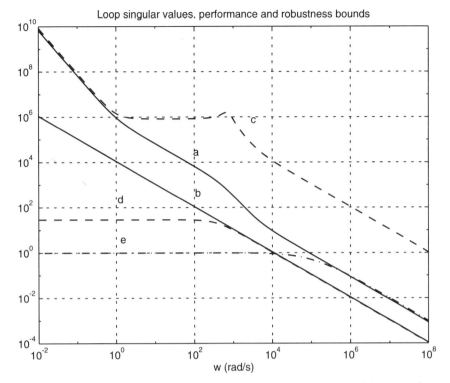

Figure 4.9. (a) $\bar{\sigma}[L(j\omega)]$, (b) $\sigma_2[L(j\omega)]$, (c) upper performance bound, (d) lower performance bound, and (e) upper robustness bound.

as follows:

$$K(s) = \begin{bmatrix} k_1 \dfrac{s+R/L}{s} & 0 & 0 & 0 \\ 0 & \dfrac{k_2 p(s)}{(s+B/J)(sp_a+1)} & \dfrac{k_3 J p(s)}{K_m(sp_a+1)(sp_b+1)} & \dfrac{k_4 J p(s)}{K_m s(sp_a+1)} \end{bmatrix}$$
(4.71)

where the poles at $-1/p_a$ and $-1/p_b$ have been added so that the controller is proper. Nominal internal stability should be verified when choosing values for the controller parameters. The open-loop matrix corresponding to this design can be seen in Figure 4.9. Figure 4.10 shows the output of $S(s)$ corresponding to the sensitivity of the angular rotation to all disturbance inputs [last row of $S(s)$]. Due to the integral action any constant disturbance of the currents i_d and i_q, or the angular velocity ω, or angular rotation θ is asymptotically rejected.

116 LOOP SHAPING

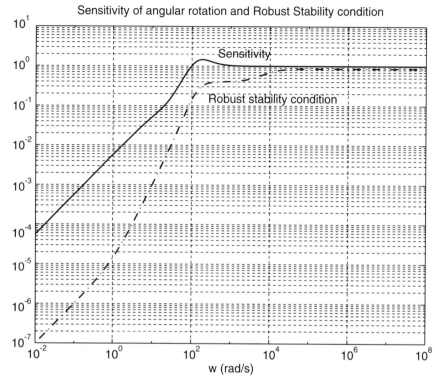

Figure 4.10. Sensitivity of angular rotation to all inputs and robust stability condition $\bar{\sigma}(W_\Delta(j\omega)T(j\omega))$.

Figure 4.11 shows a simulation of a near 200° turn at 1350 rpm. Note that constant disturbances in all outputs are rejected in steady state. The errors are 10% of the reference trajectory values. A faster constant error rejection could have been accomplished by dropping control action bounds from the performance specifications, resulting in higher loop gains. However, as we show in the sequel, this is also precluded if there exists model uncertainty, since the robust stability conditions are more stringent than the nominal performance constraints (see Figure 4.9). Finally, Figure 4.12 shows the results of a second simulation, corresponding to the case of 10% random disturbances.

Robust Stability Up to this point we have assumed that a perfect description of the motor was available. However, this is clearly an optimistic assumption in most practical applications. Now suppose that, although the low-frequency characteristics of the step motor are very well described by the above model, there is uncertainty at high frequencies. This fact should be taken into account in the design, especially if the performance objectives involve high-speed rotations. Suppose that the unmodeled dynamics are de-

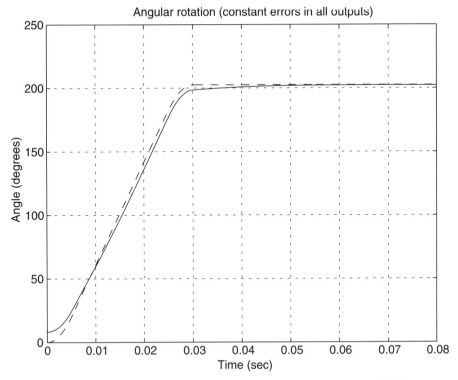

Figure 4.11. Reference (dashed) and angular rotation with 10% constant disturbances in all outputs.

scribed by means of the following output multiplicative uncertainty:

$$\mathcal{G} = \{[I + \Delta(s)W_\Delta(s)]\, G(s) \mid \|\Delta(s)\|_\infty < 1\} \quad (4.72)$$

$$W_\Delta(s) = \frac{10^{-5}s}{s\epsilon + 1}I = w_\Delta(s)I \quad (4.73)$$

The above represents a family of models that describes adequately the low-frequency characteristics[7] of the motor up to $\omega \leq 10^5$ rad/s, which in terms of motor speed represents a limit of 10^6 rpm. By using equation (4.46), we compute $[1 + |w_\Delta(j\omega)|]^{-1}$ as the upper bound for $\bar{\sigma}\,[L(j\omega)]$ (at high frequencies) imposed by model uncertainty. As seen in Figure 4.9 (curve e), this upper limit is more restrictive than the one imposed by the control action bound (curve c). Therefore the loop function will be bounded below by the

---

[7] In fact, it assumes perfect knowledge at zero frequency, and less than 10% error below 10^4 rad/s.

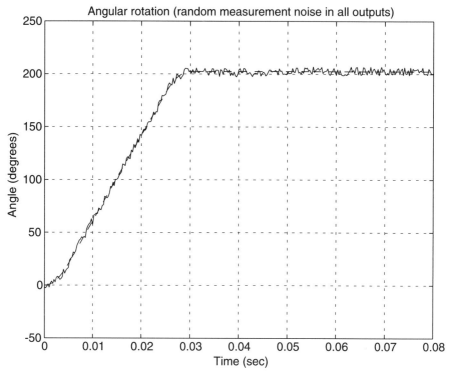

Figure 4.12. Reference (dashed) and angular rotation with 10% random disturbances in all outputs.

performance bound (curve d) and above by the robustness constraint (curve e).

A set of values for the parameters k_2 to k_4, p_a, and p_b guaranteeing internal stability and meeting the performance and robustness specifications is given by

$$k_1 = 1.1 \times 10^4, \quad k_2 = 16.04, \quad k_3 = 84.45, \quad k_4 = 0.667 \times 10^6 \quad (4.74)$$

$$p_a = 0.0018, \quad p_b = 0.0201 \quad (4.75)$$

This set of values places the roots of $p_4(s)$ (closed loop poles) at $s_1 = -1.1 \times 10^4$, $s_2 = -0.9 \times 10^4$, $s_3 = -100$, and $s_{4,5} = -100 \pm j100$.

Figure 4.10 shows a plot of $\bar{\sigma}[W_\Delta(j\omega)T(j\omega)]$. Note that since this curve is below unity at all frequencies (peak value is 0.86), robust stability is guaranteed; that is, the controller internally stabilizes all possible plants in $\mathcal{G}$. It is also possible to find the smallest uncertain block that fits the multiplicative uncertainty structure and produces instability ([20]). In this case it corresponds to the following uncertainty block:

$$\Delta_\star = - \begin{bmatrix} 0 & 0 & 0 & 0 \\ 0 & 0.1172 & 3 \times 10^{-8} + j6 \times 10^{-5} & 0 \\ 0 & 0.0073 & 2 \times 10^{-9} + j4 \times 10^{-6} & 0 \\ 0 & 1.1544 - j5 \times 10^{-4} & 3 \times 10^{-7} + j6 \times 10^{-4} & 0 \end{bmatrix} \quad (4.76)$$

which has norm greater than unity ($\bar{\sigma}(\Delta_\star) = 1.16$); that is, it does not belong to $\mathcal{G}$. A simulation with a scaled version of this "worst case" model (using $\bar{\sigma}(\Delta_\star) = 0.99$) and 10% constant disturbances in all outputs virtually coincides with the plot shown in Figure 4.11. The same can be said for the simulation in Figure 4.12.

4.7.2 Loop Shaping Q(s)

As we pointed out before, one of the limitations of the loop shaping procedure developed in Section 4.6 is that is does not automatically guarantee that the resulting controller will internally stabilize the loop. Thus the synthesis must be followed by a stability check and, if necessary, a redesign. This difficulty can be avoided by using the Youla parametrization developed in Chapter 3 to guarantee that *only* stabilizing controllers are considered. Recall that by using this parametrization all achievable closed-loop maps are transformed to affine functions of a free parameter: a stable, proper transfer matrix $Q(s)$. Therefore we can replace conditions (4.50) and (4.51) by simpler ones expressed in terms of $Q(s)$. Adding to these conditions the restriction $Q \in \mathcal{RH}_\infty$ guarantees that the corresponding controller $K(s)$ internally stabilizes the loop. This procedure is illustrated next via a simple example.

Consider a multivariable system having as nominal model

$$G_o(s) = \begin{bmatrix} \dfrac{4}{s+4} & 0 \\ \dfrac{4s(3s+16)}{(s+4)(s+8)} & \dfrac{8(s-200)}{s+8} \end{bmatrix} \quad (4.77)$$

Furthermore, assume that the actual plant G is known to belong to the following set:

$$\bar{\sigma}\left\{[G(j\omega) - G_o(j\omega)] G_o^{-1}(j\omega)\right\} \leq \tfrac{1}{2} \quad (4.78)$$

$$\forall \omega \leq 1000 \text{ rad/s} \quad (4.79)$$

The goal is to synthesize a controller that internally stabilizes all members of this set of models and such that the nominal closed loop satisfies the following performance objectives:

1. The sensitivity to output disturbances is reduced by at least 20 dB in the range $\omega \in [0, 10]$ rad/s with respect to its high-frequency value.
2. Tracking with zero steady-state error of step inputs.
3. The transfer matrix between the inputs and outputs of the closed-loop system should be diagonal and such that $y(s) = [10/(s+10)]r(s)$.

Solution Rewritting the family (4.78) as a nominal plant subject to output dynamic multiplicative uncertainty yields

$$G = (I + \Delta)G_o, \quad \bar{\sigma}\{\Delta(j\omega)\} \leq \tfrac{1}{2}, \quad \forall \omega \in [0, 1000 \text{ rad/s}] \quad (4.80)$$

Equivalently,

$$G = (I + W_\Delta \Delta)G_o, \quad \Delta \in \mathcal{BH}_\infty$$
$$W_\Delta(s) = \frac{10^{-3}s + 1}{2(10^{-6}s + 1)} I \quad (4.81)$$

where the pole at $s = 10^6$ has been added to make $W_\Delta(s)$ proper. From condition (4.19) we have that the system is robustly stable if and only if

$$\|W_\Delta(s)T(s)\|_\infty \leq 1 \quad (4.82)$$

Similarly, by using condition (4.18) together with an appropriate weighting function W_1, the nominal performance requirement reduces to

$$\|S(s)W_1(s)\|_\infty \leq 1 \quad \text{with } W_1(s) = \frac{s + 100}{2s} I \quad (4.83)$$

Note that W_1 was chosen to have a pole at the origin to force the sensitivity matrix to have a zero at $s = 0$ and, as a consequence, integral action in the loop. This is necessary and sufficient to track step reference inputs. Since the nominal plant is open-loop stable, the set of all internally stabilizing controllers can be parametrized as (see Section 3.3)

$$K(s) = Q(s)[I - G_o(s)Q(s)]^{-1}, \quad Q \in \mathcal{RH}_\infty \quad (4.84)$$

In terms of $Q(s)$ the sensitivity and complementary sensitivity matrices are given by

$$S(s) = I - G_o(s)Q(s) \quad (4.85)$$
$$T(s) = G_o(s)Q(s) \quad (4.86)$$

The next step is to search for a stable, proper $Q(s)$ satisfying conditions (4.82) and (4.83). Due to the fact that $G_o(s)$ is nonminimum phase, the plant inverse cannot be used in the controller (otherwise the internal stability

EXAMPLES

requirement will be violated). Proceeding as outlined in Section 4.6, Q is selected so that it only inverts the stable minimum phase part of $G_o(s)$. This is accomplished by selecting $Q = G_+(s)q(s)$, where $G_+(s)$ is such that $G_o(s)G_+(s)$ is an allpass transfer matrix, and where $q(s)$ is an arbitrary stable and proper transfer function. In this case an appropriate choice of G_+ is given by

$$G_+(s) = \begin{bmatrix} \dfrac{(s+4)(s-200)}{4(s+200)} & 0 \\ -\dfrac{s(3s+16)}{8(s+200)} & \dfrac{s+8}{8(s+200)} \end{bmatrix} \quad (4.87)$$

$$\Longrightarrow G_o(s)G_+(s) = \dfrac{s-200}{s+200} I$$

Note that with this choice, $q(s)$ must be selected to be strictly proper in order to get a proper Q.

The corresponding sensitivity matrices are both diagonal:

$$S(s) = \left[1 - q(s)\dfrac{s-200}{s+200}\right] I \quad (4.88)$$

$$T(s) = q(s)\dfrac{s-200}{s+200} I \quad (4.89)$$

Combining condition (4.82) with these expressions leads to the following sufficient condition for robust stability:

$$|q(\jmath\omega)| \leq 2 \left|\dfrac{\jmath\omega 10^{-6} + 1}{10^{-3}\jmath\omega + 1}\right| \quad (4.90)$$

Similarly, condition (4.83) yields the following necessary condition for nominal performance:

$$|q(\jmath\omega)| \geq 1 - \left|\dfrac{2\jmath\omega}{\jmath\omega + 100}\right| \quad (4.91)$$

Note that from (4.83) it is clear that $S(s)$ should have a zero at $s = 0$ to cancel the pole of $W_1(s)$. Equivalently, $q(s)$ must satisfy the interpolation constraint $q(0) = -1$. It can easily be verified that the following $q(s)$ satisifies all of these conditions:

$$q(s) = -\dfrac{300(s+100)}{(s+150)(s+200)} \quad (4.92)$$

Figure 4.13 shows a plot of conditions (4.90) and (4.91) and $q(\jmath\omega)$ as a function of the frequency. Since these conditions are not equivalent to (4.82)

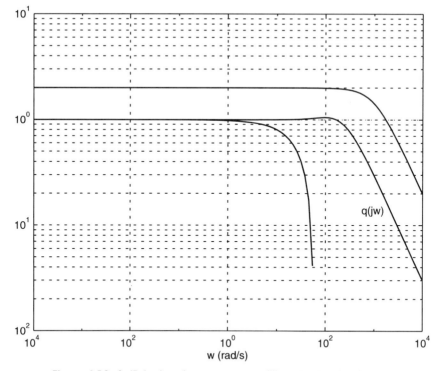

Figure 4.13. Sufficient and necessary conditions for shaping $|q(jw)|$.

and (4.83), the next step in the design is to verify whether or not the latter are satisfied. This can be accomplished through the plots shown in Figure 4.14.

Finally, the corresponding controller is obtained from Q via (4.84), yielding

$$K(s) = G_+(s) \frac{(s+200)q(s)}{s+200 - (s-200)q(s)} \tag{4.93}$$

which has integral action (recall that $q(0) = -1$) and inverts the stable minimum phase part of the plant.

Figure 4.14 shows that the nominal performance condition (4.83) is barely satisfied. Thus we can expect that satisfying the sufficient robust performance condition (4.24) will require scaling back the performance requirements. Indeed, by solving the following optimization problem,

$$\max\{\alpha \in (0,1] \mid \bar{\sigma}\left[\alpha W_1(j\omega)S(j\omega)\right] + \bar{\sigma}\left[W_\Delta(j\omega)T(j\omega)\right] \leq 1 \quad \forall \omega\} \tag{4.94}$$

it can be shown that the optimal robust performance level that can be guaranteed corresponds to $\alpha_{\max} = 0.5$ (and thus entails a 50% performance reduction).

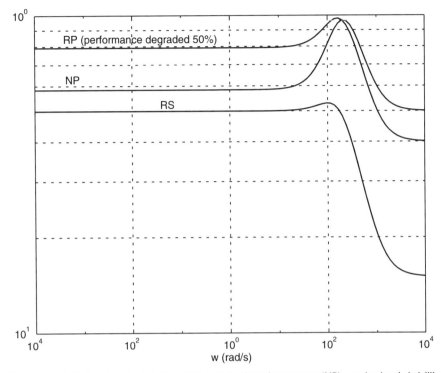

Figure 4.14. Robust performance (RP), nominal performance (NP), and robust stability (RS) conditions.

So far we have addressed only the first two performance requirements. The third can be met through the use of a prefilter $F(s)$ so that $F(s)T(s)$ achieves the desired specification. In this case, it is clear that due to the RHP zero of the plant, the specification cannot be achieved with a stable filter. Nevertheless, we can meet the requirement in terms of the *magnitude* of $F(s)T(s)$ by selecting F as follows:

$$F(s) = \frac{10}{q(s)(s+10)} \implies \bar{\sigma}\left[F(\jmath\omega)T(\jmath\omega)\right] = \left|\frac{10}{(\jmath\omega+10)}\right| \quad (4.95)$$

4.8 RELATED DESIGN PROCEDURES

The synthesis method presented in Section 4.6 provides an important step toward the generalization of SISO design methods to multivariable systems. Since the 1980s, a number of methods have been proposed based on the ideas introduced in this chapter. These methods include:

1. A combination of the LQG optimal control method with loop shaping, in order to automatically guarantee that the loop shaping takes into consideration only stabilizing controllers. The resulting method, known as as LQG/LTR (LQG with loop transfer recovery) ([101]) will be described briefly in Chapter 5. In its original formulation it was limited only to minimum phase plants, but it was later extended to the nonminimum phase case as well [294]. During the past few years, this methodology has been applied successfully to a number of practical problems (e.g., see [16]).

2. An alternative, related use of the loop shaping concept is based on transforming the energy functional (minimized by the LQG method) to the frequency domain. The weights appearing in the LQG cost functional can be related to the ones used in the loop shaping procedure to trade off robustness versus performance. The resulting optimization problem can be solved using a Wiener–Hopf approach ([258]).

3. As we indicated before, $\mathcal{H}_\infty$ optimal control (covered in Chapter 6) can be used to optimize robust stability or nominal performance or to synthesize a controller achieving both, via the following mixed sensitivity problem:

$$\min_{\text{stabilizing } K(s)} \left\| \begin{bmatrix} W_e(s)S_o(s) \\ W_\Delta(s)T_o(s) \end{bmatrix} \right\|_\infty \quad (4.96)$$

Using standard singular values properties it can be shown easily that

$$2\bar{\sigma} \begin{bmatrix} \alpha W_e(j\omega)S(j\omega) \\ W_\Delta(j\omega)T(j\omega) \end{bmatrix}$$
$$\geq \bar{\sigma}[\alpha W_e(s)S(j\omega)] + \bar{\sigma}[W_\Delta(j\omega)T(j\omega)] \quad (4.97)$$
$$\geq \bar{\sigma} \begin{bmatrix} \alpha W_e(s)S(j\omega) \\ W_\Delta(j\omega)T(j\omega) \end{bmatrix}$$

It follows then that we can attempt to satisfy the robust performance condition (4.24) by solving a scaled mixed sensitivity problem, where the performance condition is scaled by a factor α leading to

$$\min_{\text{stabilizing } K(s)} \left\| \begin{bmatrix} \alpha W_e(s)S_o(s) \\ W_\Delta(s)T_o(s) \end{bmatrix} \right\|_\infty \quad (4.98)$$

4. The optimal solution to the robust performance problem can be obtained by μ-synthesis, due to Doyle ([102]). This material will be presented in greater detail in Chapter 7.

4.9 PROBLEMS

1. Derive the necessary conditions for nominal performance and robust stability used in the loop shaping procedure.

2. Repeat Problems 3 and 4 of Chapter 2, considering that all models are MIMO.

3. A nominal linearized model of a rigid satellite is the following:

$$G(s) = \frac{1}{s^2 + 100} \begin{bmatrix} s & 10 \\ -10 & s \end{bmatrix}$$

The actuator representation is uncertain and can be described by means of a family of multiplicative dynamic uncertain models, $[I + W_\Delta(s)\Delta]$ with $\Delta \in \mathbb{C}^{2\times 2}$ bounded by unity. The performance objective consists in bounding by unity the weighted energy of the tracking error, for all possible bounded output disturbances: $\|d(s)\|_2 < 1$. Furthermore, the satellite should track with steady-state zero error, step input signals. The uncertainty and performance weights are as follows:

$$W_\Delta(s) = \frac{1}{2}\left(1 + \frac{s}{20}\right)$$

$$W_e(s) = \frac{1}{2}\left(\frac{s + 50}{s + 0.25}\right)$$

Design a controller that guarantees nominal stability and performance, and robust stability as well. What happens in the case of robust performance?

5

$\mathcal{H}_2$ OPTIMAL CONTROL

5.1 INTRODUCTION

A large number of control problems of practical importance involve designing a controller capable of stabilizing a given linear time invariant system while minimizing the worst-case response to some exogenous disturbances. This problem is relevant, for instance, for disturbance rejection, tracking, and robustness to model uncertainty. Depending on the choice of models for the input signals and on the criteria used to assess performance, this prototype problem leads to different mathematical formulations. When the exogenous disturbances are modeled as bounded energy signals and performance is measured in terms of the energy of the output, this problem leads to the well known $\mathcal{H}_\infty$ theory, explored in Chapter 6. The case where the signals involved are persistent bounded signals, with performance measured using the ℓ^∞ norm, leads to the ℓ^1 optimal control theory, formulated by Vidyasagar [315] and solved by Dahleh and Pearson [81, 83], covered in Chapter 8. In this chapter we will concentrate on the setup shown in Figure 5.1 with the exogenous input w assumed to belong to the set of signals with spectral density bounded by one. The objective is to minimize, over this class, the worst-case "size" of the output z, measured using the power seminorm.[1] Recall that the bounded spectral density to power induced norm, $\sup \|z\|_\mathcal{P}/\|w\|_\mathcal{S}$, is precisely $\|T_{zw}\|_2$, hence the $\mathcal{H}_2$ control problem name (see Appendix A).

---

[1] An alternative stochastic interpretation can be given by considering the input signal w to be white gaussian noise with unit covariance and having as design objective the minimization of the RMS value of the output, $\lim_{t \to \infty} \mathcal{E}[z^T(t)z(t)]$, where $\mathcal{E}$ denotes expectation.

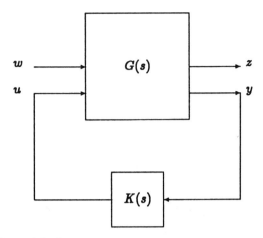

Figure 5.1. General loop structure for the $\mathcal{H}_2$ problem.

5.2 THE CLASSICAL LINEAR QUADRATIC REGULATOR PROBLEM

A *regulator* is a feedback controller designed to keep the states of a system within a neighborhood of a desirable operating point. Since we are dealing with linear systems, we can assume without loss of generality (by redefining the states if necessary) that this operating point is the origin, that is, $x = 0$. Physical considerations mandate that this behavior must be achieved using an acceptable amount of control energy since, obviously, the control energy available in any physical system is limited. Additionally, by limiting the control energy, one expects also to limit, in an indirect fashion, the *peak* value of the control input, avoiding potentially large signals that may result in damage to the plant.

This problem can be formalized by defining a performance index of the form

$$\min_{\bar{u} \in \mathcal{L}_2[0,\infty)} \int_0^\infty [x^T(t) \quad \bar{u}^T(t)] \begin{bmatrix} \bar{Q} & N \\ N^T & R \end{bmatrix} \begin{bmatrix} x(t) \\ \bar{u}(t) \end{bmatrix} dt \qquad (5.1)$$

$$\bar{Q}^T = \bar{Q} \geq 0, \quad R^T = R > 0, \quad \begin{bmatrix} \bar{Q} & N \\ N^T & R \end{bmatrix} \geq 0 \qquad (5.2)$$

Furthermore, by using the change of variable $\bar{u} = u - R^{-1} N^T x$, it is easily seen that (5.1) is equivalent to

$$\min_{u \in \mathcal{L}_2[0,\infty)} \int_0^\infty \left[x^T(t) Q x(t) + u^T(t) R u(t) \right] dt \qquad (5.3)$$

$$Q^T = Q \geq 0, \quad R^T = R > 0 \qquad (5.4)$$

where $Q = \bar{Q} - NR^{-1}N^T$. Here Q is a matrix that penalizes the deviation of the states x from the desired operating point, while R penalizes the control effort. Thus (5.3) represents a trade-off between regulation performance and control effort. A commonly used choice for the matrices Q and R is *Bryson's rule*: the matrices are chosen to be diagonal, with elements given by

$$\frac{1}{Q_{ii}} = \text{maximum acceptable value of } [x_i(t)]^2$$
$$\frac{1}{R_{ii}} = \text{maximum acceptable value of } [u_i(t)]^2 \quad (5.5)$$

However, we stress the fact that while this rule usually gives good results, there is no exact way to translate specifications given in terms of the *peak* values of x and u into specific values for Q and R. Thus several trial-and-error design iterations may be required in order to accommodate design specifications given in terms of *peak* values rather than $\mathcal{L}_2$ norms (akin to energy).

While in general the system may be subject to several classes of exogenous disturbances, we consider first the case where the system starts from an arbitrary (but deterministic) initial condition $x(0) = x_o$, and it is subject to no other exogenous disturbances during its evolution, leading to the classical LQR formulation. The effect of additional exogenous signals will be considered later in the chapter.

In the sequel, for simplicity we will normalize the control action by considering a Cholesky factorization of R, $R = V^T V$, $V > 0$ and using the change of variable $u_n = Vu$. Thus we have that $u^T(t)Ru(t) = u_n^T(t)u_n(t)$. If follows that without loss of generality we can assume that $R = I$ in (5.3). In terms of this new input, and using a factorization of $Q = C^T C$, we have that

$$\begin{aligned}
& x^T(t)Qx(t) + u_n^T(t)u_n(t) \\
&= \begin{bmatrix} x^T(t) & u_n^T(t) \end{bmatrix} \begin{bmatrix} Q & 0 \\ 0 & I \end{bmatrix} \begin{bmatrix} x(t) \\ u_n(t) \end{bmatrix} \\
&= \begin{bmatrix} x^T(t) & u_n^T(t) \end{bmatrix} \begin{bmatrix} C^T \\ D^T \end{bmatrix} \begin{bmatrix} C & D \end{bmatrix} \begin{bmatrix} x(t) \\ u_n(t) \end{bmatrix}
\end{aligned} \quad (5.6)$$

where C and D have been chosen so that $D^T [C \ D] = [0 \ I]$. Using these results, we can precisely state the classical LQR problem in the general form of Figure 5.1 as follows:

Classical Linear Quadratic Regulator Problem Given the LTI system

$$\begin{aligned}
\dot{x}(t) &= Ax(t) + Bu(t) \quad x(0) = x_0 \\
z(t) &= Cx(t) + Du(t)
\end{aligned} \quad (5.7)$$

with

$$(A, B) \text{ stabilizable} \quad (5.8)$$

$$(C, A) \text{ detectable} \quad (5.9)$$

$$C^T D = 0 \quad (5.10)$$

$$D^T D = I \quad (5.11)$$

find an optimal control law $u \in \mathcal{L}_2[0, \infty)$ that minimizes[2] $\|z\|_2^2$.

Classically, the LQR problem has been solved using either variational calculus techniques or "a completion of the square" type argument (e.g., see [61, 63]). Here, we will pursue a different approach (following after [104]) that exploits the orthogonality and norm preserving properties of certain linear operators. The main advantage of this approach is that it is easily extended to both the general $\mathcal{H}_2$ and the $\mathcal{H}_\infty$ cases. Toward this goal, we begin by giving a characterization of inner systems[3] in terms of the solution to an algebraic Riccati equation.

Lemma 5.1 *Consider the following stable system:*

$$U(s) \equiv \left[\begin{array}{c|c} A & B \\ \hline C & D \end{array} \right] \quad (5.12)$$

where the pair (A, B) is controllable. Let $X = X^T \geq 0$ denote the solution to the following Lyapunov equation:

$$A^T X + XA + C^T C = 0 \quad (5.13)$$

Then $U(s)$ is an inner system if and only if the following conditions hold:

$$\begin{aligned} D^T C + B^T X &= 0 \\ D^T D &= I \end{aligned} \quad (5.14)$$

Proof. ($\Rightarrow$) Using the state-space formulas given in Appendix B to obtain $U^\sim(s)$ and $U^\sim(s)U(s)$ yields

$$U^\sim(s)U(s) \equiv \left[\begin{array}{cc|c} A & 0 & B \\ -C^T C & -A^T & -C^T D \\ \hline D^T C & B^T & D^T D \end{array} \right] \quad (5.15)$$

---
[2] From the discussion above and the definition of the $\|.\|_2$ given in Appendix A, it should be clear at this point that $\|z\|_2^2$ is precisely given by (5.3).
[3] Recall that a LTI stable transfer matrix U is inner if $U^\sim U = I$ and that inner systems are norm preserving (Lemma 3.9).

Applying the similarity transformation

$$T = \begin{bmatrix} I & 0 \\ X & I \end{bmatrix}$$

to the realization (5.15) and using (5.13) and (5.14) yields

$$U^\sim(s)U(s) = \left[\begin{array}{cc|c} A & 0 & B \\ -XA - A^TX - C^TC & -A^T & -XB - C^TD \\ \hline D^TC + B^TX & B^T & D^TD \end{array}\right]$$

$$= \left[\begin{array}{cc|c} A & 0 & B \\ 0 & -A^T & 0 \\ \hline 0 & B^T & I \end{array}\right] \quad (5.16)$$

Finally, eliminating the nonobservable and noncontrollable modes in (5.16) yields the desired result. To show the converse, note that if the pair (A, B) is controllable, then from the first equation in (5.16) it follows immediately that $U^\sim(s)U = I$ only if $D^TD = I$ and $(D^TC + B^TX, A)$ is completely unobservable, which implies (5.14). □

Next, we introduce two preliminary results that will be used to solve the LQR problem.

Lemma 5.2 *Given the system (5.7), denote by $X \geq 0$ the solution to the following algebraic Riccati equation:*

$$A^TX + XA + C^TC - XBB^TX = 0 \quad (5.17)$$

Consider now the change of variable $v = u - Fx$, where the feedback gain F is given by

$$F = -B^TX \quad (5.18)$$

and let U denote the system in terms of this new input variable v, that is,

$$U(s) = \left[\begin{array}{c|c} A_F & B \\ \hline C_F & D \end{array}\right]$$

$$A_F = A + BF$$
$$C_F = C + DF \quad (5.19)$$

Then U is inner.

Proof. Simple algebra using the definitions of A_F and C_F and the orthogonality condition $C^TD = 0$ shows that (5.17) is equivalent to

$$A_F^TX + XA_F + C_F^TC_F = 0 \quad (5.20)$$

Moreover, from the definition of F, using again the orthogonality condition, it is easily seen that $D^T C_F + B^T X = 0$. Thus from Lemma 5.1 it follows that $U(s)$ is inner, provided that it is stable. To show stability, recall that from the properties of the Riccati equations (see Appendix C) we have that the assumption in (5.8) is a necessary and sufficient condition for the existence of a unique positive semidefinite solution $X \geq 0$ to equation (5.17). Moreover, this solution is stabilizing, that is, $\lambda(A - BB^T X) \subset \mathbb{C}_-$. □

Lemma 5.3 *Consider again the feedback gain $F = -B^T X$ and denote by $G_c(s)$ the following system:*

$$G_c(s) \stackrel{\triangle}{=} \left[\begin{array}{c|c} A_F & I \\ \hline C_F & 0 \end{array}\right] \qquad (5.21)$$

Then $U^\sim(s) G_c(s) \in \mathcal{RH}_2^\perp$.

Proof. Since $G_c(s)$ is strictly proper, we need only to show that $U^\sim G_c$ is strictly antistable. Proceeding as in Lemma 5.1, we have that

$$U^\sim(s) G_c(s) = \left[\begin{array}{cc|c} -A_F^T & -C_F^T C_F & 0 \\ 0 & A_F & I \\ \hline B^T & F & 0 \end{array}\right]$$

$$= \left[\begin{array}{cc|c} -A_F^T & 0 & -X \\ 0 & A_F & I \\ \hline B^T & 0 & 0 \end{array}\right]$$

$$= \left[\begin{array}{c|c} -A_F^T & -X \\ \hline B^T & 0 \end{array}\right] \qquad (5.22)$$

which is antistable, since A_F is stable. □

Finally, we recall (without proof) the following result from [104]:

Lemma 5.4 *Consider the system (5.7). If the control and output signals $u, z \in \mathcal{L}_2[0, \infty)$ then its states x are also in $\mathcal{L}_2[0, \infty)$. Moreover, $x(t) \to 0$ as $t \to \infty$.*

Using these results, we are ready now to give a complete solution to the LQR problem.

Theorem 5.1 *There exists a unique stabilizing control action $u(t) = F_2 x(t)$ that minimizes $\|z\|_2$ in (5.7), where $F_2 = -B^T X$ is the feedback gain defined* in (5.18). Moreover, we have that

$$\min_{u \in \mathcal{L}_2[0,\infty)} \|z(t)\|_2^2 = \|G_c(s)x_0\|_2^2 = x_0^T X x_0 \qquad (5.23)$$

where $G_c(s)$, the closed-loop system, is given by (5.21).

Proof. Proceeding as in the proof of Lemma 5.2, change the control input to $v(t) = u(t) - F_2 x(t)$. In terms of this new input we have that

$$\begin{aligned} \dot{x}(t) &= A_F x(t) + Bv(t) \quad x(0) = x_0 \\ z(t) &= C_F x(t) + Dv(t) \end{aligned} \qquad (5.24)$$

Stability of A_F implies that if $v \in \mathcal{L}_2$ then $z \in \mathcal{L}_2$. Thus we can take the Laplace transform of (5.24), obtaining

$$z(s) = G_c(s)x_0 + U(s)v(s) \qquad (5.25)$$

where $G_c(s)$ and $U(s)$ are defined in (5.21) and (5.12). Using Parseval's theorem and the fact that U is inner yields

$$\begin{aligned} \|z(t)\|_2^2 &= \|G_c(s)x_0 + U(s)v(s)\|_2^2 \\ &= \|U^\sim G_c(s)x_0 + v(s)\|_2^2 \end{aligned} \qquad (5.26)$$

Finally, using the facts that $v(s) \in \mathcal{H}_2$ and $U^\sim G_c x_o \in \mathcal{H}_2^\perp$ and that multiplication by an inner matrix is 2-norm preserving, (5.26) reduces to

$$\|z\|_2^2 = \|G_c(s)x_0\|^2 + \|v(s)\|_2^2 \qquad (5.27)$$

From this last equation it is apparent that the optimal value of $\|z\|_2$ is obtained when $v \equiv 0 \iff u(t) = F_2 x(t)$. $\square$

It is worth stressing that as a by-product of the proof we recover the well known fact that in the LQR case optimal performance among the set of all possible state-feedback stabilizing controllers is achieved by using *linear static* feedback $u = F_2 x$. This follows immediately from the fact that any other control law having $v \neq 0$ will result in a higher $\|z\|_2^2$.

5.3 THE STANDARD $\mathcal{H}_2$ PROBLEM

Consider again the block diagram of Figure 5.1. As we mentioned in the introduction, a very common control problem is to synthesize a controller that stabilizes the system and minimizes (in some sense) the "size" of the output due to a given class of input signals w. The LQR problem addressed in the last section is a special case of this problem, where the input is restricted to be of the form $w(t) = x_o \delta(t)$, with x_o arbitrary, and where it is assumed that the states are available for feedback, leading to the following state-space realization for $G(s)$:

$$G_{LQR} \equiv \left[\begin{array}{c|cc} A & I & B_2 \\ \hline C_1 & 0 & D_{12} \\ I & 0 & 0 \end{array} \right] \qquad (5.28)$$

In this section we extend these results to the case where $G(s)$ has the form

$$G \equiv \left[\begin{array}{c|cc} A & B_1 & B_2 \\ \hline C_1 & 0 & D_{12} \\ C_2 & D_{21} & 0 \end{array} \right] \qquad (5.29)$$

As before, the objective is to synthesize an internally stabilizing controller that minimizes the 2-norm of the closed-loop transfer function from w to z, $\|T_{zw}\|_2$. Recall from Appendix A that this can be given a physical interpretation in terms of minimizing the root mean square (RMS) value of the output z due to a gaussian white noise input with unit covariance. This problem can be formally stated as follows.

Standard $\mathcal{H}_2$ Problem Given the plant $G(s)$, find an internally stabilizing, proper, LTI controller that minimizes $\|T_{zw}\|_2$.

Note that we have assumed that $D_{11} = 0$ and $D_{22} = 0$. In the sequel, for the sake of simplicity, we will make the following additional assumptions:

(A1) (A, B_2) is stabilizable and (C_2, A) is detectable.
(A2) (A, B_1) is stabilizable and (C_1, A) is detectable
(A3) $C_1^T D_{12} = 0$ and $B_1 D_{21}^T = 0$.
(A4) D_{12} has full column rank with $D_{12}^T D_{12} = I$ and D_{21} has full row rank with $D_{21} D_{21}^T = I$.

The assumption $D_{11} = 0$ guarantees that the closed-loop transfer matrix is in $\mathcal{H}_2$ (recall that a real rational stable transfer matrix is in $\mathcal{H}_2$ if and only if it is strictly proper). A complete treatment of $\mathcal{H}_2$ problems with disturbance feedforward can be found in [332]. The assumption $D_{22} = 0$ is made to simplify the algebra. It can easily be removed by using the loop shifting technique discussed in Chapter 3. Assumption (A1) is clearly necessary for the system to be stabilizable via output feedback. Assumptions (A1) and (A2) together guarantee that the control and filtering Riccati equations associated with the $\mathcal{H}_2$ problem admit positive semidefinite stabilizing solutions. As we will see in the next section, they can be relaxed to G_{12} and G_{21} not having invariant zeros on the $j\omega$ axis. The orthogonality assumption (A3) is made to simplify the development. It can be relaxed at the price of the appearance of additional terms in the controller. Finally, the rank assumptions in (A4) guarantee that the $\mathcal{H}_2$ problem is nonsingular, while the normalizing assumptions ($D_{12}^T D_{12} = I$, $D_{21} D_{21}^T = I$) do not entail any loss of generality since they can always be met by redefining the inputs w and u if necessary.

Next, we introduce some notation that will be used to solve the $\mathcal{H}_2$ problem. Consider the following two Riccati equations:

$$A^T X_2 + X_2 A - X_2 B_2 B_2^T X_2 + C_1^T C_1 = 0 \tag{5.30}$$

$$A Y_2 + Y_2 A^T - Y_2 C_2^T C_2 Y_2 + B_1 B_1^T = 0 \tag{5.31}$$

Recall from Appendix C that these equations can be associated to the Hamiltonian matrices H_2 and J_2, respectively, where

$$H_2 \triangleq \begin{bmatrix} A & -B_2 B_2^T \\ -C_1^T C_1 & -A^T \end{bmatrix} \quad J_2 \triangleq \begin{bmatrix} A^T & -C_2^T C_2 \\ -B_1 B_1^T & -A \end{bmatrix} \tag{5.32}$$

Under assumptions (A2) and (A3) these matrices satisfy $H_2, J_2 \in dom(Ric)$ (see Appendix C), which in turn implies that (5.30) and (5.31) have unique solutions $X_2 = Ric(H_2) \geq 0$ and $Y_2 = Ric(Y_2) \geq 0$, respectively. Moreover, these solutions are stabilizing, that is, $A - B_2 B_2^T X_2$ and $A - Y_2 C_2^T C_2$ are Hurwitz.

With these assumptions we are ready now to give a complete solution to the $\mathcal{H}_2$ problem.

Theorem 5.2 *Under assumptions (A1)–(A4) the unique optimal $\mathcal{H}_2$ controller is given by*

$$K_{opt}(s) \equiv \left[\begin{array}{c|c} A_{FL} & L_2 \\ \hline -F_2 & 0 \end{array} \right] \tag{5.33}$$

with the corresponding optimal value of $\|T_{zw}\|_2$ given by

$$\min_{K(s) \text{stabilizing}} \|T_{zw}(s)\|_2^2 = \|G_c(s) B_1\|_2^2 + \|F_2 G_f(s)\|_2^2 \tag{5.34}$$

where

$$G_c(s) \equiv \left[\begin{array}{c|c} A_F & I \\ \hline C_{1F} & 0 \end{array}\right]$$

$$G_f(s) \equiv \left[\begin{array}{c|c} A_L & B_{1L} \\ \hline I & 0 \end{array}\right]$$

$$\begin{aligned} A_F &= A + B_2 F_2 \\ C_{1F} &= C_1 + D_{12} F_2 \\ A_L &= A + L_2 C_2 \\ A_{FL} &= A + B_2 F_2 + L_2 C_2 \\ B_{1L} &= B_1 + L_2 D_{21} \\ F_2 &= -B_2^T X_2 \\ X_2 &= \mathrm{Ric}(H_2) \\ L_2 &= -Y_2 C_2^T \\ Y_2 &= \mathrm{Ric}(J_2) \end{aligned} \qquad (5.35)$$

To prove this result we will decouple the output feedback problem into full information and output estimation subproblems, proceeding as in Chapter 3. Thus before proving the theorem in the general case, we will consider the same four special problems (FI, DF, FC, and OE) introduced in Chapter 3 as an intermediate step in obtaining the Youla parametrization.

Full Information Problem

Lemma 5.5 *Consider the FI case, that is, the case where*

$$G = \left[\begin{array}{c|cc} A & B_1 & B_2 \\ \hline C_1 & 0 & D_{12} \\ \left[\begin{smallmatrix} I \\ 0 \end{smallmatrix}\right] & \left[\begin{smallmatrix} 0 \\ I \end{smallmatrix}\right] & \left[\begin{smallmatrix} 0 \\ 0 \end{smallmatrix}\right] \end{array}\right] \qquad (5.36)$$

Then the following results hold:

1. *The optimal $\mathcal{H}_2$ controller is $K_{FI} = [F_2 \ \ 0]$.*
2. *The corresponding optimal value of the $\mathcal{H}_2$ cost is $\gamma_{opt} \stackrel{\triangle}{=} \min_{u \in \mathcal{L}_2} \|T_{zw}\|_2 = \|G_c B_1\|_2 = \mathrm{trace}(B_1^T X_2 B_1)^{1/2}$.*
3. *Given $\gamma > \gamma_{opt}$, the set of all internally stabilizing controllers such that $\|T_{zw}\|_2 \leq \gamma$ is $\{K(s) : K = [F_2 \ \ Q(s)], \ Q(s) \in \mathcal{RH}_2, \ \|Q\|_2^2 \leq \gamma^2 - \|G_c B_1\|_2^2\}$.*

Proof. As in the LQR case, define a new control input $v \triangleq u(t) - F_2 x(t)$, obtaining

$$z(s) = \left[\begin{array}{c|cc} A_F & B_1 & B_2 \\ \hline C_{1F} & 0 & D_{12} \end{array}\right] \left[\begin{array}{c} w(s) \\ v(s) \end{array}\right] \quad (5.37)$$

Or equivalently,

$$\begin{aligned} T_{zw}(s) &\equiv \left[\begin{array}{c|c} A_F & B_1 \\ \hline C_{1F} & 0 \end{array}\right] + \left[\begin{array}{c|c} A_F & B_2 \\ \hline C_{1F} & D_{12} \end{array}\right] T_{vw}(s) \\ &= G_c(s) B_1 + U(s) T_{vw}(s) \end{aligned} \quad (5.38)$$

which, combined with Lemmas 5.2 and 5.3, yields

$$\|T_{zw}(s)\|_2^2 = \|G_c(s) B_1\|_2^2 + \|T_{vw}(s)\|_2^2 \quad (5.39)$$

Clearly, this quantity is minimized by setting $v = 0$, which results in $T_{vw} \equiv 0$. This is achieved by the (unique) static controller $K_{FI} \triangleq \begin{bmatrix} F_2 & 0 \end{bmatrix}$ yielding

$$\min \|T_{zw}(s)\|_2^2 = \|G_c(s) B_1\|_2^2 = \operatorname{trace}(B_1^T X_2 B_1)^{1/2} \quad (5.40)$$

where the last equality follows from the fact that X_2 is the closed-loop observability Gramian. To show the last property, recall from Chapter 3 that the set of all stabilizing FI control actions can be obtained using controllers of the form $K(s) = [F \ Q]$, $A + B_2 F$ stable, $Q \in \mathcal{RH}_\infty$. Selecting $F = F_2$ yields

$$\|T_{zw}\|_2^2 = \|G_c B_1\|_2^2 + \|Q\|_2^2 \quad (5.41)$$

Thus $\|T_{zw}\|_2 \leq \gamma \iff Q \in \mathcal{RH}_2$, $\|Q\|_2^2 \leq \gamma^2 - \|G_c B_1\|_2^2$. □

Remarks

1. Note that the optimal controller uses *only* feedback from the states. Thus, under the assumption that $D_{11} = 0$, the FI and state-feedback (LQR) problems have the same optimal solution.
2. The optimal controller depends neither on the exogenous disturbance w nor on B_1 (a similar situation arises in the LQR case where the optimal gain does not depend on the initial condition x_o). However, the optimal value of the performance index is a function of B_1.
3. We have indeed proved a stronger result than intended. From the proof of the lemma it follows that the best possible control action (over the set of all stabilizing control actions $u \in \mathcal{L}_2$) corresponds to the control action generated by linear static state feedback.

By exploiting duality, we immediately get the following result.

Full Control Problem

Lemma 5.6 *For the full control case*

$$G_{FC}(s) \equiv \left[\begin{array}{c|ccc} A & B_1 & I & 0 \\ \hline C_1 & 0 & 0 & I \\ C_2 & D_{21} & 0 & 0 \end{array} \right] \quad (5.42)$$

the following properties hold:

1. *The optimal $\mathcal{H}_2$ controller is*

$$K_{FC} = \begin{bmatrix} L_2 \\ 0 \end{bmatrix}$$

2. *The corresponding optimal value of the $\mathcal{H}_2$ cost is $\gamma_{opt} \triangleq \min_{u \in \mathcal{L}_2} \|T_{zw}\|_2 = \|C_1 G_f\|_2 = \operatorname{trace}(C_1 Y_2 C_1^T)^{1/2}$.*
3. *Given $\gamma > \gamma_{opt}$, the set of all internally stabilizing controllers such that $\|T_{zw}\|_2 \leq \gamma$ is*

$$\left\{ K(s) : K = \begin{bmatrix} L_2 \\ Q(s) \end{bmatrix} \; Q(s) \in \mathcal{RH}_2, \; \|Q\|_2^2 \leq \gamma^2 - \|C_1 G_f\|_2^2 \right\}$$

Finally, we consider the DF and OE problems.

Disturbance Feedforward Problem

Lemma 5.7 *Consider the DF problem*

$$G_{DF}(s) \equiv \left[\begin{array}{c|cc} A & B_1 & B_2 \\ \hline C_1 & 0 & D_{12} \\ C_2 & I & 0 \end{array} \right] \quad (5.43)$$

where, as before, we make the additional assumption that $A - B_1 C_2$ is stable. Then the following properties hold:

1. $\gamma_{opt} = \min_{u \in \mathcal{L}_2} \|T_{zw}\|_2 = \|G_c B_1\|_2$
2.
$$K_{opt}(s) \equiv \left[\begin{array}{c|c} A + B_2 F_2 - B_1 C_2 & B_1 \\ \hline F_2 & 0 \end{array} \right] \quad (5.44)$$

3. *The set of all LTI controllers such that $\|T_{zw}\|_2 \leq \gamma$ is given by $K(s) = F_\ell(J_{DF}, Q)$, $Q \in \mathcal{RH}_2$, $\|Q\|_2^2 \leq \gamma^2 - \|G_c B_1\|_2^2$, where*

$$J_{DF}(s) = \left[\begin{array}{c|cc} A + B_2 F_2 - B_1 C_2 & B_1 & B_2 \\ \hline F_2 & 0 & I \\ -C_2 & I & 0 \end{array} \right] \quad (5.45)$$

THE STANDARD $\mathcal{H}_2$ PROBLEM

Proof. Recall from Chapter 3 (Lemma 3.3) that under the additional hypothesis of stability of $A - B_1 C_2$ the DF and FI problems are equivalent, in the sense that if $K_{FI} = [F \quad Q]$ stabilizes G_{FI} then $K_{DF} = F_\ell(J_{DF}, Q)$ stabilizes G_{DF} and yields the same closed-loop transfer function, that is, $F_\ell(G_{FI}, K_{FI}) = F_\ell(G_{DF}, K_{DF})$. Moreover, K_{DF} parametrizes *all* the DF stabilizing controllers. The proof follows now by combining these facts with Lemma 5.5. □

As before, the solution to the OE case follows from duality.

Output Estimation Problem

Lemma 5.8 *For the OE case where G has the form*

$$G_{OE}(s) = \left[\begin{array}{c|cc} A & B_1 & B_2 \\ \hline C_1 & 0 & I \\ C_2 & D_{21} & 0 \end{array}\right], \quad A - B_2 C_1 \text{ stable} \tag{5.46}$$

the following properties hold:

1. *The optimal $\mathcal{H}_2$ controller is*

$$K_{OE}(s) = \left[\begin{array}{c|c} A + L_2 C_2 - B_2 C_1 & L_2 \\ \hline C_1 & 0 \end{array}\right] \tag{5.47}$$

2. $\gamma_{opt} \stackrel{\triangle}{=} \min_{u \in \mathcal{L}_2} \|T_{zw}\|_2 = \|C_1 G_f\|_2$.
3. *The set of all internally stabilizing controllers such that $\|T_{zw}\|_2 \leq \gamma$ is $K(s) = F_\ell(J_{OE}, Q)$, $Q \in \mathcal{RH}_2$, $\|Q\|_2^2 \leq \gamma^2 - \|C_1 G_f\|_2^2$, where*

$$J_{OE}(s) = \left[\begin{array}{c|cc} A + L_2 C_2 - B_2 C_1 & L_2 & -B_2 \\ \hline C_1 & 0 & I \\ C_2 & I & 0 \end{array}\right] \tag{5.48}$$

Using the results of Lemmas 5.5–5.8, we can now give a proof of Theorem 5.2.

Proof of Theorem 5.2. Proceeding as in Chapter 3, make the change of input variable $u(t) = v(t) + F_2 x(t)$. This partitions the system into the two

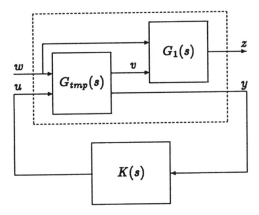

Figure 5.2. Output feedback as a cascade of an OE and a stable system.

subsystems shown in Figure 5.2, where $G_1(s)$ and G_{tmp} have the following realizations:

$$G_1(s) = \left[\begin{array}{c|cc} A + B_2 F_2 & B_1 & B_2 \\ \hline C_1 + D_{12} F_2 & 0 & D_{12} \end{array}\right]$$

$$G_{tmp}(s) = \left[\begin{array}{c|cc} A & B_1 & B_2 \\ \hline -F_2 & 0 & I \\ C_2 & D_{21} & 0 \end{array}\right]$$

(5.49)

Note that G_1 is stable and that G_{tmp} has an OE form. Moreover, $K(s)$ internally stabilizes $G(s)$ if and only if it internally stabilizes G_{tmp}. From the figure we have that

$$z(s) = \left[\begin{array}{c|cc} A_F & B_1 & B_2 \\ \hline C_{1F} & 0 & D_{12} \end{array}\right] \left[\begin{array}{c} w(s) \\ v(s) \end{array}\right]$$
$$= G_c(s) B_1 w(s) + U(s) v(s) \qquad (5.50)$$

or equivalently,

$$T_{zw}(s) = G_c(s) B_1 + U(s) T_{vw}(s) \qquad (5.51)$$

From Lemmas 5.2 and 5.3 we have that

$$\min_{K(s)\text{stab.}} \|T_{zw}(s)\|_2^2 = \|G_c(s) B_1\|_2^2 + \min_{K(s)\text{stab.}} \|T_{vw}(s)\|_2^2 \qquad (5.52)$$

But since G_{tmp} has an OE form, from Lemma 5.8 we have that the controller that minimizes this last transfer function is precisely

$$K_{OE}(s) \equiv \left[\begin{array}{c|c} A + L_2 C_2 + B_2 F_2 & L_2 \\ \hline -F_2 & 0 \end{array}\right] \qquad (5.53)$$

yielding $\min \|T_{vw}(s)\|_2 - \|F_2 G_f(s)\|_2$. It follows that

$$\min_{K(s)\text{stabilizing}} \|T_{zw}(s)\|_2^2 = \|G_c(s)B_1\|_2^2 + \|F_2 G_f(s)\|_2^2 \quad (5.54)$$

□

Remark Note that the optimal controller (5.33) clearly exhibits the well-known separation structure of the $\mathcal{H}_2$ problem. This can be made readily apparent by rewriting its state-space realization as

$$\begin{aligned}\dot{\hat{x}} &= A\hat{x} + B_2 u + L_2(C_2\hat{x} - y) \\ u &= F_2\hat{x}\end{aligned} \quad (5.55)$$

Thus the output of the controller is precisely the optimal estimate of the LQR control action $u = F_2 x$. Alternatively, the separation can be seen directly from the proof of the theorem, since the state feedback used in obtaining G_1 corresponds to the optimal LQR state feedback, and the subsystem G_{tmp} leads to an optimal output estimation problem. Note also that the optimal cost is precisely the optimal state-feedback (i.e., LQR) cost *plus* the optimal filtering cost.

5.4 RELAXING SOME OF THE ASSUMPTIONS

As we have just seen, the controller (5.33) minimizes T_{zw} under assumptions (A1)–(A4). While some of these assumptions are quite practical (e.g., (A1), required for the system to be stabilizable using output feedback), others may seem rather restrictive. In this section we address the issue of relaxing some of these assumptions, in particular, (A2) and (A3).

Recall from Appendix C that assumptions (A2) and (A3) are *sufficient* conditions for the Riccati equations (5.30) and (5.31) to have positive semi-definite stabilizing solutions. However, these assumptions are not *necessary* for this property to hold. Indeed, it can be shown (see Appendix C) that assumptions (A2)–(A4) can be relaxed to the following:

(A2′) $\begin{bmatrix} A - j\omega I & B_2 \\ C_1 & D_{12} \end{bmatrix}$ has full column rank for all ω \quad (5.56)

(A3′) $\begin{bmatrix} A - j\omega I & B_1 \\ C_2 & D_{21} \end{bmatrix}$ has full row rank for all ω \quad (5.57)

(A4′) D_{12} full column rank and D_{21} full row rank.

Assumptions (A1) and (A2′) are necessary and sufficient conditions for $H_2 \in$

$dom(Ric)$, where H_2 now denotes the following Hamiltonian matrix:

$$H_2 \triangleq \begin{bmatrix} A - B_2 R^{-1} D_{12}^T C_1 & -B_2 R^{-1} B_2^T \\ -C_1^T (I - D_{12} R^{-1} D_{12}^T) C_1 & -(A - B_2 R^{-1} D_{12}^T C_1^T) \end{bmatrix} \quad (5.58)$$

and where $R \triangleq D_{12}^T D_{12}$. Similarly, assumptions (A1) and (A3′) are necessary and sufficient for $J_2 \in dom(Ric)$, where

$$J_2 \triangleq \begin{bmatrix} (A - B_1 D_{21}^T S^{-1} C_2)^T & -C_2^T S^{-1} C_2 \\ -B_1 (I - D_{21}^T S^{-1} D_{21}) B_1^T & -(A - B_1 D_{21}^T S^{-1} C_2) \end{bmatrix} \quad (5.59)$$

and where $S \triangleq D_{21} D_{21}^T$.

Our previous results hold under these relaxed hypotheses, provided that F_2 and L_2 are suitably modified.

Theorem 5.3 *Suppose that assumptions (A1) and (A2′)–(A4′) hold. Let $X_2 \triangleq Ric(H_2)$ and $Y_2 \triangleq Ric(Y_2)$, and define*

$$\begin{aligned} F_2 &\triangleq -R^{-1}(B_2^T X_2 + D_{12}^T C_1) \\ L_2 &\triangleq -(Y_2 C_2^T + B_1 D_{21}^T) S^{-1} \end{aligned} \quad (5.60)$$

Then Lemmas 5.5–5.8 and Theorem 5.2 ho!d.

Finally, we want to comment briefly on assumption (A4′). The full rank conditions are imposed to avoid obtaining singular control or filtering problems. The treatment of such problems is beyond the scope of this book (e.g., see [298]), although we will briefly discuss a special case, the "cheap control" case, in the next section.

5.5 CLOSED-LOOP PROPERTIES

Recall that we motivated the $\mathcal{H}_2$ problem using an *optimal* control argument, where the goal was to optimize a performance index related to rejecting a class of disturbances. However, it should be clear that optimizing performance in a certain sense does not necessarily guarantee that the resulting system will be "optimal" (or even have acceptable performance) under a different set of conditions. Thus it is of interest to analyze the properties of the closed-loop system. In particular, guided by practical considerations, we are interested in assessing how sensitive the system is to modeling errors, since these errors are inevitably present in real life situations.

5.5.1 The LQR Case: Kalman's Inequality

We begin by considering the simpler LQR case shown in Figure 5.3. We will establish the surprising fact that although robustness considerations were not taken into account in the design, the resulting system exhibits, in a classical sense, excellent properties. In order to establish this fact we need the following preliminary result.

Lemma 5.9 (Kalman's Identity) *Let $L(j\omega) = F(j\omega I - A)^{-1}B$ be the state-feedback loop, where F is the optimal LQR gain. Then the following identity holds for all ω:*

$$[I + L(j\omega)]^* R [I + L(j\omega)] \\ = R + \left[(j\omega I - A)^{-1}B\right]^* Q \left[(j\omega I - A)^{-1}B\right] \quad (5.61)$$

where Q and R are the state and control weighting matrices, respectively.

Proof. See [98, 168]. □

Remark Since the second term on the right-hand side of (5.61) is positive semidefinite, it follows that

$$[I + L(j\omega)]^* R [I + L(j\omega)] \geq R \quad (5.62)$$

Moreover, recall that without loss of generality (by redefining the control input if necessary) we can always assume that $R = I$. Thus (5.62) reduces to

$$[I + L(j\omega)]^* [I + L(j\omega)] \geq I \quad (5.63)$$

which is called Kalman's inequality. Equivalently, if we define the *sensitivity function* $S(j\omega)$ in the usual way $S = [I + L(j\omega)]^{-1}$, then we have that $\|S(j\omega)\|_\infty \leq 1$ (i.e., the sensitivity is smaller than or equal to 1 at all frequencies).

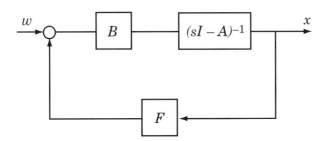

Figure 5.3. Assessing stability margins for the optimal LQR controller.

In the SISO case (5.63) reduces even further to

$$|1 + L(j\omega)| \geq 1, \quad \forall \omega \tag{5.64}$$

Recall from Chapter 2 that the left-hand side of (5.64) is the distance from a generic point on the Nyquist plot of $L(j\omega)$ to the critical point $(-1, 0)$. Thus (5.64) implies that the Nyquist plot of L is excluded from a unit disk centered at $(-1, 0)$. In turn, this implies (see Figure 5.4) that the closed-loop system has an *infinite* increasing gain margin and a decreasing gain margin of at least $\frac{1}{2}$. Additionally, exclusion from the unit disk centered at the critical point means that the *worst possible* phase margin corresponds to the situation shown in Figure 5.5, where the Nyquist plot passes through the intersection of this disk and the unit circle centered at the origin. Thus the LQR design has a phase margin of at least $60°$. Both margins indicate that, from a classical standpoint, the LQR design has excellent robustness properties.

Let us return now to the MIMO case. Recall from Table 2.1 that $\|S\|_\infty \leq 1$ is a necessary and sufficient condition for robust stability against quotient-type uncertainty, that is, $G(s) = [I + \Delta]^{-1} G_o(s)$, $\|\Delta\|_\infty \leq 1$. Thus we can expect good robustness properties against this type of uncertainty (e.g., arising from uncertainty in the location of the poles of the plant). On the other hand, it is easy to establish that $\|S\|_\infty \leq 1$ implies that $\|T\|_\infty \leq 2$. Moreover, it is not difficult to find examples where this bound is tight (e.g., see Example 3.2 in [181]). Thus the LQR design *does not* necessarily have good robustness properties against multiplicative uncertainty, since robustness can be guaranteed only for perturbations satisfying $\|\Delta\|_\infty \leq \frac{1}{2}$. Additionally, it can be shown [12] that $T(j\omega)$ has the relatively poor roll-off rate of ω^{-1}, which may not be sufficient to compensate for high-frequency unmodeled dynamics.

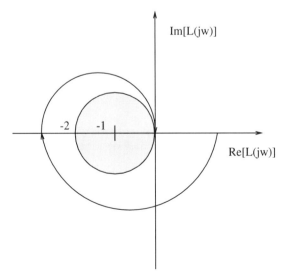

Figure 5.4. Nyquist plot illustrating the gain margin of LQ regulators.

CLOSED-LOOP PROPERTIES   145

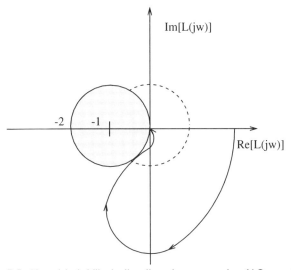

Figure 5.5. Nyquist plot illustrating the phase margin of LQ regulators.

5.5.2 Some Consequences of Kalman's Inequality

In this section we briefly discuss an important corollary of Kalman's inequality, the minimum phase property of the optimal LQR plant, and its implications for the *cheap control* case. For simplicity we will consider only the SISO case, but the results hold for general MIMO systems.

From the inequality (5.63) it follows that the sensitivity function $S(s) = [1 + f(sI - A)b]^{-1}$ is bounded real, that is, stable and such that $\|S\|_\infty \leq 1$. Assume now that $L(s) = f(sI - A)^{-1}b$ is not minimum phase. Then there exist s_o, $\mathrm{Re}(s_o) > 0$ such that $L(s_o) = 0$. It follows that $S(s_o) = 1$. Since $S(s)$ is analytic in the closed right-half plane and bounded above by 1 in the $j\omega$ axis, it follows from the *maximum modulus theorem* ([224]) that $S(s) \equiv 1$ in $\mathrm{Re}(s) \geq 0$, which implies that $f(sI - A)^{-1}b \equiv 0$. Clearly this is not possible (unless the state weight $Q = 0$). Thus we have established that the optimal loop function cannot have zeros in the open right-half plane. Note, however, that this proof does not rule out the existence of zeros on the $j\omega$ axis.

As an application of the minimum phase property, we will analyze the case where the control penalty is of the form $R = \rho I$ and $\rho \to 0$. This is usually referred to as the *cheap control* case, since very little penalty is imposed on the control effort. Intuitively, it seems clear that as $\rho \to 0$ the control action increases and that the corresponding value of the performance index (5.3) should approach 0. However, as we will see next this is true *only* in the case where the plant is minimum phase.

Setting $R = \rho I$ and factoring $Q = c^T c$, the equality (5.61) becomes

$$[1 + L(j\omega)]^* [1 + L(j\omega)] = 1 + \frac{1}{\rho} G(j\omega)^* G(j\omega) \qquad (5.65)$$

where $L(s) = f(sI - A)^{-1}b$ and $G(s) = c(sI - A)^{-1}b$. Clearly, as $\rho \to 0$ this equality implies that

$$|1 + f(j\omega I - A)^{-1}b| \to \frac{1}{\sqrt{\rho}}|c(j\omega I - A)^{-1}b| \tag{5.66}$$

Since $L(s)$ is minimum phase, it follows that its zeros are the minimum phase zeros of $G(s)$ plus the mirror image (with respect to the $j\omega$ axis) of its non-minimum phase zeros. Consider first the case where $G(s)$ is minimum phase. In this case since G and L have (asymptotically) the same zeros, it follows that as $\rho \to 0$ then $\sqrt{\rho}L(s) \to \pm G(s)$ and thus

$$c \to \pm\sqrt{\rho}f \tag{5.67}$$

Rewritting the algebraic Riccati equation (5.17) in terms of f and c we have

$$A^T X + XA + c^T c - \rho f^T f = 0 \tag{5.68}$$

From (5.67) and (5.68) it follows that as $\rho \to 0$ then $A^T X + XA \to 0$ and thus $X \to 0$. Hence

$$\lim_{\rho \to 0} J_{LQR}(x_o) = \lim_{\rho \to 0} x_o^T X x_o = 0 \tag{5.69}$$

On the other hand, if $G(s)$ is nonminimum phase then $c^T c - \rho f^T f \not\to 0$ and thus $X \to \tilde{X} \neq 0$. It follows that for nonminimum phase plants there is a nonzero LQR cost $J = x_o^T \tilde{X} x_o$ even in the case where the control action is not penalized.

5.5.3 Stability Margins of Optimal $\mathcal{H}_2$ Controllers

As we saw in Section 5.5.1, the optimal *state-feedback* $\mathcal{H}_2$ controlled system has very good robustness properties, at least in the case where the hypotheses of Section 5.2 hold (it can be shown that if R is not diagonal, then the gain margins may become arbitrarily small [181]). Motivated by these results, one would expect that some of these robustness properties will carry over to the general $\mathcal{H}_2$ case. Unfortunately, as we illustrate with the following counter-example [100], this is not true.

Example 5.1 *Consider the following system:*

$$\begin{bmatrix} \dot{x}_1 \\ \dot{x}_2 \end{bmatrix} = \begin{bmatrix} 1 & 1 \\ 0 & 1 \end{bmatrix} \begin{bmatrix} x_1 \\ x_2 \end{bmatrix} + \begin{bmatrix} 1 \\ 1 \end{bmatrix} w + \begin{bmatrix} 0 \\ 1 \end{bmatrix} u$$

$$y = \begin{bmatrix} 1 & 0 \end{bmatrix} \begin{bmatrix} x_1 \\ x_2 \end{bmatrix} + v \tag{5.70}$$

where w and v are uncorrelated white gaussian noises with covariance σ and 1, respectively. Assume that we want to minimize the following performance index:

$$J \triangleq \lim_{t\to\infty} \frac{1}{t} \mathcal{E} \left\{ \int_0^t \left[x^T Q x + u^T R u \right] dt \right\} \tag{5.71}$$

where $\mathcal{E}$ denotes the expectation operator and the weighting matrices Q and R are chosen as

$$Q \triangleq q \begin{bmatrix} 1 \\ 1 \end{bmatrix} \begin{bmatrix} 1 & 1 \end{bmatrix}, \quad q > 0, \; R = 1 \tag{5.72}$$

It is easily seen that this problem can be recast in the general $\mathcal{H}_2$ form of Section 5.3 by selecting the following realization for $G(s)$:

$$G(s) = \left[\begin{array}{cc|cc|c} 1 & 1 & \begin{bmatrix} \sqrt{\sigma} & 0 \\ \sqrt{\sigma} & 0 \end{bmatrix} & \begin{bmatrix} 0 \\ 1 \end{bmatrix} \\ 0 & 1 & & \\ \hline \begin{bmatrix} \sqrt{q} & \sqrt{q} \\ 0 & 0 \end{bmatrix} & 0 & \begin{bmatrix} 0 \\ 1 \end{bmatrix} \\ \begin{bmatrix} 1 & 0 \end{bmatrix} & \begin{bmatrix} 0 & 1 \end{bmatrix} & 0 \end{array} \right] \tag{5.73}$$

In this case equations (5.30) and (5.31) can be solved analytically yielding the following optimal gains:

$$\begin{aligned} F_2 &= f[1 \; 1] \\ L_2 &= d \begin{bmatrix} 1 \\ 1 \end{bmatrix} \\ f &\triangleq -(2 + \sqrt{4+q}) \\ d &\triangleq -(2 + \sqrt{4+\sigma}) \end{aligned} \tag{5.74}$$

Hence the optimal controller is given by

$$K(s) = \left[\begin{array}{cc|c} 1+d & 1 & -d \\ f+d & 1+f & -d \\ \hline f & f & 0 \end{array} \right] \tag{5.75}$$

To assess the gain margin of the resulting closed loop, suppose that the control action is multiplied by a scalar gain m (with nominal value 1). Closing the loop with the controller $mK(s)$ yields the following closed-loop dynamics matrix:

$$A_{cl} = \begin{bmatrix} 1 & 1 & 0 & 0 \\ 0 & 1 & mf & mf \\ -d & 0 & 1+d & 1 \\ -d & 0 & d+f & 1+f \end{bmatrix} \tag{5.76}$$

It can be shown that the characteristic polynomial of A_{cl} has the form

$$P(s) \triangleq \det(sI - A_{cl}) = s^4 + a_3 s^3 + a_2 s^2 + a_1 s + a_0 \qquad (5.77)$$

where only the last two coefficients, a_1 and a_0, are functions of m, with the following explicit expressions:

$$\begin{aligned} a_1 &= -(d+f+4) + 2(m-1)df \\ a_0 &= 1 + (1-m)df \end{aligned} \qquad (5.78)$$

Recall that a necessary condition for stability is that $a_i > 0$. But from (5.78) it can be seen that, for sufficiently large d and f, a_0 and a_1 can be rendered negative by taking $m = 1 + \epsilon$, with ϵ arbitrarily small. Thus the closed-loop system has an arbitrarily small gain margin. Note also that this situation happens precisely in cases where the driving noise gets larger (large σ) or the control weight relative to the state weight ($1/q$) gets smaller, contrary to the conjecture that either choice (large noise or small control weight) will result in improved "robustness."

Since LQR systems exhibit good robustness properties, motivated by the separation principle, one may attempt to recover these properties by making the dynamics of the observer very fast. It can be shown [101] that this approach *will not*, in general, recover the desired properties. However, they may be recovered in some special situations, most notably in the case where the plant is minimum phase. The main idea of this approach consists of introducing fictitious plant noise with a specific structure (thus giving up filter "optimality"). It can be shown that as this noise intensity approaches infinity, then the closed-loop LQG loop matrix $K(s)C(sI - A)^{-1}B$ approaches the LQR matrix $F(sI - A)^{-1}B$. This technique, first proposed in [101], is known as LQG/LTR (loop transfer recovery). In its original formulation, it was limited to *minimum phase* plants since the resulting controller inverts the plant (from the left), substituting it with the desired LQR dynamics. This fact suggests that the technique could also be used for "mildly" nonminimum phase plants, that is, cases where the norm of the transfer matrix is small at the frequency of the nonminimum phase zeros [16, 331].

Finally, it is worth noting that this technique is not restricted to the case where the transfer function recovered corresponds to the optimal LQR controller. Indeed, any return ratio that shares the right-half plane pole/zero structure of the plant can be recovered [294]. This opens the possibility of recovering some other state-feedback structure that could be applied to nonminimum phase plants. Note, however, that at this point, in addition to giving up filter optimality (by the introduction of the fictitious noise), we will also be giving up controller optimality. Thus, in this case, the method reduces to just an *ad hoc* procedure for trading off robustness versus performance. Rather than pursuing this approach further, in the next chapters we will address the issues of robust stability and performance using an $\mathcal{H}_\infty$ approach.

5.6 RELATED PROBLEMS

Before closing this chapter we want to briefly address the related *generalized* $\mathcal{H}_2$ control problem [249]. Recall that in the SISO case the $\mathcal{H}_2$ norm of a system can be given a deterministic interpretation as the $\mathcal{L}_2$ to $\mathcal{L}_\infty$ (i.e., energy to peak) induced norm. Thus, in this context, the $\mathcal{H}_2$ problem can be thought of as the problem of synthesizing a controller that stabilizes the plant and minimizes the worst possible *peak* value of the output due to exogenous signals with unit energy. In the case of MIMO systems, the two interpretations do not coincide. Specifically, assume that the output is measured using the supremum over time of the spatial 2 norm. It can be shown [320] that $\sup_{\|w\|_2=1}\{\sup_t \|z(t)\|_2\} = \sqrt{\bar{\sigma}[S(T_{zw})]}$, where $\bar{\sigma}$ denotes the maximum singular value. Thus, while the standard $\mathcal{H}_2$ problem leads to the minimization of $J_{\mathcal{H}_2} = \text{trace}(C_{cl}W_cC_{cl}^T)$, where W_c is the closed-loop controllability Gramian, the generalized $\mathcal{H}_2$ problem leads to the minimization of $J_{\text{gen}} = \bar{\sigma}(C_{cl}W_cC_{cl}^T)$. Obviously, these two indexes coincide only in the case where $C_{cl}W_cC_{cl}^T$ is a scalar, in which case Theorems 5.1 and 5.2 provide a solution also to the generalized $\mathcal{H}_2$ problem. While this approach is no longer valid in the MIMO case, some of the tools developed to solve the standard $\mathcal{H}_2$ problem can still be used, allowing for recasting the generalized problem into a linear matrix inequality form. This LMI problem can be solved efficiently using finite-dimensional convex optimization techniques.

As in the conventional $\mathcal{H}_2$ case, the optimal state-feedback controller is static, while optimal output feedback controllers have the same order as the plant and exhibit a separation structure. As an intermediate step in establishing these results we will address first the full information case.

Lemma 5.10 *The state-feedback and FI generalized $\mathcal{H}_2$ problems are equivalent in the sense that there exists a linear static state-feedback controller such that $J_{\text{gen}}^{sf} < \gamma$ if and only if there exists a linear full information controller such that $J_{\text{gen}}^{FI} < \gamma$ [243].*

Remark From this lemma it follows immediately that *dynamic* state-feedback controllers cannot improve upon the performance of static ones. Thus in the sequel we only need to consider static state-feedback laws.

Theorem 5.4 *There exists a static state-feedback controller such that the generalized $\mathcal{H}_2$ cost*

$$J_{\text{gen}} = \left[\bar{\sigma}\left(\frac{1}{2\pi}\int_{-\infty}^{\infty} T_{zw}(j\omega)T_{zw}^*(j\omega)\,d\omega\right)\right]^{1/2} < \gamma \quad (5.79)$$

if and only if the following set of linear matrix inequalities (LMIs) in the

150 $\mathcal{H}_2$ OPTIMAL CONTROL

variables Y and W has a feasible solution:

$$\begin{bmatrix} AY + YA^T + W^T B_2^T + B_2 W & B_1 \\ B_1^T & -I \end{bmatrix} < 0$$

$$\begin{bmatrix} \gamma^2 I & (CY + DW)^T \\ (CY + DW) & Y \end{bmatrix} > 0 \quad (5.80)$$

$$Y > 0$$

Moreover, in this case a solution is given by the state-feedback gain $K_o = WY^{-1}$, *with the corresponding cost given by* $J_{\text{gen}}^2 = \bar{\sigma}\left[(C + DK_o)Y(C + DK_o)^T\right].$

Proof. The proof follows from combining Theorem 3.2 in [249] with standard Schur complement arguments to recast the equations there into the LMI form. □

Next, we consider the output feedback case. As in the regular $\mathcal{H}_2$ case the solution will be obtained by reducing it to a disturbance feedforward problem through the use of an appropriate output injection.

Theorem 5.5 *Consider a plant $G(s)$ with the following state-space realization:*

$$G(s) \equiv \left[\begin{array}{c|cc} A & B_1 & B_2 \\ \hline C_1 & 0 & D_{12} \\ C_2 & D_{21} & 0 \end{array}\right] \quad (5.81)$$

and assume that the following conditions hold:

(A1) (A, B_2) *is stabilizable and* (C_2, A) *is detectable.*
(A2) D_{21} *has full row rank with* $D_{21}[B_1^T \quad D_{21}^T] = [0 \quad I].$
(A3) $\begin{bmatrix} A - j\omega I & B_1 \\ C_2 & D_{21} \end{bmatrix}$ *has full row rank for all ω.*

Then the following statements are equivalent:

1. *There exists an internally stabilizing controller $K(s)$ such that $J_{\text{gen}}[G(s), K(s)] \leq \gamma.$*
2. *There exists a static state-feedback controller K_{sf} that internally stabilizes*

the plant $G_{sf}(s)$ and such that $\bar{\sigma}\left[C_1 Y_2 C_1^T + S(G_{sf}, K_{sf})\right] \leq \gamma^2$, where

$$J_2 = \begin{bmatrix} A^T & -C_2^T C_2 \\ -B_1 B_1^T & -A \end{bmatrix}$$

$$Y_2 = Ric(J_2)$$

$$L_2 = -Y_2 C_2^T$$

$$G_{sf}(s) \equiv \left[\begin{array}{c|cc} A & -L_2 & B_2 \\ \hline C_1 & 0 & D_{12} \\ I & 0 & 0 \end{array} \right]$$

$$S(G_{sf}, K_{sf}) = \frac{1}{2\pi} \int_{-\infty}^{\infty} F_\ell[G_{sf}(j\omega), K_{sf}(j\omega)] F_\ell[G_{sf}(j\omega), K_{sf}(j\omega)]^* d\omega$$

(5.82)

Proof. By using the output injection $r = L_2 y$, the plant $G(s)$ can be decomposed into the two subsystems shown in Figure 5.6, where

$$G_{tmp}(s) \equiv \left[\begin{array}{c|cc} A + L_2 C_2 & B_1 + L_2 D_{21} \\ \hline C_1 & 0 \\ C_2 & D_{21} \end{array} \right]$$

(5.83)

$$G_1(s) \equiv \left[\begin{array}{c|cc} A & -L_2 & B_2 \\ \hline C_1 & 0 & D_{12} \\ C_2 & I & 0 \end{array} \right]$$

Note that in this decomposition $G_{tmp}(s)$ is internally stable while $G_1(s)$ corresponds to a disturbance feedforward problem. Using standard arguments it can be shown that (e.g., see [249]), *regardless of the choice of controller $K(s)$, the following orthogonality property holds for $S(G, K)$*:

$$S(G, K) = C_1 Y_2 C_1^T + S(G_1, K) \quad (5.84)$$

It follows then that there exists an internally stabilizing controller such that $J_{gen} \leq \gamma$ if and only if there exists a controller K that internally stabilizes $G_1(s)$ and such that $\bar{\sigma}\left[C_1 Y_2 C_1^T + S(G_1, K)\right] \leq \gamma^2$. By construction $G_1(s)$ has a DF structure and hence (Lemma 5.7) is equivalent to a FI problem. The proof follows now from the equivalence of the FI and SF problems shown in Lemma 5.10. □

Finally, we want to consider briefly the case where the norm used for

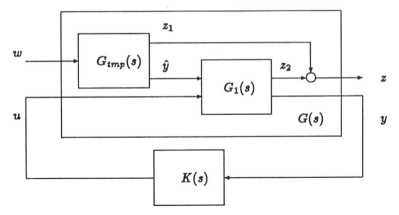

Figure 5.6. Output feedback as the combination of a DF problem and a stable system.

measuring the output is the standard $\mathcal{L}^\infty$ norm. In this case it can be shown ([320]) that

$$\sup \{\|z\|_\infty/\|w\|_2 : w \in \mathcal{L}_2[0, \infty), w \neq 0\} = \sqrt{d_{\max}[S(T_{zw})]} \qquad (5.85)$$

where $d_{\max}$ denotes the maximum diagonal entry. It follows that the same tools used to solve the generalized $\mathcal{H}_2$ problem with cost (5.79) can also be used in this case, by simply replacing $\bar{\sigma}$ by $d_{\max}$.

5.7 PROBLEMS

1. The servomotor of a tracking antenna can be approximately represented by the following second order model:

$$\dot{x}(t) = \begin{bmatrix} 0 & 1 \\ 0 & -4.6 \end{bmatrix} x(t) + \begin{bmatrix} 0 \\ 0.787 \end{bmatrix} u(t) \qquad (5.86)$$

where $u(t)$ is the input voltage and where the states are the angular position and velocity respectively.

(a) Using classical state-space tools design a controller that places the closed loop poles at $-20 \pm j20$.

(b) Find the peak value of the control action for an initial condition $x(0) = \begin{bmatrix} 0 & 1 \end{bmatrix}^T$.

(c) Find the optimal LQR controller corresponding to the choice of weights $Q = \begin{bmatrix} 1 & 0 \\ 0 & 0 \end{bmatrix}$ and $R = 2 \times 10^{-5} I$.

(d) Repeat part (b).

2. This problem illustrates some of the potential pitfalls of Optimal Control theory. Consider the following system:

$$\dot{x} = \begin{bmatrix} 1 & 1 \\ 0 & 1 \end{bmatrix} x + \begin{bmatrix} 0 \\ 1 \end{bmatrix} u$$

(a) Find the controller that minimizes:

$$J = \int_0^\infty \left(x^T Q x + u^T R u \right) dt, \quad Q = q \begin{bmatrix} 1 & 1 \\ 1 & 1 \end{bmatrix}, \quad R = 1$$

for $q = 1, 10, 100$.

(b) In each case, plot the Nyquist plot of the combined plant-controller and find the closed-loop gain and phase margins.

(c) Assume now that instead of the states the output of the plant is x_1 corrupted by a white Gaussian noise signal v and that the plant is also driven by white Gaussian noise w, that is:

$$\dot{x} = \begin{bmatrix} 1 & 1 \\ 0 & 1 \end{bmatrix} x + \begin{bmatrix} 0 \\ 1 \end{bmatrix} u + w$$

$$y = \begin{bmatrix} 1 & 0 \end{bmatrix} x + v$$

where

$$\mathcal{E} \left\{ \begin{bmatrix} w^T \\ v^T \end{bmatrix} \begin{bmatrix} w & v \end{bmatrix} \right\} = \begin{bmatrix} 1 & 0 & 0 \\ 0 & 1 & 0 \\ 0 & 0 & 1 \end{bmatrix}$$

Find the optimal $\mathcal{H}_2$ controller that minimizes

$$J = \mathcal{E} \left\{ \int_0^\infty \left(x^T Q x + u^T R u \right) dt \right\}$$

for $q = 1, 10, 100$.

(d) Repeat part (b). What conclusion can you draw about the gain and phase margins of the optimal system?

3. The issues involved in controlling systems subject to model uncertainty and multiple specifications can be illustrated by the simple system shown in Figure 5.7 consisting of two unity masses coupled by a spring with nominal value $k_o = 1.0$ but subject to uncertainty. A control force acts on body 1 and the position of body 2 is measured, resulting in a noncolocated sensor actuator problem used as a benchmark several years at the American Control Conference [46].

(a) Find a state-space model of the system.

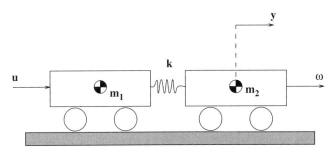

Figure 5.7. The ACC benchmark problem.

(b) Design a stabilizing controller that meets the following performance specifications: (i) the control action following a unit impulse disturbance ω acting on m_2 is constrained by $|u(t)| \leq 1$; and (ii) for the same disturbance the displacement y of m_2 has a settling time of about 15 seconds.

(c) Investigate the performance of your controller when k varies between 0.5 and 2.

4. A common performance specification is that the closed-loop system must have a prescribed degree of stability α, that is the closed-loop poles must lie to the left of the line $\text{Re}(s) = -\alpha$. Show that this can be accomplished using a LQ regulator that minimizes $\|e^{\alpha t} z(t)\|_2$ rather than $\|z(t)\|_2$.

5. Consider the optimal LQR problem for a SISO plant with $Q = cc^T$ and $R = \rho$. Let z_i, $i = 1, \ldots, p$ and p_i, $i = 1, \ldots, n$ denote the zeros and poles of $G = d(sI - A)^{-1}b$.

   (a) Show that as $\rho \to 0$ p of the n optimal closed-loop poles approach z_i if $\text{Re}(z_i) \leq 0$ or $-z_i$ if $\text{Re}(z_i) > 0$.

   (b) Show that the remaining $n - p$ poles approach infinity as the roots of $s^{2(n-p)} = (-1)^{p-n+1} \frac{\alpha}{\rho}$. This configuration is known as a *Butterworth configuration* of order $n - p$.

   (c) Show that as $\rho \to \infty$ the n optimal closed-loop poles approach p_i if $\text{Re}(p_i) \leq 0$ or $-p_i$ if $\text{Re}(p_i) > 0$.

6. Consider a discrete time system with state space realization $U(z) = D + C(zI - A)^{-1}B$.

   (a) Its conjugate is defined as $U\tilde{\ }(z) = U^T(\frac{1}{z})$. Show that if A^{-1} exists then

   $$U\tilde{\ }(z) = \left[\begin{array}{c|c} A^{-T} & -A^{-T}C^T \\ \hline B^T A^{-T} & D^T - B^T A^{-T} C^T \end{array} \right] \quad (5.87)$$

   (b) $U(z)$ is said to be inner if it is stable and $U\tilde{\ }(z)U(z) = I$. Assume that the pairs (A, B) and (A, C) are controllable and observable respec-

tively and let $X > 0$ denote the solution to the discrete-time Lyapunov equation:
$$A^T X A - X + C^T C = 0$$
Show that $U(z)$ is inner if and only if:
$$\begin{aligned} D^T C + B^T X A &= 0 \\ D^T D + B^T X B &= I \end{aligned} \qquad (5.88)$$

(c) Use the results of parts (a) and (b) to solve the discrete-time LQR problem.

6

$\mathcal{H}_\infty$ CONTROL

6.1 INTRODUCTION

As we indicated in Chapter 5, a large number of control problems of practical importance can be described using the block diagram shown in Figure 6.1. Here the goal is to synthesize an internally stabilizing controller $K(s)$ such that the worst-case output z due to a class of exogenous disturbances w is kept below a given threshold. In this chapter we consider the case where w is an $\mathcal{L}_2$ signal and where the design objective is to keep $\|z\|_2$, the energy of the output, below a given level γ. Since for LTI stable systems the $\mathcal{L}_2$ to $\mathcal{L}_2$ induced norm coincides with the $\mathcal{H}_\infty$ norm of the transfer matrix (see Appendix A), this problem is known as the $\mathcal{H}_\infty$ (sub)optimal control problem. Recall from Chapter 2 that this problem also arises in the context of achieving robust stability and robust performance in the presence of dynamic uncertainty.

The approach that we follow in the first portion of the chapter, where we give a complete solution to a class of simpler problems, follows after [104] and is similar to the approach that we pursued in Chapter 5. As there, the solution will be obtained by first solving four cases having a special structure: full information (FI), full control (FC), disturbance feedforward (DF), and output estimation (OE). The solution to the general output feedback case will then be obtained by decoupling it into a FI and an OE problem. There is, however, an important difference between the approach pursued in Chapter 5 and the one used here. While in the $\mathcal{H}_2$ case we looked for the *optimal* controller, that is, the one minimizing $\|T_{zw}\|_2$, in this chapter we shall be content finding *suboptimal* controllers, that is, controllers such that $\|T_{zw}\|_\infty < \gamma$, where γ is a prescribed level. The reasons for limiting the treatment to sub-

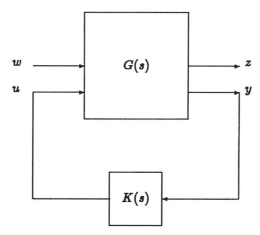

Figure 6.1. General loop structure for the $\mathcal{H}_\infty$ problem.

optimal controllers are both theoretical, since finding $\mathcal{H}_\infty$ optimal controllers is difficult [133], and practical, since these controllers may exhibit undesirable properties and their computation may lead to numerical difficulties. Nevertheless, at the end of the chapter we will briefly analyze the limiting behavior of the suboptimal controllers as $\gamma \downarrow \gamma_o$, the optimal achievable value.

6.2 THE STANDARD $\mathcal{H}_\infty$ PROBLEM

Consider a plant G with the following state-space realization:

$$G \equiv \left[\begin{array}{c|cc} A & B_1 & B_2 \\ \hline C_1 & 0 & D_{12} \\ C_2 & D_{21} & 0 \end{array}\right] \quad (6.1)$$

Then the $\mathcal{H}_\infty$ control problem can formally be stated as follows: given γ, synthesize an internally stabilizing, proper, LTI controller $K(s)$ such that the closed-loop transfer matrix T_{zw} satisfies $\|T_{zw}\|_\infty < \gamma$. Note that, as in the $\mathcal{H}_2$ case, we have assumed that $D_{11} = 0$ and $D_{22} = 0$. In the sequel, we will assume that the following additional conditions hold:

(A1) (A, B_2) is stabilizable and (C_2, A) is detectable.
(A2) (A, B_1) is stabilizable and (C_1, A) is detectable.
(A3) $C_1^T D_{12} = 0$ and $B_1 D_{21}^T = 0$.
(A4) D_{12} has full column rank with $D_{12}^T D_{12} = I$ and D_{21} has full row rank with $D_{21} D_{21}^T = I$.

Assumption (A1) is clearly necessary for the system to be stabilizable via output feedback. As in Chapter 5, assumption (A2) is made for technical reasons. Together with (A1) it guarantees that the control and filtering Riccati equations associated with a related $\mathcal{H}_2$ problem admit positive semidefinite stabilizing solutions. The rank assumptions in (A4) guarantee that the $\mathcal{H}_\infty$ problem is nonsingular, while the normalizing assumptions ($D_{12}^T D_{12} = I$, $D_{21} D_{21}^T = I$) do not entail any loss of generality since they can always be met by redefining the inputs w and u if necessary. Similarly, the orthogonality assumptions (A3) are made for simplicity. They can be relaxed at the price of introducing additional terms in the expression for the controller. Finally, the assumptions $D_{11} = 0$ and $D_{22} = 0$ are also made for the sake of simplicity. As we indicated in Chapter 5 they can be relaxed through the use of loop shifting [264]. However, the resulting formulas are considerable more complicated than the ones corresponding to the simpler case treated here.

Paralleling the development in the $\mathcal{H}_2$ case, we introduce the following two Hamiltonian matrices:

$$H_\infty \triangleq \begin{bmatrix} A & \gamma^{-2} B_1 B_1^T - B_2 B_2^T \\ -C_1^T C_1 & -A^T \end{bmatrix}$$
$$J_\infty \triangleq \begin{bmatrix} A^T & \gamma^{-2} C_1^T C_1 - C_2^T C_2 \\ -B_1 B_1^T & -A \end{bmatrix} \quad (6.2)$$

Note that in contrast with the matrices H_2 and J_2 defined in (5.32), the (1,2) blocks in H_∞ and J_∞ are not sign definite. Thus we can guarantee *a priori* neither that $H_\infty, J_\infty \in dom(Ric)$ nor that $Ric(H_\infty) \geq 0$, $Ric(J_\infty) \geq 0$. Indeed, as we will see in the sequel, these conditions are related to the existence of a suboptimal $\mathcal{H}_\infty$ controller for the given attenuation level γ. Note also in passing that as $\gamma \to \infty$ then $H_\infty \to H_2$ and $J_\infty \to J_2$.

The next result, originally from [104], gives a complete solution to the $\mathcal{H}_\infty$ problem.

Theorem 6.1 *Under assumptions (A1)–(A4) there exists an internally stabilizing controller $K(s)$ that renders $\|T_{zw}\|_\infty < \gamma$ if and only if the following conditions hold:*

1. $H_\infty \in dom(Ric)$ and $X_\infty \triangleq Ric(H_\infty) \geq 0$.
2. $J_\infty \in dom(Ric)$ and $Y_\infty \triangleq Ric(H_\infty) \geq 0$.
3. $\rho(X_\infty Y_\infty) < \gamma^2$.

In this case the set of all internally stabilizing controllers rendering $\|T_{zw}\|_\infty < \gamma$ can be parametrized as

$$K(s) = F_\ell(M_\infty, Q) \quad (6.3)$$

where $Q \in \mathcal{RH}_\infty$, $\|Q\|_\infty < \gamma$, and M_∞ has the following state-space realization:

$$M_\infty \equiv \left[\begin{array}{c|cc} A_\infty & -Z_\infty L_\infty & Z_\infty B_2 \\ \hline F_\infty & 0 & I \\ -C_2 & I & 0 \end{array}\right] \quad (6.4)$$

and where
$$\begin{aligned} A_\infty &= A + \gamma^{-2} B_1 B_1^T X_\infty + B_2 F_\infty + Z_\infty L_\infty C_2 \\ F_\infty &= -B_2^T X_\infty \\ L_\infty &= -Y_\infty C_2^T \\ Z_\infty &= \left(I - \gamma^{-2} Y_\infty X_\infty\right)^{-1} \end{aligned} \quad (6.5)$$

Remark The special controller corresponding to the choice $Q = 0$ in (6.3) is known as the "central controller." This controller has the following state-space realization:

$$K_{\text{central}} \equiv \left[\begin{array}{c|c} A_\infty & -Z_\infty L_\infty \\ \hline F_\infty & 0 \end{array}\right] \quad (6.6)$$

Hence its McMillan degree does not exceed that of the (generalized) plant $G(s)$.

As in Chapter 5, this result will be established by decoupling the output feedback problem into full information and ouput estimation subproblems. Thus the first step is to consider the same four intermediate special problems (FI, DF, FC, and OE) used there. However, before proceeding, it is instructive to analyze the structure of the central controller K_{central} to gain some insight into the problem. As in the $\mathcal{H}_2$ case, this controller exhibits a separation structure. This can be made apparent by rewriting its state-space realization as

$$\dot{\hat{x}} = A\hat{x} + B_1 \underbrace{\left(\gamma^{-2} B_1^T X_\infty \hat{x}\right)}_{w_{\text{worst}}} + B_2 u + Z_\infty L_\infty \left(C_2 \hat{x} - y\right) \quad (6.7)$$

$$u = F_\infty \hat{x}$$

We can see now that the general form of the controller does indeed resemble an observer-based controller, with a few notable differences, namely, the use of $Z_\infty L_\infty$ rather than L_∞ alone and the appearance of the $B_1\left(\gamma^{-2} B_1^T X_\infty \hat{x}\right)$ term. As we will show next $\gamma^{-2} B_1^T X_\infty x$ is a worst-case disturbance for a related FI problem. Thus this extra term can be thought of as an estimate of this worst-case disturbance based on the present estimate of the state. Similarly, $Z_\infty L_\infty$ is the optimal filter gain for estimating the optimal FI control input, $u = F_\infty x$, in the presence of the worst-case disturbance. To show this, assume that the following Riccati equation admits a solution X_∞:

$$A^T X_\infty + X_\infty A + C_1^T C_1 + X_\infty B_1 B_1^T X_\infty - X_\infty B_2 B_2^T X_\infty = 0 \quad (6.8)$$

Note that $X_\infty = Ric(H_\infty)$ when $\gamma = 1$. This can always be assumed without loss of generality by using the transformation $B_1 \to \gamma^{-1}B_1$ if necessary. Thus, in the remainder of this chapter, we will set $\gamma = 1$ for the sake of notational simplicity.

Consider now the quantity $x^T(t)X_\infty x(t)$ and its time derivative. Using (6.8) together with the orthogonality of $C_1 x$ and $D_{12} u$, it can easily be shown that

$$\begin{aligned}
\frac{d}{dt}\left[x^T(t)X_\infty x(t)\right] &= \dot{x}^T(t)X_\infty x(t) + x^T(t)X_\infty \dot{x}(t) \\
&= x^T(t)\left(A^T X_\infty + X_\infty A\right)x(t) \\
&\quad + 2w^T(t)B_1^T X_\infty x(t) + 2u^T(t)B_2^T X_\infty x(t) \\
&= -\|z\|^2 + \|u + B_2^T X_\infty x\|^2 \\
&\quad - \|B_1^T X_\infty x\|^2 + 2w^T(t)B_1^T X_\infty x(t)
\end{aligned} \qquad (6.9)$$

A completion of the squares argument yields

$$\begin{aligned}
\frac{d}{dt}\left[x^T(t)X_\infty x(t)\right] &= -\|z(t)\|^2 + \|w(t)\|^2 + \|u(t) + B_2^T X_\infty x(t)\|^2 \\
&\quad - \|w - B_1^T X_\infty x\|^2
\end{aligned} \qquad (6.10)$$

Finally, integrating this last equation, assuming that $x(0) = x(\infty) = 0$ and that $w \in \mathcal{L}_2[0, \infty)$, yields

$$\|z\|_2^2 - \|w\|_2^2 = \|u + B_2^T X_\infty x\|_2^2 - \|w - B_1^T X_\infty x\|_2^2 \qquad (6.11)$$

It follows that w_{worst} is the worst-case disturbance in the sense that it maximizes $\|z\|_2^2 - \|w\|_2^2$. On the other hand, $u = -B_2^T X_\infty x$ is the optimal control in the sense that it minimizes the same quantity in the presence of w_{worst}. Thus these values of u and w satisfy a saddle-point type condition. Note also that this suggests that when the states are available for feedback then $u = F_\infty x$ is indeed the optimal control action. As we will see in the sequel, this is precisely the case.

6.2.1 Background: Hankel and Mixed Hankel–Toeplitz Operators

In this section we provide some background material required to solve the FI $\mathcal{H}_\infty$ control problem. While this problem can also be addressed using a game-theoretic approach [136], the approach pursued here (which follows after [104]) highlights some connections between $\mathcal{H}_\infty$ control, Hankel operators, and the problem of approximating an anticausal system with a causal one. Additionally, some of the tools developed here will be used again in Chapter 9 when we deal with the problem of model reduction.

In order to introduce the Hankel operator, given a stable, strictly proper system $G(s)$,

$$G \equiv \left[\begin{array}{c|c} A & B \\ \hline C & 0 \end{array} \right] \quad (6.12)$$

and a point x_o, consider the problem of finding the minimum energy input $w(t) \in \mathcal{L}_2(-\infty, 0]$ such that if $x(-\infty) = 0$, then $x(0) = x_o$, that is,

$$\min_{w \in \mathcal{L}_2(-\infty,0]} \|w\|_2 \\ \text{subject to:} \quad x(0) = x_0, x(-\infty) = 0 \quad (6.13)$$

It is a standard result that if the pair (A, B) is controllable, then the solution to the problem is given by

$$w_{\text{opt}} = B^T e^{-A^T t} W_c^{-1} x_o \quad (6.14)$$

where $W_c > 0$ is the controllability Gramian of G. Moreover, the optimal cost is

$$\|w_{\text{opt}}\|_2^2 = x_o^T W_c^{-1} x_o \quad (6.15)$$

On the other hand, consider now the free evolution of G from a given initial condition x_o. Again, it is a standard result that

$$\|z\|_2^2 = \|Cx\|_2^2 = x_o^T W_o x_o \quad (6.16)$$

where W_o is the observability Gramian of G. Consider now the problem of finding the past input $w \in \mathcal{L}_2(-\infty, 0]$ that maximizes the energy of the future output $z(t), t \in [0, \infty)$, that is,

$$\max_{w \in \mathcal{L}_2(-\infty,0], w \neq 0} \frac{\|P_+ z\|_2^2}{\|w\|_2^2} \quad (6.17)$$

where $P_+ : \mathcal{L}_2(-\infty, \infty) \to \mathcal{L}_2[0, \infty)$ is the projection operator, that is,

$$P_+ f(t) = \begin{cases} f(t) & t \geq 0 \\ 0 & t < 0 \end{cases} \quad (6.18)$$

By combining equations (6.14) and (6.16) it follows that this problem is equivalent to

$$\max_{w \in \mathcal{L}_2(-\infty,0], w \neq 0} \frac{\|P_+ z\|_2^2}{\|w\|_2^2} = \max_{x_o \neq 0} \frac{x_o^T W_o x_o}{x_o^T W_c^{-1} x_o} = \rho(W_c W_o) \quad (6.19)$$

This last equation can be given an interpretation in terms of Hankel operators[1] as follows.

---

[1] These operators have historical importance since they were used in the early 1980s to furnish a solution to $\mathcal{H}_\infty$ control problems via a reduction to an equivalent Nehari approximation problem (see Section 6.6). A full treatment of Hankel operators is deferred until Chapter 9, where they will be discussed in the context of model reduction. Here we provide an abbreviated treatment, introducing some of the properties relevant to the connection with $\mathcal{H}_\infty$ control.

THE STANDARD $\mathcal{H}_\infty$ PROBLEM 163

Definition 6.1 *Given the system (6.12), its Hankel operator* $\Gamma_G : \mathcal{L}_2(-\infty, 0] \to \mathcal{L}_2[0, \infty)$ *is defined by*

$$z(t) = \Gamma_G v \triangleq \begin{cases} \int_{-\infty}^{0} Ce^{A(t-\tau)} Bv(\tau)\, d\tau & t \geq 0 \\ 0 & t < 0 \end{cases} \quad (6.20)$$

Thus Γ_G can be thought of as mapping the past input $v(t)$, $t \in (-\infty, 0]$ to the future output $z(t)$, $t \in [0, \infty)$, via the state $x(0)$. The adjoint operator Γ_G^* is defined by the relationship

$$\langle y, \Gamma_G v \rangle = \langle v, \Gamma_G^* y \rangle, \quad v \in \mathcal{L}_2(-\infty, 0],\ y \in \mathcal{L}_2[0, \infty) \quad (6.21)$$

It follows that $\Gamma_G^* : \mathcal{L}_2[0, \infty) \to \mathcal{L}_2(-\infty, 0]$ is given by

$$v(\tau) = \Gamma_G^* y \triangleq \begin{cases} \int_0^\infty B^T e^{A^T(t-\tau)} C^T y(t)\, dt & \tau \leq 0 \\ 0 & \tau > 0 \end{cases} \quad (6.22)$$

Next, we show that the eigenvalues of $\Gamma_G^* \Gamma_G$ coincide with the eigenvalues of $W_c W_o$. Consider a nonzero eigenvalue σ^2 of $\Gamma_G^* \Gamma_G$ and let $u \in \mathcal{L}_2(-\infty, 0]$ be the corresponding (right) eigenvector. Define $v \triangleq \sigma^{-1} \Gamma_G u$. Then we have that

$$\begin{aligned} \Gamma_G u &= \sigma v \\ \Gamma_G^* v &= \sigma u \end{aligned} \quad (6.23)$$

The pair (u, v) is called a *Schmidt pair* of Γ_G. Using the explicit expressions (6.20) and (6.22), we obtain

$$\begin{aligned} Ce^{At} x_o &= \sigma v \\ B^T e^{-A^T t} \hat{x}_o &= \sigma u \end{aligned} \quad (6.24)$$

where we defined

$$\begin{aligned} x_o &\triangleq \int_{-\infty}^{0} e^{-A\tau} Bu(\tau)\, d\tau \\ \hat{x}_o &\triangleq \int_0^\infty e^{A^T \tau} C^T v(\tau)\, d\tau \end{aligned} \quad (6.25)$$

Premultiplying the first equation in (6.24) by $e^{A^T t} C^T$ and integrating from 0 to ∞ yields

$$\int_0^\infty e^{A^T t} C^T Ce^{At} x_o\, dt = W_o x_o = \sigma \hat{x}_o \quad (6.26)$$

Similarly, premultiplying the second equation in (6.24) by $e^{-At}B$ and integrating from $-\infty$ to 0 yields

$$\int_{-\infty}^{0} e^{-At} B B^T e^{-A^T t} \hat{x}_o \, dt = W_c \hat{x}_o = \sigma x_o \qquad (6.27)$$

where we used the fact that

$$\int_{-\infty}^{0} e^{-At} B B^T e^{-A^T t} \, dt = \int_{0}^{\infty} e^{At} B B^T e^{A^T t} \, dt = W_c \qquad (6.28)$$

Combining (6.26) and (6.27) we have that

$$W_c W_o x_o = \sigma^2 x_o \qquad (6.29)$$

Thus an eigenvalue of $\Gamma_G^* \Gamma_G$ is also an eigenvalue of $W_c W_o$. Reversing the argument above using (6.14) shows that $\Gamma_G^* \Gamma_G$ and $W_c W_o$ have precisely the same eigenvalues.

Definition 6.2 *The square roots of the eigenvalues of $W_c W_o$ are called the Hankel singular values of the system $G(s)$.*

Since the $\mathcal{L}_2 \to \mathcal{L}_2$ induced norm of Γ_G is given by its largest singular value, we have that $\|\Gamma_G\|^2 = \rho(W_c W_o)$. Additionally, this can also be used to define a norm as follows.

Definition 6.3 *The Hankel norm of a stable, strictly proper system G is defined as*

$$\|G(s)\|_H = \sup_{u \in \mathcal{L}_2(-\infty, 0]} \frac{\|P_+(g * u)\|_2}{\|u\|_2} = \overline{\sigma}[\Gamma_G] = \rho^{1/2}(W_c W_o) \qquad (6.30)$$

where $g(t)$ denotes the impulse response of $G(s)$.

Next, we explore several implications of $\|G\|_\infty \leq 1$. The first one, usually known as the *bounded real lemma* [10] links this condition to a Hamiltonian matrix H and its associated Riccati equation. The second establishes a relationship between the output norm achieved by a worst-case perturbation and the solution to this Riccati equation.

Lemma 6.1 *Consider a proper, stable system $G(s) = C(sI - A)^{-1}B + D$ with $\overline{\sigma}(D) < 1$. Then $\|G\|_\infty < 1$ if and only if $H \in \text{dom}(\text{Ric})$ and $X = \text{Ric}(H) \geq 0$, where*

$$H \triangleq \begin{bmatrix} A + BR^{-1}D^T C & BR^{-1}B^T \\ -C^T(I + DR^{-1}D^T)C & -(A + BR^{-1}D^T C)^T \end{bmatrix} \qquad (6.31)$$

and $R \triangleq I - D^T D$.

Proof. Define $\Phi(j\omega) \triangleq I - \tilde{G}(j\omega)G(j\omega)$. Then

$$\|G(s)\|_\infty < 1 \iff \Phi(j\omega) > 0, \quad \forall \omega \tag{6.32}$$

It can easily be seen that a state-space realization for Φ is

$$\Phi = \left[\begin{array}{cc|c} A & 0 & -B \\ -C^TC & -A^T & C^TD \\ \hline D^TC & B^T & I - D^TD \end{array}\right] \tag{6.33}$$

Since A has no eigenvalues on the $j\omega$ axis, this realization has no uncontrollable or unobservable modes there. Then there exist a frequency ω_o and a vector u_o such that $\Phi(j\omega_o)u_o = 0$ if and only if there exists v_o such that

$$\left\{\begin{bmatrix} A & 0 \\ -C^TC & -A^T \end{bmatrix} + \begin{bmatrix} B \\ -C^TD \end{bmatrix} R^{-1}[D^TC \; B^T]\right\} v_o = j\omega_o v_o \iff$$

$$\begin{bmatrix} A + BR^{-1}D^TC & BR^{-1}B^T \\ -C^T(I + DR^{-1}D^T)C & -(A + BR^{-1}D^TC)^T \end{bmatrix} v_o = j\omega v_o \tag{6.34}$$

Therefore $\Phi(s)$ has no zeros on the $j\omega$ axis iff H has no eigenvalues there. Thus $\det[I - \tilde{G}(j\omega)G(j\omega)] \neq 0$ for all ω iff H has no eigenvalues on the $j\omega$ axis. Since $[I - \tilde{G}(j\omega)G(j\omega)] > 0$ as $\omega \to \infty$, we can conclude that $I - \tilde{G}(j\omega)G(j\omega) > 0$ all ω. It follows then that $\|G\|_\infty < 1 \iff H$ has no eigenvalues on the $j\omega$ axis. The fact that $H \in dom(Ric)$ follows now from Lemma C.2 and the fact that $(A + BR^{-1}D^TC, BR^{-1/2})$ is stabilizable since A is stable. Finally, to show that $X = Ric(H) \geq 0$, start by rewriting the corresponding Riccati equation as

$$\begin{aligned}(A + BR^{-1}D^TC)^TX + X(A + BR^{-1}D^TC) \\ + XBR^{-1}B^TX + C^T(I + DR^{-1}D^T)C &= 0 \\ \iff \\ A^TX + XA + (C^TD + XB)R^{-1}(B^TX + D^TC) + C^TC &= 0 \end{aligned} \tag{6.35}$$

Thus X is the observability Gramian for

$$\left(A, [C^T \; (XB + C^TD)R^{-1/2}]^T\right)$$

Since A is stable it follows that $X \geq 0$. □

Lemma 6.2 *Assume that $G(s)$ is strictly proper with $\|G\|_\infty < 1$ and $x(0) = x_o$. Then*

$$\sup_{w \in \mathcal{L}_2[0,\infty)} \left(\|z\|_2^2 - \|w\|_2^2\right) = x_o^T X x_o \tag{6.36}$$

where $X = Ric(H)$.

166 $\mathcal{H}_\infty$ CONTROL

Proof. Differentiating the quantity $x^T(t)Xx(t)$ along a trajectory and using the Riccati equation associated with H yields

$$\frac{d}{dt}\left[x^T(t)Xx(t)\right] = \dot{x}^T(t)Xx(t) + x^T(t)X\dot{x}(t)$$

$$= x^T(t)\left(A^TX + XA\right)x(t) + 2w^T(t)B^TXx(t)$$

$$= -\|z\|_2^2 - \|B^TXx\|_2^2 + 2w^TB^TXx \quad (6.37)$$

Completing the squares yields

$$\frac{d}{dt}\left[x^T(t)Xx(t)\right] = -\|z(t)\|^2 + \|w(t)\|^2 - \|w - B^TXx\|^2 \quad (6.38)$$

Finally, integrating this last equation yields (since $w \in \mathcal{L}_2[0, \infty)$ and thus $x \in \mathcal{L}_2[0, \infty)$)

$$\|z\|_2^2 - \|w\|_2^2 = x_o^T X x_o - \|w - B^TXx\|_2^2 \le x_o^T X x_o \quad (6.39)$$

To complete the proof note that the disturbance $w^* \triangleq B^TXx = B^TXe^{(A+BB^TX)t}x_o$ is in $\mathcal{L}_2[0, \infty)$ since $A + BB^TX$ is stable, and that for this disturbance the inequality is saturated. $\square$

We will consider now a generalization of the Hankel operator described earlier in this section. Suppose that the input to the system G is partitioned in $w = [w_1 \ w_2]$, where $w_1 \in \mathcal{L}_2(-\infty, 0]$ and $w_2 \in \mathcal{L}_2(-\infty, \infty)$, and that the matrix B is partitioned conformally in $B = [B_1 \ B_2]$. Define the space

$$\mathcal{W} \triangleq \left\{ \begin{bmatrix} w_1 \\ w_2 \end{bmatrix} : w_1 \in \mathcal{L}_2(-\infty, 0], \ w_2 \in \mathcal{L}_2(-\infty, \infty) \right\}$$

and consider the following operator $\Gamma: \mathcal{W} \to \mathcal{L}_2[0, \infty)$ defined by

$$\Gamma \begin{bmatrix} w_1 \\ w_2 \end{bmatrix} = P_+ [G_1 \ G_2] \begin{bmatrix} w_1 \\ w_2 \end{bmatrix} \quad (6.40)$$

where $P_+\mathcal{L}_2(-\infty, \infty) \to \mathcal{L}_2[0, \infty)$ denotes the projection operator. Γ can be interpreted as the sum of the Hankel operator Γ_G and the Toeplitz operator $P_+G_2P_+$. As in the case of the pure Hankel operator, we are interested in finding the worst-case perturbation, in the sense of maximizing $\|z\|_2$ and bounds on the corresponding value of the output.

Lemma 6.3 *Assume that (A, B) is controllable. Then $\sup_{w \in \mathcal{BW}} \|\Gamma w\|_2 < 1$ iff the following two conditions hold:*

(i) $H_W \in dom(Ric)$ and $W = Ric(H_W) \ge 0$ where

$$H_W \triangleq \begin{bmatrix} A & B_2 B_2^T \\ -C^TC & -A^T \end{bmatrix}.$$

(ii) $\rho(WW_c) < 1$.

THE STANDARD $\mathcal{H}_\infty$ PROBLEM

Proof. From the bounded real lemma (Lemma 6.1) we have that (i) is a necessary condition for having $\sup_{w \in \mathcal{BW}} \|\Gamma w\|_2 < 1$. Thus, to establish the desired result, we need to show that given (i), (ii) holds iff $\sup_{w \in \mathcal{BW}} \|\Gamma w\|_2 < 1$. By definition of $\mathcal{W}$, for any $w \in \mathcal{W}$ we have that

$$\|P_+z\|_2^2 - \|w\|_2^2 = \|P_+z\|_2^2 - \|P_+w\|_2^2 - \|P_-w\|_2^2$$

$$= \|P_+z\|_2^2 - \|P_+w_2\|_2^2 - \|P_-w\|_2^2 \qquad (6.41)$$

Since the last term contributes to $\|P_+z\|_2^2$ only through $x(0)$, the state at $t = 0$, and since W_c is invertible, we can use (6.14) to solve explicitly for P_-w_{worst}, where w_{worst} is the disturbance that maximizes the right-hand side of (6.41). Thus combining (6.15) and Lemma 6.2 we have that

$$\sup_{w \in \mathcal{W}} \left\{ \|P_+z\|_2^2 - \|w\|_2^2 \right\} = \sup_{w \in \mathcal{W},\, x_o} \left\{ \|P_+z\|_2^2 - \|w\|_2^2 : x(0) = x_o \right\}$$

$$= \sup_{x_o} \left\{ x_o^T W x_o - x_o^T W_c^{-1} x_o \right\} \qquad (6.42)$$

To prove the desired result, assume by contradiction that there exists $w^* \in \mathcal{BW}$ such that $\|\Gamma w^*\|_2 \geq 1$. Then $\|P_+z\|_2^2 - \|w^*\|_2^2 \geq 0$, which, together with (6.42), implies that $\rho(WW_c) \geq 1$. Conversely, if $\rho(WW_c) < 1$ then the left-hand side of (6.42) is negative for all w, which implies that $\|\Gamma w\|_2 < 1$ for all $w \in \mathcal{BW}$. $\square$

Finally, before closing this section we want to consider Γ^*, the adjoint operator of Γ. We will use this operator, together with the property that $\|\Gamma\| = \|\Gamma^*\|$, to solve the full information case. From the definition of Γ (6.40), we have that given any $z \in \mathcal{L}_2[0, \infty)$ and $w = \begin{bmatrix} w_1 \\ w_2 \end{bmatrix} \in \mathcal{W}^2$:

$$\langle z, \Gamma w \rangle = \langle z, P_+(G_1 w_1 + G_2 w_2) \rangle$$

$$= \langle z, G_1 w_1 \rangle + \langle z, G_2 w_2 \rangle = \langle P_-(\tilde{G_1} z), w_1 \rangle + \langle \tilde{G_2} z, w_2 \rangle$$

$$= \langle \Gamma^* z, w \rangle \qquad (6.43)$$

where P_- denotes the projection operator from $\mathcal{L}_2(-\infty, \infty)$ to $\mathcal{L}_2(-\infty, 0]$. It follows that

$$\Gamma^* z = \begin{bmatrix} P_-(\tilde{G_1} z) \\ \tilde{G_2} z \end{bmatrix} = \begin{bmatrix} P_- \tilde{G_1} \\ \tilde{G_2} \end{bmatrix} z \qquad (6.44)$$

With the tools introduced in this section we are ready now to give a solution to the FI $\mathcal{H}_\infty$ control problem.

---

[2] These inequalities are easily verified using the frequency-domain expression for the inner product in $\mathcal{L}^2$: $\langle x_1, x_2 \rangle = \frac{1}{2\pi} \int_{-\infty}^{+\infty} x_1^*(j\omega) x_2(j\omega)\, dw$

Full Information Problem

Lemma 6.4 *Consider the FI case, that is, the case where*

$$G = \left[\begin{array}{c|cc} A & B_1 & B_2 \\ \hline C_1 & 0 & D_{12} \\ \begin{bmatrix} I \\ 0 \end{bmatrix} & \begin{bmatrix} 0 \\ I \end{bmatrix} & \begin{bmatrix} 0 \\ 0 \end{bmatrix} \end{array}\right] \quad (6.45)$$

Then the following results hold:

1. *There exists an internally stabilizing controller such that $\|T_{zw}\|_\infty < 1$ iff $H_\infty \in dom(Ric)$ and $X_\infty = Ric(H_\infty) \geq 0$. Moreover, one such controller is $K_{FI} = [F_\infty \quad 0]$.*
2. *The following family of controllers,*

$$K_{FI}(s) : K_{FI}(s) = [F_\infty - Q(s)B_1^T X_\infty \quad Q(s)], \; Q(s) \in \mathcal{RH}_\infty, \|Q\|_\infty < 1\} \quad (6.46)$$

internally stabilizes the plant and renders $\|T_{zw}\|_\infty < 1$.

Proof. (Necessity) Assume for simplicity that the pair (C_1, A) is observable, rather than just detectable (see Problem 6.1). As in the $\mathcal{H}_2$ case, define a new control input $v \stackrel{\triangle}{=} u(t) - F_2 x(t)$, obtaining

$$T_{zw}(s) \equiv \left[\begin{array}{c|c} A_{F_2} & B_1 \\ \hline C_{1F_2} & 0 \end{array}\right] + \left[\begin{array}{c|c} A_{F_2} & B_2 \\ \hline C_{1F_2} & D_{12} \end{array}\right] T_{vw}(s)$$

$$= G_c(s)B_1 + U(s)T_{vw}(s) \quad (6.47)$$

where

$$\begin{aligned} H_2 &= \begin{bmatrix} A & -B_2 B_2^T \\ -C_1^T C_1 & -A^T \end{bmatrix} \\ X_2 &= Ric(H_2) > 0 \\ F_2 &= -B_2^T X_2 \\ A_{F_2} &= A + B_2 F \\ C_{1F_2} &= C_1 + D_{12} F_2 \end{aligned} \quad (6.48)$$

Recall (from Lemma 5.2) that U is inner. Thus there exists $U_\perp \in \mathcal{RH}_\infty$ such that $[U \quad U_\perp]$ is unitary. For instance, it can easily be shown that one such transfer matrix is

$$U_\perp \equiv \left[\begin{array}{c|c} A_{F_2} & -X_2^{-1}C_1^T D_\perp \\ \hline C_{1F_2} & D_\perp \end{array}\right] \quad (6.49)$$

where $D_\perp$ is any matrix such that $[D_{12} \ D_\perp]$ is orthogonal. Suppose that there exists an internally stabilizing controller that renders $\|T_{zw}\|_\infty < 1$. It follows then that

$$\sup_{w \in B\mathcal{L}_2[0,\infty)} \inf_{v \in \mathcal{L}_2[0,\infty)} \|z\|_2 < 1 \quad (6.50)$$

Since the 2-norm is invariant under multiplication by a unitary transfer matrix, we have that

$$\|z\|_2^2 = \|[U \ U_\perp]^\sim z\|_2^2$$
$$= \left\|\begin{bmatrix} U^\sim G_c B_1 w + v \\ U_\perp^\sim G_c B_1 w \end{bmatrix}\right\|_2^2$$
$$= \left\|\begin{bmatrix} P_+(U^\sim G_c B_1 w + v) + P_-(U^\sim G_c B_1 w) \\ U_\perp^\sim G_c B_1 w \end{bmatrix}\right\|_2^2 \quad (6.51)$$

From this equation it is apparent that the minimizing control is $v = -P_+(U^\sim G_c B_1 w)$ and that

$$\sup_{w \in B\mathcal{L}_2[0,\infty)} \inf_{v \in \mathcal{L}_2[0,\infty)} \|z\|_2 = \sup_{w \in B\mathcal{L}_2[0,\infty)} \left\|\begin{bmatrix} P_-(U^\sim G_c B_1 w) \\ U_\perp^\sim G_c B_1 w \end{bmatrix}\right\|_2^2$$
$$= \sup_{w \in B\mathcal{L}_2[0,\infty)} \left\|\begin{bmatrix} P_-(U^\sim G_c B_1) \\ U_\perp^\sim G_c B_1 \end{bmatrix} w\right\|_2^2$$
$$= \sup_{w \in B\mathcal{L}_2[0,\infty)} \|\Gamma^* w\|_2^2 \quad (6.52)$$

where Γ^* is the adjoint of the mixed Hankel–Toeplitz operator defined in (6.40). Thus we have that

$$\sup_{w \in B\mathcal{L}_2[0,\infty)} \inf_{v \in \mathcal{L}_2[0,\infty)} \|z\|_2 < 1 \iff \sup_{w \in B\mathcal{L}_2[0,\infty)} \|\Gamma^* w\|_2^2 < 1$$
$$\iff \sup_{w^* \in BW} \|\Gamma w^*\|_2^2 < 1 \quad (6.53)$$

where the last line follows from the fact that $\|\Gamma\| = \|\Gamma^*\|$, and where

$$\Gamma w^* = [P_+(B_1^T G_c^\sim U \ \ B_1^T G_c^\sim U_\perp)]\begin{bmatrix} w_1 \\ w_2 \end{bmatrix} = P_+ B_1^T G_c^\sim [U \ U_\perp]\begin{bmatrix} w_1 \\ w_2 \end{bmatrix} \quad (6.54)$$

Simple but tedious algebra using the state-space realizations of U, $U_\perp$, and G_c and the formulas in Appendix A shows that a state-space for $B_1^T G_c^\sim [U \quad U_\perp]$ is given by

$$B_1^T G_c^\sim [U \quad U_\perp] \equiv \left[\begin{array}{c|cc} A_{F_2} & B_2 & -X_2^{-1} C_1^T D_\perp \\ \hline -B_1^T X_2 & 0 & 0 \end{array}\right] \tag{6.55}$$

Combining this expression with Lemma 6.3, we have that $H_W \in dom(Ric)$ and that $W = Ric(H_W) \geq 0$, where

$$H_W = \begin{bmatrix} A_{F_2} & X_2^{-1} C_1^T C_1 X_2^{-1} \\ -X_2 B_1 B_1^T X_2 & -A_{F_2}^T \end{bmatrix} \tag{6.56}$$

Moreover, X_2^{-1} is precisely the controllability Gramian of (6.55) since it satisfies the Riccati equation

$$A_{F_2} X_2^{-1} + X_2^{-1} A_{F_2}^T + B_2 B_2^T + X_2^{-1} C_1^T C_1 X_2^{-1} = 0 \tag{6.57}$$

Thus it follows from the second part of Lemma 6.3 that $\rho(W X_2^{-1}) < 1$, or equivalently, $X_2 > W$. To complete the proof note that the Hamiltonian matrices H_∞ and H_W are related by the similarity transformation $H_\infty = T H_W T^{-1}$, where

$$T = \begin{bmatrix} -I & X_2^{-1} \\ -X_2 & 0 \end{bmatrix}$$

It follows that $H_\infty \in dom(Ric)$. Moreover, using the facts that

$$\mathcal{X}_-(H_W) = Im \begin{bmatrix} I \\ W \end{bmatrix} \tag{6.58}$$

and

$$\mathcal{X}_-(H_\infty) = T \mathcal{X}_-(H_W) = Im \begin{bmatrix} I - X_2^{-1} W \\ X_2 \end{bmatrix} \tag{6.59}$$

we have that $X_\infty = X_2(I - X_2^{-1} W)^{-1} = X_2(X_2 - W)^{-1} X_2 > 0$.

(Sufficiency) We will show first that if $H_\infty \in dom(Ric)$ and $X_\infty = Ric(H_\infty) \geq 0$, then $u = -B_2^T X_\infty x = F_\infty x$ stabilizes the system and renders $\|T_{zw}\|_\infty < 1$. To establish this, note that since $H_\infty \in dom(Ric)$, then $A + B_1 B_1^T X_\infty - B_2 B_2^T X_\infty$ is stable. Denote $A_{F_\infty} \triangleq A - B_2 B_2^T X_\infty$ and $C_{F_\infty} \triangleq C_1 + D_{12} F_\infty$. Then the Riccati equation associated with H_∞ can be rewritten in terms of A_{F_∞} and C_{F_∞} as

$$A_{F_\infty}^T X_\infty + X_\infty A_{F_\infty} + X_\infty B_1 B_1^T X_\infty + C_{F_\infty}^T C_{F_\infty} = 0 \tag{6.60}$$

where we used the fact that $D_{12}^T [C_1 \quad D_{12}] = [0 \quad I]$. Moreover, since $A - B_2 B_2^T X_\infty + B_1 B_1^T X_\infty$ is stable, a simple argument using the PBH test shows that the pair $(B_1^T X_\infty, A_{F_\infty})$ is detectable, which, together with (6.60), implies that A_{F_∞} is stable. This fact, combined with the bounded real lemma (Lemma 6.1), implies that the corresponding closed-loop transfer function

$$T_{zw}(s) \equiv \left[\begin{array}{c|c} A_{F_\infty} & B_1 \\ \hline C_{F_\infty} & 0 \end{array}\right] \tag{6.61}$$

is stable and such that $\|T_{zw}\|_\infty < 1$.

Finally, to show that the set of controllers

$$K_{FI} = \{[F_\infty - Q(s)B_1^T X_\infty \quad Q(s)], Q(s) \in \mathcal{RH}_\infty, \|Q\|_\infty < 1\}$$

is admissible, use again the change of variable $u = F_\infty x + v$. Consider now a controller from the family K_{FI}. In terms of v the control action is given by

$$v = [-Q(s)B_1^T X_\infty \quad Q(s)] \begin{bmatrix} x \\ w \end{bmatrix} = Q(s)[-B_1^T X_\infty \quad I] \begin{bmatrix} x \\ w \end{bmatrix}$$

Clearly, closing the loop around $G(s)$ with the controller K_{FI} is equivalent to closing the loop with the controller $Q(s)$ around a system having the following state-space realization:

$$P(s) \equiv \left[\begin{array}{c|cc} A_{F_\infty} & B_1 & B_2 \\ \hline C_{F_\infty} & 0 & D_{12} \\ -B_1^T X_\infty & I & 0 \end{array}\right] \tag{6.62}$$

This is illustrated in Figure 6.2, where $r = w - B_1^T X_\infty x$. Note that equation (6.60), together with Lemma 5.1, implies that P is inner. Moreover, using the formulas for system inversion given in Appendix B, we have that

$$P_{21}^{-1}(s) \equiv \left[\begin{array}{c|c} A_{F_\infty} + B_1 B_1^T X_\infty & -B_1 \\ \hline -B_1^T X_\infty & I \end{array}\right] \tag{6.63}$$

Thus $P_{21}^{-1}(s) \in \mathcal{RH}_\infty$. It follows now from Lemma 3.10 that $T_{zw} \in \mathcal{RH}_\infty$ with $\|T_{zw}\|_\infty < 1$ iff $Q \in \mathcal{RH}_\infty$, $\|Q\|_\infty < 1$. Thus the family K_{FI} has the desired properties. □

It can be shown (see [104] for details) that the family K_{FI} parametrizes the set of all admissible control actions, albeit it does not generate the set of all admissible controllers. This can easily be seen by considering a static controller of the form $[K \quad 0]$ and such that the corresponding $\|T_{zw}\|_\infty < 1$. Clearly this controller is not a member of the family K_{FI} for any finite $Q(s)$ (unless $K = F_\infty$). Recall that a similar situation arose in Chapter 3, in the

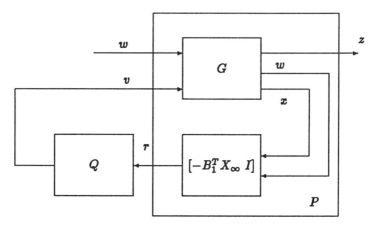

Figure 6.2. Loop transformation for proving the FI case.

context of stabilization. As there, the fact that the family K_{FI} generates all possible control actions will be sufficient for our purposes.

Remark Note that the central ($Q = 0$) FI controller uses *only* feedback from the states. Thus, as in the $\mathcal{H}_2$ case, under the assumption that $D_{11} = 0$ the FI and state feedback are, in some sense, equivalent. This is no longer true when $D_{11} \neq 0$. Note also that, contrary to what happened in the $\mathcal{H}_2$ case, the $\mathcal{H}_\infty$ controller depends on B_1 through H_∞.

By exploiting duality, we immediately get the following result.

Full Control Problem

Lemma 6.5 *For the full control case*

$$G_{FC}(s) \equiv \left[\begin{array}{c|ccc} A & B_1 & I & 0 \\ \hline C_1 & 0 & 0 & I \\ C_2 & D_{21} & 0 & 0 \end{array}\right] \tag{6.64}$$

the following results hold:

1. *There exists an internally stabilizing controller such that $\|T_{zw}\|_\infty < 1$ iff $J_\infty \in \text{dom}(\text{Ric})$ and $Y_\infty = \text{Ric}(J_\infty) \geq 0$. Moreover, one such controller is*

$$K_{FC} = \begin{bmatrix} L_\infty \\ 0 \end{bmatrix}$$

THE STANDARD $\mathcal{H}_\infty$ PROBLEM

2. The following family of controllers,

$$\left\{ K_{FC}(s) : K_{FC} = \begin{bmatrix} L_\infty - Y_\infty C_1^T Q(s) \\ Q(s) \end{bmatrix}, \ Q(s) \in \mathcal{RH}_\infty, \ \|Q\|_\infty < 1 \right\} \quad (6.65)$$

internally stabilizes G_{FC} and renders $\|T_{zw}\|_\infty < 1$.

Next, we consider the DF problem and its dual, OE.

Disturbance Feedforward Problem

Lemma 6.6 *Consider the DF problem*

$$G_{DF}(s) \equiv \left[\begin{array}{c|cc} A & B_1 & B_2 \\ \hline C_1 & 0 & D_{12} \\ C_2 & I & 0 \end{array} \right] \quad (6.66)$$

where, as before, we make the additional assumption that $A - B_1 C_2$ is stable. Then the following properties hold:

1. *There exists an internally stabilizing controller such that $\|T_{zw}\|_\infty < 1$ iff $H_\infty \in \text{dom}(\text{Ric})$ and $X_\infty = \text{Ric}(J_\infty) \geq 0$.*
2. *The set of all LTI controllers[3] such that $\|T_{zw}\|_\infty < 1$ can be parametrized as*

$$K_{OE}(s) = F_\ell(J_{DF}, Q), \quad Q \in \mathcal{RH}_\infty, \quad \|Q\|_\infty < 1 \quad (6.67)$$

where

$$J_{DF} = \left[\begin{array}{c|cc} A + B_2 F_\infty - B_1 C_2 & B_1 & B_2 \\ \hline F_\infty & 0 & I \\ -C_2 - B_1^T X_\infty & I & 0 \end{array} \right] \quad (6.68)$$

Proof. As in the $\mathcal{H}_2$ case, the proof follows from the equivalence of the FI and DF problems under the additional hypothesis of stability of $A - B_1 C_2$. Recall from Chapter 3 (Lemma 3.3) that under this additional hypothesis the FI and DF problems are equivalent in the sense that if K_{DF} stabilizes the DF plant then $K_{FI} \stackrel{\triangle}{=} K_{DF}[C_2 \ I]$ stabilizes the FI plant. Moreover, in this case $F_\ell(G_{FI}, K_{FI}) = F_\ell(G_{DF}, K_{DF})$. Conversely, if a controller $K_{FI}(s)$ is admissible for the FI plant, then $K_{DF} \stackrel{\triangle}{=} F_\ell((P_{DF}, K_{FI})$ internally stabilizes G_{DF} and $F_\ell(G_{FI}, K_{FI}) = F_\ell(G_{DF}, K_{DF})$. Item 1 and the first part of item 2 follow now from Lemma 6.4, by using the explicit expression (6.46) for K_{FI}

---

[3] In the same sense as in Chapter 3, that is, the set of controllers generating all possible control actions.

and noting that $F_\ell(P_{DF}, K_{FI}) = F_\ell(J_{DF}, Q)$. Finally, to show that the family (6.67) parametrizes all the DF controllers, proceeding as in Chapter 3, given an admissible controller $K_{DF}(s)$ define $Q \triangleq F_\ell(\hat{M}, K_{DF})$, where

$$\hat{M} = \left[\begin{array}{c|cc} A + B_1 B_1^T X_\infty & B_1 & B_2 \\ \hline -F_\infty & 0 & I \\ C_2 + B_1^T X_\infty & I & 0 \end{array} \right] \quad (6.69)$$

Using the properties of the interconnection of LFTs given in Appendix B it can easily be shown that

$$F_\ell(J_{DF}, \hat{M}) = \left[\begin{array}{cc|cc} A + B_2 F_\infty - B_1 C_2 & -B_2 F_\infty & B_1 & B_2 \\ -B_1 C_2 - B_1 B_1^T X_\infty & A + B_1 B_1^T X_\infty & B_1 & B_2 \\ \hline F_\infty & -F_\infty & 0 & I \\ -C_2 - B_1^T X_\infty & C_2 + B_1^T X_\infty & I & 0 \end{array} \right] \quad (6.70)$$

Using the similarity transformation

$$T = \begin{bmatrix} I & 0 \\ -I & I \end{bmatrix}$$

and eliminating the uncontrollable and unobservable modes from the realization yields

$$F_\ell(J_{DF}, \hat{M}) = \left[\begin{array}{cc|cc} A - B_1 C_2 & -B_2 F_\infty & B_1 & B_2 \\ 0 & A + B_2 F_\infty + B_1 B_1^T X_\infty & 0 & 0 \\ \hline 0 & -F_\infty & 0 & I \\ 0 & C_2 + B_1^T X_\infty & I & 0 \end{array} \right]$$

$$= \begin{bmatrix} 0 & I \\ I & 0 \end{bmatrix} \quad (6.71)$$

where we used the facts that $A - B_1 C_2$ is stable by assumption and $A + B_2 F_\infty + B_1 B_1^T X_\infty$ is stable since $H_\infty \in dom(Ric)$. Thus $F_\ell(J_{DF}, Q) = F_\ell\left[J_{DF}, F_\ell(\hat{M}, K_{DF})\right] = F_\ell[F_\ell(J_{DF}, \hat{M}), K_{DF}] = K_{DF}$. To complete the proof we need to show that $Q \in \mathcal{RH}_\infty, \|Q\|_\infty < 1$. To this effect consider the interconnection shown in Figure 6.3. Simple algebra using the formulas in Appendix B shows that $F_\ell(G_{DF}, J_{DF})$ has the following state-space realization (where once more we have used the similarity transformation T and eliminated uncontrollable modes exploiting the stability of $A + B_2 F_\infty$ and $A - B_1 C_2$):

$$F_\ell(G_{DF}, J_{DF}) = \left[\begin{array}{cc|cc} A & B_2 F_\infty & B_1 & B_2 \\ B_1 C_2 & A + B_2 F_\infty - B_1 C_2 & B_1 & B_2 \\ \hline C_1 & D_{12} F_\infty & 0 & D_{12} \\ C_2 & -C_2 - B_1^T X_\infty & I & 0 \end{array} \right]$$

THE STANDARD $\mathcal{H}_\infty$ PROBLEM

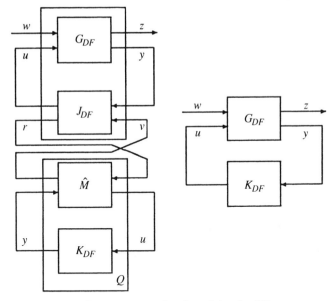

Figure 6.3. Interconnection for solving the DF case.

$$= \left[\begin{array}{cc|cc} A + B_2 F_\infty & B_2 F_\infty & B_1 & B_2 \\ 0 & A - B_1 C_2 & 0 & 0 \\ \hline C_1 + D_{12} F_\infty & D_{12} F_\infty & 0 & D_{12} \\ -B_1^T X_\infty & -C_2 - B_1^T X_\infty & I & 0 \end{array}\right]$$

$$= \left[\begin{array}{c|cc} A + B_2 F_\infty & B_1 & B_2 \\ \hline C_1 + D_{12} F_\infty & 0 & D_{12} \\ -B_1^T X_\infty & I & 0 \end{array}\right] \tag{6.72}$$

Thus it follows from Lemma 3.10 and equation (6.62) that $T_{zw} \in \mathcal{RH}_\infty$, $\|T_{zw}\|_\infty < 1 \iff Q \in \mathcal{RH}_\infty, \|Q\|_\infty < 1$. □

As before, the solution to the OE case follows from duality.

Output Estimation Problem

Lemma 6.7 *For the OE case where G has the form*

$$G_{OE} = \left[\begin{array}{c|cc} A & B_1 & B_2 \\ \hline C_1 & 0 & I \\ C_2 & D_{21} & 0 \end{array}\right], \quad A - B_2 C_1 \text{ stable} \tag{6.73}$$

the following properties hold:

1. There exists an internally stabilizing controller such that $\|T_{zw}\|_\infty < 1$ iff $J_\infty \in dom(Ric)$ and $Y_\infty = Ric(J_\infty) \geq 0$.
2. The set of all LTI controllers such that $\|T_{zw}\|_\infty < 1$ can be parametrized as

$$K_{OE}(s) = F_\ell(J_{OE}, Q), \quad Q \in \mathcal{RH}_\infty, \quad \|Q\|_\infty < 1 \tag{6.74}$$

where

$$J_{OE} = \left[\begin{array}{c|cc} A + L_\infty C_2 - B_2 C_1 & L_\infty & -B_2 - Y_\infty C_1^T \\ \hline C_1 & 0 & I \\ C_2 & I & 0 \end{array}\right] \tag{6.75}$$

6.2.2 Proof of Theorem 6.1

In this section we exploit the results of Lemmas 6.4–6.7 to prove Theorem 6.1. Motivated by the min–max problem (6.11) make the following change of variables:

$$\begin{aligned} v &= u + B_2^T X_\infty x \\ r &= w - B_1^T X_\infty x \end{aligned} \tag{6.76}$$

leading to the interconnection shown in Figure 6.4, where P and G_{tmp} have the following realizations:

$$P = \left[\begin{array}{c|cc} A + B_2 F_\infty & B_1 & B_2 \\ \hline C_1 + D_{12} F_\infty & 0 & D_{12} \\ -B_1^T X_\infty & I & 0 \end{array}\right]$$

$$G_{tmp} = \left[\begin{array}{c|cc} A_{tmp} & B_1 & B_2 \\ \hline -F_\infty & 0 & I \\ C_2 & D_{21} & 0 \end{array}\right] \tag{6.77}$$

$$A_{tmp} = A + B_1 B_1^T X_\infty$$

and where we used the fact that $D_{21} B_1^T = 0$.

Since P is inner with $P_{21}^{-1} \in \mathcal{RH}_\infty$, we have that $T_{zw} \in \mathcal{RH}_\infty$, $\|T_{zw}\|_\infty < 1 \iff F_\ell(G_{tmp}, K) \in \mathcal{RH}_\infty$, $\|F_\ell(G_{tmp}, K)\|_\infty < 1$. Note that G_{tmp} has an output estimation form. However, in order to be able to apply the OE results, we need to establish that G_{tmp} satisfies the corresponding assumptions:

1. (A_{tmp}, B_1) is stabilizable and $A_{tmp} + B_2 F_\infty$ is stable.
2. (C_2, A_{tmp}) is detectable.
3. $\begin{bmatrix} B_1 \\ D_{21} \end{bmatrix} D_{21}^T = \begin{bmatrix} 0 \\ I \end{bmatrix}$.

THE STANDARD $\mathcal{H}_\infty$ PROBLEM   177

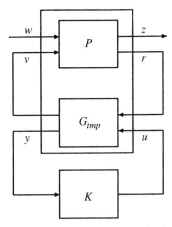

Figure 6.4. Loop transformation for reducing OF to DF.

Note that stabilizability of (A_{tmp}, B_1) follows immediately from the stabilizability of (A, B_1) using the PBH test, while stability of $A_{tmp} + B_2 F_\infty$ follows from the fact that $H_\infty \in dom(Ric)$. Finally, assumption (A3) is inherited from the corresponding assumption for the general $\mathcal{H}_\infty$ case. Thus we only need to show that the pair (C_2, A_{tmp}) is detectable. In the sequel we will establish this fact by showing that an equivalent condition holds. To obtain this condition start by considering the Hamiltonian matrix associated with the OE problem G_{tmp}:

$$J_{tmp} = \begin{bmatrix} A_{tmp}^T & F_\infty^T F_\infty - C_2^T C_2 \\ -B_1 B_1^T & -A_{tmp} \end{bmatrix} \quad (6.78)$$

Recall that as an intermediate step during the proof of the FI case we established that $H_\infty \in dom(Ric)$ and $X_\infty = Ric(H_\infty) \geq 0$ are sufficient conditions for the stability of $A - B_2 B_2^T X_\infty$. By duality we have that $J_{tmp} \in dom(Ric)$ and $Y_{tmp} = Ric(J_{tmp}) \geq 0$ are sufficient conditions for the stability of $A_{tmp} - Y_{tmp} C_2^T C_2$. The detectability of (C_2, A_{tmp}) follows now from the PBH test.

Proof of Theorem 6.1. (Sufficiency) Using the Riccati equation for X_∞, it can easily be shown that J_∞ and J_{tmp} are related by the following similarity transformation:

$$T = \begin{bmatrix} I & -X_\infty \\ 0 & I \end{bmatrix} \quad (6.79)$$

Thus $J_{tmp} \in dom(Ric) \iff J_\infty \in dom(Ric)$. Using the facts that

$$\mathcal{X}_-(J_{tmp}) = Im \begin{bmatrix} I \\ Y_{tmp} \end{bmatrix} \quad (6.80)$$

and

$$\mathcal{X}_-(J_{tmp}) = T\mathcal{X}_-(J_\infty) = T\mathrm{Im}\begin{bmatrix}I\\Y_\infty\end{bmatrix} = \mathrm{Im}\begin{bmatrix}I - X_\infty Y_\infty\\Y_\infty\end{bmatrix} \quad (6.81)$$

we have that $Y_{tmp} = Y_\infty(I - X_\infty Y_\infty)^{-1} = Z_\infty Y_\infty$. Moreover, since $\rho(X_\infty Y_\infty) < 1$ then $Y_{tmp} \geq 0$, which implies that (C_2, A_{tmp}) is detectable. Since G_{tmp} satisfies all the OE assumptions it follows from Lemma 6.7 that all controllers that stabilize G_{tmp} and render $\|T_{vr}\|_\infty < 1$ are given by

$$K = F_\ell(J, Q)$$

$$J = \left[\begin{array}{c|cc}A_{tmp} - Y_{tmp}C_2^T C_2 + B_2 F_\infty & -Y_{tmp}C_2^T & -B_2 + Y_{tmp}F_\infty^T\\\hline -F_\infty & 0 & I\\ C_2 & I & 0\end{array}\right]$$

$$= \left[\begin{array}{c|cc}A_\infty & -Z_\infty L_\infty & Z_\infty B_2\\\hline F_\infty & 0 & I\\ -C_2 & I & 0\end{array}\right] = F_\ell(M_\infty, Q) \quad (6.82)$$

where we used the definitions (6.5) and the fact that $Y_{tmp} = Z_\infty Y_\infty$.

(Necessity) Consider now an internally stabilizing controller $K(s)$ such that $\|T_{zw}\|_\infty < 1$. The controller $K(s)\begin{bmatrix}C_2 & D_{21}\end{bmatrix}$ solves the corresponding FI problem. Thus from Lemma 6.4 it follows that $H_\infty \in \mathrm{dom}(Ric)$ and $X_\infty = Ric(H_\infty) \geq 0$. From duality it follows that $J_\infty \in \mathrm{dom}(Ric)$ and $Y_\infty = Ric(H_\infty) \geq 0$. Since $X_\infty \geq 0$ we have that P is inner, with $P_{21} \in \mathcal{RH}_\infty$. Together with Lemma 3.10 this implies that

$$F_\ell(G_{tmp}, K) \in \mathcal{RH}_\infty, \quad \|F_\ell(G_{tmp}, K)\|_\infty < 1,$$

that is, the OE problem associated with G_{tmp} is solvable. It follows that the pair (C_2, A_{tmp}) is detectable. Hence G_{tmp} satisfies all the requirements for the OE case. From the output estimation results (Lemma 6.7) it follows now that $\|F_\ell(G_{tmp}, K)\|_\infty < 1 \iff J_{tmp} \in \mathrm{dom}(Ric), Y_{tmp} = Ric(J_{tmp}) \geq 0$. Using the same similarity transformation used in the first part of the proof, we have that

$$Y_{tmp} = (I - Y_\infty X_\infty)^{-1} Y_\infty \geq 0 \quad (6.83)$$

To complete the proof we need to show that $\rho(X_\infty Y_\infty) < 1$. To establish this fact consider first the case where $Y_\infty > 0$. Since $Y_\infty^{1/2} X_\infty (Y_\infty^{1/2})^T \geq 0$ it follows that if $\rho[Y_\infty^{1/2} X_\infty (Y_\infty^{1/2})^T] = 1$ there exists a right eigenvector v such that $Y_\infty^{1/2} X_\infty (Y_\infty^{1/2})^T v = v$. Define now $w = Y_\infty^{-1/2} v$. We have that $(I - X_\infty Y_\infty) w = w$ and thus $Y_\infty w = 0$ against the assumption that $Y_\infty > 0$. Hence

$\rho[Y_\infty^{1/2} X_\infty (Y_\infty^{1/2})^T] < 1 \rightarrow \rho(X_\infty Y_\infty) < 1$. Suppose now that $Y_\infty \geq 0$ and consider its singular value decomposition:

$$Y_\infty = U^T \begin{bmatrix} Y_{11} & 0 \\ 0 & 0 \end{bmatrix} U \tag{6.84}$$

where U is unitary and $Y_{11} > 0$. Using the same similarity transform on X_∞ yields

$$X_\infty = U^T \begin{bmatrix} X_{11} & X_{12} \\ X_{12}^T & X_{22} \end{bmatrix} U \tag{6.85}$$

Hence

$$Y_{tmp} = (I - Y_\infty X_\infty)^{-1} Y_\infty = U^T \begin{bmatrix} (I - Y_{11} X_{11})^{-1} Y_{11} & 0 \\ 0 & 0 \end{bmatrix} U \geq 0 \tag{6.86}$$

$$\Rightarrow \rho(X_{11} Y_{11}) < 1 \tag{6.87}$$

where the last line follows by applying the argument above to the (1,1) block of $U Y_{tmp} U^T$. The proof follows now by noting that $\rho(X_{11} Y_{11}) = \rho(X_\infty Y_\infty)$.

Remark As in the $\mathcal{H}_2$ case, the central $\mathcal{H}_\infty$ controller exhibits a separation structure. However, this structure is different from the one we observed in the $\mathcal{H}_2$ problem. Here the controller is estimating the optimal full information control gain in the presence of the worst-case disturbance w_{worst}. The gain $Z_\infty L_\infty$ corresponds to the optimal filter gain for estimating $u = F_\infty x$. The corresponding formulas in the $\mathcal{H}_2$ case are simpler since in the corresponding FI problem there is no worst-case disturbance. Moreover, in this case the problem of estimating any output, including the optimal state feedback, is equivalent to estimating the states.

6.3 RELAXING SOME OF THE ASSUMPTIONS

As in the $\mathcal{H}_2$ case it is interesting to analyze whether or not the set of assumptions (A1)–(A4) can be replaced with a set that is less restrictive. As before, assumption (A1) is necessary for the existence of stabilizing controllers, while (A4) guarantees that the problem is nonsingular. On the other hand, assumption (A2) and the orthogonality conditions can be relaxed. Recall that assumptions (A2) and (A3) are *sufficient* conditions for $H_2, J_2 \in dom(Ric)$, with $X_2 = Ric(H_2) \geq 0$ and $Y_2 = Ric(J_2) \geq 0$, properties that were used to establish the FI and FC results. However, assumptions (A2) and (A3) are not *necessary* conditions for these properties to hold. Indeed, it can be shown (see Appendix C) that these assumptions can be relaxed to the following:

(A2') $\begin{bmatrix} A - j\omega I & B_2 \\ C_1 & D_{12} \end{bmatrix}$ has full column rank for all ω \hfill (6.87)

(A3′) $\begin{bmatrix} A - j\omega I & B_1 \\ C_2 & D_{21} \end{bmatrix}$ has full row rank for all ω (6.88)

Assumptions (A1) and (A2′) are necessary and sufficient conditions for $H_2 \in dom(Ric)$ and $X_2 = Ric(H_2) \geq 0$, where H_2 denotes now the following Hamiltonian matrix:

$$H_2 \triangleq \begin{bmatrix} A - B_2 R^{-1} D_{12}^T C_1 & -B_2 R^{-1} B_2^T \\ -C_1^T (I - D_{12} R^{-1} D_{12}^T) C_1 & -(A - B_2 R^{-1} D_{12}^T C_1^T) \end{bmatrix} \quad (6.89)$$

and where $R \triangleq D_{12}^T D_{12}$. Similarly, assumptions (A1) and (A3′) are necessary and sufficient for $J_2 \in dom(Ric)$, and $Y_2 = Ric(J_2) \geq 0$, where

$$J_2 \triangleq \begin{bmatrix} (A - B_1 S^{-1} D_{21}^T C_2)^T & -C_2^T S^{-1} C_2 \\ -B_1 (I - D_{21}^T S^{-1} D_{21}) B_1^T & -(A - B_1 S^{-1} D_{21}^T C_2) \end{bmatrix} \quad (6.90)$$

with $S \triangleq D_{21} D_{21}^T$.

On the other hand, the conditions $D_{11} = 0$ and $D_{22} = 0$ can be relaxed by using a loop shifting technique (e.g., see [133] or [264]). However, the corresponding formulas become rather involved. As an alternative, in the next section we pursue a LMI-based characterization of all suboptimal $\mathcal{H}_\infty$ controllers.

6.4 LMI APPROACH TO $\mathcal{H}_\infty$ CONTROL

As we indicated in the last section some of the simplifying assumptions that we made in Section 6.2 can be relatively easily removed, albeit at the price of more complicated formulas. On the other hand, assumptions (A2′) and (A3′), namely, the fact that P_{12} and P_{21} do not have invariant zeros on the extended $j\omega$ axis, cannot be removed easily within the framework used so far. It is worth emphasizing that many problems of practical interest do actually violate these assumptions. An example will be the case where it is desired to have perfect rejection of persistent disturbances with a known frequency ω_o, including as a special case $\omega_o = 0$, that is, integral control. According to the internal mode principle, this perfect rejection necessitates including a model of the disturbance in the generalized plant. However, this will lead to zeros on the $j\omega$ axis at $\omega = \infty$.

In this section we use simple linear matrix inequality manipulations to obtain a convex LMI characterization of all suboptimal $\mathcal{H}_\infty$ controllers. This approach does not necessitate making additional assumptions on the plant, beyond the usual stabilizability and detectability assumptions (A1) and leads to convex optimization problems that can be solved efficiently. Moreover, it also gives a characterization of all reduced order controllers, although in this case the convexity property is lost, resulting in harder computational

LMI APPROACH TO $\mathcal{H}_\infty$ CONTROL

problems. The developments in this section follow after [124] and the closely related work in [158] and [275], with the main ideas being the reformulation of the bounded real lemma in terms of LMIs and the observation that since the controller parameters enter these LMIs affinely, they can be eliminated to obtain solvability conditions that depend only on the generalized plant and on two positive definite matrix variables, R and S.

To establish this result we begin by reformulating the bounded real lemma in terms of linear matrix inequalities and by establishing conditions for solvability of a certain matrix equation.

Lemma 6.8 (Bounded Real Lemma Revisited) *Consider the transfer function $G(s) = D + C(sI - A)^{-1}B$. Then the following statements are equivalent:*

1. $G(s) \in \mathcal{RH}_\infty$ and $\|G(s)\|_\infty < 1$.
2. *There exists $X = X^T > 0$ such that*

$$\begin{bmatrix} A^T X + XA & XB & C^T \\ B^T X & -I & D^T \\ C & D & -I \end{bmatrix} < 0 \qquad (6.91)$$

Proof. We will prove first that statement 1 is equivalent to $(I - D^T D) > 0$ and the existence of a positive definite solution to the following algebraic Riccati inequality (ARI):

$$A^T X + XA + (XB + C^T D)(I - D^T D)^{-1}(B^T X + D^T C) + C^T C < 0 \qquad (6.92)$$

Define the following quantities:

$$\begin{aligned} R &= (I - D^T D) \\ A_N &= A + BR^{-1}D^T C \\ B_N &= BR^{-1/2} \\ C_N &= (I + DR^{-1}D^T)^{1/2} C \end{aligned} \qquad (6.93)$$

Consider $\epsilon > 0$ small enough so that

$$\Phi_\epsilon \triangleq I - G_\epsilon^\sim(s) G_\epsilon(s) > 0 \qquad (6.94)$$

where

$$G_\epsilon \triangleq \left[\begin{array}{c|c} A & B \\ \hline C & D \\ \sqrt{\epsilon} I & 0 \end{array} \right] \qquad (6.95)$$

182 $\mathcal{H}_\infty$ CONTROL

Φ_ϵ has the following state-space realization:

$$\Phi_\epsilon = \left[\begin{array}{cc|c} A & 0 & -B \\ -C^T C - \epsilon I & -A^T & C^T D \\ \hline D^T C & B^T & I - D^T D \end{array}\right] \quad (6.96)$$

Thus the Hamiltonian matrix:

$$H_N \triangleq \begin{bmatrix} A_N & B_N B_N^T \\ -C_N^T C_N - \epsilon I & -A_N^T \end{bmatrix} \quad (6.97)$$

does not have eigenvalues on the $j\omega$ axis. Moreover, the pair (A_N, B_N) is stabilizable, since $A_N - B_N R^{-1/2} D^T C = A$ is stable. It follows then that $H_N \in \mathrm{dom}(Ric)$. Let $X = Ric(H_N)$. Then X satisfies the following Riccati equation:

$$A_N^T X + X A_N + X B_N B_N^T X + C_N^T C_N + \epsilon I = 0 \quad (6.98)$$

Since $\epsilon > 0$, this implies that

$$A_N^T X + X A_N + X B_N B_N^T X + C_N^T C_N < 0 \quad (6.99)$$

or equivalently,

$$\begin{bmatrix} A^T X + XA & XB & C^T \\ B^T X & -I & D^T \\ C & D & -I \end{bmatrix} < 0 \quad (6.100)$$

Since A is stable and the pair $(A_N, [C_N^T \; \sqrt{\epsilon}I]^T)$ is observable it follows that $X > 0$. This completes the first part of the proof. On the other hand, suppose that there exists a positive definite solution X to the ARI (6.92). Let $Q > 0$ be such that

$$A^T X + XA + (XB + C^T D) R^{-1} (B^T X + D^T C) + C^T C + Q = 0 \quad (6.101)$$

Using the similarity transformation

$$T = \begin{bmatrix} I & 0 \\ -X & I \end{bmatrix}$$

we have that

$$\Phi(s) = I - \tilde{G}(s) G(s) = \left[\begin{array}{cc|c} A & 0 & -B \\ L^T L & -A^T & XB + C^T D \\ \hline D^T C + B^T X & B^T & R \end{array}\right] \quad (6.102)$$

where

$$L \triangleq \left[(XB + C^T D) R^{-1} (B^T X + D^T C) + Q\right]^{1/2} > 0 \quad (6.103)$$

Let

$$M \triangleq \left[\begin{array}{c|c} A & B \\ \hline L & -L^{-1}(XB + C^T D) \end{array}\right] \quad (6.104)$$

Then simple algebra shows that

$$\Phi(s) = \widetilde{M(s)}M(s) + R - (B^T X + D^T C)L^{-2}(C^T D + XB) \quad (6.105)$$

Moreover, using the matrix inversion lemma it can easily be shown that $R - (B^T X + D^T C)L^{-2}(C^T D + XB) > 0$. It follows then that $\Phi(j\omega) > 0$, or equivalently $\|G\|_\infty < 1$. Stability of A follows from the Lyapunov theory and the facts that $X > 0$ and $C^T C + L^T L > 0$. Equivalence of the ARI (6.92) and the LMI (6.91) follows from Schur's complement formula. Finally, $I - D^T D > 0$ is a necessary condition for (6.91) to hold. □

Lemma 6.9 *Given a symmetric matrix* $\Psi \in \mathbb{R}^{m \times m}$ *and two matrices P and Q of column dimension m, the following statements are equivalent:*

1. *There exists some matrix* Θ *of compatible dimensions such that*

$$\Psi + P^T \Theta^T Q + Q^T \Theta P < 0 \quad (6.106)$$

2. $W_P^T \Psi W_P < 0$ *and* $W_Q^T \Psi W_Q < 0$, *where* W_P *and* W_Q *are any matrices whose columns span the null space of P and Q, respectively.*

Proof. (1 ⇒ 2) follows immediately by pre- and postmultiplying (6.106) by W_P^T and W_P or W_Q^T and W_Q. The converse can be shown through some simple, albeit lengthy, algebraic manipulations considering bases for the kernels of P and Q and invoking Schur's complement (see [124] for details). □

6.4.1 Characterization of All Output Feedback $\mathcal{H}_\infty$ Controllers

In the sequel we will assume that $D_{22} = 0$,[4] and we will consider controllers having the following state-space realization:

$$K = \left[\begin{array}{c|c} A_k & B_k \\ \hline C_k & D_k \end{array}\right] \quad (6.107)$$

where $A \in R^{k \times k}$. For ease of reference, we will combine all the controller parameters into the single variable

$$\Theta = \begin{bmatrix} A_k & B_k \\ C_k & D_k \end{bmatrix}$$

---

[4] Recall that this assumption can always be removed through a loop shifting transformation.

and we will use the following shorthand notation:

$$A_o = \begin{bmatrix} A & 0 \\ 0 & 0_k \end{bmatrix}; \quad B_o = \begin{bmatrix} B_1 \\ 0 \end{bmatrix}; \quad C_o = [C_1 \quad 0]$$

$$\mathcal{B} = \begin{bmatrix} 0 & B_2 \\ I_k & 0 \end{bmatrix}; \quad \mathcal{C} = \begin{bmatrix} 0 & I_k \\ C_2 & 0 \end{bmatrix}$$

$$\mathcal{D}_{12} = [0 \quad D_{12}]; \quad \mathcal{D}_{21} = \begin{bmatrix} 0 \\ D_{21} \end{bmatrix}$$

In terms of these variables the closed-loop matrices A_{cl}, B_{cl}, C_{cl}, and D_{cl} corresponding to the controller K can be written as

$$\begin{aligned} A_{cl} &= A_o + \mathcal{B}\Theta\mathcal{C} & B_{cl} &= B_o + \mathcal{B}\Theta\mathcal{D}_{21} \\ C_{cl} &= C_o + \mathcal{D}_{12}\Theta\mathcal{C} & D_{cl} &= D_{11} + \mathcal{D}_{12}\Theta\mathcal{D}_{21} \end{aligned} \tag{6.108}$$

Theorem 6.2 *Consider a FDLTI plant G of McMillan degree n with a minimal realization:*

$$\begin{bmatrix} z_1 \\ z_2 \\ y \end{bmatrix} = \left[\begin{array}{c|cc} A & B_1 & B_2 \\ \hline C_1 & D_{11} & D_{12} \\ C_2 & D_{21} & 0 \end{array}\right] \begin{bmatrix} w \\ u \end{bmatrix} \tag{6.109}$$

where the pairs (A, B_2) and (A, C_2) are stabilizable and detectable, respectively, and where $A \in \mathbb{R}^{n \times n}$, $D_{11} \in \mathbb{R}^{n_1 \times m_1}$, $D_{12} \in \mathbb{R}^{n_1 \times m_2}$, and $D_{21} \in \mathbb{R}^{n_2 \times m_1}$. Then there exists an internally stabilizing controller $K(s)$ with McMillan degree k that renders the closed-loop transfer function $\|T_{zw}\|_\infty < 1$ if and only if there exists a symmetric positive definite matrix $X \in R^{(n+k) \times (n+k)}$ such that

$$\begin{aligned} W_\mathcal{P}^T \Phi_X W_\mathcal{P} &< 0 \\ W_\mathcal{Q}^T \Psi_X W_\mathcal{Q} &< 0 \end{aligned} \tag{6.110}$$

where

$$\begin{aligned} \mathcal{P} &\triangleq [\mathcal{B}^T \quad 0_{(k+m_2) \times m_1} \quad \mathcal{D}_{12}^T] \\ \mathcal{Q} &\triangleq [\mathcal{C} \quad \mathcal{D}_{21} \quad 0_{(k+n_2) \times n_1}] \end{aligned} \tag{6.111}$$

where $W_\mathcal{P}$ and $W_\mathcal{Q}$ are matrices whose columns span the null spaces of $\mathcal{P}$ and $\mathcal{Q}$, respectively, and where

$$\begin{aligned} \Phi_X &\triangleq \begin{bmatrix} A_o X^{-1} + X^{-1} A_o^T & B_o & X^{-1} C_o^T \\ B_o^T & -I & D_{11}^T \\ C_o X^{-1} & D_{11} & -I \end{bmatrix} \\ \Psi_X &\triangleq \begin{bmatrix} A_o^T X + X A_o & X B_o & C_o^T \\ B_o^T X & -I & D_{11}^T \\ C_o & D_{11} & -I \end{bmatrix} \end{aligned} \tag{6.112}$$

Proof. From Lemma 6.8 we have that the controller (6.107) is admissible if and only if there exist $X \in R^{(n+k)\times(n+k)}, X > 0$ such that the following LMI holds:

$$\begin{bmatrix} A_{cl}^T X + X A_{cl} & X B_{cl} & C_{cl}^T \\ B_{cl}^T X & -I & D_{cl}^T \\ C_{cl} & D_{cl} & -I \end{bmatrix} < 0 \quad (6.113)$$

Using the explicit expressions (6.108) for the closed-loop matrices in terms of the plant and controller parameters, this LMI can be rewritten as

$$\Psi_X + \mathcal{Q}^T \Theta^T \mathcal{P}_{cl} + \mathcal{P}_{cl}^T \Theta \mathcal{Q} < 0 \quad (6.114)$$

where

$$\mathcal{P}_{cl} \triangleq [\mathcal{B}^T X \quad 0 \quad \mathcal{D}_{12}^T] \qquad \mathcal{Q} \triangleq [\mathcal{C} \quad \mathcal{D}_{21} \quad 0] \quad (6.115)$$

Next, we eliminate Θ to get a condition depending *only* on the plant. From Lemma 6.9 we have that the inequality (6.114) is solvable if and only if

$$W_{\mathcal{P}_{cl}}^T \Psi_X W_{\mathcal{P}_{cl}} < 0 \quad (6.116)$$

$$W_{\mathcal{Q}}^T \Psi_X W_{\mathcal{Q}} < 0 \quad (6.117)$$

where $W_{\mathcal{P}_{cl}}$ is a matrix spanning the null space of $\mathcal{P}_{cl}$. Since

$$\mathcal{P}_{cl} = \mathcal{P} \begin{bmatrix} X & 0 & 0 \\ 0 & I & 0 \\ 0 & 0 & I \end{bmatrix} \quad (6.118)$$

it follows that

$$W_{\mathcal{P}_{cl}} = \begin{bmatrix} X^{-1} & 0 & 0 \\ 0 & I & 0 \\ 0 & 0 & I \end{bmatrix} W_{\mathcal{P}} \quad (6.119)$$

Hence (6.116) is equivalent to

$$W_{\mathcal{P}}^T \left\{ \begin{bmatrix} X^{-1} & 0 & 0 \\ 0 & I & 0 \\ 0 & 0 & I \end{bmatrix} \Psi_X \begin{bmatrix} X^{-1} & 0 & 0 \\ 0 & I & 0 \\ 0 & 0 & I \end{bmatrix} \right\} W_{\mathcal{P}} = W_{\mathcal{P}}^T \Phi_X W_{\mathcal{P}}^T < 0 \quad (6.120)$$

□

While this theorem provides a very general condition for the solvability of the suboptimal $\mathcal{H}_\infty$ problem, these conditions are given in terms of inequalities involving both X and X^{-1}. Thus checking for the existence of an appropriate matrix X leads to a hard-to-solve nonconvex optimization problem. As we show next this difficulty can be avoided by rewriting the inequalities (6.110) in terms of two positive definite matrices R and S. As an intermediate step

186 $\mathcal{H}_\infty$ CONTROL

in this derivation we will obtain solvability conditions given in terms of two algebraic Riccati inequalities that closely resemble the two algebraic Riccati equations obtain in Section 6.2, in the special case where the assumptions (A1)–(A4) stated there hold.

Theorem 6.3 *Given a plant $G(s)$ with the minimal realization (6.109) where the pairs (A, B_2) and (A, C_2) are stabilizable and detectable, respectively, there exists an internally stabilizing controller with McMillan degree k that renders the closed-loop transfer function $\|T_{zw}\|_\infty < 1$ if and only if the following linear matrix inequalities in the variables R and S are feasible:*

$$\begin{bmatrix} N_R^T & 0 \\ 0 & I \end{bmatrix} \begin{bmatrix} AR + RA^T & RC_1^T & B_1 \\ C_1 R & -I & D_{11} \\ B_1^T & D_{11}^T & -I \end{bmatrix} \begin{bmatrix} N_R & 0 \\ 0 & I \end{bmatrix} < 0 \qquad (6.121)$$

$$\begin{bmatrix} N_S^T & 0 \\ 0 & I \end{bmatrix} \begin{bmatrix} A^T S + SA & SB_1 & C_1^T \\ B_1^T S & -I & D_{11}^T \\ C_1 & D_{11} & -I \end{bmatrix} \begin{bmatrix} N_S & 0 \\ 0 & I \end{bmatrix} < 0 \qquad (6.122)$$

$$\begin{bmatrix} R & I \\ I & S \end{bmatrix} \geq 0 \qquad (6.123)$$

where N_R and N_S are any matrices whose columns form bases of the null spaces of $[B_2^T \; D_{12}^T]$ and $[C_2 \; D_{21}]$, respectively. Moreover, the set of suboptimal controllers of order k is nonempty if and only if (6.121)–(6.123) hold for some R, S satisfying the rank constraint

$$\operatorname{rank}(I - RS) \leq k. \qquad (6.124)$$

Proof. From Theorem 6.2 we have that the set of admissible controllers is nonempty if and only if the inequalities (6.110) hold for some $X > 0$. Partition X and X^{-1} as

$$X \triangleq \begin{bmatrix} S & N \\ N^T & X_{22} \end{bmatrix}$$

$$X^{-1} \triangleq \begin{bmatrix} R & M \\ M^T & Y_{22} \end{bmatrix} \qquad (6.125)$$

where $R, S > 0 \in \mathbb{R}^{n \times n}$ and $M, N \in \mathbb{Re}^{n \times k}$. Rewriting Φ_X in (6.112) explicitly in terms of R and M yields

$$\Phi_X = \begin{bmatrix} AR + RA^T & \vdots & B_1 & RC_1^T \\ \cdots & \vdots & \cdots & \cdots \\ B_1^T & \vdots & -I & D_{11}^T \\ C_1 R & \vdots & D_{11} & -I \end{bmatrix} \qquad (6.126)$$

where $\cdots$ indicate elements that are irrelevant to the development. Note in passing that this expression is independent of Y_{22}. By rewriting $\mathcal{P}$ explicitly in terms of the plant data we have

$$\mathcal{P} = \begin{bmatrix} 0 & I_k & 0 & 0 \\ B_2^T & 0 & 0_{m_1} & D_{12}^T \end{bmatrix} \quad (6.127)$$

It follows that the bases of Ker $\mathcal{P}$ have the form

$$W_\mathcal{P} = \begin{bmatrix} W_1 & 0 \\ 0 & 0 \\ 0 & I_{m_1} \\ W_2 & 0 \end{bmatrix} \quad (6.128)$$

where

$$N_R \triangleq \begin{bmatrix} W_1 \\ W_2 \end{bmatrix} \quad (6.129)$$

is any basis of the null space of $[B_2^T \ D_{12}^T]$. Rearranging expression (6.126) and exploiting the fact that the second row of $W_\mathcal{P}$ is identically zero, we have that

$$\begin{aligned} W_\mathcal{P}^T \Phi_X W_\mathcal{P} &= \begin{bmatrix} \begin{bmatrix} W_1 \\ W_2 \\ 0 \end{bmatrix}^T & 0 \\ & I_{m1} \end{bmatrix} \begin{bmatrix} AR + RA^T & RC_1^T & B_1 \\ C_1 R & -I & D_{11} \\ B_1^T & D_{11}^T & -I \end{bmatrix} \begin{bmatrix} \begin{bmatrix} W_1 \\ W_2 \\ 0 \end{bmatrix} & 0 \\ & I_{m1} \end{bmatrix} \\ &= \begin{bmatrix} N_R^T & 0 \\ 0 & I \end{bmatrix} \begin{bmatrix} AR + RA^T & RC_1^T & B_1 \\ C_1 R & -I & D_{11} \\ B_1^T & D_{11}^T & -I \end{bmatrix} \begin{bmatrix} N_R & 0 \\ 0 & I \end{bmatrix} \end{aligned} \quad (6.130)$$

Proceeding in a similar fashion it can easily be shown that the second inequality in (6.110) is equivalent to (6.122). Finally, to establish (6.123) and the rank condition (6.124), note that using the Schur complement formula to compute the $(1,1)$ block of X^{-1} yields

$$R^{-1} = (S - NX_{22}^{-1}N^T) \iff S - R^{-1} = NX_{22}^{-1}N^T \geq 0 \quad (6.131)$$

which is equivalent to (6.123). Moreover, since $X_{22} > 0$ and $N \in \mathbb{R}^{n \times k}$ it follows that $rank(S - R^{-1}) \leq k \iff rank(I - RS) \leq k$. To show the converse note that if (6.123) and (6.124) hold then there exist $N \in R^{n \times k}$ such that

$$S - R^{-1} = NN^T \quad (6.132)$$

The construction of X and X^{-1} can then be completed by taking $X_{22} = I$, $M = -RN$, and $Y_{22} = (I + N^T RN)$. $\square$

6.4.2 Connections with the DGKF Results

In this section we explore the connections between the LMI characterization of all suboptimal $\mathcal{H}_\infty$ controllers derived in the last section with the characterization obtained in Section 6.2. Suppose that the assumptions (A1)–(A4) introduced in Section 6.2 hold. In particular, since D_{12} has full column rank and $D_{12}^T D_{12} = I$ there exists $D_\perp$ such that $[D_{12}\ D_\perp]^T \cdot [D_{12}\ D_\perp] = I$. It follows that a basis for the null space of $[B_2^T\ D_{12}^T]$ is given by

$$N_R = \begin{bmatrix} I & 0 \\ -D_{12}B_2^T & D_\perp \end{bmatrix} \quad (6.133)$$

Using this explicit expression for N_R in (6.121) and carrying out the block multiplications, exploiting the orthogonality of C_1 and D_{12} (assumption (A3)) yields

$$\begin{bmatrix} N_R^T & 0 \\ 0 & I \end{bmatrix} \begin{bmatrix} AR+RA^T & RC_1^T & B_1 \\ C_1 R & -I & D_{11} \\ B_1^T & D_{11}^T & -I \end{bmatrix} \begin{bmatrix} N_R & 0 \\ 0 & I \end{bmatrix} < 0$$

$$\Longleftrightarrow$$

$$\begin{bmatrix} AR+RA^T - B_2 B_2^T & RC_1^T D_\perp & B_1 \\ D_\perp^T C_1 R & -I & 0 \\ B_1^T & 0 & -I \end{bmatrix} < 0 \quad (6.134)$$

$$\Longleftrightarrow$$

$$AR + RA^T + B_1 B_1^T - B_2 B_2^T + RC_1^T C_1 R < 0$$

where the last line follows from using Schur's complement and the fact that $D_\perp D_\perp^T = I - D_{12} D_{12}^T$. Similarly, if D_{21} has full row rank and $D_{21}[B_1^T\ D_{21}^T] = [0\ I]$, then we have that (6.122) is equivalent to the following algebraic Riccati inequality:

$$SA + A^T S + SB_1 B_1^T S + C_1^T C_1 - C_2^T C_2 < 0 \quad (6.135)$$

Defining $X \triangleq R^{-1}$ and $Y \triangleq S^{-1}$, we have that these inequalities can be rewritten as

$$\begin{aligned} A^T X + XA + X(B_1 B_1^T - B_2 B_2^T)X + C_1^T C_1 &< 0 \\ AY + YA^T + Y(C_1^T C_1 - C_2^T C_2)Y + B_1 B_1^T &< 0 \end{aligned} \quad (6.136)$$

Rewriting (6.123) in terms of X and Y we have that $\rho(XY) \leq 1$.

Next, we show that the existence of positive definite solutions to these algebraic Riccati inequalities imply that the Hamiltonian matrices H_∞ and J_∞ defined in Section 6.2 satisfy $H_\infty \in dom(Ric)$, $J_\infty \in dom(Ric)$. Moreover, $X > X_\infty = Ric(H_\infty) \geq 0$ and $Y > Y_\infty = Ric(J_\infty) \geq 0$ and thus $\rho(X_\infty Y_\infty) < 1$. To establish these results we need the following lemma.

Lemma 6.10 *Suppose that assumptions (A1), (A2), (A3') and (A4') hold and that the following algebraic Riccati inequality*

$$A^T X + XA + X(B_1 B_1^T - B_2 B_2^T)X + C_1^T C_1 < 0 \quad (6.137)$$

admits a solution X. Then the following are true:

1. H_∞ *does not have eigenvalues on the $j\omega$ axis.*
2. *If $X > 0$ then $H_\infty \in dom(Ric)$ and $X > X_\infty \triangleq Ric(H_\infty) \geq 0$.*

Proof. Begin by noting that X is nonsingular, since if there exists $x \neq 0$ such that $x^T X = 0$ then pre- and postmultiplying (6.137) by x^T and x we obtain $x^T C_1^T C_1 x = \|C_1 x\| < 0$, which is obviously impossible. Thus we can define $R \triangleq X^{-1}$. In terms of R the ARI (6.137) can be rewritten as

$$RA^T + AR + RC_1^T C_1 R + B_1 B_1^T - B_2 B_2^T < 0 \quad (6.138)$$

To show that H_∞ does not have eigenvalues on the $j\omega$ axis, assume to the contrary that there exist ω_o and a vector $x = \begin{bmatrix} u \\ v \end{bmatrix}$ such that

$$\begin{bmatrix} A & B_1 B_1^T - B_2 B_2^T \\ -C_1^T C_1 & -A^T \end{bmatrix} \begin{bmatrix} u \\ v \end{bmatrix} = j\omega_o \begin{bmatrix} u \\ v \end{bmatrix} \quad (6.139)$$

From assumption (A2') it follows that $v \neq 0$. Pre- and postmultiplying (6.138) by v^* and v leads to the following contradiction.

$$\begin{aligned}
0 &> v^* RA^T v + v^* ARv + v^* RC_1^T C_1 Rv + v^*(B_1 B_1^T - B_2 B_2^T)v \\
&= v^* R(-C_1^T C_1 u - j\omega_o v) + (-u^* C_1^T C_1 + j\omega_o v^*)Rv \\
&\quad + v^* RC_1^T C_1 Rv + v^*(j\omega_o u - Au) \\
&= v^* RC_1^T C_1 (Rv - u) - u^* C_1^T C_1 Rv + (j\omega_o v^* - v^* A)u \\
&= v^* RC_1^T C_1 (Rv - u) - u^* C_1^T C_1 Rv + u^* C_1^T C_1 u \\
&= (Rv - u)^* C_1^T C_1 (Rv - u) \geq 0 \quad (6.140)
\end{aligned}$$

Since H_∞ cannot have eigenvalues on the $j\omega$ axis, it follows that the bases of its stable invariant subspace $\mathcal{X}_-$ have the form $\begin{bmatrix} X_1 \\ X_2 \end{bmatrix}$, where $X_1, X_2 \in R^{n \times n}$ and

$$H_\infty \begin{bmatrix} X_1 \\ X_2 \end{bmatrix} = \begin{bmatrix} X_1 \\ X_2 \end{bmatrix} \Lambda \quad (6.141)$$

where $\Lambda \in R^{n \times n}$ is a Hurwitz matrix with real entries. In order to show that $H_\infty \in dom(Ric)$ we need to show that X_2 is nonsingular. To establish this, start by expanding explicitly (6.141) to obtain

$$AX_1 + (B_1 B_1^T - B_2 B_2^T)X_2 = X_1 \Lambda \\ -C_1^T C_1 X_1 - A^T X_2 = X_2 \Lambda \qquad (6.142)$$

Premultiplying the first equation by X_2^T and the second equation by X_1^T and using the fact that[5] $X_1^T X_2 = X_2^T X_1$ yields

$$X_2^T A X_1 + X_1^T A^T X_2 + X_1^T C_1^T C_1 X_1 + X_2^T (B_1 B_1^T - B_2 B_2^T) X_2 = 0 \qquad (6.143)$$

Assume that X_2 is singular. Then there exists $x \neq 0$ such that $X_2 x = 0$. Pre- and postmultiplying (6.143) by x^T and x we have that $C_1 X_1 x = 0$. Hence $\begin{bmatrix} X_2 \\ C_1 X_1 \end{bmatrix} x = 0$. Moreover, postmultiplying the second equation in (6.142) by x we have that $X_2 \Lambda x = 0$. Repeating the same argument as before we have that

$$\begin{bmatrix} X_2 \\ C_1 X_1 \end{bmatrix} \Lambda^k x = 0, \quad k = 0, 1, \ldots \qquad (6.144)$$

It follows that the pair $(\Lambda, \begin{bmatrix} X_2 \\ C_1 X_1 \end{bmatrix})$ is not observable. Thus, from the PBH test, there exist $y \neq 0$ and λ such that

$$\begin{bmatrix} \Lambda - \lambda I \\ X_2 \\ C_1 X_1 \end{bmatrix} y = 0 \qquad (6.145)$$

Postmultiplying the first equation in (6.142) by y we have that

$$(\lambda I - A) X_1 y = 0 \qquad (6.146)$$

Assume for the time being that the pair (A, C_1) does not have any stable unobservable modes. Then $(\lambda I - A)$ is nonsingular and this last equation implies that $X_1 y = 0$. Hence $\begin{bmatrix} X_1 \\ X_2 \end{bmatrix} y = 0$ against the hypothesis that $\begin{bmatrix} X_1 \\ X_2 \end{bmatrix}$ spans the stable eigenspace of H_∞ and thus has full column rank. By interchanging the first and second row of H_∞ we have that

$$\begin{bmatrix} A^T & C_1^T C_1 \\ -(B_1 B_1^T - B_2 B_2^T) & -A \end{bmatrix} \begin{bmatrix} X_2 \\ X_1 \end{bmatrix} = -\begin{bmatrix} X_2 \\ X_1 \end{bmatrix} \Lambda \qquad (6.147)$$

---

[5] This fact can easily be established by noting that since H_∞ is a Hamiltonian matrix then $X^T S X = 0$ where $S = \begin{bmatrix} 0 & -I \\ I & 0 \end{bmatrix}$.

Let $R_\infty \triangleq X_1 X_2^{-1}$. Premultiplying this last equation by $[R_\infty \; -I]$ and postmultiplying by X_2^{-1} we have that

$$[R_\infty \; -I] \begin{bmatrix} A^T & C_1^T C_1 \\ -(B_1 B_1^T - B_2 B_2^T) & -A \end{bmatrix} \begin{bmatrix} I \\ R_\infty \end{bmatrix} =$$

$$-[R_\infty \; -I] \begin{bmatrix} I \\ R_\infty \end{bmatrix} X_2 \Lambda X_2^{-1} = 0 \quad (6.148)$$

$$\Rightarrow$$

$$R_\infty A^T + A R_\infty + B_1 B_1^T - B_2 B_2^T + R_\infty C_1^T C_1 R_\infty = 0$$

Subtracting this last equation from (6.138) we have that

$$(R - R_\infty) A^T + A(R - R_\infty) + R C_1^T C_1 R - R_\infty C_1^T C_1 R_\infty < 0 \quad (6.149)$$

or equivalently,

$$(R - R_\infty)(A + R_\infty C_1^T C_1)^T + (A + R_\infty C_1^T C_1)(R - R_\infty)$$
$$+ (R - R_\infty) C_1^T C_1 (R - R_\infty) < 0 \quad (6.150)$$

From (6.148) we have that $A^T + C_1^T C_1 R_\infty = -X_2 \Lambda X_2^{-1}$. Since Λ is stable it follows then that $A^T + C_1^T C_1 R_\infty$ is antistable. Thus, from the properties of the Lyapunov equations, it follows that $R - R_\infty < 0$, or equivalently, $R_\infty > R > 0$. Since $R_\infty > 0$, $X_\infty \triangleq R_\infty^{-1}$ is well defined and satisfies $0 < X_\infty < X$. Finally, to complete the proof, we need to remove the assumption that the pair (A, C_1) does not have any stable unobservable modes. This can be accomplished by perturbing C_1 to $\begin{bmatrix} C_1 \\ \sqrt{\epsilon} I \end{bmatrix}$, with $\epsilon > 0$ small enough so that X remains a solution of the perturbed ARI. The proof follows by noting that the stable invariant subspace of H_∞^ϵ depends continuously on ϵ and that from the discussion above we have that $0 < X_\infty^\epsilon < X$. Thus $X_\infty \triangleq \lim_{\epsilon \to 0} X_\infty^\epsilon$ is well defined and satisfies $0 \leq X_\infty \leq X$. Finally, this inequality can be strengthened to $X_\infty < X$ by replacing X by $(1 - \eta)X$ in the arguments above, with η small enough. $\square$

From this lemma it follows that whenever assumptions (A1)–(A4) (or (A1), (A2'), (A3'), and (A4)) hold then if the LMIs (6.121)–(6.123) are feasible, the conditions in Theorem 6.1 also hold. Conversely, assume that conditions in Theorem 6.1 hold. Since $H_\infty \in dom(Ric)$ and $X_\infty = Ric(H_\infty) \geq 0$, it follows that for ϵ small enough the Riccati equation

$$A^T X_\epsilon + X_\epsilon A + X_\epsilon (B_1 B_1^T - B_2 B_2^T) X_\epsilon + C_1^T C_1 + \epsilon I = 0 \quad (6.151)$$

has a stabilizing solution $X_\epsilon \geq 0$. Proceeding as before it can easily be seen that this solution cannot be singular. Thus the LMI (6.121) admits a solution

X_ϵ^{-1}. Proceeding in a similar fashion it can easily be established that if $J_\infty \in dom(Ric)$ then the LMI (6.122) admits a solution Y_ϵ^{-1}. Finally, by continuity the condition $\rho(X_\infty Y_\infty) < 1$ implies that $\rho(X_\epsilon Y_\epsilon) < 1$ or, equivalently, that X_ϵ^{-1} and Y_ϵ^{-1} satisfy the LMI (6.123).

Since the discussion above shows that whenever assumptions (A1)–(A4) (or their relaxed counterparts) hold the approach pursued in Section 6.2 and the LMI approach presented here are equivalent, one may wonder about their relative advantages. For regular $\mathcal{H}_\infty$ problems the approach of Section 6.2 is preferable since it involves solving algebraic Riccati equations, a problem with a lesser computational complexity than solving the corresponding conditions given in terms of LMIs. On the other hand, the LMI-based approach still allows for finding numerically stable solutions even in cases where Theorem 6.1 breaks down, such as singular problems or plants having invariant zeros on the $j\omega$ axis.

Finally, note in passing that the LMI approach presented in this section also parametrizes all reduced order controllers. The existence of an admissible controller having order $k < n$ can be investigated by checking whether or not the set of LMIs (6.121)–(6.123), together with the additional rank assumption (6.124), is feasible. If so, a reduced order controller can be reconstructed from the solutions R and S. Note, however, that the additional constraint (6.124) is not convex, thus resulting in an optimization problem considerably more difficult than in the case where $k \geq n$.

6.5 LIMITING BEHAVIOR

So far the treatment has been limited to suboptimal controllers. In this section we briefly examine the behavior of these controllers as $\gamma \to \gamma_{opt}$, the infimal achievable $\mathcal{H}_\infty$ norm. Note that γ_{opt} can be computed by performing a search on γ until one of the conditions in Theorem 6.1 fails. It can be shown (see [125]) that the conditions $X_\infty \geq 0$ or $Y_\infty \geq 0$ cannot fail prior to $\rho(X_\infty Y_\infty) < \gamma^2$. Thus, in the limit, one of the following two situations will arise:

1. X_∞ or Y_∞ will no longer be stabilizing, that is, either H_∞ or J_∞ will have eigenvalues on the $j\omega$ axis.
2. $\rho(X_\infty Y_\infty) = \gamma^2$.

In the first case X_∞, Y_∞, and Z_∞ have well defined limits. Hence the central controller $K_{central}$ is well defined. Moreover, assume that $H_\infty \notin dom(Ric)$ but that X_∞ is still well defined, although no longer stabilizing. It can easily be shown that if assumption (A3') holds then the pair $(A_{F_\infty}, C_{F_\infty})$ is detectable. Thus from equation (6.60) it follows that the control law $u = -B_2^T X_\infty$ is still stabilizing and renders $\|T_{zw}\|_\infty = \gamma_{opt}$. Moreover, it can also be shown that there exists no control law (even nonstabilizing) such that $\|T_{zw}\|_\infty < \gamma_{opt}$.

On the other hand, in the case where $\rho(X_\infty Y_\infty) = \gamma_{opt}^2$ then the formulas in Theorem 6.1 are no longer applicable since $I - \gamma_{opt}^{-1} Y_\infty X_\infty$ becomes singular, leading to unbounded A_∞ and $Z_\infty L_\infty$. This difficulty can be circumvented by using the following descriptor version of the controller [135, 264]:

$$(I - \gamma_{opt}^{-2} Y_\infty X_\infty)\dot{\xi} = A_c \xi + Y_\infty C_2^T y$$
$$u = -B_2^T X_\infty \xi \tag{6.152}$$

where $A_c = (I - \gamma_{opt}^{-2} Y_\infty X_\infty)[A + (\gamma_{opt}^{-2} B_1 B_1^T - B_2 B_2^T) X_\infty] - Y_\infty C_2^T C_2$. It can be shown that this system always has index 1 and defines a reduced order controller where the unbounded modes of A_c cancel at infinity. Note, however, that this formula is valid only when $I - \gamma^{-2} Y_\infty X_\infty$ is singular. Thus for values of γ close (but not equal) to γ_{opt} we still need to use the formulas in Theorem 6.1. However, since Z_∞^{-1} is nearly singular, numerical difficulties and near pole–zero cancellations in the controller may appear. These problems can be avoided by modifying the central controller with the inclusion of a feedthrough term (see [123] for details). On the other hand, note that the LMI-based approach works well even for values of γ close to the optimal.

6.6 THE YOULA PARAMETRIZATION APPROACH

In this section we consider an alternative solution to the $\mathcal{H}_\infty$ control problem based on using the Youla parametrization to recast the problem into a model matching form. In addition to having historical importance, since it provided the first solutions to the $\mathcal{H}_\infty$ problem in the early 1980s, this approach also allows for exactly solving several multiobjective control problems such as mixed $\mathcal{H}_2/\mathcal{H}_\infty$ and mixed $\ell^1/\mathcal{H}_\infty$ problems [299, 300].

Recall from Chapter 3 that the set of all stabilizing controllers can be parameterized as $K = F_\ell(J, Q)$, where Q is any arbitrary element of $\mathcal{H}_\infty$ such that $I + D_{22} Q(\infty)$ is nonsingular and where J is given in (3.34). Moreover, the corresponding set of all achievable closed-loop map is given by the following expression (affine in Q):

$$T_{zw} = T_{11} + T_{12} Q T_{21} \tag{6.153}$$

Under assumptions (A1)–(A4) (or their relaxed counterparts) there exist $D_{12,\perp}$ and $D_{21,\perp}$ such that $[D_{12} \ D_{12,\perp}]$ and $[D_{21}^T \ D_{21,\perp}^T]$ are unitary. Define now the following Hamiltonian matrices:

$$H \triangleq \begin{bmatrix} A - B_2 D_{12}^T C_1 & -B_2 B_2^T \\ -C_1^T D_{12,\perp} D_{12,\perp}^T C_1 & -(A - B_2 D_{12}^T C_1)^T \end{bmatrix}$$

$$J \triangleq \begin{bmatrix} (A - B_1 D_{21}^T C_2)^T & -C_2^T C_2 \\ -B_1 D_{21,\perp}^T D_{21,\perp} B_1^T & -(A - B_1 D_{21}^T C_2) \end{bmatrix} \tag{6.154}$$

Stabilizability of the pair $(A - B_2 D_{12}^T C_1, B_2)$ follows immediately from stabilizability of the pair (A, B_2) (just consider the feedback $\tilde{F} = D_{12}^T C_1 + F$ where F is any matrix such that $A + B_2 F$ is stable). On the other hand, if G_{12} does not have any invariant zeros on the $j\omega$ axis, then the pair $(D_{12,\perp}^T C_1, A - B_2 D_{12}^T C_1)$ cannot have unobservable modes there. To establish this fact, assume to the contrary that $(D_{12,\perp}^T C_1, A - B_2 D_{12}^T C_1)$ has an unobservable mode at $j\omega_o$. From the PBH test this is equivalent to the existence of a vector $x \neq 0$ such that

$$(A - B_2 D_{12}^T C_1)x = j\omega_o x$$
$$D_{12,\perp}^T C_1 x = 0 \tag{6.155}$$

But this implies that

$$\begin{bmatrix} A - j\omega_o I & B_2 \\ C_1 & D_{12} \end{bmatrix} \begin{bmatrix} I & 0 \\ -D_{12}^T C_1 & I \end{bmatrix} \begin{bmatrix} x \\ 0 \end{bmatrix} = 0 \tag{6.156}$$

which violates the full column rank assumption (A2'). Since $(A - B_2 D_{12}^T C_1, B_2)$ is stabilizable and $(D_{12,\perp}^T C_1, A - B_2 D_{12}^T C_1)$ does not have unobservable modes on the $j\omega$ axis, it follows from Lemma C.2 (Appendix C) that $H \in dom(Ric)$ and $X = Ric(H)$ is well defined and stabilizing. By duality the same properties hold for J and $Y = Ric(J)$. It follows that the controller and observer gains in the Youla parametrization can be selected as

$$F = -(D_{12}^T C_1 + B_2^T X)$$
$$L = -(Y C_2^T + B_1 D_{21}^T) \tag{6.157}$$

yielding the following explicit expression for T:

$$T = \begin{bmatrix} T_{11} & T_{12} \\ T_{21} & T_{22} \end{bmatrix}$$

$$= \left[\begin{array}{cc|cc} A_F & -B_2 F & B_1 & B_2 \\ 0 & A_L & B_{1L} & 0 \\ \hline C_{1F} & -D_{12} F & D_{11} & D_{12} \\ 0 & C_2 & D_{21} & 0 \end{array} \right] \tag{6.158}$$

$$A_F = A + B_2 F$$
$$A_L = A + L C_2$$
$$C_{1F} = C_1 + D_{12} F$$
$$B_{1L} = B_1 + L D_{21}$$

Using Lemma 5.1 it can easily be seen that T_{12} is inner and T_{21} is co-inner.

Define now the following transfer matrices:

$$T_{12,\perp} = \left[\begin{array}{c|c} A_F & -X^\dagger C_1^T D_{12,\perp} \\ \hline C_{1F} & D_{12,\perp} \end{array}\right]$$

$$T_{21,\perp} = \left[\begin{array}{c|c} A_L & B_{1L} \\ \hline -D_{21,\perp} B_1^T Y_2^\dagger & D_{21,\perp} \end{array}\right]$$

(6.159)

where $\dagger$ denotes the Moore–Penrose pseudoinverse. By straightfoward but somewhat tedious manipulations, using the formulas in Appendix B, it can be seen that the transfer matrices $[T_{12,\perp}(s) \quad T_{12}(s)]$ and $\left[\begin{array}{c} T_{21,\perp}(s) \\ T_{21}(s) \end{array}\right]$ are unitary. Since the $\mathcal{H}_\infty$ norm is invariant under multiplication by unitary matrices, we have that

$$\|T_{zw}\|_\infty = \|T_{11} + T_{12} Q T_{21}\|_\infty$$

$$= \left\|[T_{12,\perp} \quad T_{12}]^\sim (T_{11} + T_{12} Q T_{21}) \left[\begin{array}{c} T_{21,\perp} \\ T_{21} \end{array}\right]\right\|_\infty \quad (6.160)$$

$$= \left\|\left[\begin{array}{cc} R_{11} & R_{12} \\ R_{21} & R_{22} + Q \end{array}\right]\right\|_\infty$$

where $R \triangleq [T_{12,\perp} \quad T_{12}]^\sim T_{11} [T_{21,\perp} \quad T_{\widetilde{21}}]$ has the following state-space representation:

$$R \triangleq \left[\begin{array}{cc|cc} -A_F^T & -MB_{1L}^T & -MD_{21,\perp}^T & -MD_{21}^T \\ 0 & -A_L^T & Y^\dagger B_1 D_{21,\perp}^T & -C_2^T \\ \hline -D_{12,\perp}^T C_1 X^\dagger & D_{12,\perp}^T D_{11} B_{1,L}^T & D_{12,\perp}^T D_{11} D_{21,\perp}^T & D_{12,\perp}^T D_{11} D_{21}^T \\ B_2^T & -FY + D_{12}^T D_{11} B_{1L}^T & D_{12}^T D_{11} D_{21,\perp}^T & D_{12}^T D_{11} D_{21}^T \end{array}\right]$$

(6.161)

where $M \triangleq XB_1 + C_{1F}^T D_{11}$. Note that R is completely antistable. It follows that the $\mathcal{H}_\infty$ problem can be recast as the following approximation problem:

$$\inf_{Q_a} \left\{\|R + Q_a\|_\infty : Q_a = \left[\begin{array}{cc} 0 & 0 \\ 0 & Q \end{array}\right], \quad Q \in \mathcal{H}_\infty\right\} \quad (6.162)$$

that is, the problem of finding the best stable approximation to an unstable system, possibly subject to some additional structural constraints on Q.

Consider first the case of a SISO system. In this case both $R(s)$ and $Q(s)$ are scalar transfer functions and this problem reduces then to the classical Nehari approximation problem. The infimally achievable approximation error is given by the following result [212].

Theorem 6.4 *Let $R \in \mathcal{RH}_\infty$ and denote by Γ_R the corresponding Hankel operator introduced in Section 6.2.1. Then*

$$\inf_{Q \in \mathcal{H}_\infty^\perp} \|R + Q\|_\infty = \|\Gamma_R\| = \|R\|_H \qquad (6.163)$$

and the infimum is achieved.

Next, we indicate how to construct the optimal approximant using an algorithm originally proposed by Adamjan, Arov, and Krein [6]. To this effect assume that $R \in \mathcal{RH}_\infty^\perp$ and recast the problem into the problem of finding the best antistable approximation to a stable transfer function by noting that

$$\inf_{Q \in \mathcal{RH}_\infty} \|R + Q\|_\infty = \inf_{Q \in \mathcal{RH}_\infty^\perp} \|\tilde{R} + Q\|_\infty \qquad (6.164)$$

Theorem 6.5 *Let $G = \tilde{R} \in \mathcal{RH}_\infty$ and $\sigma_1 = \|\Gamma_G\| = \rho^{1/2}(W_c W_o)$. Consider the Schmidt pair introduced in (6.23):*

$$\begin{aligned} \Gamma_G u &= \sigma_1 v \\ \Gamma_G^* v &= \sigma_1 u \end{aligned} \qquad (6.165)$$

Then the (unique) optimal $Q = \sigma_1 V(s)/U(s) - G$, where $U(s)$ and $V(s)$ denote the Laplace transforms of u and v, respectively.

Proof. Let $H \triangleq (G + Q)U$. Then

$$\begin{aligned} \|H - \Gamma_G U\|_2^2 &= \|H\|_2^2 + \|\Gamma_G U\|_2^2 - \langle H, \Gamma_G U \rangle - \langle \Gamma_G U, H \rangle \\ &= \|H\|_2^2 + \|\Gamma_G U\|_2^2 - \langle P_+ H, \Gamma_G U \rangle - \langle \Gamma_G U, P_+ H \rangle \\ &= \|H\|_2^2 - \langle \Gamma_G U, \Gamma_G U \rangle = \|H\|_2^2 - \sigma_1^2 \|U\|_2^2 \end{aligned} \qquad (6.166)$$

where the last line follows from the fact that since $Q \in \mathcal{H}_\infty^\perp$ and $U \in \mathcal{H}_2^\perp$ then $P_+ H = P_+ GU = \Gamma_G U$. On the other hand, we have that

$$\|H\|_2^2 = \|(G + Q)U\|_2^2 \leq \|(G + Q)\|_\infty^2 \|U\|_2^2 = \sigma_1^2 \|U\|_2^2 \qquad (6.167)$$

Combining equations (6.166) and (6.167) we have that

$$\|H - \Gamma_G U\|_2^2 = \|H\|_2^2 - \sigma_1^2 \|U\|_2^2 \leq 0 \qquad (6.168)$$

Hence $H = \Gamma_G U$, or equivalently $(G + Q)U = \Gamma_G U$. From the definition of Γ_G we have that $P_+[QU] = 0$, or equivalently, $QU \in \mathcal{RH}_2^\perp$, which implies that $Q \in \mathcal{RH}_\infty^\perp$. The proof is completed by noting that since $\Gamma_G U = \sigma_1 V$ then $(G + Q)U = \sigma_1 V \iff Q = \sigma_1 V/U - G$. □

THE YOULA PARAMETRIZATION APPROACH 197

A state-space realization of the optimal Q can be obtained from the following state-space realization for the Schmidt pair (u, v) [286]:

$$U = \left[\begin{array}{c|c} -A^T & -x \\ \hline B^T & 0 \end{array}\right], \quad V = \left[\begin{array}{c|c} A & w \\ \hline C & 0 \end{array}\right] \quad (6.169)$$

where w is a right eigenvector of $W_c W_o$ associated with the eigenvalue σ_1^2 and where $x \triangleq 1/\sigma_1 W_o w$. By using the formulas in Appendix B it is easily seen that $U^\sim U = V^\sim V$. It follows then that V/U is allpass, which implies that the optimal error system $G + Q$ is also allpass.

It can be shown (see Theorem 6.3 in [132]) that the previous results also hold when R is a square transfer matrix, that is $\inf_{Q \in \mathcal{RH}_\infty^\perp} \|R + Q\|_\infty = \|R\|_H$ and $(R + Q_{\text{opt}})^\sim (R + Q_{\text{opt}}) = \sigma_1^2 I$. This case is known as the one-block approximation problem.

We return now to the four-blocks approximation problem (6.162). Let γ_o denote the optimal approximation error. Since Q is now constrained to have a specific structure, it is clear that $\gamma_o \geq \|R\|_H$, with a strict inequality holding most of the time. It should also be clear that

$$\gamma_o \geq \max\left\{\|[R_{11} \ R_{12}]\|_\infty, \left\|\begin{bmatrix} R_{11} \\ R_{21} \end{bmatrix}\right\|_\infty\right\} \quad (6.170)$$

since these are submatrices of $R + Q$, and matrix compressions do not increase the norm. Thus the relative simple procedure used to solve the one-block approximation problem cannot, in general, be used to solve problem (6.162). Nevertheless, as we show next, this problem can be reduced to an equivalent one-block problem by using an allpass embedding argument [225]. The developments in the sequel essentially follow the arguments presented in [182] and [135].

Before attempting to derive a formal solution, it is instructive to present some intuitive arguments that justify the steps to be taken. Consider a one-block Nehari approximation problem of the form

$$\gamma_o = \inf_{Q \in \mathcal{H}_\infty} \left\|\begin{bmatrix} R_{11} & R_{12} \\ R_{21} & R_{22} \end{bmatrix} + \begin{bmatrix} Q_{11} & Q_{12} \\ Q_{21} & Q_{22} \end{bmatrix}\right\|_\infty = \|R\|_H \quad (6.171)$$

Assume that the optimal solution is indeed of the form

$$Q = \begin{bmatrix} 0 & 0 \\ 0 & Q_{22} \end{bmatrix} \quad (6.172)$$

and recall that, for the one-block case, the optimal error system is allpass, that is, $(R+Q)\tilde{\ }(R+Q) = (R+Q)(R+Q)\tilde{\ } = \gamma_o^2 I$. It follows then that

$$[R_{11} \ R_{12}] \begin{bmatrix} R_{11}\tilde{\ } \\ R_{12}\tilde{\ } \end{bmatrix} = \gamma_o^2 I$$

$$[R_{11}\tilde{\ } \ R_{21}\tilde{\ }] \begin{bmatrix} R_{11} \\ R_{21} \end{bmatrix} = \gamma_o^2 I$$
(6.173)

As we will show later using an explicit state-space construction, the converse is also true in the sense that if this last equation holds then the one-block approximation problem (6.171) admits a solution of the form (6.172); that is, the one- and four-block problems are equivalent. Unfortunately, equation (6.173) does not hold in general. The key idea of the allpass embedding is that (6.173) can be forced to hold for some γ,

$$\gamma \geq \max \left\{ \|[R_{11} \ R_{12}]\|_\infty, \left\| \begin{bmatrix} R_{11} \\ R_{21} \end{bmatrix} \right\|_\infty \right\}$$

by introducing spectral factors L and N such that

$$[R_{11} \ R_{12} \ L] \begin{bmatrix} R_{11}\tilde{\ } \\ R_{12}\tilde{\ } \\ L\tilde{\ } \end{bmatrix} = \gamma^2 I$$

$$[R_{11}\tilde{\ } \ R_{21}\tilde{\ } \ N\tilde{\ }] \begin{bmatrix} R_{11} \\ R_{21} \\ N \end{bmatrix} = \gamma^2 I$$
(6.174)

provided that γ is chosen large enough. This leads to an augmented one-block problem of the form

$$\gamma = \inf_{Q \in \mathcal{H}_\infty} \left\| \begin{bmatrix} R_{11} & R_{12} & L \\ R_{21} & R_{22} & * \\ N & * & * \end{bmatrix} + \begin{bmatrix} Q_{11} & Q_{12} & Q_{13} \\ Q_{21} & Q_{22} & Q_{23} \\ Q_{31} & Q_{32} & Q_{33} \end{bmatrix} \right\|_\infty = \|R\|_H \quad (6.175)$$

which admits a solution of the form

$$Q = \begin{bmatrix} 0 & 0 & 0 \\ 0 & Q_{22} & Q_{23} \\ 0 & Q_{32} & Q_{33} \end{bmatrix}$$
(6.176)

The solution to the four-block problem is then obtained by discarding the augmentation, using the fact that matrix compressions do not increase the norm.

Next, we formalize these arguments. As before, for simplicity we will set $\gamma = 1$ and we will seek suboptimal solutions to the problem. We begin by finding necessary conditions for the existence of $Q_{22} \subset \mathcal{RH}_\infty$ such that

$$\left\|\begin{bmatrix} R_{11} & R_{12} \\ R_{21} & R_{22}+Q_{22} \end{bmatrix}\right\|_\infty < 1 \tag{6.177}$$

Assume that $R \in \mathcal{RH}_\infty^\perp$ is strictly proper with the following state-space realization:

$$R = \left[\begin{array}{c|cc} A & B_1 & B_2 \\ \hline C_1 & D_{11} & D_{12} \\ C_2 & D_{21} & 0 \end{array}\right] \tag{6.178}$$

and that there exists

$$Q_{22} = \left[\begin{array}{c|c} A_Q & B_Q \\ \hline C_Q & D_Q \end{array}\right]$$

such that (6.177) holds. Then by using standard dilation arguments (see Theorem 5.2 in [132] and Lemma 3.1 in [135]) there exist matrices B_i, C_i, B_{Q_i}, C_{Q_i} and D_{ij} such that

$$E_{aa} = \begin{array}{c} \\ p_1 \\ p_2 \\ m_1 \\ m_2 \end{array} \begin{array}{cccc} m_1 & m_2 & p_1 & p_2 \\ \left[\begin{array}{cccc} R_{11} & R_{12} & R_{13} & 0 \\ R_{21} & E_{22} & E_{23} & E_{24} \\ R_{31} & E_{32} & E_{33} & E_{34} \\ 0 & E_{42} & E_{43} & E_{44} \end{array}\right] \end{array}$$

$$= \left[\begin{array}{cc|cccc} A & 0 & B_1 & B_2 & B_3 & 0 \\ 0 & A_Q & 0 & B_Q & B_{Q_3} & B_{Q_4} \\ \hline C_1 & 0 & D_{11} & D_{12} & D_{13} & 0 \\ C_2 & C_Q & D_{21} & D_Q & D_{23} & D_{24} \\ C_3 & C_{Q_3} & D_{31} & D_{32} & D_{33} & D_{34} \\ 0 & C_{Q_4} & 0 & D_{42} & D_{43} & D_{44} \end{array}\right] \tag{6.179}$$

is allpass, that is, $E_{aa}\tilde{E}_{aa} = \tilde{E}_{aa}E_{aa} = I$, where $E_{ij} = R_{ij} + Q_{ij}$, $Q_{ij} \in \mathcal{RH}_\infty$, $R_{ij} \in \mathcal{RH}_\infty^\perp$. Let

$$E_a \triangleq \begin{bmatrix} R_{11} & R_{12} & R_{13} \\ R_{21} & E_{22} & E_{23} \\ R_{31} & E_{32} & E_{33} \end{bmatrix}$$

Since matrix compressions do not increase the norm, it follows that $\|E_a\|_\infty \le 1$. Thus from Nehari's theorem we have that $\|R_a\|_H \le 1$, where

$$R_a \triangleq \begin{bmatrix} R_{11} & R_{12} & R_{13} \\ R_{21} & R_{22} & R_{23} \\ R_{31} & R_{32} & R_{33} \end{bmatrix}$$

$$= \left[\begin{array}{c|ccc} A & B_1 & B_2 & B_3 \\ \hline C_1 & D_{11} & D_{12} & D_{13} \\ C_2 & D_{21} & * & * \\ C_3 & D_{31} & * & * \end{array}\right] \tag{6.180}$$

Note that since E_{aa} is allpass then R_{13} and R_{31} can be found without explicit knowledge of Q by solving the following spectral factorization problems:

$$\begin{aligned} R_{13}R_{\widetilde{13}} &= I - R_{11}R_{\widetilde{11}} - R_{12}R_{\widetilde{12}} > 0 \\ R_{\widetilde{31}}R_{31} &= I - R_{\widetilde{11}}R_{11} - R_{\widetilde{21}}R_{21} > 0 \end{aligned} \qquad (6.181)$$

These spectral factors can be computed using standard techniques (see Theorem 5.1 in [132]), yielding

$$\begin{aligned} D_{13}D_{13}^T &= I - D_{11}D_{11}^T - D_{12}D_{12}^T > 0 \\ D_{31}^T D_{31} &= I - D_{11}^T D_{11} - D_{21}^T D_{21} > 0 \\ B_3 &= \left(XC_1^T - [B_1 \ B_2] \begin{bmatrix} D_{11}^T \\ D_{12}^T \end{bmatrix} \right) D_{13}^{-T} \\ C_3 &= D_{31}^{-T} \left(B_1^T Y - [D_{11}^T \ D_{21}^T] \begin{bmatrix} C_1 \\ C_2 \end{bmatrix} \right) \end{aligned} \qquad (6.182)$$

where X and Y are the antistabilizing solutions to the Riccati equations:

$$XA^T + AX - [B_1 \ B_2] \begin{bmatrix} B_1^T \\ B_2^T \end{bmatrix} - B_3 B_3^T = 0 \qquad (6.183)$$

$$YA + A^T Y - [C_1^T \ C_2^T] \begin{bmatrix} C_1 \\ C_2 \end{bmatrix} - C_3^T C_3 = 0 \qquad (6.184)$$

The construction of the remaining elements in R_a, namely, R_{23}, R_{32}, and R_{33}, can now be done using standard state-space manipulations (see [135]), leading to a state-space realization of the form (6.180). Note that X and Y are the Gramians for R_a. Since $\|R_a\|_H \leq 1$ it follows that $\rho(XY) \leq 1$. Finally, it can be shown (see proposition 3.3 in [135]) that these conditions can be strengthened to $\|R_a\|_H < 1$ and $\rho(XY) < 1$. These results are summarized in the following lemma.

Lemma 6.11 Let $R = \begin{bmatrix} R_{11} & R_{12} \\ R_{21} & R_{22} \end{bmatrix} \in \mathcal{RH}_\infty^\perp$ be such that $\|[R_{11} \ R_{12}]\|_\infty < 1$ and $\left\| \begin{bmatrix} R_{11} \\ R_{21} \end{bmatrix} \right\|_\infty < 1$. If there exists $Q_{22} \in \mathcal{RH}_\infty$ such that (6.177) holds, then there exist $X \geq 0$ and $Y \geq 0$ satisfying the algebraic Riccati equations (6.183)–(6.184) and such that $\rho(XY) < 1$.

As we will see next using an explicit state-space construction, it turns out that the necessary condition for the existence of a solution to the four-block approximation problem just obtained is also *sufficient*.

Theorem 6.6 Let $R = \begin{bmatrix} R_{11} & R_{12} \\ R_{21} & R_{22} \end{bmatrix} \in \mathcal{RH}_\infty^\perp$ be such that $\|[R_{11} \ R_{12}]\|_\infty < 1$ and $\left\| \begin{bmatrix} R_{11} \\ R_{21} \end{bmatrix} \right\|_\infty < 1$. Then there exists $Q_{22} \in \mathcal{RH}_\infty$ such that (6.177) holds if

THE YOULA PARAMETRIZATION APPROACH

and only if $\rho(XY) < 1$, where $X \geq 0$ and $Y \geq 0$ are given in (6.183) and (6.184).

Proof. Necessity has already been established in Lemma 6.11. Sufficiency will be established by constructing a suitable Q. From the hypothesis it follows that

$$\bar{\sigma}([D_{11} \ D_{12}]) < 1 \quad \text{and} \quad \bar{\sigma}\left(\begin{bmatrix} D_{11} \\ D_{21} \end{bmatrix}\right) < 1$$

It follows then that

$$D = \begin{bmatrix} D_{11} & D_{12} \\ D_{21} & 0 \end{bmatrix}$$

can be embedded in a unitary matrix:

$$D_e = \begin{matrix} p_1 \\ p_2 \\ m_1 \\ m_2 \end{matrix} \begin{bmatrix} \overset{m_1}{D_{11}} & \overset{m_2}{D_{12}} & \overset{p_1}{D_{13}} & \overset{p_2}{0} \\ D_{21} & D_{22} & D_{23} & D_{24} \\ D_{31} & D_{32} & D_{33} & D_{34} \\ 0 & D_{42} & D_{43} & D_{44} \end{bmatrix} \quad (6.185)$$

For ease of reference we introduce the following notation:

$$\begin{aligned} B_e &= [B_1 \ B_2 \ B_3 \ 0] \\ C_e &= [C_1^T \ C_2^T \ C_3^T \ 0]^T \\ B_{Q_e} &= [0 \ B_{Q_2} \ B_{Q_3} \ B_{Q_4}] \\ C_{Q_e} &= [0 \ C_{Q_2}^T \ C_{Q_3}^T \ C_{Q_4}^T]^T \end{aligned} \quad (6.186)$$

and define R_a and Q_{aa} by the following state-space realizations

$$R_{aa} = \left[\begin{array}{c|c} A & B_e \\ \hline C_e & D_e \end{array}\right], \quad Q_{aa} = \left[\begin{array}{c|c} A_Q & B_{Q_e} \\ \hline C_{Q_e} & 0 \end{array}\right] \quad (6.187)$$

where A_Q, B_{Q_e}, and C_{Q_e} will be determined later. The corresponding state-space realization for the augmented error system $E_{aa} = R_{aa} + Q_{aa}$ is given precisely by (6.179). We proceed now to select Q_{aa} so that E_{aa} is allpass. Since D_e is unitary by construction, it follows from Theorem 5.1 in [132] that E_{aa} is unitary if and only if there exist X_e, and Y_e such that

$$\begin{bmatrix} A & 0 \\ 0 & A_Q \end{bmatrix} X_e + X_e \begin{bmatrix} A^T & 0 \\ 0 & A_Q^T \end{bmatrix} + \begin{bmatrix} B_e \\ B_{Q_e} \end{bmatrix} [B_e^T \ B_{Q_e}^T] = 0 \quad (6.188)$$

$$Y_e \begin{bmatrix} A & 0 \\ 0 & A_Q \end{bmatrix} + \begin{bmatrix} A^T & 0 \\ 0 & A_Q^T \end{bmatrix} Y_e + \begin{bmatrix} C_e^T \\ C_{Q_e}^T \end{bmatrix} [C_e \ C_{Q_e}] = 0 \quad (6.189)$$

$$D_e [B_e^T \ B_{Q_e}^T] + [C_e \ C_{Q_e}] X_e = 0 \quad (6.190)$$

202 $\mathcal{H}_\infty$ CONTROL

$$D_e^T [C_e \quad C_{Q_e}] + [B_e^T \quad B_{Q_e}^T] Y_e = 0 \tag{6.191}$$

$$X_e Y_e = I \tag{6.192}$$

Motivated by Lemma 8.2 in [132] we will look for a solution to these equations of the form

$$X_e \triangleq \begin{bmatrix} -X & I \\ I & YZ^{-1} \end{bmatrix}$$

$$Y_e \triangleq \begin{bmatrix} -Y & Z^T \\ Z & ZX \end{bmatrix} = X_e^{-1} \tag{6.193}$$

$$Z \triangleq I - XY$$

Note that since $\rho(XY) < 1$ then X_e and Y_e are well defined. To solve for C_{Q_e} and B_{Q_e} postmultiply equation (6.190) by $\begin{bmatrix} I & -Y \\ 0 & Z \end{bmatrix}$ to obtain

$$\begin{aligned} C_{Q_e} &= C_e X - D_e B_e^T \\ B_{Q_e} &= Z^{-T}(YB_e - C_e^T D_e) \end{aligned} \tag{6.194}$$

where we used the fact that Z is invertible and D_e is unitary. Moreover, from (6.182) it follows that

$$Z^T B_{Q_e} \begin{bmatrix} I \\ 0 \\ 0 \\ 0 \end{bmatrix} = YB_1 - C_e^T \begin{bmatrix} D_{11} \\ D_{21} \\ D_{31} \\ 0 \end{bmatrix} = 0 \tag{6.195}$$

Similarly, $[I \quad 0 \quad 0 \quad 0] C_{Q_e} = 0$. Thus B_{Q_e} and C_{Q_e} have the required structure. Next, we choose A_Q so that (6.188) holds. Premultiplying (6.188) by $[I \quad 0]$ and postmultiplying by $\begin{bmatrix} 0 \\ I \end{bmatrix}$ yields

$$\begin{aligned} A_Q &= -A^T - B_{Q_e} B_e^T \\ &= -A^T - Z^{-T}(YB_e - C_e^T D_e) B_e^T \\ &= -Z^{-T} \left[(I - YX)A^T + YB_e B_e^T - C_e^T D_e B_e^T \right] \\ &= -Z^{-T}(A^T + YAX - C_e^T D_e B_e^T) \end{aligned} \tag{6.196}$$

where the last line was obtained using the algebraic Riccati equation (6.183). Simple algebra shows that with these values for A_Q, B_{Q_e}, and C_{Q_e}, E_{aa} satisfies equations (6.188)–(6.192) and hence it is unitary. To complete the proof we need to show that $Q_{aa} \in \mathcal{RH}_\infty$. Using (6.196) it is easily seen that A_Q satisfies the following Lyapunov equation:

$$A_Q YZ^{-1} + Z^{-T} YA_Q^T + B_{Q_e} B_{Q_e}^T = 0 \tag{6.197}$$

Since $YZ^{-1} \geq 0$ and (A_Q, B_{Q_e}) is stabilizable, it follows that A_Q is Hurwitz. $\square$

6.7 PROBLEMS

1. In this problem we extend the proof of Lemma 6.4 to the case where the pair (C_1, A) is only detectable rather than observable.

   (a) Partition the states of the plant as $x = \begin{bmatrix} x_1 \\ x_2 \end{bmatrix}$ with x_2 unobservable and (C_{11}, A_{11}) observable. Assume that there exists an internally stabilizing controller $\hat{K}$ rendering $\|T_{zw}\|_\infty \leq 1$. Write down the equations for the closed-loop system in terms of x_1, x_2 and $\hat{x}$, the states of the controller.

   (b) Group these equations as follows: (i) a group containing only x_1, w and u and (ii) a group containing $x_1, x_2, \hat{x}, w$ and u. Consider the first group as a "new" plant G_{obs} and the second group as a "new" controller K_{obs}.

   (c) Write a state-space realization for G_{obs} and show that it satisfies the assumptions of the FI problem.

   (d) Show that K_{obs} stabilizes G_{obs} and renders $\|T_{zw}\|_\infty \leq 1$.

   (e) Show that if $X_{obs} > 0$ satisfies the FI Riccati equation for G_{obs} then
   $$Ric(H_\infty) = X_\infty = \begin{bmatrix} X_{obs} & 0 \\ 0 & 0 \end{bmatrix} \geq 0 \text{ exists for } G.$$ It follows that, without loss of generality (C_1, A) can be assumed observable.

2. Consider a stable strictly proper SISO transfer function $G(s)$. Let u, v denote the Schmidt pair of Γ_G defined in (6.23) and $U(s), V(s)$ the corresponding Laplace Transforms. Show that $U\tilde{U} = V\tilde{V}$. Use this result to show that V/U is allpass. [*Hint:* Use the state-space realizations (6.169).]

3. From [330], consider the following scalar $\mathcal{H}_\infty$ model-matching problem
   $$\min\{\|T_1 + T_2 Q\|_\infty : Q \in \mathcal{RH}_\infty\}$$
   where $T_1, T_2 \in \mathcal{RH}_\infty$ are where $T_2(j\omega) \neq 0$ for all ω.

   (a) Show that for the optimal Q, $T_1 + T_2 Q$ is allpass
   (b) Show that if $T_2(s)$ has N zeros in $\text{Re}(s) > 0$ then the degree of the optimal $T_1 + T_2 Q$ is at most $N - 1$.

4. Consider the problem of minimizing the $\mathcal{H}_\infty$ norm of the sensitivity for an open-loop unstable plant. Show that if the plant has an unstable pole at $s = s_1$ then the closed-loop system has a pole at $s = -s_1$. What conclusions can you draw about using $\mathcal{H}_\infty$ optimal controllers for plants with unstable poles very close to the $j\omega$ axis?

5. **Nevalinna–Pick Interpolation.** Given a set of n distinct points $\{a_1, a_2 \ldots, a_n\}$ with $\text{Re}(a_i) > 0$ and a set of values $\{b_1, b_2 \ldots, b_n\}$, $b_i \in \mathbb{C}$, the classical Nevalinna–Pick (NP) interpolation problem is to find a function $G \in \mathcal{RH}_\infty$ such that:
$$\|G(s)\|_\infty \leq 1$$
$$G(a_i) = b_i, \quad i = 1, 2 \ldots n \qquad (6.198)$$
For simplicity, assume that the sets $\{a_i\}$ and $\{b_i\}$ have complex-conjugate symmetry.

(a) Show that the NP problem is equivalent to solving the optimal FI $\mathcal{H}_\infty$ problem for the following plant:
$$\dot{x} = \begin{bmatrix} a_1 & 0 & \cdots & 0 \\ 0 & a_2 & \cdots & 0 \\ 0 & 0 & \vdots & 0 \\ 0 & 0 & 0 & a_n \end{bmatrix} x - \begin{bmatrix} b_1 \\ b_2 \\ \vdots \\ b_n \end{bmatrix} w + \begin{bmatrix} 1 \\ 1 \\ \vdots \\ 1 \end{bmatrix} u \qquad (6.199)$$
$$z = u$$

(b) Show that if the Pick matrix defined by
$$M_{i,j} = \frac{1 - a_i a_j^*}{s_i + s_j^*}$$
is positive definite then the NP problem is solvable. [*Hint:* Set up the FI Riccati equation and show that M^{-1} is its solution.]

(c) Give a LFT parametrization of all the solutions to the NP problem.

6. The purpose of this problem is to compare the performance (in the $\mathcal{H}_\infty$ sense) obtained by the optimal $\mathcal{H}_2$ controller against the optimal $\mathcal{H}_\infty$ performance. Given a function $T \in \mathcal{H}_2^\perp$ it can be shown [132] that $\|T\|_\infty / \|\Gamma_T\| \leq 2k$, where k denotes the number of unstable poles of T. Use this result to show that the following bound holds [117]:
$$\frac{\|T_1 - T_2 Q_2\|_\infty}{\|T_1 - T_2 Q_\infty\|_\infty} \leq 2k$$
where k denotes the number of zeros of $T_2(s)$ in $\text{Re}(s) > 0$ and where Q_2 and Q_∞ denote the solutions to the optimal $\mathcal{H}_2$ and $\mathcal{H}_\infty$ model matching problems, respectively.

7. $\mathcal{H}_\infty$ control tools cannot be used when the plant has poles or zeros on the $j\omega$ axis. This problem explores a technique for synthesizing suboptimal $\mathcal{H}_\infty$ controller in these cases [78].

(a) Consider the following bilinear transformation:
$$s = \frac{\bar{s} - p_1}{1 - \frac{\bar{s}}{p_2}}, \quad p_1, p_2 > 0$$

Show that this transformation maps the $j\omega$ axis on the s-plane onto a RHP circle in the $\bar{s}$ plane. Find the center of this circle and show that it intersects the real axis at $\bar{s} = p_1$ and $\bar{s} = p_2$.

(b) Show that the $j\omega$ axis in the $\bar{s}$ plane is the image of a LHP circle in the s-plane (exactly the mirror image of the $\bar{s}$-plane circle).

(c) Suppose that a controller $K(\bar{s})$ internally stabilizes a plant $P(\bar{s})$ and renders some $\|T(\bar{s})\|_\infty \leq 1$. Show that the controller $K(s)$ internally stabilizes $P(s)$ and renders $\|T(s)\|_\infty \leq 1$.

(d) Based on these results, propose an algorithm for synthesizing suboptimal $\mathcal{H}_\infty$ controllers for plants with poles or zeros on the $j\omega$ axis.

8. Use the algorithm outlined in Problem 6.6 to synthesize a controller for the ACC benchmark problem (see Problem 5.3) subject to the additional specification that the closed-loop system should be stable for all values of $k \in [0.5, 2]$. [*Hint:* Try $p_1 \sim 0.5$ and $p_2 \sim 100$.]

9. Consider the problem of synthesizing a controller for fuel injection control [177]. The nominal plant (at $T = 25°C$) has the following transfer function:

$$P_{25} = \frac{5.498s^2 + 400.7s - 444400}{s^3 + 93.72s^2 + 9530s + 121400}$$

The control should operate correctly for fuel temperatures ranging from $T = 0°C$ to $T = 60°C$. The plants corresponding to these extreme temperatures are given by:

$$P_0 = \frac{-0.01736s^2 + 493.9s - 313700}{s^3 + 98.34s^2 + 9223s + 87710}$$

$$P_{60} = \frac{4.677s^2 - 285.9s - 505300}{s^3 + 91.53s^2 + 10080s + 176200}$$

(6.200)

For the purpose of this problem we will assume that the frequency responses of all plants corresponding to $T \in [0, 60]$ are contained in the band determined by the frequency responses of the two extreme plants.

(a) Show that the family of plants can be modeled as the nominal plant P_{25} subject to multiplicative uncertainty Δ with $\|W_I \Delta\|_\infty < 1$ where

$$W_I(s) = 2.2 \frac{s + 10}{s + 100}$$

(b) Design an $\mathcal{H}_\infty$ controller that robustly stabilizes P_{25}.

(c) Plot the step response of the resulting closed-loop system for $T = [0, 25, 60]$. Is the tracking performance acceptable? Explain your results. [*Hint:* The plant is open-loop stable.]

(d) Design an $\mathcal{H}_\infty$ controller that robustly stabilizes P_{25} and renders $\|W_P S\|_\infty \le 1$ where
$$W_I(s) = 0.1\frac{s+100}{s+1}$$

(e) Repeat part (c) with the new controller.

10. The uncertain linear system
$$\dot{x}(t) = [A + w(t)DE]x \qquad (6.201)$$

where $A \in \mathbb{R}^{n \times n}$, $D \in \mathbb{R}^{n \times 1}$ and $E \in \mathbb{R}^{1 \times n}$ are known matrices and where $w(t) \in [w^-, w^+]$ represents *memoryless* time varying uncertainty is said to be *quadratically stable* if there exists a positive definite matrix $Q \in \mathbb{R}^{n \times n}$ and a scalar $\alpha > 0$ such that the derivative of the Lyapunov function $x'Qx$ along the trajectories of (6.201) satisfies:

$$\frac{d}{dt}\left(x^T Q x\right) \le -\alpha \|x\|^2 \qquad (6.202)$$

for all possible values of $w(t)$. Use the Bounded Real lemma (Lemma 6.8) to show that if $G(s) \triangleq E(sI - A)^{-1}D \in \mathcal{H}_\infty$ and $\|G(s)\|_\infty < \gamma$ then the system (6.201) is quadratically stable for $w \in [-\gamma^{-1}, \gamma^{-1}]$.

7

STRUCTURED UNCERTAINTY

7.1 INTRODUCTION

The objective of this chapter is to introduce the reader to more general uncertainty structures, which are often encountered in practical applications. There are some topics in this chapter that are still an active area of research. Therefore, in these cases, the approach will be that of a survey rather than of a textbook.

The class of model uncertainty that has been considered in Chapters 2 and 4 is *global dynamic*. This name comes from the fact that the uncertainty is attributable to all the system and describes the unknown higher-order dynamics of the plant. However, in many practical applications this type of uncertainty description "covers" in a very conservative way the real uncertainty of the plant's model. This is the case when more "structured" information on the plant is available. Hence a nonconservative analysis or synthesis procedure should take advantage of this extra information. Next, we mention some of these situations.

- Consider the total system as composed of individual subsystems, each with its own dynamic uncertainty description. Take, for example, the actuators, the system itself, and the sensors, described by the following individual sets of models:

$$\mathcal{G}_a(s) = [I + \Delta_1 W_a(s)] G_a(s); \quad \Delta_1 \in \mathbb{C}^{k_1 \times k_1}, \ \|\Delta_1\| < 1 \quad (7.1)$$

$$\mathcal{G}(s) = [I + \Delta_2 W_g(s)] G(s); \quad \Delta_2 \in \mathbb{C}^{k_2 \times k_2}, \ \|\Delta_2\| < 1 \quad (7.2)$$

$$\mathcal{G}_s(s) = [I + \Delta_3 W_s(s)] G_s(s); \quad \Delta_3 \in \mathbb{C}^{k_3 \times k_3}, \ \|\Delta_3\| < 1 \quad (7.3)$$

208 STRUCTURED UNCERTAINTY

The total interconnection yields

$$S(s) = \mathcal{G}_s(s) \cdot \mathcal{G}(s) \cdot \mathcal{G}_a(s) \tag{7.4}$$

The above set can be transformed to a LFT connection between a nominal model $\tilde{G}(s)$ and an uncertainty block in the set $\mathbf{\Delta}_{\text{struct}} \stackrel{\triangle}{=} \{\text{diag}[\Delta_1\ \Delta_2\ \Delta_3],\ \Delta_i \in \mathbb{C}^{k_i \times k_i},\ \|\Delta_i\| < 1,\ i = 1, 2, 3\}$. To this end, define the outputs from the uncertainty blocks as $\{u_i,\ i = 1, 2, 3\}$ and the outputs from the weight functions as $\{y_i,\ i = 1, 2, 3\}$ as in Figure 7.1. In addition, the input and output of set S are defined as u and y, respectively. After some algebra, we obtain the following matrix $\tilde{G}(s)$:

$$\begin{bmatrix} y_\Delta \\ y \end{bmatrix} = \begin{bmatrix} 0 & 0 & 0 & W_a G_a \\ W_g G & 0 & 0 & W_g G G_a \\ W_s G_s G & W_s G_s & 0 & W_s G_s G G_a \\ G_s G & G_s & I & G_s G G_a \end{bmatrix} \begin{bmatrix} u_\Delta \\ u \end{bmatrix} \tag{7.5}$$

$$y_\Delta \stackrel{\triangle}{=} \begin{bmatrix} y_1 \\ y_2 \\ y_3 \end{bmatrix}, \quad u_\Delta \stackrel{\triangle}{=} \begin{bmatrix} u_1 \\ u_2 \\ u_3 \end{bmatrix} \tag{7.6}$$

Therefore the set $\left\{ F_u\left[\tilde{G}(s), \Delta\right],\ \Delta \in \mathbf{\Delta}_{\text{struct}} \right\}$ is equivalent to the family of models S. This type of uncertainty is called *structured dynamic*.

If, on the other hand, the uncertainty of the plant is described as global dynamic, that is, $\left\{ F_u\left[\tilde{G}(s), \Delta\right],\ \Delta \in \mathbb{C}^{n \times n},\ \|\Delta\| < 1 \right\}$ $(n = k_1 + k_2 + k_3)$, disregarding the structural information will add more unnecessary models to the set. Hence the robustness analysis of a closed-loop system with such an uncertainty description will be conservative in general.

- In many cases, the plant has a well known mathematical model usually derived from physical equations. This is the case of some applications from mechanical, aeronautical, and astronautical engineering, where the rigid body model based on the Newton–Euler equations provides a good enough description for mild performance specifications. However, the parameters of these models may not be known. As a consequence their values are estimated either by means of classical parameter identification ([13, 184, 293]) procedures or, more recently, using set membership identification methods ([155, 204]). As with any other mathematical

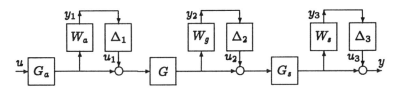

Figure 7.1. System block diagram.

INTRODUCTION

procedure involving input data from physical measurements, the nominal values of these parameters will be known within some uncertainty bounds. When these bounds are deterministic worst-case bounds, this leads to a plant representation in terms of a family of models with a mathematical fixed structure and parameters that may take values within certain specified sets. This type of uncertainty description is called *structured parametric uncertainty* or simply *parametric uncertainty*. Take, for example, the following set of models, which represents a plant with an uncertain real zero and uncertain natural frequency and damping coefficient:

$$\mathcal{G}(s,p) = \frac{s+z}{s^2 + 2\omega_n \xi s + \omega_n^2} \tag{7.7}$$

$$z \in [z_1, z_2], \quad \omega_n \in [\omega_1, \omega_2], \quad \xi \in [\xi_1, \xi_2]$$

$$p \stackrel{\triangle}{=} \begin{bmatrix} z & \omega_n & \xi \end{bmatrix}^T$$

Here the *structure* is based on the fact that only specific elements of the plant's model are uncertain (z, ω_n, ξ).

Parametric uncertainty can be present in both state-space or transfer matrix representations. In the latter case, there are basically two ways in which the uncertain parameters are handled. The first adopts the standard LFT structure with the parameters located in the upper uncertainty block Δ as in Figure 7.2. The second performs the analysis on the characteristic polynomial of the closed-loop system. In this example it is easier to put the uncertainty in the latter form in order to evaluate stability robustness:

$$P(s,p) = s^2 + \underbrace{(k + 2\omega_n \xi)}_{c_1(p)} s + \underbrace{\left(kz + \omega_n^2\right)}_{c_0(p)} \tag{7.8}$$

where k is a constant (nominal) stabilizing controller.

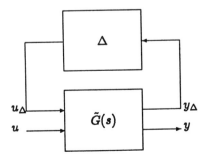

Figure 7.2. LFT uncertainty description.

- In general, both parametric and structured dynamic uncertainty appear simultaneously. This is the case, for example, of large flexible space structures. They have a well known low-frequency model described in terms of the natural vibration modes, as well as unknown higher-order dynamics. The lower-order model is represented by several second-order modes with natural frequencies and damping coefficients within real intervals, that is, parametric uncertainty. The higher-order dynamics, on the other hand, can be represented more naturally as dynamic uncertainty. This is the so-called *mixed* type uncertainty.

For all the above types of structured uncertainties, more general mathematical tools than the ones described in previous chapters should be used. In this chapter, a brief description of some of the analysis and synthesis results for these cases will be presented. Special emphasis will be placed on methods that use the standard LFT description for the uncertainty.

Recall from Chapter 2 that stability or performance robustness margins are related directly to the *type* of uncertainty present in the plant. As special cases we have mentioned the classical phase and gain margins. For structured uncertainty the same concept holds; therefore a general definition of a robustness margin should be made. Next, we present such a definition, and we leave for the rest of the chapter the different methods to compute this margin, according to the type of uncertainty structure.

7.1.1 Stability Margin

The margin of a particular property of a system (stability, performance, controllability, detectability, etc.) is always defined in terms of the distance of the nominal model, which represents the system, to the nearest model that lacks this property ([223]). To quantify this distance, a particular norm should be used. Therefore the system is more robust if this margin is larger and vice versa. Furthermore, the nearest model that lacks this property will depend on the type of uncertainty in the system. For this reason, as mentioned in Chapter 2, the concepts of uncertainty and margin computation are closely related.

We will be concerned in this subsection with the stability margin, although in many cases, as will be seen in further parts of this chapter, a simple extension to performance can be made. Furthermore, due to the fact that the stability margin is an analysis tool, we assume that a stabilizing controller for the nominal model has already been designed and closes the loop.

Characteristic Polynomial Framework A natural way to state the problem is in terms of the closed-loop characteristic polynomial (CLCP). This includes many practical applications, especially in cases where parametric uncertainty is involved. A general closed-loop polynomial for FDLTI

INTRODUCTION

systems can be defined as follows:

$$P(s,p) = s^n + c_{n-1}(p)s^{n-1} + \cdots + c_1(p)s + c_0(p), \quad p \in [a,b] \qquad (7.9)$$

where $\{c_i(\cdot), i = 1, \ldots, n\}$ are real functions of the uncertainty vector p and $p \in [a,b]$ represents $p_i \in [a_i, b_i], i = 1, \ldots, m$. Here a nominal internally stabilizing controller is assumed; therefore the nominal characteristic polynomial $P(s, p^0)$ has all its roots in $\mathbb{C}_- \triangleq \{s \in \mathbb{C}; \operatorname{Re}(s) < 0\}$. An equivalent condition for robust stability is:

$$\text{Robust stability} \iff P(s,p) \neq 0, \quad \forall p \in [a,b], \quad \forall s \in \mathbb{C}_+ \qquad (7.10)$$

where $\mathbb{C}_+ \triangleq \{s \in \mathbb{C}; \operatorname{Re}(s) \geq 0\}$.

The previous condition provides a *qualitative* answer, that is, a yes/no verification of robust stability. On the other hand, a stability margin provides a *quantitative* measure of the *degree* of stability. Obviously the former qualitative answer can also be obtained using this margin. As a consequence we seek to define a *computable* robustness margin for stability.

Checking condition (7.10) requires computing the roots of $P(s,p)$ for all possible values of $p \in [a,b]$. On the other hand (since the nominal system is stable), a robust stability margin only needs to indicate at which point the roots of $P(s,p)$ cross over from $\mathbb{C}_-$ to $\mathbb{C}_+$ as the uncertainty around the nominal set of parameters p^0 is "increased." In this situation, it seems natural that as poles move from $\mathbb{C}_-$ to $\mathbb{C}_+$, the first unstable ones reach the $j\omega$ axis before entering the interior of $\mathbb{C}_+$. In general this property is not guaranteed, however the conditions under which it holds are stated in the *boundary crossing theorem* presented next. To this end we first define a partition of the complex plane.

Definition 7.1 *An open region $\mathcal{O}_1$ and a closed one $\overline{\mathcal{O}}_2$ define a partition of $\mathbb{C}$ if and only if the following two conditions are satisfied:*

$$\mathbb{C} = \mathcal{O}_1 \cup \partial \mathcal{O}_1 \cup \mathcal{O}_2 \qquad (7.11)$$

$$\emptyset = \mathcal{O}_1 \cap \partial \mathcal{O}_2 = \mathcal{O}_1 \cap \mathcal{O}_2 = \partial \mathcal{O}_1 \cap \mathcal{O}_2 \qquad (7.12)$$

where $\mathcal{O}_2$ is the interior of the closed set $\overline{\mathcal{O}}_2$ and $\partial \mathcal{O}_1$ is the boundary of $\mathcal{O}_1$.

The conditions under which the imaginary axis crossing is achieved are as follows ([35, 47]).

Theorem 7.1 (Boundary Crossing) *Consider a polynomial $P(s,p)$ of fixed degree and continuous for all $p \in [a,b]$, an open region $\mathcal{O}_1$, and a closed one $\overline{\mathcal{O}}_2$, which form a partition of $\mathbb{C}$. Then if $P(s,a)$ has all its roots in $\mathcal{O}_1$ whereas*

$P(s, b)$ has at least one root in $\overline{\mathcal{O}}_2$, there exists at least one $x \in (a, b]$ such that $P(s, x)$ has all its roots in $\mathcal{O}_1 \cup \partial \mathcal{O}_1$ and at least one root in $\partial \mathcal{O}_1$.

Proof. Based on the theorem of Rouché, the detailed proof can be found in [47]. □

The above result can be interpreted in terms of robust stability by defining the open set $\mathcal{O}_1 \equiv \mathbb{C}_-$, the boundary $\partial \mathcal{O}_1$ as the $\jmath\omega$ axis, and the closed set $\overline{\mathcal{O}}_2 \equiv \mathbb{C}_+$. Furthermore, the assumption that $P(s, a)$ has all its roots in $\mathcal{O}_1$ means that there exists a stabilizing nominal controller, where the nominal value of the closed-loop polynomial has been assumed at $P(s, a)$. Under these conditions, the theorem states that to destabilize a nominally stable closed-loop system by moving its parameters from $a \to b$, the first *offending* set of parameters will generate an unstable pole in the boundary $\jmath\omega$ and not in the interior of $\mathbb{C}_+$. We define an offending vector x of parameters as the one that generates an unstable root of $P(s, x)$. Therefore, under these conditions, robust stability can be verified by looking only at the $s = \jmath\omega$ axis. Similarly, for discrete time systems, we define the open set $\mathcal{O}_1$ as the interior of the unit disk and $\overline{\mathcal{O}}_2$ as its complement.

This result suggests another simplification to the robust stability equivalent condition stated in (7.10). If the characteristic polynomial meets the assumptions of the previous theorem, the necessary and sufficient condition is as follows:

$$\text{Robust stability} \iff P(\jmath\omega, p) \neq 0, \quad \forall p \in [a, b], \, \forall \omega \qquad (7.13)$$

Note that this is a *maximum modulus* ([332]) type theorem for the case of parametric uncertainty.

LFT Framework In many cases, it is more convenient to structure the uncertainty as a LFT, as for additive or multiplicative dynamic uncertainty (see Chapter 2 and Appendix B). This is a natural way of evaluating robust stability when dynamic uncertainty is involved, due to the fact that in these cases the model order is not fixed. In the parametric uncertainty case the LFT setup includes only $c_i(p)$ functions that are polynomial in the parameters. Therefore the previous analysis based on the characteristic polynomial $P(s, p)$ would be more general.[1] Next, the analysis problem in a LFT setup is presented, as in Figure 7.3. Here $T(s)$ includes the nominal model, the controller, and the performance and robustness weights.

Note that, if the inputs and outputs w and z are selected conveniently and the problem is well posed, the stability of $F_u[T(s), \Delta]$ for all Δ in a prescribed

---
[1] Historically, the CLCP has been used to compute the stability margin for pure real parameter uncertainty; while the LFT formulation has been used for the structured dynamic type uncertainty, due to the above reasons.

INTRODUCTION

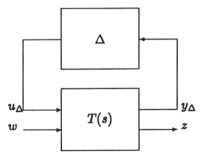

Figure 7.3. LFT for robustness analysis.

bounded set $\boldsymbol{\Delta}$ guarantees robust *internal stability* of the closed-loop system it represents. Based on the fact that nominal internal stability guarantees $T(s)$ stable—and hence the only element in $F_u[T(s), \Delta]$ that may destabilize it (assuming Δ stable or constant) is $[I - T_{11}(s)\Delta]^{-1}$—the following conditions are equivalent to robust stability:

$$F_u[T(s), \Delta] \text{ stable } \quad \forall \Delta \in \boldsymbol{\Delta} \tag{7.14}$$

$$\iff [I - T_{11}(s)\Delta]^{-1} \text{ stable } \quad \forall \Delta \in \boldsymbol{\Delta} \tag{7.15}$$

$$\iff \det[I - T_{11}(s)\Delta] \neq 0, \quad \forall \Delta \in \boldsymbol{\Delta}, \forall s \in \mathbb{C}_+ \tag{7.16}$$

The above condition can be assumed without loss of generality due to the fact that w and z can always be selected so that input–output stability is equivalent to internal stability (Problem 1).

The uncertainty structures $\boldsymbol{\Delta}$ that will be used in the sequel are

$$\boldsymbol{\Delta}_d \triangleq \left\{ \begin{bmatrix} \Delta_1 & & 0 \\ & \ddots & \\ 0 & & \Delta_m \end{bmatrix} \middle| \begin{array}{l} \Delta_i \in \mathbb{C}^{r_i \times r_i} \\ i = 1, \ldots, m \end{array} \right\} \tag{7.17}$$

for *structured dynamic uncertainty*;

$$\boldsymbol{\Delta}_p \triangleq \left\{ \begin{bmatrix} \delta_1 & & 0 \\ & \ddots & \\ 0 & & \delta_m \end{bmatrix} \middle| \begin{array}{l} \delta_i \in \mathbb{R} \\ i = 1, \ldots, m \end{array} \right\} \tag{7.18}$$

for *parametric uncertainty*; and the combination of both

$$\boldsymbol{\Delta}_M \triangleq \left\{ \begin{bmatrix} \Delta_1 & 0 \\ 0 & \Delta_2 \end{bmatrix} \middle| \Delta_1 \in \boldsymbol{\Delta}_d, \; \Delta_2 \in \boldsymbol{\Delta}_p \right\} \tag{7.19}$$

for *mixed* uncertainty descriptions. These are unbounded sets, hence we denote $\mathcal{B}\Delta_i$ as their corresponding unit (open) balls, that is, $\overline{\sigma}(\Delta_i) < 1$ and $|\delta_i| < 1$, $i = 1, \ldots, m$, for $\mathcal{B}\Delta_d$ and $\mathcal{B}\Delta_p$, respectively. The notation for closed balls is $\overline{\mathcal{B}}\Delta_i$.

As in the previous case, the robustness condition can be simplified to a test over the imaginary axis only. This *maximum modulus* type property has been proved in [55] and can be used to establish the following result:

$$\text{Robust stability} \iff \det[I - T_{11}(\jmath\omega)\Delta] \neq 0, \quad \forall \Delta \in \Delta, \, \forall \omega \quad (7.20)$$

In the definitions of the stability margins presented next, we use the LFT framework. However, in further subsections when considering specific classes of uncertainty structures, each corresponding type of formulation will be used.

Stability Margin Definition The next two definitions take into account the previous discussion concerning the stability margins. These are formulated so that direct necessary and sufficient conditions for robust stability in the case of general types of uncertainty can be stated.

Definition 7.2 *The* multivariable stability margin k_m *([257]) is defined as*

$$k_m[M(\jmath\omega)] \triangleq \inf\{k \in (0, \infty) | \det[I - kM(\jmath\omega)\Delta] = 0, \text{ for some } \Delta \in \mathcal{B}\Delta\}$$

Definition 7.3 *The* structured singular value μ_Δ *([102]) is defined as*

$$\mu_\Delta[M(\jmath\omega)] \triangleq \left[\inf_{\Delta \in \Delta}\{\overline{\sigma}(\Delta) | \det[I - M(\jmath\omega)\Delta] = 0\}\right]^{-1}$$

and $\mu_\Delta[M(\jmath\omega)] = 0$ *if* $\det[I - M(\jmath\omega)\Delta] \neq 0$ *for all* $\Delta \in \Delta$.

From these definitions it is clear that $k_m[M(\jmath\omega)] = \mu_\Delta^{-1}[M(\jmath\omega)]$ when the uncertainty structure Δ is the same. From both definitions, the following necessary and sufficient conditions for robust stability can be stated.

Lemma 7.1 *The following statements are equivalent:*

1. *The family of closed-loop systems* $\{F_u[T(s), \Delta], \Delta \in \mathcal{B}\Delta\}$ *of Figure 7.3 is robustly stable.*
2. $\quad\quad\quad\quad\mu_\Delta[T_{11}(\jmath\omega)] \leq 1 \quad \text{for all } \omega \in \mathbb{R} \quad\quad\quad (7.21)$
3. $\quad\quad\quad\quad k_m[T_{11}(\jmath\omega)] \geq 1 \quad \text{for all } \omega \in \mathbb{R} \quad\quad\quad (7.22)$

Here the set Δ may be any of the previously defined uncertainty structures.[2] In further sections we replace $T_{11}(s)$ by $M(s)$ for simplicity.

---
[2] Note that we have considered the open ball in Δ, which produces the above nonstrict inequalities. The case of closed balls is more difficult to evaluate (see [332]).

As noted in the above discussion, these quantities have been defined to provide directly the equivalent robustness condition for these rather general types of uncertainty. The hard problem is the computation of these margins for all the previous uncertainty structures. In fact, it can be proved that the calculation of the exact stability margin (or even an approximation with guaranteed *a priori* bounds) of an uncertain model with a general parametric uncertainty structure is a NP-hard problem ([60, 248]). To illustrate this point, note that this computation involves the mapping of a polynomial function $P(s, p)$ for all parameters in the set. For parameters defined within real intervals, this set is an m-dimensional hypercube at each $s = j\omega$, and for the general case its image produces a nonconvex set. The computation of the image of the vertices alone involves 2^m evaluations of the function. In addition, there is no way to determine, for a general $P(s, p)$, a subset of these vertices, polynomial in m, which may define this image. It is well known that this is a NP-hard problem ([62]).

Under this strong constraint on the computational aspect of the problem, it seems that research should be pointed at least in two directions. First, looking for exact (or approximate with guaranteed bounds) polynomial time analysis of parametric uncertainty structures, which may not be general but can accommodate *relevant* practical applications; and second, studying approximate methods (branch and bound, heuristics) that can bound in polynomial time the stability margin for general cases and, although they may not have guaranteed *a priori* error bounds, work reasonably well in practical situations ([326]). The former approach has been pursued by the research area briefly described in Section 7.3 of this chapter, which seeks extensions to practical cases of analytical or tractable computational results. The formulation is in terms of the characteristic polynomial $P(s, p)$ of the closed-loop system. The latter approach is based on the structured singular value applied in a LFT setup and will be explained in the last section of this chapter.

The formulation used for pure dynamic uncertainty has no computational hard constraints and will be presented in the next section in terms of the structured singular value in a LFT setup. In addition, the use of these margins as analysis and synthesis tools will be illustrated.

7.2 STRUCTURED DYNAMIC UNCERTAINTY

7.2.1 Computation

In this section we restrict the uncertainty to the set Δ_d defined before; therefore the definition of μ will be considered with respect to such a structure. Historically, this problem has been developed in [102] using a LFT framework. An excellent tutorial on this subject can be found in [220].

One of the reasons why this margin has been defined as the inverse of the destabilizing perturbation is that in this form it naturally generalizes the

216 STRUCTURED UNCERTAINTY

notions of maximum singular value and spectral radius.[3] This can be explained by means of the following two examples.

Example 7.1 Consider the set $\Delta_\ell \triangleq \{\delta I;\ \delta \in \mathbb{C}\} \subset \Delta_d$. By definition we have $\det[I - M(\jmath\omega)\delta] = 0$ for $\delta = \lambda_i^{-1}[M(\jmath\omega)]$, therefore the minimum $|\delta_\star| = |\lambda_1|^{-1}$, where λ_1 is the maximum magnitude eigenvalue, that is, $\rho[M(\jmath\omega)] = |\lambda_1|$. Hence for the uncertainty structure Δ_ℓ, we have

$$\mu_{\Delta_\ell}[M(\jmath\omega)] = \rho[M(\jmath\omega)] \tag{7.23}$$

Example 7.2 Take the set $\Delta_u \triangleq \{\Delta \in \mathbb{C}^{r \times r}\}$, which includes the structure Δ_d when $r = r_1 + \cdots + r_m$ in (7.17). This is the global dynamic uncertainty case considered in Chapters 2 and 4. Therefore, according to the robust stability condition in that case, the following holds:

$$\mu_{\Delta_u}[M(\jmath\omega)] = \overline{\sigma}[M(\jmath\omega)] \tag{7.24}$$

In fact, we can select the smallest perturbation as $\Delta_\star = V\Sigma_\Delta U^*$, $\Sigma_\Delta = \operatorname{diag}[\sigma_1^{-1}, 0, \ldots, 0]$, where $M(\jmath\omega) = U\Sigma V^*$ and $\overline{\sigma}(\Sigma) = \sigma_1$. It is simple to verify that $\det[I - M(\jmath\omega)\Delta_\star] = 0$.

From the above examples and the fact that $\Delta_\ell \subset \Delta_d \subset \Delta_u$, we obtain

$$\inf_{\Delta \in \Delta_\ell} \overline{\sigma}(\Delta) \geq \inf_{\Delta \in \Delta_d} \overline{\sigma}(\Delta) \geq \inf_{\Delta \in \Delta_u} \overline{\sigma}(\Delta) \tag{7.25}$$

$$\implies \rho[M(\jmath\omega)] \leq \mu_{\Delta_d}[M(\jmath\omega)] \leq \overline{\sigma}[M(\jmath\omega)] \tag{7.26}$$

This inequality presents the structured singular value as a generalization of the maximum singular value and the spectral radius. These measures will be instrumental in computing tight bounds on μ, although a further step needs to be performed due to the fact that the above bounds may be conservative in general.

Example 7.3 Consider the constant matrix case (or, equivalently, a system "frozen" at a given frequency) $R = M(\jmath\omega)$:

$$R = \begin{bmatrix} 0 & 0 \\ 1 & 0 \end{bmatrix} \tag{7.27}$$

which has spectral radius $\rho(R) = 0$ and maximum singular value $\overline{\sigma}(R) = 1$. Take the following uncertainty structure to compute $\mu(\cdot)$:

$$\Delta = \left\{ \begin{bmatrix} \delta_1 & 0 \\ 0 & \delta_2 \end{bmatrix},\ \delta_i \in \mathbb{C} \right\} \tag{7.28}$$

---

[3] The spectral radius is defined as $\rho(A) \triangleq \max |\lambda_i(A)|$, where $\lambda_i(\cdot)$ are the eigenvalues of A.

Then $\det(I - R\Delta) = 1$ for all $\Delta \in \mathbf{\Delta}$ and, by definition, $\mu(R) = 0$. Hence $\rho(R) = \mu_\Delta(R) = 0 \leq \overline{\sigma}(R) = 1$, which by scaling can be made arbitrarily conservative.

Example 7.4 *Take instead the constant matrix*

$$R = \begin{bmatrix} 0.5 & -0.5 \\ 0.5 & -0.5 \end{bmatrix} \quad (7.29)$$

with $\rho(R) = 0$ and $\overline{\sigma}(R) = 1$, and the same uncertainty structure of the previous example. Replacing $\Delta \in \mathbf{\Delta}$ we obtain

$$\det(I - R\Delta) = 1 - \tfrac{1}{2}(\delta_1 - \delta_2) = 0 \quad (7.30)$$

for $\Delta_\star = \begin{bmatrix} 1 & 0 \\ 0 & -1 \end{bmatrix}$. Therefore $\mu_\Delta(R) = 1$ and hence $\rho(R) = 0 \leq \mu_\Delta(R) = \overline{\sigma}(R) = 1$, which by scaling can be made arbitrarily conservative.

To improve these bounds in order to compute μ, the following properties will be used. For simplicity we drop the argument in the system matrix $M(s)$ and the uncertainty set $\mathbf{\Delta}_d$ as the index of the structured singular value.

1. $\mu(\alpha M) = |\alpha|\mu(M)$, $\alpha \in \mathbb{C}$. This is similar to the scaling property of norms (see Appendix A) and is due to the fact that

$$\det(I - \alpha M \Delta) = \det[I - M(\alpha \Delta)] \quad (7.31)$$

and $\overline{\sigma}(\alpha \Delta) = |\alpha|\overline{\sigma}(\Delta)$.

2. $\mu(A \cdot B) \leq \overline{\sigma}(A)\mu(B)$. This is based on the facts that $\det(I - A \cdot B \cdot \Delta) = \det[I - B(\Delta \cdot A)]$ and $\overline{\sigma}(\Delta A) \leq \overline{\sigma}(\Delta) \cdot \overline{\sigma}(A)$.

3. Consider a general uncertainty set $\mathbf{\Delta}$, with respect to which μ is defined. Then:

$$\mu(M) = \max_{\Delta \in \overline{\mathcal{B}}\mathbf{\Delta}} \rho(M\Delta) \quad (7.32)$$

This can be used as an alternative definition for $\mu(\cdot)$. From the scaling property of μ, it follows that we need to prove the above condition only for the two following cases:

$$\max_{\Delta \in \overline{\mathcal{B}}\mathbf{\Delta}} \rho(M\Delta) = 1 \iff \mu(M) = 1 \quad (7.33)$$

which holds by definition, and

$$\det(I - M\Delta) \neq 0, \quad \forall \Delta \in \mathbf{\Delta} \tag{7.34}$$

$$\iff \mu(M) = 0 \tag{7.35}$$

$$\iff \sup_{\Delta \in \mathbf{\Delta}} \rho(M\Delta) < 1 \tag{7.36}$$

$$\iff \sup_{\Delta \in \overline{B}\mathbf{\Delta}} \rho(M\Delta) = 0 \tag{7.37}$$

Otherwise, if the last equation is not equal to zero, it may be scaled in order to contradict (7.36). Note that $\mu_{\mathbf{\Delta}_d}(\cdot)$ is continuous due to the continuity of the spectral radius and the *max* functions and the fact that $\overline{B}\mathbf{\Delta}$ is compact.

4. $\mu(MU) = \mu(M)$ for $U \in \mathcal{U}$, the latter defined as follows:

$$\mathcal{U} \triangleq \{U \in \mathbf{\Delta}_d, \ U_i \text{ unitary}; \ i = 1, \ldots, m\} \tag{7.38}$$

This can be proved with simple arguments (Problem 2).

5. $\mu(DMD^{-1}) = \mu(M)$ for $D \in \mathcal{D}$, the latter defined as follows:

$$\mathcal{D} \triangleq \left\{ \begin{bmatrix} d_1 I_1 & & 0 \\ & \ddots & \\ 0 & & d_m I_m \end{bmatrix} \;\middle|\; \begin{array}{l} I_i : \text{identity in } r_i \\ d_i > 0 \end{array} \right\} \tag{7.39}$$

which can also be proved using the definition of μ (Problem 3).

Equation (7.26) can be used to improve the bounds as follows:

$$\max_{U \in \mathcal{U}} \rho(MU) \leq \max_{\Delta \in \overline{B}\mathbf{\Delta}} \rho(M\Delta) = \mu(M) \leq \inf_{D \in \mathcal{D}} \overline{\sigma}(DMD^{-1}) \tag{7.40}$$

Here the equality follows from property 3. Moreover, tighter bounds can be obtained using the following two results from [102].

Theorem 7.2

$$\max_{U \in \mathcal{U}} \rho(MU) = \mu(M) \tag{7.41}$$

Proof. The argument is based on the fact that for $\mu(M) = 1$ there exists a $U \in \mathcal{U}$ such that $\det(I - MU) = 0$, which in turn implies $\rho(MU) \geq 1$. Property (7.40) completes the proof. See [102] for details. □

The fact that the optimum in (7.40) of the spectral radius over $\bar{\mathcal{B}}\Delta$ and $\mathcal{U}$ coincide means that the optimal uncertainty Δ in this problem can always be made unitary.

Theorem 7.3

$$\inf_{D \in \mathcal{D}} \bar{\sigma}(DMD^{-1}) = \mu(M) \tag{7.42}$$

for $m \leq 3$ in the uncertainty structure Δ_d defined in (7.17)

Proof. The proof exceeds the scope of this book and thus will not be presented here. It can be found in [102]. □

The computation of the structured singular value by means of Theorem 7.2 involves a nonconvex optimization problem; thus there is no guarantee of reaching the global optimum. On the other hand, the optimization problem in Theorem 7.3 is convex ([259]) as will be proved next.

Lemma 7.2 *The set $\{\bar{\sigma}(D^{1/2}MD^{-1/2}) \leq \gamma, \ D \in \mathcal{D}\}$, for $\gamma \geq 0$ is convex in D.*

Proof. The following are all equivalent:

$$\bar{\sigma}(D^{1/2}MD^{-1/2}) \leq \gamma$$
$$\iff \bar{\lambda}(D^{-1/2}M^*DMD^{-1/2}) \leq \gamma^2$$
$$\iff D^{-1/2}M^*DMD^{-1/2} - \gamma^2 I \leq 0$$
$$\iff M^*DM - \gamma^2 D \leq 0$$

which is convex in $D \in \mathcal{D}$ ([220]). □

The latter is a linear matrix inequality (LMI), which can be computed efficiently ([59, 126]). From Theorem 7.3 we have that the upper bound is tight only in the case of three or less uncertainty blocks. However, extensive computational tests have shown that it does not exceed 15% in the case of pure dynamic uncertainty ([113, 219]).

7.2.2 Analysis and Design

Robust Performance Analysis The condition presented in Lemma 7.1 generalizes the robust stability theorems in Chapters 2 and 4, which apply only to the global dynamic uncertainty case, that is, $m = 1$ in the definition of Δ_d. In addition ([102]), a necessary and sufficient condition for robust performance can also be obtained. In fact, we will show next that robust performance for a family of models with a structured dynamic uncertainty set Δ_d can also be interpreted as robust stability of a different uncertainty set (see Chapter 2).

Based on the performance definition of Chapters 2 and 4, and using the condition for nominal performance based on the induced norm, the following conditions can be stated for the system in Figure 7.3:

$$\text{Robust performance} \iff \|z\|_2 \leq 1 \quad \forall \|w\|_2 < 1, \quad \forall \Delta \in \Delta_d \quad (7.43)$$

$$\iff \|F_u[T(s), \Delta]\|_\infty \leq 1, \quad \forall \Delta \in \Delta_d \quad (7.44)$$

From the robust stability condition obtained in Chapter 4 for plants described by a set of models with *global* dynamic uncertainty, it follows that equation (7.44) is necessary and sufficient for the system $\{F_u[T(s), \Delta], \Delta \in \Delta_d\}$ to be robustly stable against $\{\tilde{\Delta} \in \mathbb{C}^{r \times r}, \overline{\sigma}(\tilde{\Delta}) < 1\}$, connected between input w and output z (see Figure 7.4).

Due to the fact that an interconnection of LFTs remains a LFT, Figure 7.4 can be transformed to Figure 7.5. Hence the new uncertainty structure can be redefined as follows:

$$\tilde{\boldsymbol{\Delta}} \triangleq \left\{ \begin{bmatrix} \Delta & 0 \\ 0 & \tilde{\Delta} \end{bmatrix} \begin{array}{c} \Delta \in \Delta_d \\ \tilde{\Delta} \in \mathbb{C}^{r \times r} \end{array} \right\} \quad (7.45)$$

The above uncertainty set is qualitatively the same as Δ_d, but with an additional uncertainty block, usually called the *performance block*. However, the analysis tool remains the same, the structured singular value, but now with a slightly different uncertainty structure. Therefore, due to the equivalence between robust performance and structured robust stability depicted in

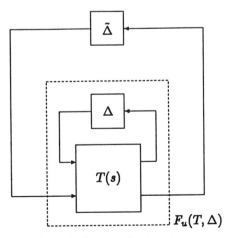

Figure 7.4. Equivalence between robust performance and stability.

STRUCTURED DYNAMIC UNCERTAINTY   221

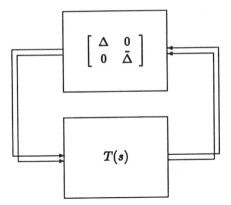

Figure 7.5. Robust performance analysis.

Figures 7.3, 7.4, and 7.5, equivalent conditions for the former can be established as follows.

Theorem 7.4 *The set of models $T(s, \Delta)$ of Figure 7.4 has robust performance if and only if*

$$\mu_{\tilde{\Delta}}[T(j\omega)] \leq 1, \ \forall \omega \in \mathbb{R} \tag{7.46}$$

There should be no confusion in the notation concerning the uncertainty structure and the system to which μ is applied. Recall that in order to simplify notation, we have made $M(s) = T_{11}(s)$; therefore it is clear that $\mu_{\Delta_d}[M(j\omega)] \neq \mu_{\tilde{\Delta}}[T(j\omega)]$. In addition, $\mu_{\Delta_d}[T(j\omega)]$ or $\mu_{\tilde{\Delta}}[M(j\omega)]$ cannot be computed due to a dimensional incompatibility between the system and the uncertainty structure.

In Chapter 4 a sufficient condition for robust performance of sets of models with global dynamic uncertainty was obtained. Since (7.46) is necessary and sufficient, the following holds in the two-block case ($m = 2$ when defining $\boldsymbol{\Delta}_d$):

$$\mu_{\tilde{\Delta}}[T(j\omega)] \leq \overline{\sigma}[T_{11}(j\omega)] + \overline{\sigma}[T_{22}(j\omega)] \tag{7.47}$$

when $T_{11}(s)$ and $T_{22}(s)$ represent the robust stability and nominal performance transfer matrices, respectively. Furthermore, in the SISO case (Chapter 2) the above inequality turns into an equality (Problem 4).

Synthesis   When disturbances and errors are measured by energy content ($\|\cdot\|_2$) and uncertainty blocks by the equivalent induced norm ($\|\cdot\|_\infty$) for $\Delta \in \mathcal{RH}_\infty$ and $\overline{\sigma}(\cdot)$ for $\Delta \in \mathbb{C}^{r \times r}$), two design procedures have been presented so far:

$$\inf_{\substack{\text{stabilizing} \\ K(s)}} \|F_\ell[P(s), K(s)]\|_\alpha, \quad \alpha \in \{2, \infty\} \tag{7.48}$$

The case $\alpha = 2$ is the LQG or $\mathcal{H}_2$ optimal control problem addressed in Chapter 5. The case $\alpha = \infty$ leads to $\mathcal{H}_\infty$ optimal control covered in Chapter 6. It guarantees nominal internal stability and solves in an optimal way either robust stability for global dynamic uncertainty or nominal performance. The latter considers the more general case of norm-bounded disturbances. It can also solve, although not optimally, the robust performance problem known as *mixed sensitivity*.

A more general procedure that seeks a controller that solves the optimal robust performance problem, in the case of structured dynamic uncertainty and norm-bounded disturbances, is known as μ-synthesis. It can be stated as the following optimization problem:

$$\inf_{\substack{\text{stabilizing} \\ K(s)}} \sup_{s=j\omega} \mu_{\Delta_d} \{F_\ell [P(s), K(s)]\} \tag{7.49}$$

Due to the fact that the structured singular value is computed via its bounds, the following suboptimal (in the case of $m \leq 3$ it is optimal) problem should be solved:

$$\inf_{\substack{\text{stabilizing} \\ K(s)}} \inf_{D(s) \in \mathcal{D}} \|D(s) F_\ell [P(s), K(s)] D(s)^{-1}\|_\infty \tag{7.50}$$

A suboptimal solution to the latter problem can be obtained through the $D - K$ *iteration* algorithm. This procedure "freezes" one optimization variable at a time, say, $K_i(s)$, and alternatively solves the other (convex) optimization problem, say, $D_{i+1}(s)$.

Several remarks are in order:

- The bound obtained in equation (7.42) holds for each $s = j\omega$, therefore in that case $D \in \mathcal{D}$ is a constant matrix. On the other hand, the scaling matrices $D(s)$ in equation (7.50) are functions of s; hence they should be obtained at each frequency and fitted by a rational matrix function, for example, spline interpolation.
- For a fixed $D(s)$ matrix, equation (7.50) is a weighted $\mathcal{H}_\infty$ problem that is convex and has a well known set of solutions. In particular, for the case when $D(s)$ equals the identity matrix, it is the $\mathcal{H}_\infty$ optimal control solution to the problem. In this sense the $D - K$ iteration can be interpreted as an algorithm that solves scaled $\mathcal{H}_\infty$ problems at each iteration.
- The order of the controller depends on the order of the augmented system, which in this case includes the weights $D(s)$. Therefore the order of the matrix function $D(s)$ that fits the optimal scales at each frequency should be made as low as possible. Otherwise, at each step, the order of the controller will increase significantly.
- For a fixed controller $K(s)$, the optimization has been shown to be convex; therefore a global minimum $D(j\omega)$ can be found efficiently at each frequency point.

However, it is well known that the combination of two convex problems is not necessarily convex. As a consequence, if a fixed controller $K_0(s)$ has been obtained and if by solving the optimal D-scale problem and interpolating it we obtain $D_1(s)$, and so on, there is no guarantee that the optimals $K_{opt}(s)$ and $D_{opt}(s)$ will be reached. A way to partially solve this is by checking both upper and lower bounds on μ.

In practice, many challenging application problems have been solved efficiently using commercial software products ([20, 126, 262]), which implement this procedure, although research in this area is still active. Some examples will be presented in the last chapter of the book.

7.3 PARAMETRIC UNCERTAINTY

7.3.1 Introduction

System models with parametric uncertainty originate either from physical laws or are the outcome of parametric identification algorithms. In the former case, the parameters may have a physical interpretation; in the latter they usually do not. In general, parameter identification algorithms fit a given mathematical structure that represents a certain physical phenomenon to a set of experimental data from this same phenomenon. This is accomplished through an optimization procedure with the parameters as its variables. The classical stochastic approach has been studied extensively ([13, 184, 293]), particularly in the area of adaptive control. More recently, set membership theory has been applied to model identification ([155, 204]). In both approaches, not only is a set of parameters obtained, but also error bounds for each parameter. These bounds can be described stochastically (*soft* bounds) or deterministically (*hard* bounds).

In any case, in this section we assume a given mathematical FDLTI structure in the frequency domain with unknown fixed parameters lying within real intervals.

$$\begin{cases} G(s,p), & p = [\,p_1 \; \cdots \; p_m\,]^T \\ p_i \in [a_i, b_i], & i = 1, \ldots, m \end{cases} \quad (7.51)$$

This is consistent with the deterministic approach pursued in robust control.

Once the above given family of models is assumed, the next step is to analyze the robust stability of this set for a given controller. A natural way to do this is by evaluating the set of roots of the characteristic polynomial of such a family, as mentioned at the beginning of this chapter:

$$P(s,p) = s^n + c_{n-1}(p)s^{n-1} + \cdots + c_1(p)s + c_0(p), \quad p \in [a,b] \quad (7.52)$$

According to the boundary crossing theorem of Section 7.1.1, under certain conditions robust stability can be verified by looking only at the $j\omega$ axis. To this end, we assume that the controller internally stabilizes at least a model with a (nominal) set of parameters, say, $p = p^0 \in [a, b]$. In addition, the functions $c_i(p)$, $i = 1, \ldots, m$, are continuous and do not vanish for $p \in [a, b]$. This accommodates many practical applications of interest. In this case, robust stability can be tested using the stability margins defined in Section 7.1.1, computed at each $s = j\omega$.

In particular, this structure includes many cases that may be recast in a LFT format as in Figure 7.2, with $\Delta \in \Delta_p$ defined previously. These are cases in which the characteristic polynomial coefficients $c_i(p)$ depend on the parameter vector p polynomially, for all $i = 1, \ldots, m$. To this end a shifting and scaling of the uncertain parameters should be made:

$$p_i = p_i^0 + h_i \cdot \delta_i \tag{7.53}$$

$$p_i^0 = \tfrac{1}{2}(a_i + b_i) \tag{7.54}$$

$$h_i = \tfrac{1}{2}(b_i - a_i) \tag{7.55}$$

Note that $p_i \in [a_i, b_i]$ corresponds to $\delta_i \in [-1, 1]$, and the nominal set of parameters to $\delta_i = 0$, $i = 1, \ldots, m$. The values of p_i^0 and h_i can be incorporated into the nominal model $\tilde{G}(s)$. This is illustrated by the following example.

Example 7.5 *Consider the following set of models:*

$$\mathcal{G}(s, z, \alpha) = \frac{s + z}{(s + \alpha)(s + \beta)}, \quad z \in [z_1, z_2], \quad \alpha \in [\alpha_1, \alpha_2] \tag{7.56}$$

which has an uncertain real pole and zero. This uncertainty structure can be transformed to fit the LFT framework as follows:

$$\tilde{\mathcal{G}}(s, \Delta) = \frac{(s + p_1^0) + h_1 \delta_1}{(s + \beta)(s + p_2^0)} \cdot \frac{1}{\left[1 + \delta_2 \left(\dfrac{h_2}{s + p_2^0}\right)\right]}, \quad \Delta = \begin{bmatrix} \delta_1 & 0 \\ 0 & \delta_2 \end{bmatrix}$$

$$\delta_i \in [-1, 1], \ i = 1, 2$$

$$p_1^0 = \tfrac{1}{2}(z_1 + z_2), \quad p_2^0 = \tfrac{1}{2}(\alpha_1 + \alpha_2)$$

$$h_1 = \tfrac{1}{2}(z_2 - z_1), \quad h_2 = \tfrac{1}{2}(\alpha_2 - \alpha_1)$$

The first term represents additive uncertainty and the second one quotient type uncertainty (see Chapter 2). Note that in this case the weights are

$$W_1(s) = \frac{h_1}{(s + \beta)(s + p_2^0)}, \quad W_2(s) = \frac{-h_2}{s + p_2^0} \tag{7.57}$$

Defining the inputs and outputs (u_i, y_i), $i = 1, 2$, as the outputs of the uncertainties and weights as in Figure 7.1, we obtain the following for the system of Figure 7.2:

$$\tilde{G}(s) = \begin{bmatrix} 0 & 0 & \dfrac{h_1}{(s+\beta)(s+p_2^0)} \\ \dfrac{-h_2}{s+p_2^0} & \dfrac{-h_2}{s+p_2^0} & \dfrac{-h_2(s+p_1^0)}{(s+\beta)(s+p_2^0)^2} \\ 1 & 1 & \dfrac{s+p_1^0}{(s+\beta)(s+p_2^0)} \end{bmatrix} \quad (7.58)$$

with (u, y) the input and output to the system. The uncertainty structure is the set Δ_p as defined previously, now with $m = 2$ parameters (δ_1, δ_2). Therefore $\tilde{\mathcal{G}}(s, \Delta) = F_u[\tilde{G}(s), \Delta]$.

7.3.2 Research Directions

The objective of this subsection is to provide a rough general idea of the main approaches pursued in this area of robust control. Due to the fact that this is still a very active research subject, the rest of this chapter will present only the basic results, from which many analytical and algorithmical tools for robustness analysis have been derived. A complete and very up-to-date treatment of this prolific area of research can be found in the books by Ackermann ([5]), Barmish ([35]) and Bhattacharyya and co–authors ([47]).

Most of the research effort related to the robustness analysis of systems with parameter uncertainty has taken place since the beginning of the 1980s. However, this problem had already been considered before by several researchers: Neimark ([213]), Horowitz ([152]), Siljak ([285]), Zadeh and Desoer ([328]), and more recently Ackermann ([4]). Horowitz templates gave origin to the *value set* approach ([35]), which uses the complex map of the characteristic polynomial at each $s = j\omega$ to evaluate robustness against parametric uncertainty. The book by Zadeh and Desoer introduced the *mapping theorem*, which has been used extensively in computational solutions to the analysis problem.

The more recent history of this area is closely related with the coefficient functions $c_i(p)$ of the closed-loop characteristic polynomial $P(s, p)$

$$P(s, p) = s^n + c_{n-1}(p)s^{n-1} + \cdots + c_1(p)s + c_0(p), \quad p \in [a, b]$$

As mentioned before, when considering only parametric uncertainty, this is a general framework that includes the LFT formulation as a special case. The complexity of the aforementioned functions of the uncertain parameters $c_i(p)$ determines the computational complexity of the solution. In addition, the type of functions considered leads to two clearly different research

Table 7.1. Coefficient functions, LFT structure, value sets and analysis results

CLCP Framework $c_i(p)$	LFT Framework Δ_p Structure	Value Set	Result
$c_i = p_i$	$M(s)$ rank 1	Rectangle	Kharitonov
Affine	$M(s)$ rank 1	Polytope	Edge theorem
Multilinear	Independent δ_i	Nonconvex	Analytical,
	$M(s)$ general	$P(\omega, \mathcal{H}) \subset \text{co}[P(\omega, \mathcal{V})]$	computational
Polynomial	Repeated δ_i	Nonconvex	Computational
	$M(s)$ general	$P(\omega, \mathcal{H}) \not\subset \text{co}[P(\omega, \mathcal{V})]$	

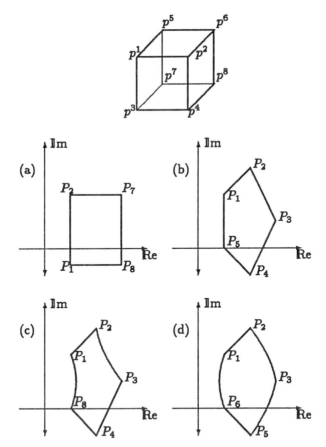

Figure 7.6. (a) Interval, (b) affine, (c) multilinear, and (d) polynomial. Here $P_i = P(j\omega, p^i)$ with $p^i \in \mathbb{R}^m$.

approaches. This is illustrated in Table 7.1, which classifies the different tools according to these functions (see also Figure 7.6).

The third row of this table, that is, the case of multilinear dependence of the coefficients with the parameters, establishes a boundary between two different research approaches. The first one seeks a small number of distinguished models or test set of models ([35]), which may solve the analysis problem, hopefully with a small amount of computations. These are known as the *Kharitonov* type results, which start from the independent coefficient case and extend to particular multilinear dependence in the unknown parameters. The second approach starts directly from the general multilinear dependence case and generalizes to polynomial functions $c_i(p)$. Many of these methods are based on the mapping theorem and have a clear computational basis. As mentioned before, the general parametric analysis is NP-hard; therefore the procedures that are able to compute the stability margin exactly (or with guaranteed bounds) have exponential time complexity. Next, we comment in some detail on both these approaches.

Kharitonov-like Results The simplest case is when the parameters are the coefficients of the polynomial themselves, leading to the so-called *interval polynomials*. This also includes situations where the coefficients are functions of independent sets of parameters, since this essentially reduces to the case $c_i = p_i$. As will be shown in the sequel, in this case the image of $P(s, p)$ for all $p \in [a, b]$, defined as its *value set*, is a rectangle in $\mathbb{C}$ (see Figure 7.6a). The four vertices of this rectangle correspond to the only distinguished models that are necessary (and sufficient) to establish robust stability. This goes back to the 1978 result of Kharitonov ([172]), which was introduced to the control community in 1983 ([27, 48]). In addition, the graphical result in [307] allows the computation of the stability margin for this class of uncertainty (see also [122]).

The first relevant generalization of the previous result considers the case where the coefficients are affine functions of the parameters. Therefore these coefficients are no longer independent, as in the previous case. The CLCP can be transformed to

$$P(s,p) = \sum_{i=0}^{n} c_i(p)s^i = v_0(s) + \sum_{i=1}^{m} p_i v_i(s) \qquad (7.59)$$

where $v_i(s)$ are known polynomials. The image in $\mathbb{C}$ at each ω is a convex combination of certain vertex polynomials $\{P_1(j\omega), \ldots, P_\ell(j\omega)\}$, $\ell \leq 2^m$. This is a subset of the polynomials that correspond to the 2^m vectors $p \in [a, b]$ of the extreme values of all intervals. Hence the value set is a polytope in the complex plane (see Figure 7.6b). The first solution to this problem appeared in 1988 ([36]) and is known as the *edge theorem*. The test for robust stability consists of a search over the CLCPs corresponding to all the exposed edges of the value set. If these are stable, the whole set is stable. Further simplifications of this result can be found in [31, 283] and [68]. The latter considers the characteristic polynomial of an *interval plant*, defined as a transfer function that has interval polynomials as its numerator and denominator, the set of parameters of each of them being independent. From the LFT perspective, it has been shown in [73] that both the interval and affine parameter dependence of the characteristic polynomials are special cases of $\det[I - M(s)\Delta]$, $\Delta \in \Delta_p$, when $M(s)$ is a rank 1 matrix, as noted in Table 7.1.

The next natural step is to consider multilinear coefficient functions. The general multilinear case can be analyzed at the expense of an exponential growth in computations with the number of parameters. It is also well known ([35]) that there is no "edge" result in this case, that is, verifying the stability of the CLCPs of the boundary of the value set is not enough for the stability of the whole set. The worst parameter vector, in terms of stability, may be in the interior of the parameter set. In fact, interval matrices, realizable transfer functions, and random CLCPs may produce such examples ([35, 265]).

However, there are results for specific multilinear structures such as the ones in [33, 71], which consider the CLCP corresponding to the series connections of interval plants. A particular case of this structure would have the following characteristic polynomial:

$$P(s, p^1, q^1, p^2, q^2) = N_1(s, p^1) \cdot N_2(s, p^2) + D_1(s, q^1) \cdot D_2(s, q^2)$$

produced by the series connection of two plants of the form

$$g_i(s, p^i, q^i) = \frac{N(s, p^i)}{D(s, q^i)} = \frac{s^m + p^i_{m-1} s^{m-1} + \cdots + p^i_1 s + p^i_0}{s^n + q^i_{n-1} s^{n-1} + \cdots + q^i_1 s + q^i_0}$$

$$p^i \in \mathbb{R}^m, \quad q^i \in \mathbb{R}^n, \quad p^i_j \in \left[a^i_j, b^i_j\right], \quad q_k \in \left[c^i_k, d^i_k\right]$$

$$i = 1, 2, \quad j = 0, \ldots, m-1, \quad k = 0, \ldots, n-1$$

The results in [88, 111, 112, 270] provide polynomial time computation for systems with products and ratios of affine uncertainty structures, as follows:

$$g(s, p) = \frac{\prod_{i=1}^{\ell} f_i(s, p)}{\prod_{j=\ell+1}^{n} f_j(s, p)}, \quad p \in \mathbb{R}^m, \quad p_j \in [a_j, b_j], \quad j = 1, \ldots, m \quad (7.60)$$

where the $f_i(s, p)$, $i = 1, \ldots, n$, represent polytopes of polynomials. This includes the case of products of uncertain poles and zeros, for example. The general case of multilinear dependence with the parameters may have a nonconvex value set in $\mathbb{C}$. Nevertheless, it can be proved that this value set (denoted as $P(\omega, \mathcal{H})$ in Table 7.1) is included in the convex polytope constructed with the image of the 2^m vertices of the uncertainty set $\{P(s, p), p_i = a_i \text{ or } b_i, i = 1, \ldots, m\}$ (defined as the *convex hull* and denoted as co$[P(\omega, \mathcal{V})]$ in Table 7.1, see also Figure 7.6c). This will be explained in further detail when introducing the mapping theorem. Additional results in this area include conditions under which the value set is a polytope, as if the uncertain coefficients were affine functions ([11, 150, 241]). This covers part of the third row of Table 7.1.

When the CLCP has a polynomial dependence on the parameters, its value set cannot be included in the convex hull of the vertex polynomials, as in the previous case (see Figure 7.6d). The model in (7.60) with numerator and denominator raised to given powers can be included as a special case of this uncertainty structure. A general setup that allows the analysis of complicated CLCPs in a polynomial computation time is the *tree structured decomposition* ([30, 112]).

Computational Results The computational approach is mainly based on the mapping theorem, which will be proved further along this section. A first step in this research direction was proposed in [254] based on the aforementioned theorem. It was followed by the algorithm in [92], which computes exactly the stability margin for systems in a LFT framework with multilinear parameter dependency. Note that the $c_i(\Delta)$ are multilinear functions when the characteristic polynomial is det $[I - M(s)\Delta]$, $\Delta \in \Delta_p$ and the parameters in Δ are independent.

A more general uncertainty structure is the case of repeated parameters in $\Delta \in \Delta_p$. This corresponds to CLCPs that have a polynomial dependence with the uncertain parameters. This fact suggests a simplification of polynomial dependence to multilinear dependence by adding equality constraints to the repeated parameters. Therefore, essentially the same procedures extend to the polynomial case ([266, 280]).

All of the above algorithms are based on a *branch and bound* method over the two-dimensional value sets in the complex plane for each frequency ω. A further simplification was introduced in [281], which applies the Routh–Hurwitz procedure to eliminate the frequency dependency, therefore eliminating the ω-sweep for the computation of the stability margin. Furthermore, the value sets are reduced to one-dimensional intervals, which significantly simplifies the computations. Many other branch and bound procedures compute exactly the stability margin in these general cases ([17, 112, 250, 312]). However, the exact computation using these algorithms can be performed only for a small number of parameters, due to the combinatorial explosion with the number of parameters. There is yet another algorithmic approach that seeks an approximate tractable computation of the stability margin in the framework of the structured singular value ([326, 327]). The latter will be presented in the context of mixed parametric/dynamic uncertainty at the end of this chapter.

The above description of this area of research is by no means complete and has only an introductory objective. A complete and up-to-date treatment of this subject can be found in [5, 35, 47]. In the rest of this section we present only the basic results, which initiated both of the approaches mentioned before: Kharitonov and the mapping theorems.

7.3.3 Kharitonov's Theorem

This result is the starting point of a prolific research area, which seeks the stability analysis of systems with parameter uncertainty based on the evaluation of a finite (hopefully small) number of distinguished models in the class.

The original result was presented in [172] and introduced in [27, 48] to the area of control systems. It considers the simplest case of uncertain characteristic polynomials, that is, *interval* polynomials. Although this is seldom the case in practice, it generated a fruitful research area with generalizations of this theorem, which could be applied to practical cases.

Consider the following family of closed-loop characteristic polynomials:

$$P(s,c) = s^n + c_{n-1}s^{n-1} + \cdots + c_1 s + c_0 \tag{7.61}$$

$$c = \begin{bmatrix} c_0 \\ \vdots \\ c_{n-1} \end{bmatrix} \in \mathcal{H}$$

where $\mathcal{H}$ is an n-dimensional rectangle, as follows:

$$\mathcal{H} \triangleq \{x \in \mathbb{R}^n \mid x_i \in [a_i, b_i], \quad 0 \notin [a_i, b_i], \quad i = 0, \ldots, n-1\} \tag{7.62}$$

The following notation will be used in the sequel to represent certain polynomials in the previous set, by means of their coefficients;

$$E_1 \triangleq \begin{bmatrix} b_0 & a_2 & b_4 & a_6 & \cdots & b_n \end{bmatrix}^T \tag{7.63}$$

$$E_2 \triangleq \begin{bmatrix} a_0 & b_2 & a_4 & b_6 & \cdots & a_n \end{bmatrix}^T \tag{7.64}$$

$$O_1 \triangleq \begin{bmatrix} b_1 & a_3 & b_5 & a_7 & \cdots & a_{n-1} \end{bmatrix}^T \tag{7.65}$$

$$O_2 \triangleq \begin{bmatrix} a_1 & b_3 & a_5 & b_7 & \cdots & b_{n-1} \end{bmatrix}^T \tag{7.66}$$

$$E^\star \triangleq \begin{bmatrix} c_0^\star & c_2^\star & c_4^\star & c_6^\star & \cdots & c_n^\star \end{bmatrix}^T \tag{7.67}$$

$$O^\star \triangleq \begin{bmatrix} c_1^\star & c_3^\star & c_5^\star & c_7^\star & \cdots & c_{n-1}^\star \end{bmatrix}^T \tag{7.68}$$

where we have assumed without loss of generality that n is a multiple of 4. For example, a polynomial can be denoted as $P(s, E_1, O^\star)$.

Lemma 7.3 *The image in $\mathbb{C}$ of the polynomial $P(\jmath\omega, c)$ for a given frequency ω and all values $c \in \mathcal{H}$ is a rectangle with vertices $P(\jmath\omega, E_1, O_1)$, $P(\jmath\omega, E_1, O_2)$, $P(\jmath\omega, E_2, O_1)$, and $P(\jmath\omega, E_2, O_2)$.*

Proof. The set of characteristic polynomials (7.61) can be decomposed as follows:

$$P(\jmath\omega, c) = \operatorname{Re}[P(\jmath\omega, c)] + \jmath\operatorname{Im}[P(\jmath\omega, c)] \qquad (7.69)$$

$$\operatorname{Re}[P(\jmath\omega, c)] = c_0 - c_2\omega^2 + c_4\omega^4 - c_6\omega^6 + \cdots \qquad (7.70)$$

$$\operatorname{Im}[P(\jmath\omega, c)] = c_1\omega - c_3\omega^3 + c_5\omega^5 - c_7\omega^7 + \cdots \qquad (7.71)$$

Hence the real and imaginary parts of $P(\jmath\omega, \cdot)$ are functions of independent subsets of parameters, and therefore their maximum and minimum values are independent as well. Due to the linearity of $P(\jmath\omega, \cdot)$ in the parameters,

$$\operatorname{Re}[P(\jmath\omega, E_2, O)] \leq \operatorname{Re}[P(\jmath\omega, c)] \leq \operatorname{Re}[P(\jmath\omega, E_1, O)]$$

$$\operatorname{Im}[P(\jmath\omega, E, O_2)] \leq \operatorname{Im}[P(\jmath\omega, c)] \leq \operatorname{Im}[P(\jmath\omega, E, O_1)]$$

As a consequence, the four extreme vertices of the image are

$$P(\jmath\omega, E_1, O_1), \; P(\jmath\omega, E_1, O_2), \; P(\jmath\omega, E_2, O_1), \; P(\jmath\omega, E_2, O_2) \qquad (7.72)$$

□

A basic element in the proof of Kharitonov's theorem is the *interlacing* property derived from the Hermite–Biehler theorem. This is a property of *Hurwitz* polynomials, defined as the ones with all their roots in $\mathbb{C}_-$. To this end note that any polynomial with real coefficients can be put in the form $P(s) = f(s^2) + sg(s^2)$, where f includes all even coefficients and g all odd coefficients of $P(s, c)$. In addition, a necessary condition for a polynomial to be Hurwitz is to have all of its coefficients non-zero and with the same sign. This will be assumed in the sequel.

Theorem 7.5 (Hermite–Biehler) *A polynomial $P(s) = f(s^2) + sg(s^2)$ with real coefficients is Hurwitz if and only if the roots $\{x_1^f, \ldots, x_m^f\}$ of $f(x)$ and $\{x_1^g, \ldots, x_m^g\}$ of $g(x)$ are all real and negative and satisfy the interlacing property*[4]:

$$x_1^g < x_1^f < x_2^g < x_2^f < \cdots < x_m^f < 0$$

Proof. See [129] for a detailed proof. □

When the functions f and g have the above property they are called a *positive pair* ([129]).

---
[4] According to the order of $P(s)$, the degree of $g(s)$ may be $(m-1)$.

232 STRUCTURED UNCERTAINTY

Note that the functions $f(s^2)$ and $sg(s^2)$ at $s = \jmath\omega$ correspond to the real and imaginary parts of $P(\jmath\omega, c)$. Furthermore, a negative root of $f(x)$ corresponds to an imaginary root of $f(s^2)$, and similarly for g. Next, we present the main result of this subsection. The proof is simpler than the original one in [172] and can be found in [322]. Alternative proofs have been presented in [54, 69, 207].

Theorem 7.6 (Kharitonov) *The set of interval polynomials in (7.61) is Hurwitz if and only if the following four elements of the set are Hurwitz:*

$$P_1(s) \triangleq P(s, E_1, O_1) \qquad (7.73)$$

$$P_2(s) \triangleq P(s, E_1, O_2) \qquad (7.74)$$

$$P_3(s) \triangleq P(s, E_2, O_1) \qquad (7.75)$$

$$P_4(s) \triangleq P(s, E_2, O_2) \qquad (7.76)$$

Proof. Necessity is trivial, due to the fact that all four polynomials belong to set (7.61).

Sufficiency is based on the following argument. If $P(\jmath\omega, E^\star, O_1)$ and $P(\jmath\omega, E^\star, O_2)$ are Hurwitz for a certain vector of even parameters $E^\star$, then any element of the set $P(\jmath\omega, E^\star, O)$ is Hurwitz, where O represents the set of all odd coefficients in $\mathcal{H}$. This holds because

$$P(\jmath\omega, E^\star, O_1) = f(-\omega^2, E_\star) + \jmath\omega g(-\omega^2, O_1) \qquad (7.77)$$

$$P(\jmath\omega, E^\star, O_2) = f(-\omega^2, E_\star) + \jmath\omega g(-\omega^2, O_2) \qquad (7.78)$$

being Hurwitz imply that f and g are positive pairs. From Theorem 7.5, $P(\jmath\omega, E^\star, O_1)$ and $P(\jmath\omega, E^\star, O_2)$ alternatively intersect the real and imaginary axes m times, the latter being the degree of f and g.[5] Because g depends only on the odd coefficients, which affect only the imaginary part of set $P(\jmath\omega, E^\star, O)$, we have

$$\mathrm{Im}\,[P(\jmath\omega, E^\star, O_2)] \le \mathrm{Im}\,[P(\jmath\omega, E^\star, O)] \le \mathrm{Im}\,[P(\jmath\omega, E^\star, O_1)]$$

By continuity, all members of this set will intersect alternatively the real and imaginary axes m times. Hence, from Theorem 7.5, all of the set $P(\jmath\omega, E^\star, O)$ is Hurwitz.

With a similar argument it can be proved that if $P(\jmath\omega, E_1, O^\star)$ and $P(\jmath\omega, E_2, O^\star)$ are Hurwitz, all elements in $P(\jmath\omega, E, O^\star)$ are Hurwitz as well. Here E represents the set of all even coefficients in $\mathcal{H}$.

---

[5] Or m and $(m-1)$ in case $g(\cdot)$ has the least degree.

Combining the above arguments, we have

$$\left.\begin{array}{l}P(\jmath\omega, E_1, O_1)\\ P(\jmath\omega, E_1, O_2)\end{array}\right\} \text{Hurwitz} \Rightarrow P(\jmath\omega, E_1, O) \text{ Hurwitz} \quad (7.79)$$

$$\left.\begin{array}{l}P(\jmath\omega, E_2, O_1)\\ P(\jmath\omega, E_2, O_2)\end{array}\right\} \text{Hurwitz} \Rightarrow P(\jmath\omega, E_2, O) \text{ Hurwitz} \quad (7.80)$$

and finally,

$$\left.\begin{array}{l}P(\jmath\omega, E_1, O)\\ P(\jmath\omega, E_2, O)\end{array}\right\} \text{Hurwitz} \Rightarrow P(\jmath\omega, E, O) \text{ Hurwitz} \quad (7.81)$$

equivalent to set (7.61) being Hurwitz. □

7.3.4 Mapping Theorem

The computational approach to the analysis problem can be stated in terms of the value set of the CLCP, represented by $P(s, p)$, $p \in [a, b]$, in the general case, or by $\det[I - M(s)\Delta]$, $\Delta \in \Delta_p$ in a LFT framework. In any case, the objective is to compute the value set in $\mathbb{C}$ (or rather its boundary) to check whether or not it includes the origin $z = 0$. In such a case $P(\jmath\omega^z, p^z) = 0$ for a certain combination of parameters $p^z \in \mathcal{H}$, which means that the system has an unstable pole at $s = \jmath\omega^z$. As noted in Figure 7.6, except for the interval and affine cases, the boundaries of these value sets cannot be defined with a finite number of points, that is, the vertices. Nevertheless, in the case where $P(s, p)$ is multilinear in p, the mapping theorem provides an important simplification, which allows the bounding of the value set.

To this end, consider the set of CLCPs $\{P(s, p), p \in [a, b]\}$ and define

$$\mathcal{H} \triangleq \{p \in \mathbb{R}^m, \quad p_i \in [a_i, b_i], \quad i = 1, \ldots, m\} \quad (7.82)$$

$$\mathcal{V} \triangleq \{p_i = a_i \quad \text{or} \quad p_i = b_i, \quad i = 1, \ldots, m\} \quad (7.83)$$

as the parameter domain $\mathcal{H}$ and the set of its 2^m vertices, respectively. Note that here $\mathcal{H}$ is an m-dimensional rectangle that contains all possible uncertain parameters p, and should not be confused with the domain of coefficients defined in the previous section.

We are particularly interested in the image of all vertices in $\mathcal{V}$ through the function $P(\jmath\omega, \cdot)$, which we denote as $P(\jmath\omega, \mathcal{V})$. In this case, the convex hull $\text{co}[P(\jmath\omega, \mathcal{V})]$[6] is defined by r vertices with $r \leq 2^m$, as seen in Figure 7.6b. In the case of any multilinear complex function, the following result holds ([328]).

---
[6] The convex hull, $\text{co}(\mathcal{S})$ of a set $\mathcal{S}$ is the smallest convex set that contains $\mathcal{S}$

Theorem 7.7 (Mapping Theorem) *The image of any multilinear complex function $\mathcal{F}: \mathcal{H} \to \mathbb{C}$ is contained in the convex hull of the image of the vertices of $\mathcal{H}$, that is, $\mathcal{F}(\mathcal{H}) \subset \mathrm{co}[\mathcal{F}(\mathcal{V})]$.*

Proof. First, we show that the map of $\mathcal{F}$ along a coordinate axis of set $\mathcal{H}$ is a line in $\mathbb{C}$. This is clear considering that a coordinate axis in $\mathcal{H}$ fixes all parameters but one, that is, $p_i = x_i$ for all $i = 1, \ldots, m, i \neq j$, and $p_j \in [a_j, b_j]$. Therefore by the multilinearity of $\mathcal{F}$, the image is affine in p_j.

Next, consider any point $p^\star \in \mathcal{H}$ and a line that contains it and is parallel to the first axis of $\mathcal{H}$. This line intersects the opposite $(n-1)$-dimensional "faces" of $\mathcal{H}$ at $p^{b\star}$ and $p^{a\star}$, where

$$p^{b\star} = [\, b_1 \ \ p_2^\star \ \ \cdots \ \ p_m^\star \,]^T \tag{7.84}$$

$$p^{a\star} = [\, a_1 \ \ p_2^\star \ \ \cdots \ \ p_m^\star \,]^T \tag{7.85}$$

The new points $p^{b\star}$ and $p^{a\star}$ can be considered as generic points of the $(n-1)$-dimensional faces of $\mathcal{H}$ mentioned before. The same procedure can be applied to each of them, to generate four new points, each in a different $(n-2)$-dimensional face of $\mathcal{H}$. These are

$$p^{bb\star} = [\, b_1 \ \ b_2 \ \ p_3^\star \ \ \cdots \ \ p_m^\star \,]^T \tag{7.86}$$

$$p^{ba\star} = [\, b_1 \ \ a_2 \ \ p_3^\star \ \ \cdots \ \ p_m^\star \,]^T \tag{7.87}$$

$$p^{ab\star} = [\, a_1 \ \ b_2 \ \ p_3^\star \ \ \cdots \ \ p_m^\star \,]^T \tag{7.88}$$

$$p^{aa\star} = [\, a_1 \ \ a_2 \ \ p_3^\star \ \ \cdots \ \ p_m^\star \,]^T \tag{7.89}$$

This procedure can be applied m times and ends up with the set of 2^m vertices. Each of the segments $\overline{p^{b\star}p^{a\star}}$, $\overline{p^{bb\star}p^{ba\star}}$, $\overline{p^{ab\star}p^{aa\star}}$, and so on defines a linear image in $\mathbb{C}$ due to the fact that they are parallel to the coordinate axes of $\mathcal{H}$. In addition, the image $\mathcal{F}(p^\star)$ is contained in the segment $\mathcal{F}(\overline{p^{b\star}p^{a\star}})$. The extremes of the latter are contained in the segments $\mathcal{F}(\overline{p^{bb\star}p^{ba\star}})$ and $\mathcal{F}(\overline{p^{ab\star}p^{aa\star}})$, respectively, and so on. Therefore the image of the first point $\mathcal{F}(p^\star)$ (which is generic) is contained in the convex hull of the image of the vertices, that is, $\mathrm{co}[\mathcal{F}(\mathcal{V})]$. $\square$

In particular, this can be applied to the complex CLCP $P(\jmath\omega, p)$ over the set of parameters $p \in \mathcal{H}$.

The above theorem provides an important tool for assessing robustness of a parametric family of models. A computational procedure applied to the image of the vertices $P(\jmath\omega, \mathcal{V})$ determines, at each frequency ω, a sufficient condition for robust stability when this polytope does not contain the origin. This is due to the fact that $P(\jmath\omega, \mathcal{V})$ includes the actual value set. However, this only provides a qualitative answer: stability/instability.

If, instead, we seek a quantitative solution to the problem by means of the computation of the stability margin, the above theorem provides an upper bound in the case of μ_{Δ_p} or a lower bound in the case of k_m. To compute either one of these margins we should scale the parameter domain $\mathcal{H}$ by $k \in (0, \infty)$ until its image $P(j\omega, k\mathcal{H})$ intersects the origin. This scaling is exactly the stability margin (see Definitions 7.2 and 7.3). To this end, the following property is useful, the proof being left as an exercise (Problem 5).

Lemma 7.4 *The following holds when scaling a multilinear complex function $\mathcal{F} : \mathcal{H} \to \mathbb{C}$ by real numbers $k_2 > k_1 > 0$:*

$$\mathcal{F}(k_2 \cdot \mathcal{H}) \supset \mathcal{F}(k_1 \cdot \mathcal{H}) \tag{7.90}$$

$$\text{co}\left[\mathcal{F}(k_2 \cdot \mathcal{H})\right] \supset \text{co}\left[\mathcal{F}(k_1 \cdot \mathcal{H})\right] \tag{7.91}$$

There are still certain issues that may limit the above result. The first one is the fact that the mapping theorem applies only to multilinear functions. The second one is the fact that it is useful only to bound the value set at each frequency ω; therefore the procedure should be repeated over a grid of frequencies in $\omega \in [0, \infty)$.

Next, we present further simplifications to the above problems. The first one transforms a set of parameters that appear polynomially in $P(s, p)$ to a new set that appears multilinearly and has equality constraints. Therefore the above theorem can be applied. To this end, we consider for simplicity and without loss of generality the LFT framework. In the case of polynomial parametric dependence, this setup is equivalent to the CLCP provided that Δ_p is allowed to have repeated parameters. In the sequel we denote these Δs as Δ_{pr}. It is easy to establish the equivalence between the following uncertainty sets:

$$\Delta_{pr} \triangleq \begin{cases} \text{diag}(\overbrace{\delta_1, \ldots, \delta_1}^{n_1}, \ldots, \overbrace{\delta_i, \ldots, \delta_i}^{n_i}, \ldots, \overbrace{\delta_{m_r}, \ldots, \delta_{m_r}}^{n_{m_r}}) \\ \delta_i \in \mathbb{R}, \quad i = 1, \ldots, m_r \end{cases} \tag{7.92}$$

and Δ_p of dimension M_R, restricted by the following equality constraints;

$$\begin{cases} \delta_1 = \ldots = \delta_{n_1} \\ \vdots \\ \delta_{j+1} = \ldots = \delta_{j+n_i}, \quad j = \sum_{k=1}^{i-1} n_k \\ \vdots \\ \delta_{j+1} = \ldots = \delta_{j+n_{m_r}}, \quad j = \sum_{k=1}^{m_r-1} n_k \\ \delta_i \in \mathbb{R}, \quad i = 1, \ldots, M_R = \sum_{k=1}^{m_r} n_k \end{cases} \tag{7.93}$$

The (closed) ball $\overline{\mathcal{B}}\Delta_{pr}$ is defined for $|\delta_i| \leq 1$, $i = 1, \ldots, m_r$.

A second simplification consists in the elimination of the frequency search. This not only reduces the computational burden but also solves the problem of gridding the $j\omega$ axis, especially due to the fact that the stability margin for pure real parametric uncertainty can be discontinuous ([32, 221, 284]). One way to perform this simplification is by applying the Routh–Hurwitz criteria to evaluate the stability of the CLCP. In this way, n positivity conditions should be verified to guarantee stability, n being the order of the polynomial $P(s, p)$ in the variable s. In general, these inequalities have polynomial dependence with the unknown parameters but are real functions, independent from $j\omega$. Therefore the mapping theorem can be applied along with the previous simplification over the n positivity constraints in order to verify the intersection with the origin. This eliminates the frequency search and transforms the two-dimensional mapping of $P(s, p)$ to n one-dimensional ones, which simplifies considerably the software and reduces the computation time. In fact, the convex hull of each of these functions is a real interval, which coincides with the actual value set of the same function (Problem 6), that is, $\mathcal{F}(\mathcal{H}) \equiv \mathrm{co}[\mathcal{F}(\mathcal{V})]$ when $\mathcal{F} : \mathcal{H} \to \mathbb{R}$. The increase in the number of functions to be evaluated is not a problem, because it is $\mathcal{O}(n)$, where n is independent of the number of parameters; as a consequence, it does not add complexity to the algorithmic solution. However, in this transformation of the problem, a symbolic manipulation procedure should be used in the case of large number of parameters ([281]). Another efficient method to eliminate the ω-search, which directly deals with the state space realization of the plant in a LFT framework, can be found in [284].

Finally, applying the mapping theorem with these simplifications in mind, a program can be constructed to scale the parameter set $\mathcal{H}$ until its image $f(\mathcal{H})$[7] reaches the origin. The value of the scaling parameter at this point is the stability margin. To this end several branch and bound procedures have been proposed ([17, 92, 250, 266, 281, 312, 326]). These iteratively compute upper and lower bounds of the stability margin and, by different mechanisms, partition the parameter domain and compute bounds for the new subdomains. Finally, with the aid of certain criteria to eliminate some of these subdomains, the procedure continues until the bounds converge to a predetermined error. One of the important issues in the implementation of these programs is the increment in the number of regions after each partition, which depend on the criteria adopted to eliminate some of these subdomains ([265, 325, 326]). Another important point is the computation time of the bounds at each subdomain, which in many cases seems to dominate the total computational time, for reasons we will explain next. In addition, it is important also to determine the "offending" parameter vector: this is the one that maps the scaled set $k_m \mathcal{H}$ to the origin. Here k_m is the multivariable stability margin defined previously.

---

[7] This function could be either $P(\omega, \mathcal{H})$ at each frequency or the n real-valued Routh–Hurwitz functions.

In any case, the first step that maps the image of the vertices involves 2^m computations of a function $f(\mathcal{H})$, one for each vertex of set $\mathcal{H}$, m being the number of parameters. This is related to the problem being NP-hard. Therefore these algorithms will work reasonably well only for a small number of uncertain parameters, say, 10 to 20. Nevertheless, this may include many practical cases of interest, as, for example, the robustness analysis of aircrafts with uncertain aerodynamic coefficients. (See Chapter 11 and [265] for the analysis of the X-29 aircraft.)

With the approach considered in this section, the only way to achieve tractable computational time would be with the use of some extra *a priori* knowledge of the model structure. This *a priori* information can be used to eliminate a number of vertices of $\mathcal{H}$, in order to decrease the number of computations of $f(\mathcal{H})$ to a polynomial function of m. For example, with $m = 10$ parameters the computation of all vertices involves 1024 evaluations of $f(\cdot)$, and an $\mathcal{O}(m^3)$ algorithm would involve roughly the same number. Instead, for $m = 20$, the image of the vertices involves approximately 10^6 computations of $f(\cdot)$ and an $\mathcal{O}(m^3)$ algorithm only 8000 evaluations. This extra information should therefore reduce the number of vertices to be evaluated by two orders of magnitude in this last case.

Alternatively, only approximate algorithms with no guarantee of convergence to predefined error bounds can be constructed. These procedures could also use a branch and bound approach but with polynomial time computation at each subdomain. Therefore if the increase in subdomains remains polynomial, the complete computation is tractable, allowing for solving problems with fewer than 100 parameters. In this case the objective is not a precise answer, but a fast one. It seems that a good approximation (10–20% error) can be achieved in many practical problems. These algorithms will be presented in the next section in the context of mixed parametric/dynamic uncertainty structures.

7.4 MIXED TYPE UNCERTAINTY

7.4.1 Introduction

From the practical point of view, uncertainty is present in the plant in both the dynamics and the parameters. Usually the lower frequency part of the plant may have a well known model structure, with parameters computed using identification methods. The upper-frequency portion may be unknown but bounded. This is typical for large mechanical flexible structures ([121]), for which well known second-order models fit the lower-frequency portion and where high-frequency oscillations, which have less detailed information, may be covered with dynamic uncertainty. The parameters, natural frequencies, and damping coefficients are estimated within real bounds.

Furthermore, even for a very well known mathematical model (fixed

238 STRUCTURED UNCERTAINTY

order), which may not have dynamic uncertainty, it has been shown in a previous section that performance can be inserted as an extra dynamic uncertainty *performance* block (see Figure 7.4).

In addition, as mentioned in the last section, the pure parametric uncertainty case can be discontinuous in the problem data ([32, 284]). As shown in [221], a regularization of the parameter uncertainty, by adding some phase to each parameter, may prevent the stability margin from being discontinuous. In fact, it has been shown that adding at least one complex uncertainty block, for example the performance block, to the problem may eliminate the discontinuity as well. This is important when considering the gridding of the ω axis to compute the stability margin, either $\mu(j\omega)$ or $k_m(j\omega)$. Therefore from a practical point of view, it seems natural to attempt the analysis and design considering a mixed type uncertainty structure.

From the point of view of computational complexity, due to the fact that the computation of the stability margin for the pure parametric case is NP-hard, it seems reasonable to expect the same limitation in the mixed case. Nevertheless, the addition of complex dynamic uncertainty does not increment the complexity.

In the area of mixed parametric dynamic uncertainty several results have appeared since 1988; among others we can mention [28, 70, 149, 151, 282, 318] for the combination of parametric and global dynamic uncertainty and [115, 267, 325, 326, 327] for the general case. In can be shown that in the mixed case, the minimization of the upper bound of μ still leads to a convex problem ([267, 325]). Therefore the addition of parametric uncertainty does not destroy the convexity properties of the structured pure dynamic case.

In the LFT framework, one approach to deal with this type of uncertainty structure is by partitioning the uncertainty matrix into two parts as indicated in Figure 7.7. Therefore, for the system $M(s)$ to be robustly stable for the mixed type structure Δ_M, it is necessary and sufficient to have ([267]):

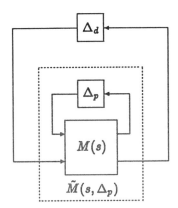

Figure 7.7. LFT approach for combined uncertainty.

1. $\tilde{M}(s,\Delta_p) \triangleq \Gamma_u[M(s),\Delta_p]$ stable for all $\Delta_p \in \mathcal{B}\Delta_p$.
2. $\mu_{\Delta_d}\left[\tilde{M}(\jmath\omega,\Delta_p)\right] \leq 1$ for all $\Delta_p \in \mathcal{B}\Delta_p$ and all ω.

Condition 1 is a necessary condition and establishes that the "nominal" model $\tilde{M}(s,\Delta_p)$ should be robustly stable, when only considering parametric uncertainty. Under certain conditions it can be verified by checking whether or not

$$\mu_{\Delta_p}[M_{11}(\jmath\omega)] \leq 1 \qquad (7.94)$$

for all ω. If this holds, then condition 2, which represents the stability margin with respect to the pure complex case, is necessary and sufficient. As before, the latter is checked via its bounds. Thus, a sufficient condition for robust stability in the mixed case is:

$$\inf_{D \in \mathcal{D}} \overline{\sigma}\left[D\tilde{M}(\jmath\omega,\Delta_p)D^{-1}\right] \leq 1, \quad \forall \omega \qquad (7.95)$$

for all $\Delta_p \in \mathcal{B}\Delta_p$. Here $\mathcal{D}$ is the set of scaling matrices introduced in Section 7.2.1. Equation (7.95) is also a margin computation with respect to pure parametric uncertainty. In this way, we can transform the problem of mixed uncertainty into two stability margin computations with pure parametric uncertainty. This also supports the fact that the mixed case is NP-hard, due exclusively to the pure parametric case. As a consequence, the same branch and bound algorithms constructed for the parametric case ([267]) could be used in the mixed case as well, bearing in mind the limitations due to the computational complexity of the problem.

On the other hand, for a large number of uncertain parameters (up to 100) there is no other way to approach the problem but by bounding the stability margin in polynomial time, although these bounds will have no guarantee of reaching a prescribed error. A branch and bound algorithm that proceeds along these lines is available in a commercial package ([20, 325]). Next, we present the basic results that are used in such an algorithm.

7.4.2 Mixed μ

This section follows the research effort aimed to construct bounds computable in polynomial time for the stability margin in the case of mixed type uncertainty. For further details the reader is referred to [38, 216, 302, 325, 326, 327]. The uncertainty set for mixed parametric–structured dynamic uncertainty has been defined in (7.19). For this same structure, here we also specify the repeated parametric uncertainty case, already defined in (7.92), and the repeated complex scalar case defined as follows:

$$\boldsymbol{\Delta}_{dr} \triangleq \begin{cases} \text{diag}(\overbrace{\delta_1,\ldots,\delta_1}^{k_1},\ldots,\overbrace{\delta_i,\ldots,\delta_i}^{k_i},\ldots,\overbrace{\delta_{m_c},\ldots,\delta_{m_c}}^{k_{m_c}}) \\ \delta_i \in \mathbb{C}, \quad i=1,\ldots,m_c \end{cases} \quad (7.96)$$

Therefore the mixed uncertainty structure is as follows:

$$\boldsymbol{\Delta}_M \triangleq \begin{cases} \text{block diag}(\Delta_1,\Delta_2,\Delta_3) \\ \Delta_1 \in \boldsymbol{\Delta}_{pr} \\ \Delta_2 \in \boldsymbol{\Delta}_{dr} \\ \Delta_3 \in \boldsymbol{\Delta}_d \end{cases} \quad (7.97)$$

Next, define the following sets of matrices:

$$\mathcal{Q} \triangleq \left\{ \Delta \in \boldsymbol{\Delta}_M \;\middle|\; \begin{array}{l} \delta_i \in \{1,-1\}, \quad i=1,\ldots,m_r \\ \delta_j^*\delta_j = 1, \quad j=m_r+1,\ldots,m_r+m_c \\ \Delta_k^*\Delta_k = I, \quad k=m_r+m_c+1,\ldots,m \end{array} \right\}$$

$$\mathcal{D} \triangleq \begin{cases} \text{block diag}(D_1,\ldots,D_{m_r+m_c},d_{m_r+m_c+1}I,\ldots,d_m I) \\ D_i^* = D_i > 0, \quad D_i \in \mathbb{C}^{n_i \times n_i}, \quad i=1,\ldots,m_r \\ D_i^* = D_i > 0, \quad D_i \in \mathbb{C}^{k_i \times k_i}, \quad i=m_r+1,\ldots,m_c \\ d_j > 0, \quad d_j \in \mathbb{R}, \quad j=m_r+m_c+1,\ldots,m \end{cases}$$

$$\mathcal{G} \triangleq \begin{cases} \text{block diag}(G_1,\ldots,G_{m_r},0,\ldots,0) \\ G_i^* = G_i \in \mathbb{C}^{n_i \times n_i}, \quad i=1,\ldots,m_r \end{cases}$$

$$\widehat{\mathcal{D}} \triangleq \begin{cases} \text{block diag}(D_1,\ldots,D_{m_r+m_c},d_{m_r+m_c+1}I,\ldots,d_m I) \\ \det(D_i) \neq 0, \quad D_i \in \mathbb{C}^{n_i \times n_i}, \quad i=1,\ldots,m_r \\ \det(D_i) \neq 0, \quad D_i \in \mathbb{C}^{k_i \times k_i}, \quad i=m_r+1,\ldots,m_r+m_c \\ d_j \neq 0, \quad d_j \in \mathbb{C}, \quad j=m_r+m_c+1,\ldots,m \end{cases}$$

$$\widehat{\mathcal{G}} \triangleq \begin{cases} \text{block diag}(g_1,\ldots,g_{M_R},0,\ldots,0) \\ g_i \in \mathbb{R}, \quad i=1,\ldots,M_R \end{cases}$$

Here M_R has been defined in (7.93) and 0 and I are the null and identity matrices, respectively, of appropiate dimensions so that the elements of the previous sets may be multiplied by the ones in $\boldsymbol{\Delta}_M$. In addition, define

MIXED TYPE UNCERTAINTY

$\rho_R(A) \triangleq \max\{|\lambda_i|, \lambda_i \text{ is a real eigenvalue of } A\}$ and the structured singular value as in Section 7.1.1, now with respect to Δ_M.

Property 5 of Section 7.2.1 still holds for this more general structure. In addition, using the definition of μ, the following is also true (see property 3 in Section 7.2.1):

$$\mu_{\Delta_M}(M) = \max_{\Delta \in \overline{B}\Delta_M} \rho_R(M\Delta) \tag{7.98}$$

It is clearly easier to search over the boundary of the set $\overline{B}\Delta_M$ than over the whole set. For this reason, taking into account that the boundary $\mathcal{Q} \subset \overline{B}\Delta_M$, the following bounds are obtained:

$$\max_{Q \in \mathcal{Q}} \rho_R(MQ) \leq \mu_{\Delta_M}(M) \leq \inf_{D \in \mathcal{D}} \overline{\sigma}(DMD^{-1}) \tag{7.99}$$

Next, two theorems are presented that prove that the lower bound is tight and a generalization of the upper bound can be made tight as well, in the case of less than three complex blocks.

Theorem 7.8

$$\max_{Q \in \mathcal{Q}} \rho_R(MQ) = \mu_{\Delta_M}(M) \tag{7.100}$$

Proof. See [325]. □

Theorem 7.9 *Let:*

$$\alpha_\star = \inf_{\substack{D \in \mathcal{D} \\ G \in \mathcal{G}}} [\min\{\alpha \in \mathbb{R} \mid [M^*DM + \jmath(GM - M^*G) - \alpha D] \leq 0\}] \tag{7.101}$$

Then if $\alpha_\star \leq 0$, $\mu_{\Delta_M} = 0$, *otherwise*

$$\mu_{\Delta_M} \leq \sqrt{\alpha_\star} \tag{7.102}$$

Proof. See [115]. □

The lower bound is a nonconvex optimization, as with the pure dynamic uncertainty structure. The upper bound is a convex optimization problem due to the fact that it is a LMI in both G and D. Therefore the global minimum can be computed efficiently with the use of commercial software packages ([20, 126]).

Note that in the case of only nonrepeated structured dynamic uncertainty we have $m_r = m_c = 0$, which imply $\mathcal{G} = \{0\}$. Therefore the previous theorem reduces to

$$\alpha_\star = \inf_{D \in \mathcal{D}} [\min\{\alpha \mid [M^*DM - \alpha D] \leq 0\}] \tag{7.103}$$

which can be shown is equivalent to the upper bound in equation (7.40), Section 7.2.1 (see Theorem 7.3 and Lemma 7.2).

In [326] for computational purposes the following alternative characterization of the upper bound was introduced:

Theorem 7.10 *Consider a constant matrix $M \in \mathbb{C}^{r \times r}$ and a real scalar $\beta > 0$; then there exist matrices $D \in \mathcal{D}$ and $G \in \mathcal{G}$ such that*

$$\bar{\lambda}\left[M^*DM + j(GM - M^*G) - \beta^2 D\right] \leq 0 \qquad (7.104)$$

if and only if there exist matrices $\widehat{D} \in \widehat{\mathcal{D}}$ and $\widehat{G} \in \widehat{\mathcal{G}}$ such that

$$\bar{\sigma}\left[(I + \widehat{G}^2)^{-1/4}\left(\frac{1}{\beta}\widehat{D}M\widehat{D}^{-1} - j\widehat{G}\right)(I + \widehat{G}^2)^{-1/4}\right] \leq 1 \qquad (7.105)$$

From this theorem it is clear that the minimization problem to compute μ_{Δ_M} can be carried out alternatively as the computation of the minimal $\beta > 0$, such that

$$\inf_{\substack{\widehat{D} \in \widehat{\mathcal{D}} \\ \widehat{G} \in \widehat{\mathcal{G}}}} \bar{\sigma}\left[(I + \widehat{G}^2)^{1/4}\left(\frac{1}{\beta}\widehat{D}M\widehat{D}^{-1} - j\widehat{G}\right)(I + \widehat{G}^2)^{-1/4}\right] \leq 1 \qquad (7.106)$$

Although the above problem is no longer convex in $\widehat{D}$ and $\widehat{G}$, the former enters the problem as in the pure complex case, the latter is now a real diagonal matrix, and the minimization of the norm $\bar{\sigma}(\cdot)$ offers certain numerical advantages. It seems that a good choice is initially to solve the previous minimization, which provides an initial guess for the LMI problem in (7.101). This has been implemented in commercial packages ([20]) that compute the mixed μ stability margin.

7.5 PROBLEMS

1. Show, in a loop block diagram example, the signals w and z that guarantee that stability of $T_{zw}(s)$ is equivalent to nominal internal stability. In the same way show for which signals this is not the case.

2. Prove that $\mu(MU) = \mu(M)$ for $U \in \mathcal{U}$, with

$$\mathcal{U} \triangleq \{U \in \Delta_C, \text{ for } U_i \text{ unitary } i = 1, \ldots, n\}$$

3. Prove that $\mu(DMD^{-1}) = \mu(M)$ for $D \in \mathcal{D}$, with

$$\mathcal{D} \triangleq \left\{\begin{bmatrix} d_1 I_1 & & 0 \\ & \ddots & \\ 0 & & d_n I_n \end{bmatrix} \middle| \begin{array}{l} I_i : \text{identity in } r_i \\ d_i \in \mathbb{R}_+ \end{array}\right\}$$

4. Prove that (7.47) holds when the uncertainty structure is global dynamic and $T_{11}(s)$ and $T_{22}(s)$ represent the transfer matrices to be tested for robust stability and nominal performance, respectively. Verify that equality holds in the SISO case.

5. Prove Lemma 7.4.

6. Consider a real polynomial $P(p)$ that is multilinear in $p \in \mathcal{H}$. Prove that $\text{co}[P(\mathcal{V})] \equiv P(\mathcal{H})$.

8

$\mathcal{L}^1$ CONTROL

8.1 INTRODUCTION

Up to now, most of the material in this book has been devoted to the case where the signals involved are $\mathcal{L}_2$ signals. Thus, in this context, it is natural to assess performance in terms of the $\mathcal{L}_2$-induced norm (i.e., the $\mathcal{H}_\infty$ norm) and to consider the effects of $\mathcal{L}_2$ bounded model uncertainty. However, in several situations of practical relevance the $\mathcal{L}_2$ norm is usually not the most adequate to capture the features of the problem. Examples of this situation are cases involving time-domain specifications and persistent, unknown but bounded exogenous signals, such as when it is desired to minimize the worst tracking error in the presence of persistent, bounded noise, while, at the same time, keeping the control action below a given value. Another common situation where the need for a worst-case time-domain norm arises is the case where the plant model being used for synthesis has been obtained from the linearization of a nonlinear plant at a given operating point. In this case the states of the plant should be kept confined to a region around this nominal condition where the linearized model is known to be accurate enough.

While conceivably these problems could be addressed by using the $\mathcal{H}_2$ and $\mathcal{H}_\infty$ methodologies discussed in Chapters 5 and 6, this will entail the use of weighting functions in an attempt to recast the time-domain specifications in terms of the $\mathcal{L}_2$ norm of a suitably weighted output. Thus we can expect at best a lengthy trial-and-error type process without guarantee of success. Rather than pursuing this approach, in this chapter we will develop tools to handle directly cases involving persistent bounded signals and performance specifications given in terms of peak time-domain values, that is, the $\mathcal{L}_\infty$ norm.

Consider the prototype control loop used in Chapters 5 and 6 and shown again in Figure 8.1. As before, the goal is to synthesize an internally stabilizing controller $K(s)$ such that the worst-case value of the output $z(t)$ (measured now in terms of its peak time-domain value) to persistent exogenous disturbances $w(t)$, $|w(t)| \leq 1$, is optimized. Denoting by $\|.\|_{\mathcal{L}_\infty \to \mathcal{L}_\infty}$ the $\mathcal{L}_\infty$-induced norm of a given mapping, this problem can be stated precisely as

$$\min_{K \text{ stabilizing}} \|\mathcal{F}_l(G,K)\|_{\mathcal{L}_\infty \to \mathcal{L}_\infty} \tag{8.1}$$

Since for LTI stable systems the $\mathcal{L}_\infty$ to $\mathcal{L}_\infty$-induced norm coincides with the $\mathcal{L}_1$ norm of the impulse response matrix (see Appendix A), this problem is known as the $\mathcal{L}_1$ control problem.

The problem of rejecting persistent disturbances has been considered as far back as the early 1970s, [39, 40, 131]. These papers addressed the problem of finding a static state-feedback control, possibly under control input constraint, guaranteeing the permanence of the state in a given time-dependent set, in the presence of unknown but bounded persistent disturbances. The problem was solved by finding a sequence of sets (the reduced target tube) in which the state could be confined by means of an appropriate control action. While some progress was made in this direction, this line of research was abandoned, probably in view of the computational complexity involved in implementing these controllers, not compatible with the computer technology available then, in all but the simplest cases.

Interest in this problem arose again in the early and mid-1980s, when it was shown that in the case of linear time invariant plants it could be recast into an equivalent $\mathcal{L}_1$ model matching problem and solved using linear programming [26, 81, 82, 315]. The approach that we pursue in the first part of this chapter follows closely after the work of Dahleh and Pearson ([81, 82]). For reasons

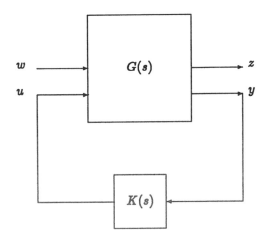

Figure 8.1. General loop structure for the $\mathcal{L}_1$ problem.

that will become clear latter, we also will depart from the approach taken so far and consider discrete, rather than continuous, time systems first. The continuous time case will be reduced to an equivalent discrete time problem using the simple change of variable proposed in [41].

Before proceeding to solve the $\mathcal{L}_1$ control problem, we explore the connections between $\mathcal{L}_1$ norm minimization and robust stability.

8.2 ROBUST STABILITY REVISITED

Up to the present point, the only class of dynamic model uncertainty that we have considered has been linear time invariant with bounded $\mathcal{L}_2$-induced norm. In this context we developed in Chapter 2 (for the SISO case) and Chapter 4 (for MIMO systems) *necessary and sufficient* conditions for $\mathcal{L}_2$ stability of the form

$$\boxed{\text{Robust stability}} \iff \|M_{11}\|_\infty \leq 1 \tag{8.2}$$

Next, we want to extend the class of admissible perturbations to include nonlinear, possibly time varying operators. To this end we begin by introducing the concept of causal operators.

Definition 8.1 *An operator $M: \ell_n^p \to \ell_m^p$ is causal if*

$$P_k M = P_k M P_k \quad \text{for all } k \tag{8.3}$$

where P_k denotes the standard truncation operation, that is,

$$P_k(x_o, x_1 \ldots x_k \ldots) = (x_o, x_1 \ldots x_k, 0, 0 \ldots) \tag{8.4}$$

If

$$P_k M = P_k M P_{k-1} \quad \text{for all } k \tag{8.5}$$

then M is said to be strictly causal.

That is, an operator is causal if its present output does not depend on future inputs. If, in addition, the present output does not depend on the present input, then it is strictly causal.

Definition 8.2 *A mapping $H: \ell^p \to \ell^p$ is said to be ℓ^p stable if:*

1. *whenever $x \in \ell^p$ then $Hx \in \ell^p$, and*
2. *there exist finite constants k and b such that*

$$\|Hx\|_p \leq k\|x\|_p + b, \quad \forall x \in \ell^p$$

248 $\mathcal{L}^1$ CONTROL

Note that in the case of arbitrary nonlinear operators both conditions are required. For instance, the operator $H(x)$ defined by $(Hx)(i) = x(i)^2$ maps ℓ^∞ into itself. However, the second condition fails since clearly no suitable constants k and b can be found.

When $p = \infty$, Definition 8.2 reduces to the usual bounded-input bounded-output (BIBO) stability. If, in addition, we restrict our attention to linear time invariant mappings, a necessary and sufficient condition for ℓ^∞ stability is given by the following well known result.

Lemma 8.1 *Consider a LTI operator $H: \ell^\infty \to \ell^\infty$ and let $\{h_i\}$ denote its impulse response. Then H is ℓ^∞ stable if and only if $\{h_i\} \in \ell^1$.*

Proof. (Sufficiency) For any $x \in \ell^\infty$ we have that (see Appendix A)

$$\|Hx\|_\infty \leq \|H\|_{\ell^\infty \to \ell^\infty} \|x\|_\infty \\ = \|h\|_1 \|x\|_\infty \tag{8.6}$$

Therefore $Hx \in \ell^\infty$ and condition 2 in Definition 8.2 is satisfied with $k = \|h\|_1$ and $b = 0$.

(Necessity) Assume that $h \notin \ell^1 \Rightarrow \sum_{i=0}^\infty |h_i| = \infty$. Consider the following input $x = \{x(i)\}$, $x(i) = \text{sign}[h(N - i)]$. In this case we have that $|(Hx)(N)| = \sum_{i=0}^N |h_i| \to \infty$. Thus $Hx \notin \ell^\infty$. □

Define now the class of nonlinear, causal perturbations with bounded induced ℓ^p norm as[1]

$$\mathcal{BNLTV} = \left\{ \Delta: \ell^p \to \ell^p, \text{ causal}: \sup_{x \neq 0} \frac{\|\Delta * x\|_p}{\|x\|_p} \triangleq \|\Delta\|_{\ell^p \text{ induced}} < 1 \right\} \tag{8.7}$$

where $*$ denotes convolution. Our goal is to assess internal stability of the interconnection shown in Figure 8.2 when Δ is restricted to some appropriately chosen subset $\mathcal{S} \subseteq \mathcal{BNLTV}$. The first step is to extend the concept of internal stability introduced in Chapter 3.

Definition 8.3 *The feedback loop shown in Figure 8.2 is internally ℓ^p stable if whenever the inputs $[u_1(s), u_2(s)]$ are in ℓ^p the corresponding outputs $[e_1(s), e_2(s)]$ are also in ℓ^p, and, in addition, there exist finite constants k and b such that*

$$\|e_1\|_p \leq k(\|u_1\|_p + \|u_2\|_p) + b \\ \|e_2\|_p \leq k(\|u_1\|_p + \|u_2\|_p) + b \tag{8.8}$$

---
[1] Without loss of generality we take this bound to be 1.

ROBUST STABILITY REVISITED

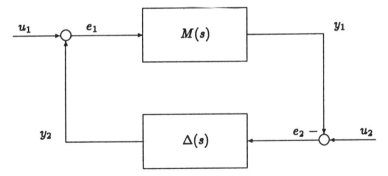

Figure 8.2. Prototype loop interconnection.

A *sufficient* condition for stability is given by the well known small gain theorem.

Theorem 8.1 *Consider the interconnection shown in Figure 8.2. Assume that $M: \ell^p \to \ell^p$ is ℓ^p stable and that the loop is well posed for all $\Delta \in \mathcal{BNLTV}$. If $\|M\|_{\ell^p \to \ell^p} \leq 1$, then the interconnection is stable for all $\Delta \in \mathcal{BNLTV}$.*

Proof. From the block diagram of Figure 8.2 we have that

$$\|e_1\|_p \leq \|\Delta\|_{\ell^p \to \ell^p} \|e_2\|_p + \|u_1\|_p$$
$$\|e_2\|_p \leq \|M\|_{\ell^p \to \ell^p} \|e_1\|_p + \|u_2\|_p \qquad (8.9)$$

Eliminating $\|e_2\|_p$ yields

$$\|e_1\|_p \leq \frac{1}{1 - \|\Delta\|_{\ell^p \to \ell^p} \|M\|_{\ell^p \to \ell^p}} (\|\Delta\|_{\ell^p \to \ell^p} \|u_2\|_p + \|u_1\|_p)$$
$$< \frac{1}{1 - \|\Delta\|_{\ell^p \to \ell^p} \|M\|_{\ell^p \to \ell^p} \to \ell^p} (\|u_2\|_p + \|u_1\|_p) \qquad (8.10)$$

The proof follows now from Definition 8.3 and the fact that for any admissible Δ, we have that $1 - \|\Delta\|_{\ell^p \to \ell^p} \|M\|_{\ell^p \to \ell^p} > 0$. □

In the case of *linear* operators, the small gain condition guarantees that the loop is well posed. However, this is not necessarily true in the case of general $\mathcal{NLTV}$ operators. The following lemma provides a *sufficient* condition for well posedness in this case.

Lemma 8.2 *Consider the interconnection shown in the block diagram of Figure 8.2, where M is a proper linear system and Δ is a $\mathcal{NLTV}$ operator. If Δ is strictly causal then the interconnection is well posed.*

Proof. Since M is proper and Δ is strictly causal, it follows that the operators $M\Delta$ and ΔM are both strictly causal. Thus given the input sequences u_1 and u_2, the equations

$$\begin{aligned} e_1 &= u_1 - \Delta M e_1 + \Delta u_2 \\ e_2 &= u_2 - M u_1 - M \Delta e_2 \end{aligned} \quad (8.11)$$

can be solved recursively to obtain the (unique) solution e_1, e_2. □

Given that the small gain theorem provides only a sufficient condition for stability, a question that arises is whether or not this condition is conservative. In the case of ℓ^2 stability, since the set $\mathcal{BNLTV}$ includes all perturbations $\Delta \in \mathcal{BH}_\infty$, it follows from equation (8.2) that the small gain is both necessary and sufficient for stability. Following along these lines, it is natural to investigate whether or not necessity holds in other situations and, in particular, in the case of ℓ^∞ stability. As we show next the answer is positive, provided that the set of admissible perturbations is enlarged to include either linear time varying or nonlinear operators.

8.2.1 Robust Stability Under LTV Perturbations

Consider a linear time varying causal operator $L: \ell^p \to \ell^p$ defined by its kernel $L(i, j)$, that is, if $y = Lu$ then

$$y(i) = \sum_{k=0}^{i} L(i, k) u(k) \quad (8.12)$$

An alternative representation of this operator is via an infinite-dimensional lower block triangular matrix $\mathcal{L}$:

$$\mathcal{L} = \begin{bmatrix} L(0,0) & \cdots & \cdots & \cdots & \cdots \\ L(1,0) & L(1,1) & \cdots & \cdots & \cdots \\ \vdots & \vdots & \ddots & \cdots & \cdots \\ L(k,0) & L(k,1) & \cdots & L(k,k) & \cdots \\ \vdots & \vdots & \vdots & \vdots & \ddots \end{bmatrix} \quad (8.13)$$

Clearly a necessary and sufficient condition for the operator L to have bounded ℓ^∞-induced norm is

$$\sup_k \|\mathcal{L}(k,:)\|_1 \leq M < \infty \quad (8.14)$$

where $\mathcal{L}(k,:)$ denotes the kth row of the infinite matrix $\mathcal{L}$ and $\|.\|_1$ denotes the standard ℓ^1 norm for matrices. Thus the space of all bounded, causal, LTV operators is isomorphic to the space of matrices of the form $\mathcal{L}$ subject to (8.14). Next, we exploit this characterization to show how to construct a bounded LTV operator that maps two given ℓ^∞ sequences.

ROBUST STABILITY REVISITED   251

Lemma 8.3 *Consider two ℓ^∞ sequences $y = \{y(i)\}$, $\xi = \{\xi(i)\}$. Assume that*

$$\|P_k y\|_\infty \geq \|P_k \xi\|_\infty \quad \forall k \qquad (8.15)$$

where P_k denotes the projection operator. Then there exists a causal operator $\Delta \in \mathcal{BLTV}$ such that $\xi = \Delta y$.

Proof. Assume that $y \neq 0$ (otherwise (8.15) implies that $\xi \equiv 0$ and the proof is trivial). Define

$$\begin{aligned} i_1 & \quad \text{smallest integer such that } y(i_1) \neq 0 \\ i_2 & \quad \text{smallest integer such that } |y(i_2)| > |y(i_1)| \\ & \vdots \\ i_n & \quad \text{smallest integer such that } |y(i_n)| > |y(i_{n-1})| \end{aligned} \qquad (8.16)$$

Consider the following operator defined in terms of its matrix representation:

$$\Delta = \begin{bmatrix} \ddots & & & & & & \\ & \frac{\xi(i_1)}{y(i_1)} & & & & & \\ & \vdots & 0 & & & & \\ & \vdots & & \ddots & & & \\ & \frac{\xi(i_2-1)}{y(i_1)} & 0 & \cdots & 0 & & \\ & & & & \frac{\xi(i_2)}{y(i_2)} & & \\ & & & & \vdots & 0 & \\ & & & & \vdots & & \ddots \\ & & & & \frac{\xi(i_3-1)}{y(i_2)} & 0 & \cdots & 0 \\ & & & & & & & \frac{\xi(i_3)}{y(i_3)} \\ & & & & & & & \vdots & \ddots \end{bmatrix} \qquad (8.17)$$

By construction Δ is a LTV causal operator such that $\xi = \Delta y$. In addition, since each row of Δ has at most one nonzero element it is easily seen that

$$\|\Delta\|_{\ell^\infty \to \ell^\infty} = \sup_k \frac{\|P_k \xi\|_\infty}{\|P_k y\|_\infty} \leq 1 \qquad (8.18)$$

and thus $\Delta \in \mathcal{BLTV}$. □

With this result we are ready to show now that the small gain theorem is both necessary and sufficient for ℓ^∞ stability provided that the set of admis-

252 $\mathcal{L}^1$ CONTROL

sible perturbations is enlarged to $\mathcal{BLTV}$. This fact was originally established in [84]. The simpler proof that we provide here follows after [86] and consists of two parts: (i) construction of a suitable input signal resulting in an unbounded output; and (ii) contruction of an admissible perturbation that maps this output to the input. Combination of (i) and (ii) results in a loop where a bounded input generates an unbounded internal signal, thus violating internal stability.

Theorem 8.2 *Consider the loop interconnection of Figure 8.2. Assume that M is a LTI stable system and that the loop is well posed. Then the interconnection is stable for all $\Delta \in \mathcal{BLTV}$ if and only if*

$$\|M\|_{\ell^\infty \to \ell^\infty} \leq 1 \tag{8.19}$$

Proof. Sufficiency follows from the small gain theorem. To establish necessity assume that $\|M\|_{\ell^\infty \to \ell^\infty} \geq \gamma > 1$. We will show that in this case there exists a destabilizing $\Delta \in \mathcal{BLTV}$. Proceeding as outlined before, we first construct an input sequence u_2 and a signal e_1 such that

1. $\lim_{k \to \infty} \|P_k e_1\|_\infty \to \infty$.
2. $\dfrac{1}{\gamma}\|P_k e_2\|_\infty \geq \|P_k e_1\|_\infty$ for all k.

For simplicity in the sequel we will assume that M has a finite impulse response of length N.[2] Select the first N elements of the sequence e_1 so that $\|P_{N-1} e_1\|_\infty = 1$ and $|y_1(N-1)| \geq \gamma$. Select the corresponding sequence u_2 as $u_2(i) = -\gamma \, sign(y_1(i))$. With these choices we have that

$$\begin{aligned} |e_2(k)| &= \gamma + |y_1(k)| \geq \gamma, \quad k = 0, 1, \ldots, N-2 \\ |e_2(N-1)| &= \gamma + |y_1(N-1)| \geq 2\gamma \end{aligned} \tag{8.20}$$

Select now the next N elements of the sequence e_1 so that $\|P_{2N-1} e_1\|_\infty = 2$ and $|y_1(2N-1)| \geq 2\gamma$, with $u_2(i) = -\gamma \, sign(y_1(i))$. With this choice we have now that $\|P_k(e_2)\|_\infty \geq \gamma \|P_k e_1\|_\infty$, $k \leq 2N - 1$ and $|e_2(2N-1)| \geq 3\gamma$. Continuing along these lines, pick the mth chunk of the sequence e_1 so that $\|P_{mN-1}(e_1)\|_\infty = m$ and $|y(mN-1)| \geq m\gamma$. It is clear that with these choices both conditions 1 and 2 above are satisfied. From Lemma 8.3 it follows that there exists a causal operator $\Delta \in \mathcal{BLTV}$ such that $e_1 = \Delta e_2$. Moreover, by selecting the sequences so that $e_1(0) = 0$ and $e_2(0) \neq 0$ (e.g., by setting $u_2(1) = \gamma$) it follows that Δ is strictly causal and thus the loop is well posed. To

---

[2] This assumption can be removed by using the fact that since M is stable, for any $\epsilon > 0$ there exists N such that $\sum_{i=N}^{\infty} |M_i| < \epsilon$, where M_i denote the Markov parameters of M (see [84] for details).

complete the proof we need to show that this operator Δ destabilizes the loop in Figure 8.2. This follows immediately from the fact that for this choice of Δ the bounded input signal u_2 results in an unbounded internal signal e_1. □

8.2.2 Stability Under Time Invariant Perturbations

In the last section we have shown that the small gain theorem is necessary and sufficient for ℓ^∞ stability under linear time varying perturbations. Next, we show that this is also true for the time invariant case, *provided* that the perturbations are allowed to be *nonlinear* causal mappings. The proof is similar to the proof of Theorem 8.2 except that the destabilizing perturbation that maps e_2 to e_1 is selected to be NLTI rather than LTV. To this effect, denote by S_i the "shift-right by i" operator, and given any sequence $f = \{f(k)\} \in \ell^\infty$ consider the following mapping:

$$(\Delta f)(k) = \begin{cases} e_2(k-i), & \text{if } P_k f = P_k S_i e_1 \text{ for some } i \\ 0, & \text{otherwise} \end{cases} \quad (8.21)$$

It is easily seen that this perturbation maps e_1 to e_2. Moreover, since by construction $\gamma^{-1}\|P_k e_2\|_\infty \geq \|P_k e_1\|_\infty$, it follows that $\|\Delta\|_{\ell^\infty \to \ell^\infty} \leq \gamma^{-1} < 1$. Finally, it can also be shown that $S_1 \Delta = \Delta S_1$ and thus Δ is time invariant. Combining this result with Theorem 8.2 yields the following corollary.

Corollary 8.1 *The condition $\|M\|_{\ell^\infty \to \ell^\infty} \leq 1$ is necessary and sufficient for the stability of the the loop interconnection of Figure 8.2 for all causal Δ with $\|\Delta\|_{\ell^\infty \to \ell^\infty} < 1$ when Δ is either a LTV or NLTI mapping.*

Finally, we address the issue of the conservativeness of the small gain theorem for systems subject to LTI perturbations. Consider an ℓ^∞ stable LTI system H and let $h(i)$ denote its impulse response. Since H is ℓ^∞ stable, the sequence $\{h_i\} \in \ell^1$ and therefore $H(z) = \sum_{i=0}^\infty h(i) z^{-i}$ is analytic on the exterior of the unit disk and continuous on the unit circle. It follows that $H(z) \in \mathcal{H}_\infty$. Moreover, for any ω

$$|H(e^{j\omega})| \leq \sum_{i=0}^\infty |h_i| = \|h\|_1 \quad (8.22)$$

Hence $\|H\|_{\mathcal{H}_\infty} \leq \|h\|_1$. Thus, for LTI systems, ℓ^∞ stability implies ℓ^2 stability. The converse, however, is not true as illustrated by the following example.

Example 8.1 *Consider a LTI system having as z transform $H(z) = \exp\{z/(1-z)\}$. It can be shown that*

$$\|H\|_{\mathcal{H}_\infty} = \operatorname*{ess\,sup}_{|z|=1} \left|\exp\left(\frac{z}{1-z}\right)\right| = \exp(-0.5) \quad (8.23)$$

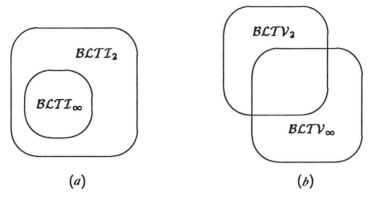

Figure 8.3. Comparison of ℓ^∞- and ℓ^2-induced unit balls: (a) LTI case and (b) LTV case.

Therefore $H(z) \in \mathcal{H}_\infty$ and the system is ℓ^2 stable. On the other hand, since $H(z)$ is not continuous on the unit circle its Laurent series expansion does not converge absolutely there. Hence $h(i) \notin \ell^1$ and the system is not ℓ^∞ stable.

From this discussion it follows that in the case of LTI perturbations the ℓ^∞-induced unit ball $\mathcal{BLTI}_\infty$ is strictly contained inside the ℓ^2-induced unit ball $\mathcal{BLTI}_2$ as illustrated in Figure 8.3. Thus condition (8.19) is more conservative than (8.2). Moreover, given any n, it is relatively simple to construct a LTI stable operator $H \in \overline{\mathcal{BLTI}_2}$ satisfying

$$\|H\|_{\ell^\infty \to \ell^\infty} = (2n+1)\|H\|_{\mathcal{H}_\infty} = (2n+1) \tag{8.24}$$

For example, consider a system having the following z transform:

$$G(z) = \prod_{i=1}^{n} \frac{(1-\alpha^i)z - 1}{z - (1-\alpha^i)}, \quad 0 < \alpha < 1 \tag{8.25}$$

Since $G(z)$ is allpass, $\|G\|_{\mathcal{H}_\infty} = 1$. On the other hand, it is easily seen (by using partial fraction expansions) that as $\alpha \to 0$ then $\|G\|_1 \to 2n+1$.

Since the gap between the $\|.\|_{\ell^\infty \to \ell^\infty}$ and $\|.\|_{\mathcal{H}_\infty}$ norms can be made arbitrarily large, (8.19) can be arbitrarily conservative when used for LTI systems. This situation no longer holds in the case of LTV systems since in that case the sets $\mathcal{BLTV}_2$ and $\mathcal{BLTV}_\infty$ are no longer comparable (see Figure 8.3).

8.3 A SOLUTION TO THE SISO ℓ^1 CONTROL PROBLEM

Consider again the ℓ^1 control problem (8.1). Assume that the plant G satisfies:

(A1) (A, B_2) stabilizable.
(A2) (A, C_2) detectable.

A SOLUTION TO THE SISO ℓ^1 CONTROL PROBLEM

By using the Youla parametrization introduced in Chapter 3, the problem can be recast in the following model matching form:

$$\inf_{Q \in \ell^1} \|T_{11} - T_{12} * Q * T_{21}\|_1 \tag{8.26}$$

where the operators $T_{i,j} \in \ell^\infty$ are fixed LTI operators related to the plant, and where $*$ denotes convolution. Consider first the case where the plant is SISO, that is, all the signals involved are scalars. In this case (8.26) reduces to

$$\inf_{Q \in \ell^1} \|T_1 - T_2 * Q\|_1 \tag{8.27}$$

where T_1 and T_2 are finite-dimensional, ℓ^∞ stable LTI operators.

Recall that a similar technique was used in Section 6.6 as a first step in solving the $\mathcal{H}_\infty$ optimal control problem and also underlies the approach pursued in Section 5.3 to solve the $\mathcal{H}_2$ problem. However, there is an important distinction between these cases: since $\mathcal{H}_2$ is a Hilbert space, the optimal approximation problem can be solved by a simple projection.[3] On the other hand, neither $\mathcal{H}_\infty$ nor ℓ^1 is an inner product space and thus the solution becomes more involved. In the $\mathcal{H}_\infty$ case, it was obtained from Nehari's theorem. As we show next, in the ℓ^1 case, the model matching problem can be solved by recasting it into an equivalent linear programming problem through the use of duality theory (briefly covered in Appendix A).

To simplify the exposition, in the sequel we use the following notation. Given an operator $H: \ell^\infty \to \ell^\infty$, we denote its z transform by $\hat{H}(z)$ and its impulse response sequence by $h = \{h(i)\}$. $\mathcal{R}\ell^1$ denotes the subspace of ℓ^1 formed by sequences having a real rational z transform. With a slight abuse of notation, we sometimes use the shorthand $\|\hat{H}(z)\|_1$ to denote $\|H\|_1$, that is, the ℓ^1 norm of $\{h(i)\}$, and $\hat{H}(z) \in \mathcal{R}\ell^1$ (or $H \in \mathcal{R}\ell^1$) to indicate that $\{h(i)\} \in \mathcal{R}\ell^1$.

Consider now a normed linear space X. Its dual space X^* is defined as the space of all bounded linear functionals on X. Given $x \in X$ and $r \in X^*$, $\langle x, r \rangle$ denotes the value of the linear functional r at the point x. It is a standard result in functional analysis (see Appendix A for details) that if $X = \ell^p$, $1 \leq p < \infty$, then $X^* = \ell^q$, where $1/p + 1/q = 1$ (in the case where $p = 1$, we take $q = \infty$). Moreover, every linear functional on ℓ^p, $1 \leq p < \infty$, has a unique representation of the form

$$\langle x, r \rangle = \sum_{i=1}^{\infty} \eta_i x_i \tag{8.28}$$

where $\eta = \{\eta_i\} \in \ell^q$. Note that $(\ell^1)^* = \ell^\infty$. However the dual of ℓ^∞ is not ℓ^1.[4]

---

[3] This was the approach pursued in Lemma 5.5, where the Youla parametrization was selected to render $\hat{T}_2$ an inner function. The optimal Q is then given by the stable portion of $\hat{T}_2^\sim \hat{T}_1$.
[4] This can easily be established by noting that while ℓ^1 is separable, ℓ^∞ is not. Indeed, $\ell^1 = c_o^*$, where c_o denotes the subspace of ℓ^∞ formed by sequences converging to zero.

At this point we also introduce the following two additional assumptions:

(A3) $\hat{T}_2(z) \neq 0$ for all $|z| = 1$.
(A4) All zeros of $\hat{T}_2(z)$ in $|z| > 1$ have multiplicity 1.

Assumption (A4) is made for convenience, in order to simplify the subsequent developments. We will briefly indicate how to relax it at the end of the section. On the other hand, assumption (A3) is required for the solution to the problem to be well behaved. As shown in [317], if this assumption fails the optimal approximation is not a continuous function of the problem data.

Lemma 8.4 *Assume that $\hat{T}_2(z) \in \mathcal{R}\ell^1$ has n single zeros $z_1, z_2, \ldots, z_n$, $|z_i| > 1$, and no zeros on $|z| = 1$. Then, given $R \in \mathcal{R}\ell^1$, there exists $Q \in \mathcal{R}\ell^1$ such that $R = T_2 * Q$ if and only if*

$$\hat{R}(z_i) = 0, \quad i = 1, 2, \ldots, n \quad (8.29)$$

Proof. (Necessity) If $Q \in \mathcal{R}\ell^1$ then $\hat{Q}(z)$ is analytic in $|z| > 1$. Thus $\hat{T}_2(z_i)\hat{Q}(z_i) = 0$, which implies (8.29).

(Sufficiency) Since $R = T_2 * Q \in \mathcal{R}\ell^1$, the only situation where $Q \notin \ell^1$ is the case where one of the unstable zeros of $\hat{T}_2(z)$, say, z_i, cancels an unstable pole of $\hat{Q}(z)$ at that location. However, since all the nonminimum phase zeros of $\hat{T}_2$ have multiplicity 1, this implies that $\hat{T}_2(z_i)\hat{Q}_2(z_i) \neq 0$, contradicting (8.29). □

Let

$$S = \left\{ R \in \mathcal{R}\ell^1 : \hat{R}(z_i) = 0, \quad i = 1, 2, \ldots, n \right\} \quad (8.30)$$

In terms of this set the ℓ^1 model matching problem can be recast as the following minimum norm problem:

$$\inf_{R \in S} \|T_1 - R\|_1 \quad (8.31)$$

Note that this problem has the same form as the minimum norm problems set in general ℓ^p spaces discussed in Appendix A. As we show there, these problems can be solved by exploiting the following duality principle stating the equivalence of the original problem (the primal) and an optimization problem set in the dual space (the dual problem).

Theorem 8.3 [187] *Consider a real normed linear space X, its dual X^*, a subspace $M \subseteq X$, and its annihilator $M^\perp \subseteq X^*$. Let $x^* \in X^*$ be at a distance μ from $M^\perp$. Then*

$$\mu = \min_{r^* \in M^\perp} \|x^* - r^*\| = \sup_{x \in \overline{B}M} \langle x, x^* \rangle$$

where the minimum is achieved for some $r_0^* \in M^\perp$. If the supremum on the right is achieved for some $x_0 \in M$, then $\langle x^* - r_0^*, x_0 \rangle = \|x^* - r_0^*\| \cdot \|x_0\|$ (i.e., $x^* - r_0^*$ is aligned with x_0).

A special case of the above theorem is the case when M is finite dimensional. In this situation, the supremum on the right will always exist, and hence both problems have solutions. As we show next this is precisely the case for problem (8.31).

In order to apply Theorem 8.3 to this problem we need to identify the spaces X and X^* and the corresponding subspaces M and $M^\perp$. Clearly, in this case, we need $X^* = \ell^1$ and $S = M^\perp$. Thus we should select $X = c_o$, the subspace of ℓ^∞ formed by sequences converging to zero, since $c_o^* = \ell^1$ [187]. With these choices, given any $r = \{r(i)\} \in M^\perp \subset \ell^1$ and $x = \{x(i)\} \in M \subset c_o$, we have that

$$\langle r, x \rangle = \sum_{i=0}^{\infty} r(i) x(i) \tag{8.32}$$

The only remaining task is to characterize M. To this effect consider the nonminimum phase zeros of $\hat{T}_2(z)$, z_i, and define the following sequences:

$$\begin{aligned} \mathcal{Z}_j^R &= \operatorname{Re}\left\{1, z_j^{-1}, \ldots, z_j^{-n}, \ldots\right\} \\ \mathcal{Z}_j^I &= \operatorname{Im}\left\{0, z_j^{-1}, \ldots, z_j^{-n}, \ldots\right\} \end{aligned} \tag{8.33}$$

In terms of these sequences M can be characterized as follows.

Lemma 8.5 *Let*

$$M = \operatorname{Span}\left\{\mathcal{Z}_j^R, \mathcal{Z}_j^I\right\}$$

then $S = M^\perp$.

Proof. For any element $R \in S$ we have that

$$\hat{R}(z_i) = \sum_{k=0}^{\infty} r(k) z_i^{-k} = 0 \tag{8.34}$$

where $r = \{r(i)\}$ denotes the impulse response sequence of R. Thus the following holds for all i:

$$\begin{aligned} \operatorname{Re}\left[\hat{R}(z_i)\right] &= \langle r, \mathcal{Z}_i^R \rangle = 0 \\ \operatorname{Im}\left[\hat{R}(z_i)\right] &= \langle r, \mathcal{Z}_i^I \rangle = 0 \end{aligned} \tag{8.35}$$

Hence $\langle r, m \rangle = 0$ for all $m \in M$, which implies that $S \subseteq M^\perp$. A similar argument shows that if some sequence $r \in \ell^1$ satisfies $\langle r, m \rangle = 0$ for all $m \in M$,

258 $\mathcal{L}^1$ CONTROL

then its z transform $\hat{R}$ satisfies $\hat{R}(z_i) = 0$, $i = 1, 2, \ldots, n$. This implies that $M^\perp \subseteq S$, thus establishing the desired result. □

Combining Theorem 8.3 with Lemma 8.5 yields the following result.

Theorem 8.4 *Let $\hat{T}_2(z)$ have n simple zeros z_i outside the unit disk and no zeros on the unit circle. Then*

$$\mu_0 = \min_{Q \in \ell^1} \|\hat{T}_1(z) - \hat{T}_2(z)\hat{Q}(z)\|_1$$

$$= \max_{\alpha_j} \left\{ \sum_{i=1}^{n} \alpha_i \mathrm{Re}\left[\hat{T}_1(z_i)\right] + \sum_{i=1}^{n} \alpha_{i+n} \mathrm{Im}\left[\hat{T}_1(z_i)\right] \right\} \quad (8.36)$$

subject to

$$|x(k)| = \left| \sum_{i=1}^{n} \alpha_i \mathrm{Re}\left[z_i^{-k}\right] + \sum_{i=1}^{n} \alpha_{i+n} \mathrm{Im}\left[z_i^{-k}\right] \right| \leq 1, \quad \forall\, k = 0, 1, \ldots \quad (8.37)$$

Furthermore, denote by $\hat{\Phi}(z)$ the resulting optimal closed-loop system, by $\phi_i \in \ell^1$ the corresponding impulse response sequence, and by $\tilde{x} \in \ell^1$ the optimal solution to the dual problem. Then $\hat{\Phi}$ has the following properties:

(a) $\phi(k) = 0$ whenever $|\tilde{x}(k)| < 1$.
(b) $\phi(k)\tilde{x}(k) \geq 0$.
(c) $\sum_{i=1}^{\infty} |\phi(i)| = \mu_0$.
(d) $\sum_{i=1}^{\infty} \phi(i) z_k^{-i} = \hat{T}_1(z_k), \quad k = 1, \ldots, n$.

Proof. Let $t = \{t(i)\}$ denote the impulse response sequence of T_1. Then

$$\begin{aligned} \langle \mathcal{Z}_j^R, t \rangle &= \sum_{i=0}^{\infty} t(i) \mathrm{Re}\left[z_j^{-i}\right] = \mathrm{Re}\left[\hat{T}_1(Z_j)\right] \\ \langle \mathcal{Z}_j^I, t \rangle &= \sum_{i=0}^{\infty} t(i) \mathrm{Im}\left[z_j^{-i}\right] = \mathrm{Im}\left[\hat{T}_1(z_j)\right] \end{aligned} \quad (8.38)$$

From Lemma 8.5 we have that any $x \in M$ can be written as

$$x = \sum_{i=1}^{n} \alpha_i \mathcal{Z}_i^R + \sum_{i=1}^{n} \alpha_{i+n} \mathcal{Z}_j^I \quad (8.39)$$

Thus for each $x \in M$, we have that

$$\langle x, t \rangle = \sum_{i=1}^{n} \alpha_i \mathrm{Re}[\hat{T}_1(z_i)] + \sum_{i=1}^{n} \alpha_{i+n} \mathrm{Im}[\hat{T}_1(z_i)] \quad (8.40)$$

A SOLUTION TO THE SISO ℓ^1 CONTROL PROBLEM

Equation (8.36) follows now immediately from Theorem 8.3 and Lemma 8.5. The constraint (8.37) is simply a restatement of $\|\tilde{x}\| \leq 1$. Next, we establish the properties of the optimal solution. Properties (a), (b), and (c) all follow from the alignment condition in Theorem 8.3. Here this condition takes the form

$$\langle \phi, \tilde{x} \rangle = \|\phi\|_1 \|\tilde{x}\|_\infty = \mu_o$$
$$\Updownarrow \qquad\qquad\qquad\qquad (8.41)$$
$$\sum_{i=0}^{\infty} \phi(i)\tilde{x}(i) = \|\phi\|_1 \|\tilde{x}\|_\infty$$

Clearly this condition holds if and only if (a), (b), and (c) hold. Finally, to establish (d) note that $\Phi - T_1 \in S$. Thus $\hat{\Phi}(z_i) - \hat{T}_1(z_i) = 0$, $i = 1, 2, \ldots, n$. □

Example 8.2 *Consider the simple SISO stabilization problem shown in Figure 8.4, where the goal is to design a controller $K(s)$ that internally stabilizes the loop for all $\Delta \in \gamma \mathcal{BLTV}$, where γ is a scaling factor to be maximized. Assume that the nominal plant has the following transfer function:*

$$G_o(z) = \frac{z}{z^2 + a^2}, \quad a > 1$$

The problem can be solved by designing $K(s)$ such that it minimizes $\|T_{y_\Delta u_\Delta}\|_1$. The first step is to invoke the Youla parametrization to recast $T_{y_\Delta u_\Delta}$ as an affine function of $Q \in \ell^1$. Since the plant is open-loop unstable, we need to use the general form of the parametrization, obtained in Theorem 3.4. Begin by factoring the plant as

$$G_o = N(z)M(z)^{-1}, \quad N = \frac{1}{z}, \quad M = \frac{z^2 + a^2}{z^2} \qquad (8.42)$$

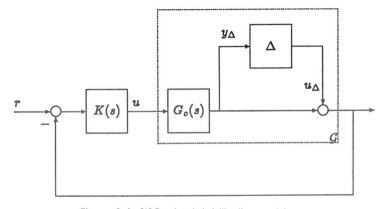

Figure 8.4. SISO robust stabilization problem.

It can easily be checked that the left coprime factors

$$U_o = -\frac{a^2}{z}, \quad V_o = 1 \tag{8.43}$$

satisfy the Bezout equation $NU_o + MV_o = 1$. Hence the set of all controllers stabilizing G_o can be parametrized as

$$K = (U_o - MQ)(V_o + NQ)^{-1}, \quad Q \in \ell^1 \tag{8.44}$$

with the corresponding complementary sensitivity function given by

$$T_{y_\Delta u_\Delta} = N(U_o - MQ) = \frac{1}{z}\left(-\frac{a^2}{z} - \frac{z^2+a^2}{z^2}Q\right) \tag{8.45}$$

Note that the ℓ^1 norm is invariant under multiplication by $1/z$ (since this corresponds to a unit delay). It follows that minimizing $\|T_{y_\Delta u_\Delta}\|_1$ is equivalent to the following ℓ^1 optimal control problem:

$$\mu_o = \min_{Q \in \ell^1} \|T_1(z) - T_2(z)Q(z)\|_1 \tag{8.46}$$

where

$$T_1(z) = -\frac{a^2}{z}, \quad T_2(z) = \frac{z^2+a^2}{z^2} \tag{8.47}$$

Since T_2 has two nonminimum phase zeros at $z = \pm ja$, from Theorem 8.4 we have that problem (8.46) is equivalent to

$$\max_{\alpha_i}(\alpha_3 - \alpha_4)a$$

subject to (8.48)

$$\left|\alpha_1 \mathrm{Re}[(ja)^{-k}] + \alpha_2 \mathrm{Re}[(-ja)^{-k}] + \alpha_3 \mathrm{Im}[(ja)^{-k}] + \alpha_4 \mathrm{Im}[(-ja)^{-k}]\right| \leq 1, \quad k = 0, 1, \ldots$$

Clearly this is equivalent to

$$\max_{\beta_i} \beta_2 a$$

subject to (8.49)

$$\left|\beta_1 a^{-k}\right| \leq 1, \quad k = 0, 2, \ldots$$

$$\left|\beta_2 a^{-k}\right| \leq 1, \quad k = 1, 3, \ldots$$

where we have defined $\beta_1 = \alpha_1 + \alpha_2$ and $\beta_2 = \alpha_3 - \alpha_4$. The solution to this problem is $\beta_1 = 0$, $\beta_2 = a$. Thus the minimum achievable norm is $\mu_o = a^2$. Next, we construct the optimal solution as follows. Note that the constraints in (8.44) saturate only for $k = 1$. From property (a) in Theorem 8.4 it follows that $\phi(k) = 0$ for all $k \neq 1$. Combining this fact with property (d) yields $\phi(1) = -a^2$. Hence

$$\Phi(z) = -\frac{a^2}{z} \Rightarrow Q_{opt} = 0 \tag{8.50}$$

Finally, replacing $Q_{opt} = 0$ in (8.44) yields

$$K_{opt} = U_o V_o^{-1} = -a^2/z$$
$$T_{y_\Delta u_\Delta} = -a^2/z^2 \tag{8.51}$$

Hence this controller guarantees robust stability against all perturbations having $\|\Delta\|_{\ell^\infty \to \ell^\infty} < a^{-2}$.

In principle, problem (8.36) is a semi-infinite linear programming problem, since it involves a finite number of variables ($\alpha_i, i = 1, \ldots, 2n$) and an infinite number of constraints (8.37). However, in the example above these constraints were active only at a finite number of points. As a consequence, the resulting optimal closed-loop system has a finite impulse response (FIR) and the problem can be recast into a finite-dimensional optimization form. As we show next these are general features of SISO ℓ^1 optimal solutions.

Theorem 8.5 *There exists $N > 0$ such that $\phi(k) = 0$, $\forall k \geq N$.*

Proof. For simplicity assume that all the nonminimum phase zeros of $\hat{T}_2(z)$ are real. For a given m, consider the following matrix:

$$F_m = \begin{bmatrix} 1 & \cdots & 1 \\ z_1^{-1} & \cdots & z_n^{-1} \\ \vdots & \vdots & \vdots \\ z_1^{-(m-1)} & \cdots & z_n^{-(m-1)} \end{bmatrix} \tag{8.52}$$

Since F_m is a Vandermonde matrix, it has full column rank for all $m \geq n$ and therefore its left inverse F_m^{-1} is well defined. Let $\alpha = [\alpha_1 \ \cdots \ \alpha_n]$. In terms of F and α the constraints (8.37) can be rewritten as

$$\|F_\infty \alpha\|_\infty \leq 1 \tag{8.53}$$

Next, we show that there exists N such that if the first N constraints in (8.37) are satisfied, then for all $k > N$ we have that

$$\left| \sum_{i=1}^n \alpha_i z_i^{-k} \right| = \left| F_\infty^{(k)} \alpha \right| < 1 \tag{8.54}$$

where $F_\infty^{(k)}$ denotes the kth row of F_∞. Note that

$$\left|\sum_{i=1}^{n} \alpha_i z_i^{-k}\right| \leq \|\alpha\|_1 \max_i |z_i^{-k}| \qquad (8.55)$$

A bound on $\|\alpha\|_1$ for all α satisfying (8.37) can be found as follows. Assume that the first n constraints in (8.37) are satisfied. Then

$$\alpha = F_n^{-1} F_n \alpha$$
$$\Rightarrow$$
$$\|\alpha\|_\infty \leq \|F_n^{-1}\|_1 \|F_n \alpha\|_\infty \leq \|F_n^{-1}\|_1 \qquad (8.56)$$
$$\Rightarrow$$
$$\|\alpha\|_1 \leq n \|\alpha\|_\infty \leq n \|F_n^{-1}\|_1$$

It follows that if N is chosen such that

$$n \|F_n^{-1}\|_1 \max_i |z_i^{-N}| < 1 \qquad (8.57)$$

then (8.54) is satisfied for all $k \geq N$. The proof of the theorem follows now from property (a) in Theorem 8.4. $\square$

Corollary 8.2 *Let $b = [T_1(z_1), T_1(z_2), \ldots, T_1(z_n)]^T$. Then problem (8.36)–(8.37) is equivalent to the following finite-dimensional linear programming problem:*

$$\mu_0 = \min_{\phi^+(i) \geq 0, \phi^-(i) \geq 0} \sum_{i=1}^{N} \phi^+(i) + \phi^-(i)$$

subject to
$$\qquad (8.58)$$
$$F_N^T (\Phi_N^+ - \Phi_N^-) = b$$

where

$$\Phi_N^+ = \begin{bmatrix} \phi^+(1) \\ \vdots \\ \phi^+(N) \end{bmatrix}, \quad \Phi_N^- = \begin{bmatrix} \phi^-(1) \\ \vdots \\ \phi^-(N) \end{bmatrix} \qquad (8.59)$$

Proof. The proof follows from combining Theorems 8.4 and 8.5 with standard linear programming arguments. $\square$

8.3.1 Properties of the Solution

In the last section we have shown that the optimal ℓ^1 (SISO) control problem can be solved by recasting it into a model matching form that can be solved via finite-dimensional linear programming. In this section we explore some

A SOLUTION TO THE SISO ℓ^1 CONTROL PROBLEM

of the properties of the solution to this matching problem. We will show that, contrary to the $\mathcal{H}_2$ and $\mathcal{H}_\infty$ cases, this solution is not unique and its order cannot be *a priori* bounded in terms of the order of the plant. In fact, we will illustrate with a simple example that even in the state-feedback case (where optimal static controllers exist in both the $\mathcal{H}_2$ and $\mathcal{H}_\infty$ cases), the solution may be dynamic, with arbitrarily high order.

We first address the issue of the nonuniqueness of the solution to the model matching problem (8.27). Consider the following example [203].

Example 8.3 *Suppose that $T_1(z)$ and $T_2(z)$ are given by*

$$T_1(z) = \frac{49z^2(3z+2)^2}{(4z-1)^2(31z-30)^2}, \quad T_2(z) = \frac{(2-z)(3-z)}{z^2} \tag{8.60}$$

Applying Theorem 8.4 we have that the dual problem is given by

$$\max_{\alpha_1,\alpha_2} \frac{\alpha_1}{4} + \frac{\alpha_2}{9}$$

subject to
$$|\alpha_1 2^{-k} + \alpha_2 3^{-k}| \leq 1, \quad k = 0, 1, \ldots \tag{8.61}$$

Note that the left-hand side of the constraint corresponding to the value $k=2$ coincides precisely with the objective function. Thus the maximum value of the objective is 1, and the constraint corresponding to $k=2$ is active. It can easily be seen that the constraints corresponding to $k=0$ and $k=1$ are also active, while all the constraints corresponding to $k \geq 3$ are inactive. Moreover, the corresponding dual functional is of form $r = \{-1, 1, 1, *, *, \ldots\}$, where the $*$ denotes elements with absolute value less than 1. From the alignment conditions in Theorem 8.4 we have that the optimal approximation has the form

$$\phi = \phi_o + \frac{\phi_1}{z} + \frac{\phi_2}{z^2} \tag{8.62}$$

where $\phi_o \leq 0$ and $\phi_1, \phi_2 \geq 0$. From the interpolation constraints at $z=2$ and at $z=3$ we have that

$$\begin{aligned}\phi(2) &= \phi_o + \frac{\phi_1}{2} + \frac{\phi_2}{4} = \frac{1}{4} \\ \phi(3) &= \phi_o + \frac{\phi_1}{3} + \frac{\phi_2}{9} = \frac{1}{9}\end{aligned} \tag{8.63}$$

Finally, the constraint $\|\phi\|_1 = 1$ yields

$$\|\phi\|_1 = -\phi_o + \phi_1 + \phi_2 = 1 \tag{8.64}$$

264 $\mathcal{L}^1$ CONTROL

The system of equations (8.63) and (8.64) admits infinite solutions that can be parametrized as

$$\phi_o = \alpha$$
$$\phi_1 = -5\alpha \qquad (8.65)$$
$$\phi_2 = 1 + 6\alpha$$

where α is a real such that $-\frac{1}{6} \leq \alpha \leq 0$.

Next, we show that, *even in the full state-feedback case*, the order of the controller can be arbitrarily high. To this effect we need to introduce the following preliminary result [93] that provides a sufficient condition for the order of the closed-loop system to be bounded by the number of unstable zeros of the plant.

Lemma 8.6 *Let n denote the number of unstable zeros of T_2, $\{z_i\}$. The optimal approximation Φ is a FIR of degree $n-1$ if*

$$\sum_{i=1}^{n-1} |a_i| < |a_o| - 1 \qquad (8.66)$$

where $\Pi_{j=1}^n (z - z_j) = z^n + a_{n-1} z^{n-1} + \cdots + a_1 z + a_o$

Consider now a full state-feedback ℓ^1 problem where the control, exogenous disturbance, and performance variable are all scalars and where the generalized plant has the following realization:

$$G = \left[\begin{array}{c|cc} A & b_1 & b_2 \\ \hline c_1 & 0 & d_{12} \\ I & 0 & 0 \end{array} \right], \quad A \in R^{n \times n} \qquad (8.67)$$

where the pair (A, b_2) is controllable. By using the Youla parametrization introduced in Chapter 3 and formulas (3.41) with controller gain f and observer gain $-A$, we have that the ℓ^1 problem reduces to the form (8.26) with

$$T_{11} = \left[\begin{array}{c|c} A_f & Ab_1 \\ \hline c_f & c_1 b_1 \end{array} \right] z^{-1}, \qquad T_{21} = z^{-1} b_1$$

$$T_{12} = \left[\begin{array}{c|c} A_f & b_2 \\ \hline c_f & d_{12} \end{array} \right] \qquad (8.68)$$

where $A_f = A + b_2 f$ and $c_f = c_1 + d_{12} f$. Suppose that G_{12} has r nonminimum phase zeros $z_1, \ldots, z_r$ and select f to place $n - r$ poles exactly at the location of the minimum phase zeros of G_{12} and the remaining r at the origin. With this choice T_{12} becomes a FIR of order r, that is,

$$T_{12} = d_{12} + \sum_{i=0}^{r-1} \frac{c_f A_f^i b_2}{z^{i+1}}$$

Since the pair (A, b_2) is controllable, it follows that $c_f A_f^n = 0$ for $n \geq r$. Thus with this choice of f, T_{11} has also a finite impulse response (of order $r+1$). It follows that the model matching problem (8.26) is equivalent to

$$\min \|\phi(z)\|_1 = \min \left\| z^{-(r+1)} \left[c_1 \Pi_{i=1}^r (z - \xi_i) + c_2 \Pi_{j=1}^r (z - z_j) \tilde{q}(z) \right] \right\|$$
(8.69)
$$= \min \|\tilde{T}_1 + \tilde{T}_2 \tilde{q}\|_1$$

where

$$\tilde{T}_1 = c_1 z^{-r} \Pi_{i=1}^r (z - \xi_i)$$

$$\tilde{T}_2 = c_2 z^{-r} \Pi_{j=1}^r (z - z_j)$$
(8.70)

$$\tilde{q} = q b_1$$

c_1 and c_2 are constants, and where we used the fact that the ℓ^1 norm is invariant under multiplication by z^{-1}.

Lemma 8.7 *If the nonminimum phase zeros of G_{12} satisfy condition (8.66) then the state-feedback gain f is ℓ^1 optimal.*

Proof. From Lemma 8.6 we have that the optimal ϕ is a FIR of order $r - 1$. Since $\tilde{T}_1$ and $\tilde{T}_2$ in (8.69) are both FIRs of order r, it follows that $\tilde{q}(z)$ must be a constant $\tilde{q}_o$ and such that

$$0 = \tilde{\phi}_r = c_f A_f^{r-1}(Ab_1 + b_2 \tilde{q}_o)$$
$$= c_f A_f^{r-1}(A + b_2 q_o) b_1$$
$$= c_f A_f^r b_1 - c_f A_f^{r-1} b_2 (f - q_o) b_1 = -c_f A_f^{r-1} b_2 (f - q_o) b_1$$
(8.71)

since $c_f A_f^r = 0$. Obviously $q_o = f$ satisfies this condition. Thus, the controller $k = F_\ell(J, f) = f$ is optimal. □

Example 8.4 [87] *Consider the following third-order system, where $\kappa > 0$ is a parameter:*

$$G(z) = \left[\begin{array}{ccc|cc} 2.7 & -23.5 & 4.6 & 1 & 1 \\ 1 & 0 & 0 & 0 & 0 \\ 0 & 1 & 0 & 0 & 0 \\ 1 & -2.5 & \kappa & 0 & 0 \end{array} \right]$$
(8.72)

Table 8.1. Optimal l^1 norm for Example 8.4. N_κ denotes the order of the optimal ℓ^1 controller

κ	N_κ	$\|\Phi\|_1$
2	2	4.21
1.51	9	3.04
1.501	16	3.01
1.500	∞	3.00
1.499	0	2.50

The zeros of G_{12} are given by the roots of $z^2 - 2.5z + \kappa$. Thus, for $\kappa > 3.5$, condition (8.66) holds and the optimal ℓ^1 state-feedback controller is static ($f = [-2.7 \quad 23.5 \quad -4.6]$). On the other hand, for $1.5 < \kappa < 3.5$, the optimal state-feedback controller is dynamic. As $\kappa \downarrow 1.5$, one of the nonminimum phase zeros moves toward the unit circle, while the controller order $N_\kappa \uparrow \infty$ and the optimal cost $\|\Phi\|_1 \downarrow 3$. Table 8.1 shows the optimal ℓ^1 norm and the order of the corresponding controller for different values of κ. If $\kappa < 1.5$, G_{12} has only one nonminimum phase zero. Therefore condition (8.66) is trivially satisfied and the optimal controller becomes static again. It is also worth noting the abrupt change in the value of the optimal ℓ^1 norm as this zero moves across the unit circle.

8.3.2 The MIMO Case

As we have shown in the last section, the SISO ℓ^1 optimal control problem leads to an ℓ^1 model matching problem that can be solved via finite-dimensional linear programming. A key element in this derivation is the characterization of the annihilator space $M^\perp$ obtained in Lemma 8.5, used to establish that the dual problem is finite-dimensional. While the general idea of recasting the problem into a model matching form and exploiting duality is still valid in the MIMO case, here the problem is considerably more involved.

Consider the MIMO model matching problem (8.26). Proceeding as in Section 8.3, the first step is to obtain an equivalent to Lemma 8.4: that is, given $R \in \mathcal{R}\ell^1_{n_z \cdot n_w}$, $T_{12} \in \mathcal{R}\ell^1_{n_z \cdot n_u}$, and $T_{21} \in \mathcal{R}\ell^1_{n_y \cdot n_w}$, find necessary and sufficient conditions for the existence of $Q \in \mathcal{R}\ell^1_{n_u \cdot n_y}$ such that $R = T_{12} * Q * T_{21}$. Recall that in the SISO case these conditions had the form $R(z_i) = 0$, where z_i denote the unstable zeros of $T_{12}(z) * T_{21}(z)$. However, in the MIMO case these interpolation constraints must take into account, in addition to preservation of the unstable zero structure of T_{12} and T_{21}, the *directional* information associated with these zeros. This leads to two sets of constraints: (i) the zero interpolation conditions (essentially a generalization of the SISO conditions) and (ii) the rank interpolation conditions [87].

Consider first the so-called one-block case, where the plant has as many controls as regulated outputs and as many measurements as exogenous disturbances, that is, $n_u = n_z$ and $n_y = n_w$. These problems are the closest MIMO equivalent to the SISO case, sharing many of its properties. For instance, it can be shown that these problems require considering only the zero interpolation constraints. This results in a primal problem having a finite number of constraints and in a dual problem having only finitely many variables. Moreover, by exploiting the alignment conditions it can be shown that, as in the SISO case, the optimal ℓ^1 closed-loop system Φ has a finite impulse response. It follows that the one-block MIMO case can be reduced to a finite-dimensional linear programming problem, proceeding as in Theorem 8.5 and Corollary 8.2 (see [81]).

Problems having more disturbances than measurements or more regulated variables than control inputs ($n_w > n_y$ or $n_z > n_u$) are known as multiblock (or bad rank) problems. Here both sets of interpolation conditions are required, with the rank interpolation constraints leading to infinite-dimensional linear programs in both the primal and dual spaces. One can attempt to obtain approximate solutions to these problems either by constraining the closed loop to have (a sufficiently large) finite support (the finitely many variables approximation) or by retaining only a finite set of constraints in the primal problem (the finitely many equations approximation). Alternatively, one can attempt to embed the problem into a one-block problem by augmenting the matrices T_{12} and T_{21} with pure delays. The solution to the original problem is then obtained through a compression, by simply discarding the additional inputs. A complete treatment of the multiblock MIMO ℓ^1 control problem is beyond the scope of this book (interested readers are referred to [87] for details). Rather, in the next section we will introduce a method for obtaining controllers that minimize the peak time-domain value of the euclidean norm. While these controllers are suboptimal for the ℓ^1 problem, they have advantages in terms of both their order (bounded by the order of the plant) and the computational complexity of the synthesis process.

8.4 APPROXIMATE SOLUTIONS

As we have shown in Section 8.3.1, ℓ^1 optimal control theory may result in large-order controllers, even in the state-feedback case.[5] This feature limits the applicability of the theory in the case of relatively fast plants having a limited amount of computational power available (e.g., see Section 11.1). In principle, this problem can be solved by applying some of the model reduction techniques covered in Chapter 9 to the optimal ℓ^1 controller. However, this model reduction step is far from trivial. Table 8.2 shows the results obtained when attempting to reduce the order of the controller corresponding to $\kappa =$

---

[5] This disadvantage is shared by other widely used design methods, such as μ-synthesis, that will also produce controllers with no guaranteed complexity bound.

Table 8.2. Performance of reduced-order controllers, $\kappa = 1.501$

N	16	15	14	12	11
$\|\Phi\|_1$	3.01	3.08	3.62	4.71	Unstable

1.501 in Example 8.4 using balanced truncation. While the controller can easily be reduced to 15 order, attempting to reduce the order below 12 yields controllers that do not stabilize the plant.

As an alternative to model reduction, low-order suboptimal controllers can be obtained by minimizing an upper bound of the ℓ^1 norm, obtained from geometrical considerations. As we show in the sequel, computing this upper bound only entails solving a LMI optimization problem, followed by a one-dimensional quasi-convex optimization. Both problems can be solved efficiently with existing algorithms [3, 59]. Proceeding along these lines, suboptimal controllers can be synthesized by minimizing this bound, rather than the actual cost. As shown in [3] for the continuous time case and in [64] for the discrete time case, this approach has the advantage that, in the state-feedback case, the resulting controller is *static*, while in the output-feedback case it has the *same order as the plant*.

8.4.1 An Upper Bound of the ℓ^1 Norm

As a first step in synthesizing these suboptimal controllers, we introduce a geometrical interpretation of the ℓ^1 norm, based on the concept of invariant sets.

Definition 8.4 *Consider the discrete time dynamic system*

$$x(k+1) = Ax(k) + Bw(k) \tag{8.73}$$

where $x(k) \in R^n$ and $w(k) \in R^q$, $\|w(k)\|_\infty \leq 1$. A set P is said to be positively invariant for this system if for all $x \in P$ we have $Ax + Bw \in P$.

Definition 8.5 *Consider the system (8.73). Given a sequence*

$$w = \{w(0), w(1), \ldots\}$$

and an initial condition x_o, denote by $\phi(k, x_o, w(\cdot))$ the corresponding trajectory. The origin-reachable set R_∞ is defined as $R_\infty \triangleq \{\xi : \xi = \phi(k, 0, w)\}$ for some finite time k and some sequence $w(k), \|w(k)\|_\infty \leq 1$.

It can easily be shown that the set R_∞ is the smallest invariant set containing the origin in its interior. Moreover, consider a stable system having a

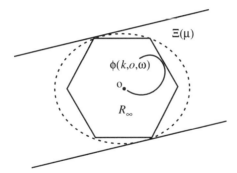

Figure 8.5. Geometric interpretation of the ℓ^1 norm and its upper bound.

state-space realization

$$G = \left[\begin{array}{c|c} A & B \\ \hline C & D \end{array}\right] \quad (8.74)$$

and define the following set:

$$\Xi(\mu) = \{\xi \colon |C\xi| \le \mu \bar{1} - \delta\} \quad (8.75)$$

where $\bar{1} \doteq [\,1\ \ 1\ \ \cdots\ \ 1\,]^T \in R^p$ and $\delta \in R^p$ is the vector whose ith component is given by $\delta_i \triangleq \|D_i\|_1$. Then $\|G\|_1 \le \mu$ if and only if $R_\infty \subseteq \Xi(\mu)$ [43].

This result can be used to synthesize (sub)optimal ℓ^1 controllers by selecting a performance level μ and then finding the largest invariant set S (along with the corresponding control action) contained in $\Xi(\mu)$ (see [43] for details). This approach has been used successfully to synthesize static nonlinear optimal ℓ^1 controllers.[6] However, attempting to proceed in a similar way to synthesize *linear* controllers leads to nondifferentiable, nonconvex optimization problems. This difficulty can be circumvented by bounding R_∞ by an invariant ellipsoid, and finding the smallest $\hat{\mu}$ such that this ellipsoid fits in $\Xi(\mu)$, as illustrated in Figure 8.5. This idea, originally proposed in [276], and used in the context of peak norm minimization in [3, 64] is formalized next.

8.4.2 The ⋆ Norm

Consider again the space ℓ_n^∞ of bounded vector sequences. Instead of the usual ℓ^∞ norm $\|h\|_\infty \triangleq \sup_i \|h_i(k)\|_\infty$, we will equip it with the norm $\|h\|_{\infty,e} \doteq \sup_k \{h^T(k)h(k)\}^{1/2}$, that is, the supremum over time of the pointwise

---

[6] Optimal controllers can be synthesized by finding the smallest μ such that the set S is nonempty.

270 $\mathcal{L}^1$ CONTROL

euclidean norm of the vector $h(k)$. In the sequel we denote the resulting space as $\ell^{\infty,e}$ to distinguish it from the usual ℓ^∞. Given a bounded linear operator $H: \ell_q^{\infty,e} \to \ell_p^{\infty,e}$, we denote by $\|H\|_{1,e}$ the operator norm induced by $\|.\|_{\infty,e}$, that is,

$$\|H\|_{1,e} \triangleq \sup_{\|v\|_{\infty,e} \leq 1} \|H*v\|_{\infty,e}$$

Note that for scalar signals this norm coincides with the usual ℓ^∞-induced norm, while in the general case we have

$$\frac{1}{\sqrt{q}}\|H\|_1 \leq \|H\|_{1,e} \leq \sqrt{p}\|H\|_1$$

Motivated by the discussion in the last section, we consider next the problem of computing the tightest invariant ellipsoid-based upper bound of $\|.\|_{1,e}$.

Lemma 8.8 *Consider the stable proper discrete time system:*

$$G = \left[\begin{array}{c|c} A & B \\ \hline C & D \end{array}\right] \tag{8.76}$$

Given α, $0 < \alpha < 1$, *define*

$$V(\alpha) = \inf_{\sigma>0, Q>0} \left\{ \eta \text{ such that } \begin{bmatrix} \sigma Q^{-1} & 0 & C^T \\ 0 & (\eta - \sigma)I & D^T \\ C & D & I \end{bmatrix} > 0 \right\} \tag{8.77}$$

where $Q > 0$ satisfies the following linear matrix inequality:

$$\begin{bmatrix} -\left(Q - \frac{1}{\alpha}BB^T\right) & AQ \\ QA^T & (\alpha-1)Q \end{bmatrix} \leq 0 \tag{8.78}$$

Then the following properties hold:

1. *The ellipsoid* $\{x: x^T Q^{-1} x \leq 1\}$ *is an invariant set for G.*
2. $\|G\|_{1,e} \leq V(\alpha)^{1/2}$.
3. $V(\alpha)$ *is quasiconvex for* $\alpha \in (0, \alpha_{\max})$, *where* $\alpha_{\max} \triangleq 1 - \rho^2(A)$.

In order to prove Lemma 8.8 we need the following preliminary result [3].

Lemma 8.9 *Given $P > 0$, then*

$$\max_{\substack{x^T P x \leq 1 \\ w^T w \leq 1}} \|Cx + Dw\|_2^2$$

$$= \inf \left\{ \eta : \exists \sigma > 0 \text{ such that } \begin{bmatrix} \sigma P & 0 & C^T \\ 0 & (\eta-\sigma)I & D^T \\ C & D & I \end{bmatrix} > 0 \right\} \tag{8.79}$$

APPROXIMATE SOLUTIONS

Proof of Lemma 8.8. Part 1: Since A is stable, it follows that for every $\alpha \in (0, \alpha_{\max})$ there exists a symmetric matrix $Q > 0$ satisfying

$$\frac{1}{1-\alpha} AQA^T - Q + \frac{1}{\alpha} BB^T \leq 0 \qquad (8.80)$$

Define the set $V_Q \triangleq \{x : x^T Q^{-1} x \leq 1\}$. Then, from Section 4.3.2 in [276], it follows that for all $w(k)$, $x(k)$ such that $w^T(k)w(k) \leq 1$ and $x^T(k)Q^{-1}x(k) \leq 1$, $x(k+1)$ satisfies

$$x(k+1)^T \Gamma^{-1} x(k+1) \leq 1$$

where

$$\Gamma = \left(\frac{1}{1-\alpha}\right) AQA^T + \frac{1}{\alpha} BB^T$$

Since (8.80) implies $Q^{-1} \leq \Gamma^{-1}$, it follows that $x(k+1)^T Q^{-1} x(k+1) \leq 1$. Hence V_Q is an invariant set.

Part 2: Since $\mathcal{R}_\infty \subseteq V_Q$, it follows that

$$\|G\|_{1,e} \leq \sup_{\substack{x \in V_Q \\ w^T w \leq 1}} \|Cx + Dw\|_2 \qquad (8.81)$$

The proof follows now from Lemma 8.9.

Part 3: For simplicity assume that the pair (A, B) is controllable. Let

$$U_\alpha = \inf_{\sigma > 0} \left\{ \eta \text{ such that } \begin{bmatrix} \sigma Q_\alpha^{-1} & 0 & C^T \\ 0 & (\eta - \sigma)I & D^T \\ C & D & I \end{bmatrix} > 0 \right\}$$

where Q_α is the solution to the discrete Lyapunov equation

$$\frac{1}{1-\alpha} AQ_\alpha A^T - Q_\alpha + \frac{1}{\alpha} BB^T = 0 \qquad (8.82)$$

From (8.77) it follows that $V(\alpha) \leq U_\alpha$. If $V(\alpha) < U_\alpha$, there exists η such that $V(\alpha) < \eta < U_\alpha$. Thus, from (8.77), it follows that there exist $Q > 0$ and $\sigma > 0$ such that $[1/(1-\alpha)]AQA^T - Q + (1/\alpha)BB^T \leq 0$ and

$$\begin{bmatrix} \sigma Q^{-1} & 0 & C^T \\ 0 & (\eta - \sigma)I & D^T \\ C & D & I \end{bmatrix} > 0 \qquad (8.83)$$

272 $\mathcal{L}^1$ CONTROL

It is easy to show that $0 < Q_\alpha \leq Q$, and

$$\begin{bmatrix} \sigma Q_\alpha^{-1} & 0 & C^T \\ 0 & (\eta - \sigma)I & D^T \\ C & D & I \end{bmatrix} \geq \begin{bmatrix} \sigma Q^{-1} & 0 & C^T \\ 0 & (\eta - \sigma)I & D^T \\ C & D & I \end{bmatrix} > 0 \qquad (8.84)$$

Hence $U_\alpha \leq \eta$, which contradicts the assumption $\eta < U_\alpha$. Therefore $V(\alpha) = U_\alpha$.

To prove that $V(\alpha)$ is quasiconvex is equivalent to showing that the sublevel set $S_\eta = \{\alpha \in (0, \alpha_{\max}) : V(\alpha) < \eta\}$ is convex for all real η. Note that $\alpha \in S_\eta$ if and only if $\alpha \in (0, \alpha_{\max})$ and there exists a real number $\sigma > 0$ such that

$$\begin{bmatrix} \sigma Q_\alpha^{-1} & 0 & C^T \\ 0 & (\eta - \sigma)I & D^T \\ C & D & I \end{bmatrix} > 0 \qquad (8.85)$$

Pre- and postmultiplying by

$$\begin{bmatrix} \frac{1}{\sqrt{\sigma}} Q_\alpha & 0 & 0 \\ 0 & \sqrt{\sigma}I & 0 \\ 0 & 0 & \sqrt{\sigma}I \end{bmatrix} \qquad (8.86)$$

yields the equivalent condition

$$M(\alpha, \sigma) \triangleq \begin{bmatrix} Q_\alpha & 0 & Q_\alpha C^T \\ 0 & f_\eta(\sigma)I & \sigma D^T \\ CQ_\alpha & \sigma D & \sigma I \end{bmatrix} > 0, \quad \text{where } f_\eta(\sigma) \triangleq \sigma(\eta - \sigma)$$

Given any $\alpha_1, \alpha_2 \in S_\eta$, and $\lambda \in (0, 1)$, define $\bar{\alpha} \triangleq \lambda \alpha_1 + (1 - \lambda)\alpha_2$. Suppose that there exist σ_1 and σ_2 such that $M(\alpha_1, \sigma_1) > 0$ and $M(\alpha_2, \sigma_2) > 0$. Then, for any $\lambda \in (0, 1)$, $\lambda M(\alpha_1, \sigma_1) + (1 - \lambda)M(\alpha_2, \sigma_2) > 0$ or, equivalently,

$$\begin{bmatrix} \bar{Q} & 0 & \bar{Q}C^T \\ 0 & (\lambda f_\eta(\sigma_1) + (1-\lambda)f_\eta(\sigma_2))I & \bar{\sigma}D^T \\ C\bar{Q} & \bar{\sigma}D & \bar{\sigma}I \end{bmatrix} > 0 \qquad (8.87)$$

where $\bar{Q} = \lambda Q(\alpha_1) + (1 - \lambda)Q(\alpha_2)$ and $\bar{\sigma} = \lambda \sigma_1 + (1 - \lambda)\sigma_2$. Note that the function $f_\eta(\sigma) = \sigma(\eta - \sigma)$ is *concave* in σ, that is, $f_\eta(\bar{\sigma}) \geq \lambda f_\eta(\sigma_1) + (1-\lambda)f_\eta(\sigma_2)$. Therefore

$$\begin{bmatrix} \bar{Q} & 0 & \bar{Q}C^T \\ 0 & f_\eta(\bar{\sigma})I & \bar{\sigma}D^T \\ C\bar{Q} & \bar{\sigma}D & \bar{\sigma}I \end{bmatrix} > 0 \qquad (8.88)$$

Multiplying both sides by

$$\begin{bmatrix} \bar{Q}^{-1} & 0 & 0 \\ 0 & I & 0 \\ 0 & 0 & I \end{bmatrix} \tag{8.89}$$

yields

$$\begin{bmatrix} \bar{Q}^{-1} & 0 & C^T \\ 0 & f_\eta(\bar{\sigma})I & \bar{\sigma}D^T \\ C & \bar{\sigma}D & \bar{\sigma}I \end{bmatrix} > 0 \tag{8.90}$$

Using the explicit expression for the solution of the Lyapunov equation (8.82)

$$Q_\alpha = \sum_{n=0}^{\infty} \frac{1}{\alpha(1-\alpha)^n} A^n B B^T (A^T)^n$$

it can easily be established that Q_α is a convex function of α. Thus $Q(\bar{\alpha})^{-1} \geq \bar{Q}^{-1}$ and

$$\begin{bmatrix} Q(\bar{\alpha})^{-1} & 0 & C^T \\ 0 & f_\eta(\bar{\sigma})I & \bar{\sigma}D^T \\ C & \bar{\sigma}D & \bar{\sigma}I \end{bmatrix} \geq \begin{bmatrix} \bar{Q}^{-1} & 0 & C^T \\ 0 & f_\eta(\bar{\sigma})I & \bar{\sigma}D^T \\ C & \bar{\sigma}D & \bar{\sigma}I \end{bmatrix} > 0 \tag{8.91}$$

Hence $\bar{\alpha} \in S_\eta$, which implies that S_η is convex on α. Thus $V(\alpha)$ is quasiconvex on the interval $(0, \alpha_{\max})$. □

This lemma suggests that an upper bound of $\|G\|_{1,e}$ can be computed by minimizing $V(\alpha)$. Following the approach in [3] we will define this upper bound as the $\star$ norm of G, that is,

$$\|G\|_\star \stackrel{\triangle}{=} \inf_{0<\alpha<1} V(\alpha)^{1/2} \tag{8.92}$$

It can easily be shown that the $\|.\|_\star$ satisfies all the properties of a norm. Moreover, computing its value only entails solving a LMI optimization problem and the scalar minimization of the function $V(\alpha)$ in the interval $(0, k)$.

8.4.3 Full State Feedback

In this section we consider the problem of synthesizing full state feedback controllers that minimize the $\star$ norm of the closed-loop system. The main result shows that these controllers are *static* and can be found by combining a LMI optimization with a scalar optimization in $(0, 1)$.

274 $\mathcal{L}^1$ CONTROL

Theorem 8.6 *Assume that G has the following state-space realization:*

$$G = \left[\begin{array}{c|cc} A & B_1 & B_2 \\ \hline C_1 & 0 & D_{12} \\ I & 0 & 0 \end{array}\right] \qquad (8.93)$$

where the pair (A, B_2) is stabilizable. Then the following statements are equivalent:

1. *There exists a finite-dimensional, full state-feedback internally stabilizing LTI controller such that $\|T_{zw}\|_\star \leq \gamma$.*
2. *There exists a static control law $u = Kx$ such that $\|T_{zw}\|_\star \leq \gamma$.*
3. *There exists a scalar α, $0 < \alpha < 1$, such that the following LMIs (in Q and V) are feasible:*

$$\begin{bmatrix} Q & QC_1^T + V^T D_{12}^T \\ C_1 Q + D_{12} V & \gamma^2 I \end{bmatrix} > 0$$

$$\begin{bmatrix} -\left(Q - \dfrac{1}{\alpha} B_1 B_1^T\right) & AQ + B_2 V \\ QA^T + V^T B_2^T & (\alpha - 1)Q \end{bmatrix} \leq 0 \qquad (8.94)$$

In this case, an internally stabilizing static controller such that $\|T_{zw}\|_\star \leq \gamma$ is given by $K = VQ^{-1}$.

Proof. $(1 \Rightarrow 3)$ Suppose that there exists a finite-dimensional LTI controller K that internally stabilizes the system and renders $\|T_{zw}\|_\star < \gamma$. Assume a realization of K is given by

$$K = \left[\begin{array}{c|c} A_k & B_k \\ \hline C_k & D_k \end{array}\right]$$

The corresponding realization of T_{zw} is given by

$$T_{zw} = \left[\begin{array}{cc|c} A + B_2 D_k & B_2 C_k & B_1 \\ B_k & A_k & 0 \\ \hline C_1 + D_{12} D_k & D_{12} C_k & 0 \end{array}\right] \doteq \left[\begin{array}{c|c} \bar{A} & \bar{B} \\ \hline \bar{C} & 0 \end{array}\right] \qquad (8.95)$$

From Lemma 8.8, there exist a scalar α and a symmetric matrix $Q > 0$ such that

$$\frac{1}{1-\alpha} \bar{A} Q \bar{A}^T - Q + \frac{1}{\alpha} \bar{B} \bar{B}^T \leq 0$$

and

$$\bar{C} Q \bar{C}^T < \gamma^2 I$$

By Schur complements, this is equivalent to

$$\begin{bmatrix} -Q + \frac{1}{\alpha} \bar{B}\bar{B}^T & \bar{A}Q \\ Q\bar{A}^T & (\alpha - 1)Q \end{bmatrix} \leq 0 \qquad (8.96)$$

$$\begin{bmatrix} Q^{-1} & \bar{C}^T \\ \bar{C} & \gamma^2 I \end{bmatrix} > 0 \qquad (8.97)$$

Partition Q as

$$\begin{bmatrix} Q_{11} & Q_{12} \\ Q_{12}^T & Q_{22} \end{bmatrix}$$

Since $Q > 0$, it follows that $Q_{11} > 0$. Multiplying on the left-hand side of (8.96) by

$$\begin{bmatrix} [I \ 0] & 0 \\ 0 & [I \ 0] \end{bmatrix}$$

and on the right by its transpose yields

$$\begin{bmatrix} -Q_{11} + \frac{1}{\alpha} B_1 B_1^T & AQ_{11} + B_2 V \\ Q_{11} A^T + V^T B_2^T & (\alpha - 1)Q_{11} \end{bmatrix} \leq 0 \qquad (8.98)$$

where $V = D_k Q_{11} + C_k Q_{12}^T$. Pre- and postmultiplying (8.97) by

$$\begin{bmatrix} Q & 0 \\ 0 & I \end{bmatrix}$$

one obtains

$$\begin{bmatrix} Q & Q\bar{C}^T \\ \bar{C}Q & \gamma^2 I \end{bmatrix} > 0 \qquad (8.99)$$

Finally, multiplying on the left by

$$\begin{bmatrix} [I \ 0] & 0 \\ 0 & I \end{bmatrix}$$

and on the right by its transpose yields

$$\begin{bmatrix} Q_{11} & Q_{11} C_1^T + V^T D_{12}^T \\ C_1 Q_{11} + D_{12} V & \gamma^2 I \end{bmatrix} > 0 \qquad (8.100)$$

$(3 \Rightarrow 2$ and $3 \Rightarrow 1)$ Suppose there exist $Q > 0$ and V that satisfy the LMI (8.94). With the controller given by $u = VQ^{-1}x$, the closed-loop system becomes

$$T_{zw} = \begin{bmatrix} A + B_2 V Q^{-1} & B_1 \\ C_1 + D_{12} V Q^{-1} & 0 \end{bmatrix} = \begin{bmatrix} A_f & B_1 \\ C_f & 0 \end{bmatrix} \qquad (8.101)$$

276 $\mathcal{L}^1$ CONTROL

Table 8.3. Optimal versus static-feedback ℓ^1 norm for Example 8.4

κ	N_{ℓ^1}	$\|\Phi\|_1$	$\|\Phi_{\text{static}}\|_1$	Gap	γ
2	2	4.21	4.65	10%	5.2
1.51	11	3.04	3.48	14%	3.91
1.501	16	3.01	3.46	15%	3.90
1.500	∞	3.00	3.46	15%	3.90

Since Q and V satisfy the LMI (8.94), it can easily be verified that

$$\frac{1}{1-\alpha}A_f Q A_f^T - Q + \frac{1}{\alpha}B_1 B_1^T \leq 0 \quad \text{and} \quad C_f Q C_f^T - \gamma^2 I < 0$$

The proof follows now from Lemma 8.8. □

Next, we illustrate these results by synthesizing $\star$-norm controllers for the system used in Example 8.4 and comparing their performance against that of the optimal ℓ^1 controller. Table 8.3 shows a comparison of the optimal ℓ^1 norm corresponding to different values of κ versus the ℓ^1 norm achieved by the *static* optimal $\star$-norm controller. It is worth noting that the gap remains constant at about 15%, even when the order of the optimal controller approaches ∞.

8.4.4 All Output Feedback Controllers for Optimal $\star$-Norm Problems

In this section we establish necessary and sufficient conditions for the existence of γ-suboptimal $\star$-norm output-feedback controllers.

Theorem 8.7 *Consider a discrete time FDLTI plant G of McMillan degree n with a minimal realization:*

$$G = \left[\begin{array}{c|cc} A & B_1 & B_2 \\ \hline C_1 & 0 & D_{12} \\ C_2 & D_{21} & D_{22} \end{array}\right] \quad (8.102)$$

Assume that the pairs (A, B_2) and (A, C_2) are stabilizable and detectable, respectively. The suboptimal $\star$-norm control problem with parameter γ, that is, $\|T_{zw}\|_\star < \gamma$, is solvable if and only if there exist pairs of symmetric matrices (R, S) in $\mathbb{R}^{n \times n}$ and a real number $\sigma > 0$ such that the following inequalities *are feasible:*

$$W_1^T \left(\frac{1}{1-\alpha} ARA^T - R + \frac{1}{\alpha} B_1 B_1^T \right) W_1 < 0$$

$$N_Q^T \begin{bmatrix} A^T SA - (1-\alpha)S & A^T SB_1 \\ B_1^T SA & -\alpha I + B_1^T SB_1 \end{bmatrix} N_Q < 0$$

$$\begin{bmatrix} R & I \\ I & S \end{bmatrix} \geq 0 \qquad (8.103)$$

$$W_2^T (\sigma I - C_1 R C_1^T) W_2 > 0$$

$$\begin{bmatrix} S & C_1^T \\ C_1 & \sigma I \end{bmatrix} > 0$$

$$\gamma^2 - \sigma > 0$$

where W_1, N_Q, and W_2 are any matrices whose columns span the null spaces of B_2^T, $[C_2 \ D_{21}]$, and D_{12}^T, respectively. Moreover, the set of γ-suboptimal controllers of order k is nonempty if and only if the LMIs (8.103) hold for some R, S, which further satisfy the rank constraint rank $(I - RS) \leq k$.

Proof. Assume, without loss of generality, that $D_{22} = 0$.[7] Proceeding as in Section 6.4, given a controller with state-space realization:

$$K = \left[\begin{array}{c|c} A_k & B_k \\ \hline C_k & D_k \end{array} \right] \qquad (8.104)$$

combine its parameters into the single variable

$$\Theta = \begin{bmatrix} A_k & B_k \\ C_k & D_k \end{bmatrix}$$

As before, in the sequel we use the following shorthand notation:

$$A_o = \begin{bmatrix} A & 0 \\ 0 & 0_k \end{bmatrix}; \qquad B_o = \begin{bmatrix} B_1 \\ 0 \end{bmatrix}; \qquad C_o = [C_1 \ 0];$$

$$\mathcal{B} = \begin{bmatrix} 0 & B_2 \\ I_k & 0 \end{bmatrix}; \qquad \mathcal{C} = \begin{bmatrix} 0 & I_k \\ C_2 & 0 \end{bmatrix}$$

$$\mathcal{D}_{12} = [0 \ D_{12}]; \qquad \mathcal{D}_{21} = \begin{bmatrix} 0 \\ D_{21} \end{bmatrix}$$

---

[7] Recall that this assumption can always be removed through a loop shifting transformation.

In terms of these variables the closed-loop matrices A_{cl}, B_{cl}, C_{cl}, and D_{cl} corresponding to the controller K can be written as

$$\begin{aligned} A_{cl} &= A_o + \mathcal{B}\Theta\mathcal{C}, & B_{cl} &= B_o + \mathcal{B}\Theta\mathcal{D}_{21} \\ C_{cl} &= C_o + \mathcal{D}_{12}\Theta\mathcal{C}, & D_{cl} &= \mathcal{D}_{12}\Theta\mathcal{D}_{21} \end{aligned} \tag{8.105}$$

From Lemma 8.8 we have that the $\star$ norm of the closed-loop transfer function is less than γ if there exists $X_c > 0$ in $R^{(n+k)(n+k)}$ such that

$$\begin{bmatrix} \sigma X_c & 0 & C_{cl}^T \\ 0 & (\gamma^2 - \sigma)I & D_{cl}^T \\ C_{cl} & D_{cl} & I \end{bmatrix} > 0 \tag{8.106}$$

$$\begin{bmatrix} -X_c^{-1} & A_{cl} & B_{cl} \\ A_{cl}^T & (\alpha - 1)X_c & 0 \\ B_{cl}^T & 0 & -\alpha I \end{bmatrix} < 0 \tag{8.107}$$

Rewriting these inequalities in terms of the matrices (8.105) yields

$$\begin{aligned} \Phi + P^T \Theta Q + Q^T \Theta^T P &< 0 \\ \Psi + \mathcal{P}^T \Theta \mathcal{Q} + \mathcal{Q}^T \Theta^T \mathcal{P} &> 0 \end{aligned} \tag{8.108}$$

where

$$\begin{aligned} \Phi &= \begin{bmatrix} -X_c^{-1} & A_0 & B_0 \\ A_0^T & (\alpha - 1)X_c & 0 \\ B_0^T & 0 & -\alpha I_{m_1} \end{bmatrix} \\ P &= \begin{bmatrix} 0 & I_k & 0_{k \times (n+k)} & 0_{k \times m_1} \\ B_2^T & 0 & 0 & 0 \end{bmatrix} \\ Q &= \begin{bmatrix} 0_{k \times (n+k)} & 0 & I_k & 0 \\ 0 & C_2 & 0 & D_{21} \end{bmatrix} \\ \Psi &= \begin{bmatrix} \sigma X_c & 0 & C_0^T \\ 0 & (\gamma^2 - \sigma)I_{m_1} & 0 \\ C_0 & 0 & I_{n_1} \end{bmatrix} \\ \mathcal{P} &= \begin{bmatrix} 0_{k \times n} & 0_k & 0_{k \times m_1} & 0 \\ 0 & 0 & 0 & D_{12}^T \end{bmatrix} \\ \mathcal{Q} &= \begin{bmatrix} 0 & I_k & 0 & 0_{k \times n_1} \\ C_2 & 0 & D_{21} & 0 \end{bmatrix} \end{aligned} \tag{8.109}$$

From Lemma 6.9 it follows that (8.108) is feasible in Θ if and only if

$$W_P^T \Phi W_P < 0; \qquad W_Q^T \Phi W_Q < 0$$
$$W_P^T \Psi W_P > 0; \qquad W_Q^T \Psi W_Q > 0 \qquad (8.110)$$

where W_P, W_Q, $W_\mathcal{P}$, and $W_\mathcal{Q}$ denote bases of Ker P, Ker Q, Ker $\mathcal{P}$, and Ker $\mathcal{Q}$, respectively.

It can easily be seen that W_P has the form

$$W_P = \begin{bmatrix} W_1 & 0 & 0 \\ 0 & 0_{k(n+k)} & 0 \\ 0 & I_{n+k} & 0 \\ 0 & 0 & I_{m_1} \end{bmatrix} \qquad (8.111)$$

where W_1 denotes any basis of the null space of B_2^T. Proceeding as in Theorem 6.3, partition X_c and X_c^{-1} as

$$X_c = \begin{bmatrix} S & N \\ N^T & X_{22} \end{bmatrix}; \qquad X_c^{-1} = \begin{bmatrix} R & M \\ M^T & Y_{22} \end{bmatrix} \qquad (8.112)$$

In terms of these matrices the condition $W_P^T \Phi W_P < 0$ reduces to

$$M_P^T \begin{pmatrix} -R & A & 0 & B_1 \\ A^T & & & 0 \\ 0 & (\alpha-1)X_c & & 0 \\ B_1^T & 0 & 0 & -\alpha I_{m_1} \end{pmatrix} M_P < 0 \qquad (8.113)$$

where

$$M_P \triangleq \begin{bmatrix} W_1 & 0 & 0 \\ 0 & I_{n+k} & 0 \\ 0 & 0 & I_{m_1} \end{bmatrix} \qquad (8.114)$$

Using Schur's complement formula we have that this is equivalent to

$$X_c > 0$$
$$0 > W_1^T \left(\frac{1}{1-\alpha} ARA^T - R + \frac{1}{\alpha} B_1 B_1^T \right) W_1 \qquad (8.115)$$

By proceeding in a similar fashion, it can be established that the inequality $W_Q \Phi W_Q < 0$ is equivalent to

$$N_Q^T \begin{bmatrix} A^T SA - (1-\alpha)S & A^T SB_1 \\ B_1^T SA & -\alpha I + B_1^T SB_1 \end{bmatrix} N_Q < 0 \qquad (8.116)$$

where N_Q denotes any basis of the null space of $[C_2 \ D_{21}]$. The condition $X_c \in R^{(n+k)(n+k)} > 0$ is equivalent to R, S satisfying

$$\begin{bmatrix} R & I \\ I & S \end{bmatrix} \geq 0 \quad \text{and} \quad rank\,(I - RS) \leq k$$

(see the proof of Theorem 6.3).

Note that W_P has the form

$$W_P = \begin{bmatrix} 0 & I_{n+k} & 0 \\ 0 & 0 & I_{m_1} \\ W_2 & 0 & 0 \end{bmatrix}$$

where W_2 is any basis of the null space of D_{12}^T. Hence $W_P^T \Psi W_P > 0$ reduces to

$$M_Q^T \begin{bmatrix} \sigma X_c & 0 & \begin{bmatrix} C_1^T \\ 0 \end{bmatrix} \\ 0 & (\gamma^2 - \sigma)I_{m_1} & 0 \\ [C_1 \ 0] & 0 & I_{n_1} \end{bmatrix} M_Q > 0 \qquad (8.117)$$

where

$$M_Q \triangleq \begin{bmatrix} 0 & I_{n+k} & 0 \\ 0 & 0 & I_{m_1} \\ W_2 & 0 & 0 \end{bmatrix} \qquad (8.118)$$

After rearranging rows and columns this can be rewritten as

$$\bar{M}_Q^T \begin{bmatrix} I_{n_1} & 0 & [C_1 \ 0] \\ 0 & (\gamma^2 - \sigma)I_{m_1} & 0 \\ \begin{bmatrix} C_1^T \\ 0 \end{bmatrix} & 0 & \sigma X_c \end{bmatrix} \bar{M}_Q > 0 \qquad (8.119)$$

where

$$\bar{M}_Q \triangleq \begin{bmatrix} W_2 & 0 & 0 \\ 0 & I_{m_1} & 0 \\ 0 & 0 & I_{n+k} \end{bmatrix} \qquad (8.120)$$

Using Schur complements once more, this is equivalent to

$$\begin{bmatrix} W_2^T & 0 \\ 0 & I_{m_1} \end{bmatrix} \begin{bmatrix} I_{n_1} - \frac{1}{\sigma}C_1 R C_1^T & 0 \\ 0 & (\gamma^2 - \sigma)I_{m_1} \end{bmatrix} \begin{bmatrix} W_2 & 0 \\ 0 & I_{m_1} \end{bmatrix} > 0 \qquad (8.121)$$

Carrying out the block multiplication explicitly yields

$$W_2^T \left(\sigma I - C_1 R C_1^T \right) W_2 > 0; \quad \gamma^2 - \sigma > 0 \qquad (8.122)$$

Similarly, from $W_Q^T \Psi W_Q > 0$, one can obtain that

$$\begin{bmatrix} S & C_1^T \\ C_1 & \sigma I \end{bmatrix} > 0$$

To recap, if $X_c > 0$ of dimension $n + k$ solves (8.106) and (8.107), then the inequalities (8.103) hold for some R, S, and σ. Conversely, if the set (8.103) admits a solution (R, S, σ), then a matrix $X_c > 0$ satisfying the inequalities (8.106) and (8.107) can be reconstructed from R, S. □

Remark Theorem 8.7 implies that if $\|T_{zw}\|_\star < \gamma$ is achieved by some controller of order $k \geq n$, there exists a controller of order n also rendering $\|T_{zw}\|_\star < \gamma$. It follows that in the output-feedback case the optimal $\star$ norm can always be achieved with controllers having the same order as the generalized plant.

The LMI-based approach introduced in Theorem 8.7 is also useful for synthesizing reduced-order controllers. These γ-suboptimal controllers of order $k < n$ correspond to pairs of (R, S) satisfying (8.103) and the additional rank constraint $rank(I - RS) = k$. Note that this additional constraint is non-convex in R and S, making the problem harder to solve.

Example 8.5 *Assume that the generalized plant has the following realization:*

$$G = \begin{bmatrix} 2.7 & -23.5 & 4.6 & 1 & 1 \\ 1 & 0 & 0 & 0 & 0 \\ 0 & 1 & 0 & 0 & 0 \\ \hline 1 & -2.5 & \kappa & 1 & 0 \\ 1 & 0 & 0 & 0.1 & 1 \end{bmatrix} \quad (8.123)$$

Table 8.4 shows a comparison of the optimal ℓ^1 closed-loop norm, $\|\Phi\|_{\ell_1}$, versus $\|\Phi_{out}\|_{\ell_1}$, the ℓ^1 norm achieved by the third-order output-feedback optimal $\star$-norm controller, for different values of the parameter κ. As before, when $\kappa \downarrow 1.5$, $N_{\ell^1} \uparrow \infty$, while the maximum gap between the optimal and suboptimal costs remains below 9%.

Table 8.4. Optimal versus output-feedback ℓ^1 norm: general case

κ	N_{ℓ^1}	$\|\Phi\|_{\ell_1}$	$\|\Phi_{out}\|_\star$	$\|\Phi_{out}\|_{\ell_1}$	Gap
2	14	12.07	16.29	12.88	7%
1.51	21	8.698	12.54	9.310	7%
1.5001	26	8.471	12.47	9.190	8%
1.5000	∞	8.465	12.46	9.187	9%

8.5 THE CONTINUOUS TIME CASE

In this section we briefly examine the continuous time counterpart of problem (8.26). While we follow a similar approach to that used in Section 8.3, there are some fundamental differences between the continuous and discrete time cases. Noteworthy, the continuous time problem leads, even in the SISO case, to a semi-infinite linear programming problem. Moreover, the resulting controller (and closed-loop system) is infinite-dimensional and has a nonrational Laplace transform.

In the sequel we use the following notation: R_+ denotes the set of non-negative real numbers. $\mathcal{L}^\infty(R_+)$ denotes the space of measurable functions $f(t)$ equipped with the norm: $\|f\|_\infty = \operatorname{ess\,sup}_{R_+} |f(t)|$. $\mathcal{L}^1(R_+)$ denotes the space of Lebesgue integrable functions on R_+ equipped with the norm $\|f\|_1 = \int_0^\infty |f(t)|\, dt < \infty$. AM denotes the space of all purely atomic measures on R_+, that is,

$$AM = \left\{ h, h(t) = \sum_{i=0}^\infty h_i \delta(t - t_i),\ \{h_i\} \in \ell^1 \right\}$$

equipped with the norm $\|h\|_{AM} = \sum_{i=0}^\infty |h_i|$. A denotes the space whose elements have the form

$$h = h^L(t) + \sum_{i=0}^\infty h_i^I \delta(t - t_i)$$

where $h^L(t) \in \mathcal{L}_1(R_+)$, $\{h_i^I\} \in \ell^1$, and $t_i \geq 0$ (i.e., $A = AM \times \mathcal{L}_1(R_+)$), equipped with the norm $\|h\|_A \triangleq \|h^L\|_{L_1} + \|h^I\|_{\ell_1}$. $\hat{A}$ denotes the space of Laplace transforms of elements in A with the norm defined as

$$\|H\|_{\hat{A}} = \|h\|_A$$

where $H(s)$ is the Laplace transform of $h(t)$. $B(R_+)$ denotes the space of bounded functions on R_+, equipped with the norm $\|f\|_B = \sup_{t \in R_+} |f(t)|$.

It can be shown [82] that $A^* = B(R_+) \times \mathcal{L}^\infty(R_+)$. Each element $(f_1, f_2) \in A^*$ defines a bounded linear functional on A, with its value given by

$$\langle (h_{AM}, h^L), (f_1, f_2) \rangle = \sum_{i=0}^\infty h_i f_1(t_i) + \int_0^\infty h^L(t) f_2(t)\, dt$$

8.5.1 Solution Via Duality

Proceeding as in Section 8.3, the first step toward applying duality is to identify the sets M and $M^\perp$. To that effect, we need the following continuous time equivalent of Lemma 8.4.

Lemma 8.10 *Assume that $T_2(s) \in \hat{A}$ has n single zeros $z_1, z_2, \ldots, z_n$, $\mathbb{Re}(z_i) > 0$, and no zeros on the $j\omega$ axis. Then, given $R \in \hat{A}$, there exists $Q \in \hat{A}$ such that $R = T_2 * Q$ if and only if*

$$R(z_i) = 0, \quad i = 1, 2, \ldots, n \quad (8.124)$$

Let

$$S = \left\{ R \in \hat{A} : R(z_i) = 0, \quad i = 1, \ldots, n \right\} \quad (8.125)$$

with this definition the continuous time $\mathcal{L}^1$ problem can be reduced (via the Youla parametrization) to

$$\mu_o = \inf_{R \in S} \|T_1 + R\|_A \quad (8.126)$$

We will solve this model matching problem by exploiting the dual of Theorem 8.3. In this case it is clear that $X = A$ and $S = M$. Moreover, let

$$\begin{aligned} f_j(t) &= \mathbb{Re}\left\{e^{-z_j t}\right\} \\ g_j(t) &= \mathbb{Im}\left\{e^{-z_j t}\right\} \end{aligned} \quad (8.127)$$

It can easily be shown that in terms of these sequences $M^\perp$ is given by

$$M^\perp = \text{Span}\left\{(f_j, f_j), (g_j, g_j)\right\}$$

Combining these results with the dual of Theorem 8.3 yields [82] the following.

Theorem 8.8 *Let $T_2(s)$ have n zeros z_i in the open right-half plane and no zeros on the jw axis. Then*

$$\mu_0 = \max_{\alpha_j} \left[\sum_{i=1}^n \alpha_i \mathbb{Re}\{T_1(z_i)\} + \sum_{i=1}^n \alpha_{i+n} \mathbb{Im}\{T_1(z_i)\} \right] \quad (8.128)$$

subject to

$$|r(t)| = \left| \sum_{i=1}^n \alpha_i \mathbb{Re}\{e^{-z_i t}\} + \sum_{i=1}^n \alpha_{i+n} \mathbb{Im}\{e^{-z_i t}\} \right| \leq 1, \quad \forall\, t \in R_+ \quad (8.129)$$

Furthermore, the following facts hold: (i) the optimal extremal functional $r^*(t)$ equals 1 at only finite points: $t_1, \ldots, t_m$; (ii) an optimal solution $\Phi(s) = T_1(s) - T_2(s)Q(s)$ to problem (8.126) always exists; and (iii) the optimal ϕ has the following form:

$$\phi = \sum_{i=1}^m \phi_i \delta(t - t_i), \; t_i \in \mathcal{R}_+, \quad m \text{ finite} \quad (8.130)$$

and satisfies the following conditions:

(a) $\phi_i r^*(t_i) \geq 0$.
(b) $\sum_{i=1}^{m} |\phi_i| = \mu_0$.
(c) $\sum_{i=1}^{m} \phi_i e^{-z_k t_i} = T_1(z_k), \quad k = 1, \ldots, n$.

Remark It can be shown [82] that the constraints (8.129) need to be satisfied only for all $t \leq t_{\max}$, where $t_{\max}$ is finite and can be determined *a priori*. Even so, there are still infinite constraints, and therefore the dual problem is a semi-infinite linear programming problem.

Example 8.6 *Consider the following unstable, nonminimum phase plant*

$$P(s) = \frac{s-1}{s-2}$$

The control objective is to minimize $\|\Phi\|_1 = \|PC(1+PC)^{-1}\|_1$. *By using the Youla parametrization, this problem can be recast into the form (8.27) with*

$$T_1 = 3\frac{s-1}{s+1}, \qquad T_2 = \frac{(s-1)(s-2)}{(s+1)^2} \qquad (8.131)$$

The dual problem is given by

$$\mu_o = \max_{\alpha_1, \alpha_2} \alpha_2 \qquad (8.132)$$

subject to

$$|r(t, \alpha)| = \left|\alpha_1 e^{-t} + \alpha_2 e^{-2t}\right| \leq 1 \quad \text{for all } t \in R_+ \qquad (8.133)$$

By differentiating (8.133) it can easily be shown that $r(t, \alpha)$ achieves local extrema at the points $t = \tau$, where

$$e^{-\tau} = -\frac{\alpha_1}{2\alpha_2}, \quad \alpha_2 \neq 0 \qquad (8.134)$$

with the corresponding value $r(\tau, \alpha) = -\alpha_1^2/4\alpha_2$. Thus problem (8.132) is equivalent to

$$\mu_o = \max_{\alpha_1, \alpha_2} \alpha_2 \qquad (8.135)$$

subject to

$$|\alpha_1 + \alpha_2| \leq 1$$
$$\left|\frac{\alpha_1^2}{4\alpha_2}\right| \leq 1 \qquad (8.136)$$
$$\alpha_1 \leq 0$$
$$\alpha_2 > 0$$

The solution to this last problem is given by

$$\alpha_1 = -(2 + 2\sqrt{2}); \qquad \alpha_2 = 3 + 2\sqrt{2} \qquad (8.137)$$

The constraint (8.133) saturates only at two points, $t = 0$ and $t = \tau = -\ln(1 + \sqrt{2})/(3 + 2\sqrt{2})$. Thus the optimal closed-loop system has the form

$$\phi = \phi_o \delta(t) + \phi_1 \delta(t - \tau)$$

Finally, the values of ϕ_o and ϕ_1 can be obtained from the interpolation constraints at $s = 1$ and $s = 2$ yielding

$$\phi_o = \frac{1 + \sqrt{2}}{\sqrt{2}}; \qquad \phi_1 = -\frac{3 + 2\sqrt{2}}{\sqrt{2}}$$

with $\|\phi_o\|_A = \alpha_2 = 5.8284$. The corresponding controller is given by

$$K_{\mathcal{L}^1} = \frac{(s-2)(1.7071 - 4.1213e^{-0.8814s})}{(s-1)(-0.7071 + 4.1213e^{-0.8814s})}$$

8.5.2 Rational Approximations to the Optimal $\mathcal{L}_1$ Controller

From equation (8.130) it follows that, unlike in the discrete-time case, the $\mathcal{L}_1$-optimal controller is irrational even if the plant is rational. Prompted by the difficulty in physically implementing a controller with a nonrational transfer function, a rational approximation method was developed in [41]. In the sequel we will briefly review this result, based on the observation that the $\mathcal{L}_1$ norm of a continuous time system is (uniformly) upper bounded by the ℓ^1 norm of its discretization using the forward Euler approximation.

Definition 8.6 *Consider the continuous time system $G(s)$. Its Euler approximating system (EAS) is defined as the following discrete time system:*

$$G^E(z, \tau) = \left[\begin{array}{c|c} I + \tau A & \tau B \\ \hline C & D \end{array}\right] \qquad (8.138)$$

From this definition it is easily seen that we can obtain the EAS of $G(s)$ by the simple variable transformation $s = (z - 1)/\tau$, that is,

$$G^E(z, \tau) = G\left(\frac{z-1}{\tau}\right)$$

On the other hand, for any given τ we can relate a discrete time system to a continuous system by the inverse transformation $z = 1 + \tau s$. It is obvious that the discrete time system is, in fact, the EAS of the continuous time system obtained in this form.

Next, we recall the main result of [41] showing that the $\mathcal{L}^1$ norm of a stable transfer function is bounded above by the ℓ^1 norm of its Euler approximating system (EAS). Moreover, this bound can be made arbitrarily tight by taking the parameter τ in (8.138) small enough. This result is the basis for the approximation procedure proposed in [41].

Theorem 8.9 *Consider a continuous time system with rational Laplace transform $\Phi(s)$ and its EAS, $\Phi^E(z,\tau)$. If $\Phi^E(z,\tau)$ is asymptotically stable, then $\Phi(s)$ is also asymptotically stable and such that*

$$\|\Phi(s)\|_1 \leq \|\Phi^E(z,\tau)\|_1$$

Conversely, if $\Phi(s)$ is asymptotically stable and such that $\|\Phi(s)\|_1 \triangleq \mu_c$, then for all $\mu > \mu_c$ there exists $\tau^* > 0$ such that for all $0 < \tau \leq \tau^*$, $\Phi^E(z,\tau)$ is asymptotically stable and such that $\|\Phi^E(z,\tau)\|_1 \leq \mu$.

Theorem 8.10 [41] *Consider a strictly decreasing sequence $\tau_i \to 0$, and define*

$$\mu_i \triangleq \inf_{\text{stabilizing } K} \|\Phi^E_{cl}(z,\tau_i)\|_1$$

where $\Phi^E_{cl}(z,\tau_i)$ denotes the closed-loop transfer function. Then the sequence μ_i is nonincreasing and such that $\mu_i \to \mu_0$, the optimal $\mathcal{L}_1$ cost.

Corollary 8.3 *A suboptimal rational solution to the $\mathcal{L}_1$ optimal control problem for continuous time systems, with cost arbitrarily close to the optimal cost, can be obtained by solving a discrete time ℓ^1 optimal control problem for the corresponding EAS. Moreover, if $K(z)$ denotes the optimal ℓ_1 compensator for the EAS, the suboptimal $\mathcal{L}_1$ compensator is given by $K(\tau s + 1)$.*

Example 8.7 Consider again Example 8.6. For $\tau = 0.1$ the EAS is given by

$$G^E(z, 0.1) = \left[\begin{array}{c|cc} 1.2 & 0 & 0.1 \\ 1 & 0 & 1 \\ \hline -1 & 1 & -1 \end{array}\right] \quad (8.139)$$

and the Youla parametrization with $F = -2.9091$ and $L = 0.3667$ yields

$$T = \left[\begin{array}{cc|cc} 0.9091 & 0.2909 & 0 & 0.0909 \\ 0 & 0.8333 & 0.3667 & 0 \\ \hline -1.9091 & 2.9091 & 0 & 0.9091 \\ 0 & -0.8333 & 0.8333 & 0 \end{array}\right]$$

$$J = \left[\begin{array}{c|cc} 1.6091 & -0.3667 & -0.2424 \\ \hline -2.9091 & 0 & 0.9091 \\ -1.5909 & 0.8333 & 0.7576 \end{array}\right]$$

(8.140)

Hence we have that

$$T_1 = \frac{176(z-1.1)}{125(1.1z-1)(1.2z-1)}$$
$$T_2 = \frac{(z-1.1)(z-1.2)}{(1.1z-1)(1.2z-1)}$$
(8.141)

Solving for the optimal ℓ^1 compensator yields optimal cost $\mu_d = 6.184$ and optimal error:

$$\phi(z) = 1.8414 - 4.3423z^{-9}$$

The corresponding optimal Q and compensator K_{EAS} are given by

$$Q(z) = 2.4309 - 0.0525z^{-1} + 0.0607z^{-2} + 0.2089z^{-3} + 0.4004z^{-4}$$
$$+ 0.6542z^{-5} + 0.9554z^{-6} + 1.3458z^{-7} + 1.8343z^{-8} - 3.2895z^{-9}$$
(8.142)

$$K_{EAS} = \mathcal{F}_l(J, Q)$$

Finally, the transformation $z = \tau s + 1$ yields the corresponding compensator for the continuous time system.

Figure 8.6 shows $\|\phi\|_A$ versus τ. Note that as $\tau \to 0$, $\|\phi\|_A \to \mu_o$. Moreover, the error $\to 0$ as $\mathcal{O}(\tau)$ [316].

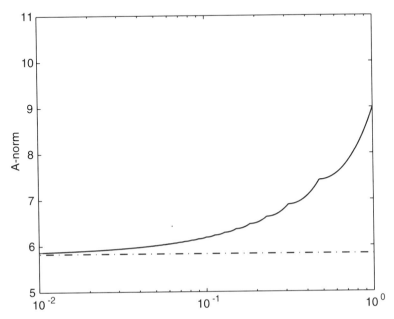

Figure 8.6. Optimal cost obtained using the EAS method versus τ.

8.6 PROBLEMS

1. Consider the uncertain linear system

$$x(k+1) = [A_o + w(k)DE]\,x(k)$$

$$A_o = 0.5 * \begin{bmatrix} 1 & 1 \\ -1 & 1 \end{bmatrix}; \quad D = \begin{bmatrix} 0 \\ 1 \end{bmatrix}; \quad E = [1\ 0] \qquad (8.143)$$

$$|w(k)| < w_{\max}$$

where $w(k)$ represents memoryless (arbitrarily fast) time varying uncertainty.

(a) Use Theorem 8.2 to show that if $w_{\max} = 0.6$ then the system (8.143) is BIBO stable.

(b) Define $G(z) \triangleq E(zI - A_o)^{-1}D$ and recall from Problem 6.10 that if $w_{\max} \leq (\|G\|_\infty)^{-1}$ then there exist $P > 0$ such that $V(x) = x^T P x$ is a Lyapunov function for the system (8.143). Use this result to reduce the conservatism in the estimate of the maximum value of $w_{\max}$ guaranteeing stability.

(c) Consider now the following *polyhedral* Lyapunov function:

$$V_p(x) = \max_{1 \leq i \leq 4} \left| \frac{(Fx)_i}{\gamma_i} \right| \qquad (8.144)$$

where

$$F = \begin{bmatrix} 0 & 1 \\ 1 & 0 \\ 1 & 1 \\ 1 & -0.3390 \end{bmatrix}, \quad \gamma = \begin{bmatrix} 1 \\ 1 \\ 1.98 \\ 0.6712 \end{bmatrix} \qquad (8.145)$$

Show that if $w_{\max} \leq 0.975$ then $V_p(x)$ is a Lyapunov function for the system (8.143). In fact, using similar techniques it can be shown that (8.143) is stable *if and only if* $w_{\max} \leq 1$. (See [42] for a thorough treatment of polyhedral Lyapunov functions).

2. The purpose of this problem is to investigate ℓ^1 control problems with zeros on the unit circle [317]. Consider the SISO ℓ^1 model-matching problem:

$$\mu_o = \inf_{Q \in \ell^1} \|T_1(z) - T_2(z)Q\|_1$$

Assume that $T_2(z)$ has two complex conjugate zeros $z_1 = e^{j\theta}$ and $z_2 = e^{-j\theta}$ on the unit circle and no other unstable zeros.

(a) Show that

$$\mu_o = \inf_{Q \in \ell^1} \|T_1 - T_2 * Q\|_1$$
$$= \max_{\alpha_i} \left[\alpha_1 \mathrm{Re}\{\hat{T}_1(z_1)\} + \alpha_2 \mathrm{Im}\{\hat{T}_1(z_2)\} \right] \qquad (8.146)$$

subject to:
$$\left|r(k)\right| = \left|\alpha_1 \cos(k\theta) - \alpha_2 \sin(k\theta)\right| \leq 1 \qquad (8.147)$$
$$k = 0, 1, 2, \ldots$$

(b) Show that for the case $\theta = \gamma\pi$ where γ is a rational number the infimum in the primal problem is always achieved and the optimization problem can be solved via finite-dimensional linear programming.

(c) Show that if $\theta = \gamma\pi$ where γ is an irrational number, there exist values of $\hat{T}_1$ such that the infimum in the primal problem can not be achieved.

(d) Show that when γ is irrational, the set of α satisfying the constraints (8.147) S_α does not depend on the exact value of γ (i.e., S_α is the same for all irrational γ). Denote this as S_α^I. Similarly, let $S_\alpha^R(\gamma)$ denote the constrained area of the dual problem for the case where $\gamma = m/n$. Show that $S_\alpha^R(\gamma)$ does not depend on m and that

$$S_\alpha^R(\gamma) \supset S_\alpha^I, \quad \text{for any of rational } \gamma \qquad (8.148)$$

(e) Use these results to conclude that the optimal approximation error μ_o does not change continuously when the zeros of $T_2(z)$ move continuously on the unit circle.

3. Consider the pitch axis control of a forward-swept wing X29 aircraft [95]. A simplified model of the plant is given by:

$$P(s) = \frac{(s+3)}{(s+10)(s-6)} \frac{20}{(s+20)} \frac{(s-26)}{(s+26)} \qquad (8.149)$$

We are interested in synthesizing a controller to minimize $\|W(s)S(s)\|_1$ where S denotes the closed-loop sensitivity function and where

$$W(s) = \frac{(s+1)}{(s+.001)} \qquad (8.150)$$

Assume that the plant is discretized via a zero-order hold at the inputs and sampling at $T_s = \frac{1}{30}$.

(a) Design a controller to minimize $\|W_d(z)S(z)\|_1$ where the weighting function $W_d(z)$ is obtained from $W(s)$ using a bilinear transformation with $T_s = 30^{-1}$.

(b) Use a CAD package to set up a simulation diagram where the continuous time plant (8.149) is connected via sample- and hold-devices to the discrete-time controller found in part (b). Find the impulse response of the resulting sampled-data system and compare it against the discrete impulse response of the system obtained by connecting the discretized version of (8.149) and the controller found in (b).

290   $\mathcal{L}^1$ CONTROL

4. Consider the following discrete time plant P:

$$P = \begin{bmatrix} A & B_1 & B_2 \\ \hline C_1 & D_{11} & D_{12} \\ C_2 & D_{21} & 0 \end{bmatrix} \quad (8.151)$$

where (A, B_2) and (A_2, C_2) are stabilizable and detectable.

(a) Proceeding as in Section 8.4.4 show that the suboptimal $\star$-norm control problem with parameter γ is solvable, if and only if, there exist pairs of symmetric matrices (R, S) in $\mathbb{R}^{n \times n}$ and a real number $\sigma > 0$ so that the following inequalities are feasible [66]:

$$W_1^T \left(\frac{1}{1-\alpha} ARA^T - R + \frac{1}{\alpha} B_1 B_1^T \right) W_1 < 0$$

$$N_Q^T \begin{bmatrix} A^T S A - (1-\alpha)S & A^T S B_1 \\ B_1^T S A & -\alpha I + B_1^T S B_1 \end{bmatrix} N_Q < 0$$

$$\begin{bmatrix} R & I \\ I & S \end{bmatrix} \geq 0 \quad (8.152)$$

$$\begin{bmatrix} W_2^T & 0 \\ 0 & I \end{bmatrix} \begin{bmatrix} I - \frac{1}{\sigma} C_1 R C_1^T & D_{11} \\ D_{11}^T & (\gamma^2 - \sigma)I \end{bmatrix} \begin{bmatrix} W_2 & 0 \\ 0 & I \end{bmatrix} > 0$$

$$N_Q^T \begin{bmatrix} \sigma S - C_1^T C_1 & -C_1^T D_{11} \\ -D_{11}^T C_1 & (\gamma^2 - \sigma)I - D_{11}^T D_{11} \end{bmatrix} N_Q > 0$$

where W_1, N_Q, W_2 are any matrices whose columns form bases of the null spaces of B_2^T, $[C_2 \; D_{21}]$, and D_{12}^T, respectively.

(b) Show that if $D_{11} = 0$ then the inequalities (8.152) reduce to the LMIs (8.103)

5. Consider the optimal closed-loop system found in Problem 3.

(a) Find the worst case peak of the control action when the input to the system is a persistent signal with magnitude bounded by one. What is the worst-case input signal?

(b) The peak control action can be reduced by synthesizing a controller $K(z)$ that minimizes

$$\left\| \begin{matrix} W_d(z)S(z) \\ 0.01 K(z)S(z) \end{matrix} \right\|_1$$

This leads to a two-block MIMO ℓ^1 problem. Find a suboptimal solution to this problem using the inequalities (8.152) and compare its performance against that of the controller found in Problem 3.

6. In order to obtain a rational controller, the X29 plant in Problem 3 was discretized using sample-and-hold elements, and the resulting discrete-time problem solved using ℓ^1 theory. However, this approach could result in significant intersampling ripple. To avoid this, use the EAS technique to synthesize a suboptimal, continuous-time rational $\mathcal{L}^1$ controller for the plant (8.149) and the weight (8.150).

9

MODEL ORDER REDUCTION

9.1 INTRODUCTION

Computers play an important role in the modeling, analysis, and design of modern control systems. Thus, it is necessary to understand not only their advantages but also the new problems that appear with their use. These problems will be considered in this chapter. In particular we will concentrate on problems that concern state-space realizations of FDLTI systems, since this is how they are represented in a computer.

It is easy to understand in other fields the problems to which we are referring. Take, for example, matrix inversion. A nonsingular matrix can be *theoretically* well defined by means of its determinant. However, from a *numerical* point of view the situation is not as clear, that is, a badly conditioned matrix can be *theoretically* nonsingular but *numerically* singular, as will be seen in the next example. This is a simple consequence of using truncated rational numbers in a computer instead of real numbers, as in the supporting theory.

Example 9.1 *Consider the following matrix:*

$$A = \begin{bmatrix} -1 & 1 & \cdots & \cdots & 1 \\ 0 & -1 & 1 & \cdots & \vdots \\ \vdots & \ddots & \ddots & \ddots & 1 \\ 0 & 0 & \cdots & 0 & -1 \end{bmatrix} \in \mathbb{R}^{n \times n} \qquad (9.1)$$

Its eigenvalues, $\lambda_1 = \cdots = \lambda_n = -1$, and its determinant, $\det(A) = (-1)^n$, indicate that in theory it is a nonsingular matrix. Nevertheless, if we multiply it by vector $u = [1 \ \ 1/2 \ \ \cdots \ \ 1/2^{n-1}]^T$ we obtain $Au = -2^{1-n}[1 \ \ \cdots \ \ 1]^T$. As the size of A increases, $\|Au\|_2$ tends to zero. This implies that numerically this is a "near singular" matrix. The latter is confirmed by the fact that $\underline{\sigma}(A) \propto 2^{1-n}$.

Therefore a theoretical zero may be represented computationally by a very small (nonzero) number that depends on the existing computing technology. The problem can be solved by quantifying the meaning of "small" and "large" in a computational numerical sense. In this example, it is well known that this can be accomplished using the singular values of the matrix. The smallest singular value is equivalent to the measure of the smallest perturbation, which may transform a nonsingular matrix into singular. This provides a "measure" of singularity. Therefore in a computational world, a matrix is *near* or *far* from singularity depending on this value, a *quantitative* property. Instead, theoretically, it can be either singular or nonsingular, a *qualitative* distinction.

The same problem appears when we work with FDLTI dynamical systems, represented in a computer by a state-space realization. The basic theory of minimal realizations ([165, 166]) suffers from the problem of *structural instability*. This means that "small" numerical perturbations can change the properties of a system. The properties we are interested in are controllability, observability, stabilizability, and detectability. As in the case of matrix singularity, measures of each of these properties can be computed, using the singular values of specific matrices (see [67, 108, 171, 202, 223, 269, 311]). As in the matrix inversion case, each of these values measures the minimal numerical perturbation that cancels a particular property. For example, ρ_s provides the smallest complex perturbation $(\delta A, \delta B)$ on the stabilizable system (A, B) such that it is no longer stabilizable. It can be computed as follows ([269]):

$$\rho_s = \inf_{s \in \mathbb{C}_+} \underline{\sigma} \{ [(A - sI) \ B] \} \quad (9.2)$$

This and other similar measures quantify a property of a system but do not distinguish separately each of its states.

In this chapter, we present a way to distinguish the states of a given system according to a property of interest. For example, we might want to order the states according to their controllability, with the first state being the "most" controllable and the last one the "least" controllable. This provides a realization useful when the less controllable states need to be eliminated. From all possible realizations of a certain system whose order is fixed, it is often convenient to select the one having its states ordered by their relative controllability and observability. This is the case when we need to verify how "minimal" a realization is from the numerical point of view, or when performing model reduction.

The latter problem is of fundamental practical importance. The following is a list, by no means extensive, of the need for model order reduction.

- The series, parallel, or feedback interconnection of systems (see Appendix B), even when these are represented by state-space minimal realizations, may produce a nonminimal one.
- Many design methodologies (e.g., μ-synthesis, ℓ^1 optimal control), result in high-order controllers.
- The state-space realizations of time-delayed systems or systems described by partial differential equations are infinite-dimensional. There are many methods that generate (high-order) finite-dimensional approximations. Thus sometimes a further model reduction step needs to be performed.
- In many cases, when working with a high-dimensional model, the numerical errors may distort the analysis or simulation results.
- Because of computation time, limited memory, or simplification of the controller hardware, it is important to have a small number of states in its realization.

In all the above cases, it is also important to have a measure of the error between the nominal and reduced-order model, which can be considered as model uncertainty in a robust control context.

9.2 GEOMETRY OF STATE-SPACE REALIZATIONS

9.2.1 Controllable/Unobservable Spaces

Consider the following FDLTI realization:

$$\dot{x}(t) = Ax(t) + Bu(t), \quad x(0) = x_0 \qquad (9.3)$$

$$y(t) = Cx(t) \qquad (9.4)$$

Define in $\mathbb{R}^n$ the controllable and unobservable subspaces as follows ([61]).

Definition 9.1 *The controllable subspace X_C contains all final states corresponding to all piecewise continuous inputs $u \in \mathcal{U}$ and zero initial conditions, $x_0 = 0$.*

Definition 9.2 *The unobservable subspace $X_{\overline{O}}$ contains all initial conditions $x_0 \in \mathbb{R}^n$, which produce null outputs, $y(t) \equiv 0$.*

It can be proved that $X_C \equiv \mathbf{R}(e^{At}B) \equiv \mathbf{R}(W_c)$, and $X_{\overline{O}} = \mathbf{N}(Ce^{At}) \equiv \mathbf{N}(W_o)$ (see [61], Lemma 1, page 75). Here W_c and W_o are the controllability and observability Gramians, respectively. This has been represented graphically in Figure 9.1.

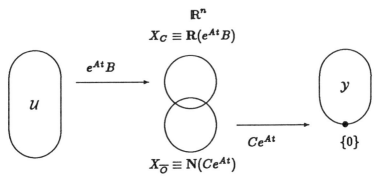

Figure 9.1. Controllable and unobservable spaces.

The state space $\mathbb{R}^n$ can be divided as follows:

$$\mathbb{R}^n = X_{CO} \oplus X_{CO}^{\perp} \tag{9.5}$$

$$X_{CO} = X_C \cap (X_C \cap X_{\overline{O}})^{\perp} \tag{9.6}$$

with X_{CO} the minimal subspace needed to describe the input–output behavior of the system. Therefore each state can be represented as the sum of two orthogonal vectors—one in X_{CO}, the other in $X_{CO}^{\perp}$.

Note that we have not defined the "observable subspace," but instead $(X_C \cap X_{\overline{O}})^{\perp}$. This is due to the fact that the set of observable states is not a vector space. Take, for example, two observable initial states $x_a(0)$ and $x_b(0)$ with zero input ($u(t) \equiv 0$) and outputs $y(t)$ and $-y(t)$, respectively. The sum of both initial conditions $x_0 = x_a(0) + x_b(0)$ gives a null output and, as a consequence, is not observable. In general, the sum of observable states of the form $x_\alpha(0) = x_r + \eta_1$ and $x_\beta(0) = -x_r + \eta_2$, where $y(t) = Ce^{At}x_r \neq 0$ and $\eta_1, \eta_2 \in X_{\overline{O}}$, may not be observable.

The subspace $X \triangleq (X_C \cap X_{\overline{O}})^{\perp}$ contains the zero element as well as observable states. It also contains unobservable and uncontrollable states, as we prove next. Divide the space $\mathbb{R}^n$ in X and $X^{\perp}$, and consider nonzero states $x_\alpha \in X^{\perp}$ and $x_\beta \in X$. For any state $x = x_\alpha + x_\beta$, we have $Ce^{At}x = Ce^{At}x_\beta$. If $Ce^{At}x_\beta \neq 0$, then it is observable. On the other hand, if $Ce^{At}x_\beta = 0$, then $x_\beta \in X_{\overline{O}}$ and $x_\beta \notin X_C$ because it cannot be in X and $X^{\perp}$ simultaneously unless $x_\beta = 0$. Therefore all unobservable states in X are also uncontrollable. As a consequence X_{CO} includes only controllable and observable states.

The subspace X_{CO} contains all states of a minimal realization of the system and therefore completely characterizes the input–output behavior from the state-space standpoint. Related with this input–output behavior, we can also describe both subspaces X_C and $X_{\overline{O}}$ in terms of input signals, as follows.

- Define $x^\ell(t)$, $\ell = 1, \ldots, p$, as the state response to input vectors $u_\ell(t) = e_\ell \delta(t)$, $\ell = 1, \ldots, p$, where e_ℓ is the ℓth column of the identity matrix in $\mathbb{R}^{p \times p}$ and $\delta(t)$ is Dirac's delta distribution. Then X_C can be defined as the smallest dimension subspace that contains the range of all outputs, that is, $\mathbf{R}[X(t); t > 0]$, with

$$X(t) = \begin{bmatrix} x^1(t) & \cdots & x^p(t) \end{bmatrix} \tag{9.7}$$

- Define $y^\ell(t)$, $\ell = 1, \ldots, m$, as the output response to initial condition vectors $x^\ell(t) = e_\ell$, $\ell = 1, \ldots, m$, with e_ℓ the ℓth column of the identity matrix in $\mathbb{R}^{m \times m}$. Then $X_{\overline{O}}$ can be defined as the largest dimension subspace contained in the null space of all outputs, that is, $\mathbf{N}[Y(t); t > 0]$, with

$$Y(t) = \begin{bmatrix} y^1(t) & \cdots & y^m(t) \end{bmatrix} \tag{9.8}$$

9.2.2 Principal Components

In this subsection we describe the geometrical characteristics of the state space in terms of the input-to-state transformation and state-to-output transformation. For simplicity we define an operator and its corresponding Gramian, which can be interpreted as any of the latter transformations. In particular, the input-to-state transformation is related to the controllability Gramian and the state-to-output transformation to the observability Gramian.

Consider a piecewise continuous function $F : \mathbb{R} \to \mathbb{R}^{n \times n}$ and its associated Gramian:

$$W = \int_{t_0}^{t_1} F(t) F^T(t) \, dt \tag{9.9}$$

The above matrix is symmetric and positive semidefinite (prove); therefore it has nonnegative real eigenvalues. As a consequence we can select the following set of orthonormal eigenvectors (Problem 1):

$$W = V \Sigma^2 V^T \tag{9.10}$$

$$\Sigma^2 = \begin{bmatrix} \sigma_1^2 & & 0 \\ & \ddots & \\ 0 & & \sigma_n^2 \end{bmatrix} \tag{9.11}$$

where the eigenvalues have been ordered for convenience ($\sigma_1 \geq \sigma_2 \geq \cdots \geq \sigma_n$) and V is a unitary matrix whose columns are the corresponding eigenvectors. To clarify, we present next specific examples of the above and the state-space interpretation of the corresponding Gramians.

Example 9.2 *Consider the problem of reaching a final (controllable) state $x_0 \neq 0$ at time $t_1 = 0$, starting from zero initial conditions at $t_0 \to -\infty$ with a minimum energy input signal $u \in \mathcal{L}_2(-\infty, 0)$:*

$$\min_{u \in \mathcal{L}_2(-\infty,0)} \|u\|_2 \quad \text{such that } x(0) = x_0 \tag{9.12}$$

This problem is similar to the following least squares one in a finite-dimensional space:

$$\min_{u \in \mathbb{R}^m} \|u\|_2 \quad \text{subject to } x_0 = Mu \in \mathbb{R}^n \tag{9.13}$$

with $M \in \mathbb{R}^{n \times m}$ and $m > n$. Denote by M^* the adjoint operator of M (in this case, the transpose). Using the geometrical fact that $\mathbb{R}^m$ can be separated into the direct sum of $\mathbf{R}(M^*)$ and $\mathbf{N}(M)$, it is clear that the minimum norm u_{opt} should be in the range of M^*. The fact that $\mathbb{R}^n$ can be separated into the direct sum of $\mathbf{R}(M)$ and $\mathbf{N}(M^*)$ implies

$$u_{opt} = M^*(x_r + x_n) = M^* x_r; \quad x_r \in \mathbf{R}(M), \ x_n \in \mathbf{N}(M^*) \tag{9.14}$$

$$\implies x_0 = M u_{opt} = M M^* x_r \tag{9.15}$$

It is easy to prove that $\mathbf{R}(M) \equiv \mathbf{R}(MM^*)$ as follows. If $y \in \mathbf{R}(MM^*) \Rightarrow y \in \mathbf{R}(M)$ because $y = MM^* x_1 = M x_2$ for $x_2 = M^* x_1$. On the other hand, if $y \notin \mathbf{R}(MM^*) \Rightarrow y \in \mathbf{N}(MM^*)$, therefore $MM^* y = 0$. For $z = M^* y$ we obtain $z \in \mathbf{N}(M)$ and $z \in \mathbf{R}(M^*)$, that is, $z = 0$. This in turn implies that y belongs to the kernel of M^* and as a consequence is not in the range of M.

When M has full rank, this fact combined with equation (9.15) yields:

$$u_{opt} = M^*(MM^*)^{-1} x_0 \tag{9.16}$$

$$\|u_{opt}\|_2^2 = x_0^T (MM^*)^{-1} x_0 \tag{9.17}$$

where $M^*(MM^*)^{-1}$ is known as the pseudoinverse of M.

A similar solution can be obtained when the input space is $\mathcal{L}_2(-\infty, 0)$ (infinite-dimensional), although the operator is different. In this case we define the infinite-dimensional operator M as

$$x_0 = x(0) = M(u) \triangleq \int_{-\infty}^{0} e^{-A\tau} B u(\tau) \, d\tau \tag{9.18}$$

$$= \int_{0}^{\infty} e^{At} B u(-t) \, dt \tag{9.19}$$

Therefore in (9.9) $F(t) = e^{At} B$ and all eigenvalues of A should be in the open left-half plane for W to exist. In this case W is exactly the controllability

GEOMETRY OF STATE-SPACE REALIZATIONS

Gramian W_c. The minimum energy input is

$$u_{opt}(t) = \underbrace{B^T e^{A^T t}}_{M^*} \underbrace{W_c^{-1}}_{(MM^*)^{-1}} x_0 \qquad (9.20)$$

$$\|u_{opt}\|_2^2 = x_0^T W_c^{-1} x_0 \qquad (9.21)$$

In this case $B^T e^{A^T t} W_c^{-1}$ is the pseudoinverse of the operator M. It is clear from the above equation that if W_c^{-1} is "large" ($\underline{\sigma}[W_c]$ "small"), certain states will be reached only by high-energy inputs, being u_{opt} the optimal one. This gives a quantitative measure of controllability, since the most controllable states can be reached with the least input energy, and vice versa. In the limit, as W_c tends to singularity,

$$\underline{\sigma}[W_c] \to 0 \qquad (9.22)$$

$$\Rightarrow \|u_{opt}\|_2 \to \infty \qquad (9.23)$$

$$\Rightarrow x_0 \to \text{noncontrollable} \qquad (9.24)$$

Example 9.3 *The dual problem is as follows. Using the output equation (9.4) with $u(t) \equiv 0$, $t > 0$, we want to find the initial condition x_0 that maximizes the energy of the output:*

$$\max_{\|x_0\|_2 \le 1} \|y\|_2 \quad \text{such that} \quad \begin{cases} \dot{x}(t) = Ax(t), \; x(0) = x_0 \\ y(t) = Cx(t) \end{cases} \qquad (9.25)$$

In this case the transformation can be defined as $y(t) \triangleq M(x_0) = Ce^{At}x_0$. Using duality and the solution of the last example, we obtain

$$\|y_{opt}\|_2^2 = x_0^T W_o x_0 \qquad (9.26)$$

As before, a quantitative measure of observability can be obtained from the "size" of the Gramian W_o. If it is near singularity ($\underline{\sigma}[W_o] \to 0$), no matter how large certain initial states may be, the output energy will tend to zero, that is, this x_0 will tend to be nonobservable.

Next we represent the function $F(t)$ in terms of the orthonormal basis formed by the eigenvectors of the Gramian W.

$$F(t) = \sum_{i=1}^{n} v_i f_i^T(t) \qquad (9.27)$$

$$f_i^T(t) \triangleq v_i^T F(t) \qquad (9.28)$$

Vector functions $v_i f_i^T(t)$ are called the *principal components* and $f_i^T(t)$ are defined as the *functional components* ([209]). The following property is useful as an interpretation of the optimal projection and "energy" distribution of this operator (for simplicity we use $t_0 = 0$ from now on).

Lemma 9.1 *The set $\{f_1(t), \ldots, f_n(t)\}$ is an orthogonal basis of functions. The norm of each element of the set is related to the singular values of the Gramian W, as follows:*

$$\int_0^{t_1} f_i^T(t) f_j(t)\, dt = \begin{cases} 0, & \forall\, i \neq j \\ \sigma_i^2, & \forall\, i = j \end{cases} \tag{9.29}$$

If $t_1 \to \infty$ and the following integral converges, then it coincides with the norm $\|F\|_2^2$:

$$\int_0^\infty \operatorname{trace}\left[F(t) F^T(t)\right] dt = \sum_{i=1}^n \sigma_i^2 \tag{9.30}$$

For each t, the argument of the integral is defined as the Frobenius matrix norm $\|F(t)\|_F^2$.

Proof.

$$\begin{aligned}
\int_0^{t_1} f_i^T(t) f_j(t)\, dt &= v_i^T \left[\int_0^{t_1} F(t) F^T(t)\, dt\right] v_j \\
&= v_i^T V \Sigma^2 V^T v_j \\
&= \sigma_i^2 v_i^T v_j \\
&= \begin{cases} 0, & \forall\, i \neq j \\ \sigma_i^2, & \forall\, i = j \end{cases}
\end{aligned} \tag{9.31}$$

which proves (9.29). Since the trace of a matrix is the sum of its eigenvalues, we obtain

$$\begin{aligned}
\operatorname{trace}(W) &= \sum_{i=1}^n \sigma_i^2 = \int_0^\infty \operatorname{trace}\left[F(t) F^T(t)\right] dt \\
&= \|F\|_2^2
\end{aligned} \tag{9.32}$$

□

The above result can be interpreted as follows. The total "energy" of the operator $F(t)$, that is, $\|F\|_2$, is distributed spacially according to the directions given by the constant eigenvectors v_i. The relative importance of each direction is quantified by the eigenvalues σ_i. Therefore the Gramian W, through its eigenvectors and values plays an important role in the geometrical interpretation of $F(t)$: in particular, as will be seen next, in the description of the controllable and unobservable subspaces and also in the input–output priority of each state (see Section 9.4).

To this end define the following subspace of $\mathbb{R}^n$:

$$S_F \triangleq \mathbf{R}(W) \tag{9.33}$$

which can be spanned by vectors v_i corresponding to all $\sigma_i > 0$. Assume without loss of generality, that the whole space can be spanned, that is, $S_F \equiv \mathbb{R}^n$. Then the optimal projection satisfies the following property.

Lemma 9.2 *Take $1 \leq r \leq n$ and $\hat{F}(t)$ any piecewise continuous function satisfying* dim $\{S_{\hat{F}}\} = r$, *Then*

$$\min_{\hat{F}} \int_0^\infty \|F(t) - \hat{F}(t)\|_F^2 \, dt = \sum_{i=r+1}^n \sigma_i^2 \tag{9.34}$$

$$\min_{\hat{F}} \int_0^\infty \overline{\sigma}^2 \left[F(t) - \hat{F}(t) \right] dt = \sigma_{r+1}^2 \tag{9.35}$$

The optimal projection in both cases is

$$\hat{F}(t) = \sum_{i=1}^r v_i f_i^T(t) \tag{9.36}$$

Proof. It can be found as Proposition 4 in [209]. □

As in Example 9.2, using the fact that $\mathbf{R}(M) \equiv \mathbf{R}(MM^*)$, it can be shown that S_F coincides with the image of $F(t)$ as a convolution operator (see [61], page 75):

$$S_F = \left\{ z \in \mathbb{R}^n \, ; \, z = \int_0^t F(t - \tau) u(\tau) \, d\tau, \quad t \in [0, t_1] \right\} \tag{9.37}$$

To obtain a geometrical interpretation, take the set $S \subset S_F$ generated by normalized inputs:

$$S \triangleq \left\{ z = \int_0^t F(t - \tau) u(\tau) \, d\tau \, , \, \|u\|_2 \leq 1, \quad t \in [0, t_1] \right\} \tag{9.38}$$

Next, we prove the main result of this section, which quantifies the relative weight of each direction v_i in $\mathbb{R}^n$ in terms of the values of σ_i. Specifically we show that S is a hyperellipsoid generated by vectors $\sigma_i v_i$.

Lemma 9.3 *The set S defined in (9.38) satisfies*

$$S \equiv \{z \in \mathbb{R}^n | z = \alpha \hat{z}, \quad \hat{z} \in \hat{S}, \quad \alpha \in [0,1]\} \tag{9.39}$$

where

$$\hat{S} \equiv \{\hat{z} \in \mathbb{R}^n | \hat{z} = V\Sigma p, \quad \|p\| = 1\} \tag{9.40}$$

is the boundary of the hyperellipsoid whose principal axes are $\sigma_i v_i$.

Proof. Take $\sigma_1 \geq \sigma_2 \geq \cdots \geq \sigma_n > 0$ in (9.11). Then there exists $q \triangleq V\Sigma^{-1}p$ such that $\hat{z} = Wq = V\Sigma p \in \hat{S}$, with $\|p\| = 1$. Therefore the input signal

$$u(t) = F^T(\tau - t)q \tag{9.41}$$

$$\|u\|_2^2 = q^T \left[\int_0^\tau F(\tau - t)F^T(\tau - t)\,dt\right] q = q^T W q$$

$$= \|p\|^2 = 1 \tag{9.42}$$

takes the state $z(0) = 0$ to $z(\tau) = \hat{z}$ $\Rightarrow$ $\hat{z} \in S$ (Problem 2). This proves that $\hat{S} \subseteq S$.

In Example 9.2 we have proved that the above input $u(t)$ is the minimum energy one[1] satisfying $\hat{z} = \int_0^\tau F(\tau - t)u(t)\,dt$. Due to linearity and $\|u\|_2 = 1$, this means that there is no $\|u\|_2 \leq 1$ that generates a final state outside the boundary of the hyperellipsoid S; therefore this boundary is $\hat{S}$. □

From the above results we can note the following:

Notes

1. If $\hat{z} = \sigma_i v_i$, then $q = W^{-1}\hat{z} = \sigma_i^{-1} v_i$ which in turn implies

$$u(t) = \frac{F^T(\tau - t)v_i}{\sigma_i} = \frac{f_i(\tau - t)}{\sigma_i} \tag{9.43}$$

Therefore the set $\{\sigma_i v_i\}$ generates the reachable states in $\mathbb{R}^n$ and the functions $\{f_i/\sigma_i\}$ generate the orthonormal set of corresponding input signals.

2. When $F(t) = e^{At}B$, the boundary represents the reachable states starting from $x(0) = 0$ with normalized inputs $\|u\|_2 = 1$. The boundary of the hyperellipsoid in the dual case—$F(t) = e^{A^T t}C^T$—represents the initial

---

[1] Take the final state $\hat{z}$ as x_0 and $W_c = W$. Then $\|u_{opt}\|_2^2 = q^T W q = x_0^T W_c^{-1} x_0 = \hat{z}^T W^{-1} \hat{z}$.

GEOMETRY OF STATE-SPACE REALIZATIONS 303

conditions from which normalized outputs $\|y\|_2 = 1$ can be obtained, with zero input.

In each of these cases, the values (σ_1/σ_n) reflect the condition number of control and observation, respectively. By this we mean the relative controllability and observability with respect to all directions.

3. The specific coordinates and magnitudes for controllability and observability depend on the Gramians, which in turn depend on the specific set of selected states; that is they are not invariant under state transformations. Therefore it is important to select a particular state transformation that can be useful to reflect the relative controllability and observability "measures" simultaneously. To this end, note the following fact.

A state transformation changes the system matrices as follows:

$$\left[\begin{array}{c|c} A & B \\ \hline C & D \end{array}\right] \xrightarrow{T} \left[\begin{array}{c|c} TAT^{-1} & TB \\ \hline CT^{-1} & D \end{array}\right] \tag{9.44}$$

The Gramians (W_c, W_o) are changed to $(TW_cT^T, T^{-T}W_oT^{-1})$, respectively. Except for special cases (unitary T), the Gramian eigenvalues will change under state transformations. Nevertheless, the eigenvalues of W_cW_o remain unchanged, since the transformed product is $TW_cW_oT^{-1}$.

Example 9.4 *Consider the following realization of the system* $1/(s+2)(s+1)$:

$$\dot{x}(t) = \begin{bmatrix} -1 & 0 \\ 0 & -2 \end{bmatrix} x(t) + \begin{bmatrix} 10^{-6} \\ -10^6 \end{bmatrix} u(t) \tag{9.45}$$

$$y(t) = \begin{bmatrix} 10^6 & 10^{-6} \end{bmatrix} x(t) \tag{9.46}$$

For this realization, the Gramians are

$$W_c = \begin{bmatrix} 0.5 \times 10^{-12} & \frac{1}{3} \\ \frac{1}{3} & 0.25 \times 10^{12} \end{bmatrix} \tag{9.47}$$

$$W_o = \begin{bmatrix} 0.25 \times 10^{12} & \frac{1}{3} \\ \frac{1}{3} & 0.5 \times 10^{-12} \end{bmatrix} \tag{9.48}$$

with similar condition numbers in the order of $\tilde{\kappa}_c \approx \tilde{\kappa}_o \approx 10^{12}$. *It is clear that for this realization x_1 is "almost" uncontrollable and x_2 "almost" unobservable. With the following state transformation,*

$$\tilde{x}(t) = \begin{bmatrix} 10^6 & 0 \\ 0 & 10^{-6} \end{bmatrix} x(t) \tag{9.49}$$

we obtain the following realization:

$$\dot{\tilde{x}}(t) = \begin{bmatrix} -1 & 0 \\ 0 & -2 \end{bmatrix} \tilde{x}(t) + \begin{bmatrix} 1 \\ -1 \end{bmatrix} u(t) \qquad (9.50)$$

$$y(t) = \begin{bmatrix} 1 & 1 \end{bmatrix} \tilde{x}(t) \qquad (9.51)$$

with a better "balance" between both condition numbers, which have decreased to $\tilde{\kappa}_c = \tilde{\kappa}_o = 38$. Clearly, this new realization has much better numerical properties. From the geometrical point of view, the highly eccentric ellipsoids of controllability and observability in $\mathbb{R}^2$ have now a more "spherical" description.

9.3 HANKEL SINGULAR VALUES

9.3.1 Continuous Systems

As indicated in the last section, it is useful to represent the dynamic system in terms of a particular set of states, which "balances" the quantitative measures of controllability and observability. In this way it is easy to order the states from the most controllable/observable to the least one. This is useful from the point of view of model reduction because this order provides the relative priority of each state in terms of its input–output effect.

The quantitative measure of relative controllability/observability should depend now on both Gramians W_c and W_o. In particular, these measures should be a property of the system, that is, they should be independent of the specific realization selected. According to the notes mentioned in the last section, the eigenvalues of the product of Gramians are invariant under state transformations. This is coherent with the definition of the Hankel singular values of Chapter 6 (Definition 6.2), as the square roots of the eigenvalues of the product $W_c W_o$ (or $W_o W_c$), indicated as

$$(W_c W_o) v_i = \left(\sigma_i^H \right)^2 v_i, \quad v_i \neq 0, \ i = 1, \ldots, n \qquad (9.52)$$

For convenience, they are usually ordered as follows $\sigma_1^H \geq \sigma_2^H \geq \cdots \geq \sigma_n^H$.

In [209] these values are called *second-order modes*. They represent the fundamental measures of gain and complexity of a LTI system ([132]). The preferred state realization, which we seek for, should make both Gramians equal and diagonal (see Section 9.4) such that

$$W_c = W_o = \begin{bmatrix} \sigma_1^H & & 0 \\ & \ddots & \\ 0 & & \sigma_n^H \end{bmatrix} \qquad (9.53)$$

Hankel Operator As mentioned in Section 6.2.1, the Hankel singular values can be interpreted in terms of the singular values of the Hankel operator introduced in Definition 6.1. This operator provides an interesting relation between the input–output gains of a system and the controllability and observability of its states. It is easily seen that Definition 6.1 is equivalent to:

$$y(t) = \Gamma_G v \triangleq \begin{cases} \int_0^\infty Ce^{A(t+\tau)} Bv(\tau)\, d\tau, & t \geq 0 \\ 0 & t < 0, \end{cases} \qquad (9.54)$$

where $v \in \mathcal{L}_2[0, \infty)$.

The interpretation of this operator in terms of inputs and outputs is as follows. Take the input $u(t) = v(-t)$, $t < 0$, therefore the input–output transformation can be performed in terms of the Hankel operator: $y(t) = (\Gamma_G v)(t) = (g * u)(t)$. This is the convolution of a system whose input is defined for negative times (in $\mathcal{L}_2(-\infty, 0]$) and its output for positive times. The final state at $t = 0$ for input $u(t)$ is

$$x(0) = \int_{-\infty}^0 e^{-A\tau} Bu(\tau)\, d\tau \qquad (9.55)$$

It corresponds to the initial state for the unforced system whose output is $y(t)$, $t > 0$. This operator transforms past inputs into future outputs through the state $x(0)$.

The Hankel singular values can be defined either way, as in Chapter 6 (Definition 6.2) or in terms of the Hankel operator. Furthermore, it can be shown that the Smith–McMillan order of a system is equal to the number of positive Hankel singular values.

Recall that the Hankel norm of a system $G(s)$ has been defined (Definition 6.3) as the largest singular value of its corresponding Hankel operator $\|G(s)\|_H = \bar{\sigma}[\Gamma_G]$. The Hankel norm can be interpreted as an induced norm between past inputs and future outputs [see equation (6.19)]:

$$\|G(s)\|_H = \max_{\substack{u \in \mathcal{L}_2(-\infty,0] \\ u \neq o}} \frac{\|P_+(g * u)\|_2}{\|u\|_2} \qquad (9.56)$$

which also equals $\sqrt{\rho(W_c W_o)}$.

Note that the Hankel operator, its singular values, and the Hankel norm are all independent of the constant D term of the realization. This is the reason why, without loss of generality, we have considered only strictly proper systems.

Next, we present the results for discrete time systems. In turns out that if the latter is obtained as a bilinear transformation of a continuous system, the Hankel singular values and Hankel norm remain unchanged.

9.3.2 Discrete Systems

Consider the discrete time, linear, shift invariant, asymptotically stable system:

$$x[(k+1)T] = A_d x(kT) + B_d u(kT) \tag{9.57}$$

$$y(kT) = C_d x(kT) + D_d u(kT) \tag{9.58}$$

Since $|\lambda_i(A_d)| < 1, \forall\, i = 1, \ldots, n$, the Gramians can be defined as

$$W_c^d \triangleq \sum_{k=0}^{\infty} A_d^k B_d B_d^T \left(A_d^k\right)^T \tag{9.59}$$

$$W_o^d \triangleq \sum_{k=0}^{\infty} \left(A_d^k\right)^T C_d^T C_d A_d^k \tag{9.60}$$

Both are obtained also as solutions of the discrete Lyapunov equations:

$$W_c^d - A_d W_c^d A_d^T - B_d B_d^T = 0 \tag{9.61}$$

$$W_o^d - A_d^T W_o^d A_d - C_d^T C_d = 0 \tag{9.62}$$

respectively.

Similar to the continuous case, the Hankel singular values are defined as follows.

Definition 9.3 *Consider the asymptotically stable system*

$$G_d(z) \equiv \left[\begin{array}{c|c} A_d & B_d \\ \hline C_d & 0 \end{array}\right] \tag{9.63}$$

with controllability and observability Gramians W_c^d and W_o^d, respectively. The set of Hankel singular values of $G_d(s)$ are the square roots of the eigenvalues of the product $W_c^d W_o^d$, indicated as

$$(W_c^d W_o^d) v_i = \left(\sigma_i^H\right)^2 v_i, \quad v_i \neq 0 \tag{9.64}$$

For convenience they are usually ordered as follows $\sigma_1^H \geq \sigma_2^H \geq \cdots \geq \sigma_n^H$.

Next, we show that the equivalent continuous system, obtained from a bilinear transform of the above, preserves the controllability and observability Gramians. Therefore the Hankel operator, singular values, and norm are preserved as well. Also, due to the fact that the McMillan degree can be defined in terms of the Hankel singular values, the order of both discrete and continuous systems is the same.

The bilinear function is[2]

$$s = \frac{z-1}{z+1} \quad (9.65)$$

$$z = \frac{1+s}{1-s} \quad (9.66)$$

which maps the interior of the unit disk $|z| < 1$ into the open left-half plane $\mathbb{Re}(s) < 0$. Applying this transformation to $G_d(z)$, we obtain

$$G(s) = G_d\left(z = \frac{1+s}{1-s}\right) \quad (9.67)$$

and for the state-space realization matrices (Problem 3)

$$A = (I + A_d)^{-1}(A_d - I) \quad (9.68)$$

$$B = \sqrt{2}(I + A_d)^{-1}B_d \quad (9.69)$$

$$C = \sqrt{2}C_d(I + A_d)^{-1} \quad (9.70)$$

$$D = D_d - C_d(I + A_d)^{-1}B_d \quad (9.71)$$

Next, pre- and postmultiply the continuous controllability Lyapunov equation by $(I + A_d)$ and $(I + A_d^T)$ as follows:

$$(I + A_d)(AW_c + W_cA^T + BB^T)(I + A_d^T) \quad (9.72)$$

By substituting (9.68) and (9.69) we obtain equation (9.61) (Problem 3). Its unique solution therefore coincides with the continuous case, that is, $W_c = W_c^d$. Similarly, we can prove that $W_o = W_o^d$.

Furthermore, if we evaluate both systems in the boundary of their respective stability regions ($|z| = 1$ and $s = j\omega$), we obtain from the bilinear transform

$$G_d(e^{j\theta}) = G\left[j\tan\left(\frac{\theta}{2}\right)\right] \quad (9.73)$$

This indicates that both frequency responses coincide in $\theta \in [0, \pi)$, and therefore their respective infinity norms. Due to the fact that the McMillan degree can be defined in terms of the Hankel singular values, the order of both discrete and continuous systems is the same.

---

[2] For simplicity consider the sampling time $T = 2$.

From the above, we conclude that any model approximation of systems $G(s)$ and $G_d(z)$ of order n by others $G_r(s)$ and $G_r^d(z)$ of order $r < n$ will have the following property. If the approximation errors $E(s) = G(s) - G_r(s)$ and $E^d(z) = G^d(z) - G_r^d(z)$ are measured in terms of the norms $\|\cdot\|_\infty$ and $\|\cdot\|_H$, they will coincide in the continuous and discrete cases. Furthermore, both continuous and discrete reduced-order models will be related by the bilinear transform as with the full-order models.

Hankel Operator In the discrete time case, the Hankel operator can be defined as follows.

Definition 9.4 *The Hankel operator or Hankel matrix of the discrete time system $G_d(z)$, described in state space by (A_d, B_d, C_d), is the infinite matrix*

$$H \triangleq \begin{bmatrix} C_d B_d & C_d A_d B_d & C_d A_d^2 B_d & \cdots \\ C_d A_d B_d & C_d A_d^2 B_d & \cdots & \cdots \\ C_d A_d^2 B_d & \vdots & \ddots & \cdots \\ \vdots & \vdots & \vdots & \ddots \end{bmatrix} \qquad (9.74)$$

Notes

1. The elements of the Hankel matrix, $C_d A_d^k B_d$, are called the *Markov parameters*.
2. The matrices used in the Kalman test of controllability and observability are, respectively,

$$K_C \triangleq \begin{bmatrix} B_d & A_d B_d & A_d^2 B_d & \cdots \end{bmatrix}, \qquad K_O \triangleq \begin{bmatrix} C_d \\ C_d A_d \\ C_d A_d^2 \\ \vdots \end{bmatrix} \qquad (9.75)$$

It is easy to verify that $H = K_O K_C$; therefore the order of the system $G_d(z)$ is $\mathrm{rank}(H)$. This is due to the fact that that both controllability and observability are determined by the ranks of K_C and K_O, respectively, and the order of the system coincides with the order of the smallest completely controllable and observable realization.

In [61] (page 111) the McMillan degree is defined also as the rank of H, which in turn can be determined as the number of positive singular values $\sigma_i(H)$, which coincide with the Hankel singular values σ_i^H, as will be proved next.

3. As in the case of continuous systems, the Hankel singular values coincide with the singular values of the Hankel operator. This is clear

from the following:

$$\begin{aligned}\sigma_i^2(H) &\triangleq \lambda_i(H^T H) = \lambda_i(K_C^T K_O^T K_O K_C) \\ &= \lambda_i(K_C K_C^T K_O^T K_O) \\ &= \lambda_i(W_c^d W_o^d) = \left(\sigma_i^H\right)^2 \end{aligned} \quad (9.76)$$

In the above we have used the fact that $W_c^d = K_C K_C^T$ and $W_o^d = K_O^T K_O$, which can be obtained from equations (9.59) and (9.60).

4. As before, the Hankel operator can be interpreted in terms of input–output signals. The terminal state $x(0) \neq 0$ due to past inputs $u(kT)$, $k = -1, -2, \ldots$, is given by:

$$x(0) = K_C \begin{bmatrix} u(-T) \\ u(-2T) \\ \vdots \end{bmatrix} \quad (9.77)$$

We can also compute the outputs for positive times $y(kT)$, $k = 0, 1 \ldots$, due to the initial condition $x(0) \neq 0$ and no input:

$$\begin{bmatrix} y(0) \\ y(T) \\ \vdots \end{bmatrix} = K_O x(0) \quad (9.78)$$

The transformation mapping past inputs ($k < 0$) to future outputs ($k > 0$) can be obtained by combining the above equations:

$$\begin{bmatrix} y(0) \\ y(T) \\ \vdots \end{bmatrix} = \underbrace{K_O K_C}_{H} \begin{bmatrix} u(-T) \\ u(-2T) \\ \vdots \end{bmatrix} \quad (9.79)$$

As before, the Hankel singular values $\sigma_i^H \equiv \sigma_i(H)$ provide a measure of the relative controllability/observability of each state $x(0)$. Geometrically, we can represent the above equation as a hyperellipsoid with each axis representing an eigenvector of H with a relative magnitude given by σ_i^H. Therefore the relative controllability/observability of $x(0)$ depends on its direction relative to the principal axes of the hyperellipsoid.

Finally, we define the Hankel norm of a discrete time system.

Definition 9.5 *The Hankel norm of a stable system $G_d(s)$ is the largest singular value of its corresponding Hankel operator: $\|G_d(s)\|_H = \overline{\sigma}(H)$ (or equivalently the largest Hankel singular value σ_1^H).*

9.4 MODEL REDUCTION

9.4.1 Introduction

The purpose of this chapter is to provide a model reduction procedure such that, given a model $G(s)$ of MacMillan degree n, we obtain a reduced model $G_r(s)$ with the same number of inputs and outputs but with r states, $r < n$. We are also interested in having a bound on the approximation error $\|G(s) - G_r(s)\|_\alpha$ in terms of a meaningful norm α. By meaningful we mean that the specific norm chosen should have relevance from the modeling and control point of view.

Take, for example, the *Hankel norm* defined in the latter section. It provides a measure of the most controllable/observable state. This is fundamental for model reduction, its main objective being to discard the less relevant states from the input–output perspective, that is, the less controllable/observable states. This is also an important measure of the "minimality" of a realization from the numerical point of view.

The importance of the *infinity norm*, particularly in robust control, can be explained quite simply. If we use a reduced-order model to design a controller, the approximation error can be interpreted as model uncertainty. In particular, it fits naturally as (global) dynamic uncertainty. As we have seen in Chapters 2 and 4, the robust stability conditions for this type of uncertainty are based on the $\mathcal{H}_\infty$ norm.

In face of the results we have so far (see Lemma 9.2), the truncation of states that correspond to small Hankel singular values provides an optimal projection in the *2-norm*.

In this section we describe the balanced-truncation model reduction method, due to Moore ([209]), and provide bounds for the approximation error in terms of the infinity and Hankel norms. Although this procedure does not solve the optimal problem,

$$\min_{r<n} \|G(s) - G_r(s)\|_\alpha, \quad \alpha = H, \infty \qquad (9.80)$$

it guarantees an approximation error that is very close to the optimal, with a smaller computational effort. The optimal Hankel norm approximation problem has been solved in [132]. An extension of the balanced-truncation method and the optimal Hankel norm approximation applied to the approximation of infinite dimensional systems can be found in [79].

9.4.2 Hankel Operator Reduction

For simplicity, consider the discrete time system described by matrices (A_d, B_d, C_d, D_d) with Hankel operator H of rank n. Due to the fact that the rank of the latter is defined as the order of the model, we can state the model order reduction problem in terms of this operator as follows:

$$\min_{r<n} \|H - H_r\|_\alpha \tag{9.81}$$

where H_r is a Hankel operator of rank r. This problem applied to finite-dimensional matrices was one of the early uses of the singular value decomposition. The solution is due to Mirsky ([208]) and is easy to prove.

Theorem 9.1 *Take* $M \in \mathbb{C}^{m \times m}$ *with a singular value decomposition* $M = U\Sigma V^*$, $\Sigma = \text{diag}[\sigma_1, \ldots, \sigma_n, 0, \ldots, 0]$ *such that* $\sigma_1 \geq \cdots \geq \sigma_n > 0$. *Also, consider a unitarily invariant matrix norm, that is,* $\|M\|_\alpha = \|UM\|_\alpha = \|MU\|_\alpha$, *with unitary* U. *Then*

$$\inf\left\{\|M - \hat{M}\|_\alpha, \ \mathbf{R}(\hat{M}) \leq r\right\} = \left\|\begin{bmatrix} \sigma_{r+1} & & 0 \\ & \ddots & \\ 0 & & \sigma_n \end{bmatrix}\right\|_\alpha \tag{9.82}$$

The optimal approximation is

$$\hat{M}_\star = U \begin{bmatrix} \sigma_1 & & & & & \\ & \ddots & & & 0 & \\ & & \sigma_r & & & \\ & & & 0 & & \\ & 0 & & & \ddots & \\ & & & & & 0 \end{bmatrix} V^* \tag{9.83}$$

Proof. See [208]. □

In principle, the problem (9.81) is not yet solved, because $\hat{M}_\star$ may not be a Hankel matrix. In 1971 Adamjan et al., ([7]) proved that Mirsky's theorem can be extended to infinite matrices, which correspond to Hankel operators of SISO discrete time dynamic systems. Furthermore, the optimal solution remains unchanged, even with $\hat{M}$ restricted to being a Hankel operator.

The following are examples of unitarily invariant matrix norms:

- $\|M\|_{2 \to 2} = \bar{\sigma}(M)$ (matrix norm induced by the euclidean norm).
- $\|M\|_T = \sum_{i=1}^r \sigma_i(M)$ (trace norm).
- $\|M\|_F^2 = \text{trace}(MM^*) = \sum_{i,j} |m_{ij}|^2 = \sum_{i=1}^r \sigma_i^2(M)$ (Frobenius norm).

The Hankel norm of a discrete system is a unitarily invariant norm of the **Hankel operator**, because it is based on the maximum singular value of H. Therefore we can conclude that the optimal solution in terms of the Hankel operator H can be obtained by truncating the set of smallest singular values, $\{\sigma_i^H, i = r+1, \ldots, n\}$, and results in

$$\min_{r<n} \bar{\sigma}(H - H_r) = \sigma_{r+1}(H) = \sigma_{r+1}^H \qquad (9.84)$$

To obtain the optimal Hankel reduced-order model we should search for a system $G_r^d(z)$ that achieves

$$\min_{r<n} \|G^d(z) - G_r^d(z)\|_H = \sigma_{r+1}(H) = \sigma_{r+1}^H \qquad (9.85)$$

The optimal system $G_r^d(z)$ should have the optimal H_r as its Hankel operator. This problem is solved in [132], both for the Hankel and infinity norms for continuous FDLTI systems. The parametrization of all optimal Hankel norm reduced models was used in one of the first algorithmic solutions to the $\mathcal{H}_\infty$ optimal control problem ([103]).

The objective of this section is to indicate that the truncation of the states that have the smallest Hankel singular values has an important conceptual significance, in terms of model reduction. We have also stressed along these lines that there is a strong relation between the Hankel operator and the system, specifically the rank of the former is the order of the latter.

Note that if we truncate the states in the *system* instead of in the Hankel operator as above, these results do not apply. This is because the truncation operation in the system realization is not a linear procedure with respect to the Hankel singular values. In other words, the Hankel singular values of the difference of a system and its balanced-truncated version are, in general, not equal to the Hankel singular values of the difference of their Hankel operators (Problem 4). This means that the truncation procedure on the system will not provide the optimal Hankel norm solution. Nevertheless, if the realization of this system is balanced, the reduced-order model, although suboptimal, has a near optimal approximation error.

9.4.3 Balanced Realizations

Definition 9.6 *The following minimal, asymptotically stable realization,*

$$\left[\begin{array}{c|c} A_b & B_b \\ \hline C_b & D_b \end{array} \right] \qquad (9.86)$$

is called input normal, output normal, internally balanced, respectively, if and only if the controllability and observability Gramians satisfy the following:

- $W_o = \Sigma^2, \; W_c = I.$
- $W_c = \Sigma^2, \; W_o = I.$
- $W_c = W_o = \Sigma = \mathrm{diag}\begin{bmatrix} \sigma_1^H & \cdots & \sigma_n^H \end{bmatrix}.$

Given a minimal realization of a system,

$$\left[\begin{array}{c|c} A & B \\ \hline C & D \end{array}\right] \qquad (9.87)$$

it is always possible to obtain an input normal, output normal, or internally balanced realization by means of nonsingular state transformations T_{in}, T_{on}, and T_{ib}, respectively.

Consider a transformation matrix T_{ib} as defined above, which takes $x(t)$ of system (9.87) to $x_{ib}(t) = T_{ib}x(t)$ of an internally balanced realization. It can be shown (Problem 5) that it relates to the other transformations as follows:

$$T_{in} = \Sigma^{-1/2} T_{ib} \qquad (9.88)$$

$$T_{on} = \Sigma^{1/2} T_{ib} \qquad (9.89)$$

For this reason we concentrate on the transformation matrix T_{ib} and on the properties of balanced realizations.

Lemma 9.4 *Consider the system (9.87) with state $x(t)$ and a nonsingular state transformation $\tilde{x}(t) = Tx(t)$. The optimization problem,*

$$\min_T \max \left[\kappa_C(T), \kappa_O(T)\right] = \frac{\tilde{\sigma}_1^H}{\tilde{\sigma}_n^H} \qquad (9.90)$$

is achieved by $T = T_{ib}$. Here $\kappa_C(T)$ and $\kappa_O(T)$ denote the condition numbers of the Gramians $\tilde{W}_c$ and $\tilde{W}_o$ of the transformed realization and $(\tilde{\sigma}_1^H, \tilde{\sigma}_n^H)$ are the maximum and minimum Hankel singular values.

Proof. By definition we have

$$\kappa_C(T) = \frac{\overline{\sigma}(\tilde{W}_c)}{\underline{\sigma}(\tilde{W}_c)}, \qquad \kappa_O(T) = \frac{\overline{\sigma}(\tilde{W}_o)}{\underline{\sigma}(\tilde{W}_o)} \qquad (9.91)$$

Using the following singular values properties,

$$\overline{\sigma}(XY) \leq \overline{\sigma}(X)\overline{\sigma}(Y) \qquad (9.92)$$

$$\underline{\sigma}(XY) \geq \underline{\sigma}(X)\underline{\sigma}(Y) \qquad (9.93)$$

and rearranging, we obtain

$$\kappa(\tilde{W}_c \tilde{W}_o) = \frac{\overline{\sigma}(\tilde{W}_c \tilde{W}_o)}{\underline{\sigma}(\tilde{W}_c \tilde{W}_o)} = \left(\frac{\tilde{\sigma}_1^H}{\tilde{\sigma}_n^H}\right)^2 \qquad (9.94)$$

$$\leq \frac{\overline{\sigma}(\tilde{W}_c)}{\underline{\sigma}(\tilde{W}_c)} \cdot \frac{\overline{\sigma}(\tilde{W}_o)}{\underline{\sigma}(\tilde{W}_o)} = \kappa_C(T)\kappa_O(T) \qquad (9.95)$$

314 MODEL ORDER REDUCTION

where we have used in (9.94) the symmetry of $\tilde{W}_c$ and $\tilde{W}_o$ to obtain $\sigma_i^2(\tilde{W}_c\tilde{W}_o) = \lambda_i(\tilde{W}_c\tilde{W}_o^2\tilde{W}_c) = \lambda_i^2(\tilde{W}_c\tilde{W}_o)$. Furthermore, we have:

$$\max[\kappa_C(T), \kappa_O(T)] \geq \sqrt{\kappa_C(T)\kappa_O(T)} \geq \left(\frac{\tilde{\sigma}_1^H}{\tilde{\sigma}_n^H}\right) \tag{9.96}$$

Therefore the minimum is achieved by $T = T_{ib}$ for which $\kappa_C(T_{ib}) = \kappa_O(T_{ib}) = (\tilde{\sigma}_1^H/\tilde{\sigma}_n^H)$. □

This property of balanced realizations indicates that they offer the best compromise in terms of conditioning of controllability and observability measures. Graphically, this means that they tend to make both controllability and observability hyperellipsoids the least eccentric (most "spherical"). Therefore this selection of states equilibrates the relation between state feedback and observation. They also provide the directions for which the controllability and observability measures coincide with the Hankel singular values, which has an input–output significance.

It can be proved (Problem 6) that the transformation that takes any realization to its internally balanced form is nonunique. This indicates that there can be a family of internally balanced realizations of a system.

9.4.4 Balanced Truncation

An important property of balanced realizations is the fact that a truncated model preserves balancing and stability. This indicates that truncated-balanced realizations is an effective model reduction methodology. Before proving the main result, we present a useful intermediate property.

Lemma 9.5 *Consider a realization (A, B) that satisfies the following Lyapunov equation for some positive definite P:*

$$AP + PA^T + BB^T = 0 \tag{9.97}$$

The system is asymptotically stable if and only if (A, B) is controllable. Otherwise (either uncontrollable or unstable) the unstable eigenvalues of A are only on the $j\omega$ axis.

Proof. Consider the eigenvalue and left eigenvector pair (λ_i, z_i^*) of A. Pre- and postmultiply the Lyapunov equation by z_i^* and z_i, respectively:

$$z_i^* AP z_i + z_i^* PA^T z_i = (\lambda_i + \bar{\lambda}_i) z_i^* P z_i = -z_i^* BB^T z_i \tag{9.98}$$

Since $z_i^* P z_i > 0$ and $-z_i^* BB^T z_i = -\|B^T z_i\|_2^2 \leq 0$, we obtain

$$\text{Re}(\lambda_i) \leq 0 \tag{9.99}$$

If (A, B) is not controllable, using the Hautus test we have

$$z_i^* \begin{bmatrix} A - \lambda_i I & B \end{bmatrix} = 0 \iff \begin{cases} B^T z_i = 0 \\ z_i^* A = \lambda_i z_i^* \end{cases} \tag{9.100}$$

and using (9.98), we obtain $\mathrm{Re}(\lambda_i) = 0$; that is, A is unstable with imaginary unstable eigenvalues.

Now assume A is not stable and again, using (9.99), all unstable eigenvalues are located in the $\jmath\omega$ axis, say, at $\lambda_i = \jmath\omega_i$. From (9.98) we obtain $\|B^T z_i\|_2 = 0$, which implies (A, B) is not completely controllable. □

By duality we can prove a similar result for observability using the corresponding Lyapunov equation. Next, we prove the main result.

Theorem 9.2 *Consider a minimal, asymptotically stable balanced realization with states $x(t) \in \mathbb{R}^n$, $t \geq 0$:*

$$G(s) \equiv \left[\begin{array}{c|c} A & B \\ \hline C & 0 \end{array}\right] \tag{9.101}$$

with Gramians $W_c = W_o = \Sigma = \mathrm{diag}\begin{bmatrix} \sigma_1^H & \cdots & \sigma_n^H \end{bmatrix}$. Divide the state vector into two subvectors as follows:

$$x(t) = \begin{bmatrix} x_1(t) \\ x_2(t) \end{bmatrix} \quad \text{with } x_i(t) \in \mathbb{R}^{n_i},\ i = 1, 2,$$

such that $n_1 + n_2 = n$, and partition the system and Gramians accordingly.

$$A = \begin{bmatrix} A_{11} & A_{12} \\ A_{21} & A_{22} \end{bmatrix}, \quad B = \begin{bmatrix} B_1 \\ B_2 \end{bmatrix}$$

$$C = \begin{bmatrix} C_1 & C_2 \end{bmatrix} \tag{9.102}$$

$$\Sigma = \begin{bmatrix} \Sigma_1 & 0 \\ 0 & \Sigma_2 \end{bmatrix} \tag{9.103}$$

Assume the Gramians of both subsystems do not share common diagonal elements. Under these conditions both realizations,

$$G_i(s) \equiv \left[\begin{array}{c|c} A_{ii} & B_i \\ \hline C_i & 0 \end{array}\right], \quad i = 1, 2 \tag{9.104}$$

are minimal, internally balanced, and asymptotically stable.

Proof. Separating both Lyapunov equations into two subsystems, we obtain for the first block-diagonal entries

$$\Sigma_1 A_{11} + A_{11}^T \Sigma_1 + C_1^T C_1 = 0 \tag{9.105}$$

$$A_{11} \Sigma_1 + \Sigma_1 A_{11}^T + B_1 B_1^T = 0 \tag{9.106}$$

316 MODEL ORDER REDUCTION

and for the lower-left block entries

$$A_{21}\Sigma_1 + \Sigma_2 A_{12}^T + B_2 B_1^T = 0 \tag{9.107}$$

$$A_{12}^T \Sigma_1 + \Sigma_2 A_{21} + C_2^T C_1 = 0 \tag{9.108}$$

Both (9.105) and (9.106) imply that subsystem (A_{11}, B_1, C_1) is balanced. Assume it is not stable; therefore by the last lemma A_{11} has at least one imaginary eigenvalue, say, at $\lambda_i = j\omega$. Then

$$(A_{11} - j\omega I) V = 0 \tag{9.109}$$

$$V^* \left(A_{11}^T + j\omega I \right) = 0 \tag{9.110}$$

where the columns of V span the right nullspace of $(A_{11} - j\omega I)$.

In (9.105) and (9.106) add and subtract $j\omega\Sigma_1$. Then pre- and postmultiply the former by V^* and V, respectively, and the latter by $V^*\Sigma_1$ and $\Sigma_1 V$, respectively. Then:

$$V^*\Sigma_1(A_{11} - j\omega I)V + V^*(A_{11}^T + j\omega I)\Sigma_1 V$$
$$= -V^* C_1^T C_1 V = 0 \tag{9.111}$$

$$\Rightarrow C_1 V = 0 \Rightarrow (A_{11}^T + j\omega I)\Sigma_1 V = 0 \tag{9.112}$$

$$V^*\Sigma_1(A_{11} - j\omega I)\Sigma_1^2 V + V^*\Sigma_1^2(A_{11}^T + j\omega I)\Sigma_1 V \tag{9.113}$$

$$= -V^*\Sigma_1 B_1 B_1^T \Sigma_1 V = 0 \tag{9.114}$$

$$\Rightarrow B_1^T \Sigma_1 V = 0 \Rightarrow (A_{11} - j\omega I)\Sigma_1^2 V = 0 \tag{9.115}$$

Therefore the columns of $\Sigma_1^2 V$ are included in the nullspace spanned by V and we can always find a matrix $\overline{\Sigma}_1$ such that $\Sigma_1^2 V = V\overline{\Sigma}_1^2$. In particular, V can be selected so that $\overline{\Sigma}_1$ is diagonal with a subset of the diagonal entries of Σ_1.

Now postmultiply (9.107) by $\Sigma_1 V$ and (9.108) by V to get

$$A_{21}\Sigma_1^2 V + \Sigma_2 A_{12}^T \Sigma_1 V = 0 \tag{9.116}$$

$$A_{12}^T \Sigma_1 V + \Sigma_2 A_{21} V = 0 \tag{9.117}$$

which in turn implies

$$A_{21} V \overline{\Sigma}_1^2 = \Sigma_2^2 A_{21} V \tag{9.118}$$

Because $\overline{\Sigma}_1$ and Σ_2 do not share common entries, it follows that $A_{21}V = 0$. Therefore rearranging equation (9.118) and the eigenvalue equation for A_{11},

$$\begin{bmatrix} A_{11} & A_{12} \\ A_{21} & A_{22} \end{bmatrix} \begin{bmatrix} V \\ 0 \end{bmatrix} - j\omega \begin{bmatrix} V \\ 0 \end{bmatrix} \tag{9.119}$$

which contradicts the stability of the original system. Therefore (A_{11}, B_1, C_1) is asymptotically stable. Similarly, we can prove the stability and balance of system (A_{22}, B_2, C_2) as well. Using the previous lemma, both subsystems are controllable and observable, therefore minimal. □

The above result is useful as a model reduction method. The objective is to separate the full model $G(s)$ into two subsystems $G_r(s)$ and $G_{n-r}(s)$. Care must be taken to order the states according to their Hankel singular values, such that $\sigma_1^H \geq \cdots \sigma_r^H \geq \sigma_{r+1}^H \cdots \geq \sigma_n^H$. In this way, the truncated model with the first r states $G_r(s)$ contains the first r Hankel singular values. Therefore the *most* controllable and observable states are contained in this reduced system, which are more relevant in an input–output sense.

The assumption that both subsystems should not share common Hankel singular values is critical. The following example ([240]) indicates that, otherwise, one of the subsystems could be uncontrollable, unobservable, or unstable. Consider the balanced system

$$\dot{x}(t) = \begin{bmatrix} -1 & 1 \\ -1 & 0 \end{bmatrix} x(t) + \begin{bmatrix} 1 \\ 0 \end{bmatrix} u(t)$$
$$y(t) = \begin{bmatrix} 1 & 0 \end{bmatrix} x(t)$$

Both Hankel singular values are 0.5 and, clearly, the subsystem $(0,0,0)$ is neither controllable, nor observable, nor stable. From the practical point of view it is not a problem, due to the fact that, generically, any matrix has distinct eigenvalues. Also, for model reduction, it is important to eliminate the subsystem that has very small Hankel singular values ($\sigma_r^H \gg \sigma_{r+1}^H$), which is not this case. This last point is important when considering the model reduction error.

It can be proved ([132]) that the approximation error for the balanced and truncated method, defined as the norm of the residual system $G_{n-r}(s)$, is given by

$$\|G(s) - G_r(s)\|_\alpha \leq 2 \sum_{r+1}^n \sigma_i^H, \quad \alpha = H, \infty \tag{9.120}$$

for both the Hankel and infinity norms, when all Hankel singular values are distinct (otherwise we may have an unstable error). Although this error is not minimal, it is close to the optimal solution. In [132] the family of optimal reduced-order models, measured in terms of the Hankel norm approximation, is obtained. The corresponding minimum approximation error is

$$\min_{r<n} \|G(s) - G_r(s)\|_H = \sigma_{r+1}^H \tag{9.121}$$

318 MODEL ORDER REDUCTION

It is possible to find a D_r matrix for the Hankel optimal reduced-order model $G_r(s)$ such that

$$\|G(s) - [G_r(s) + D_r]\|_\infty = \sum_{r+1}^{n} \sigma_i^H \qquad (9.122)$$

This is so because the Hankel norm does not depend on the constant D term of the realization.

Therefore the balanced truncation algorithm produces a solution suboptimal up to a factor of 2 in the infinity norm, with less computational work.

9.5 ALGORITHMS

Moore's algorithm ([209]) for balancing an asymptotically stable system

$$G(s) \equiv \left[\begin{array}{c|c} A & B \\ \hline C & 0 \end{array} \right] \qquad (9.123)$$

is as follows.

1. Compute the controllability Gramian W_c from the corresponding Lyapunov equation.
2. Compute its Jordan form with ordered eigenvalues (largest first) and orthogonal eigenvectors:

$$W_c = V_C \Sigma_C^2 V_C^* \qquad (9.124)$$

This is used to form the transformation matrix $T_1 \stackrel{\triangle}{=} \Sigma_C^{-1} V_C^*$, which produces the new system matrices $\tilde{A} = T_1 A T_1^{-1}$, $\tilde{B} = T_1 B$, $\tilde{C} = C T_1^{-1}$. The new controllability Gramian is $\tilde{W}_c = I$.

3. Compute the eigenvalue decomposition of the new observability Gramian as before, $\tilde{W}_o = \tilde{V}_O \tilde{\Sigma}_O^2 \tilde{V}_O^*$. This is used to form the transformation matrix $T_2 \stackrel{\triangle}{=} \tilde{\Sigma}^{1/2} \tilde{V}_O^*$, which produces the balanced realization:

$$A_b = T_2 \tilde{A} T_2^{-1} = T_2 T_1 A T_1^{-1} T_2^{-1} \qquad (9.125)$$

$$B_b = T_2 \tilde{B} = T_2 T_1 B \qquad (9.126)$$

$$C_b = \tilde{C} T_2^{-1} = C T_1^{-1} T_2^{-1} \qquad (9.127)$$

It is not difficult to verify that this is an internally balanced realization and that the transformation matrix $T = T_2 T_1$ is nonsingular under certain assumptions (Problem 8).

Another version of this algorithm was developed by Laub [180] and Glover [132].

1. Solve the corresponding Lyapunov equations to compute the Gramians W_c and W_o.
2. Compute the Cholesky form $W_o = R^*R$ (also known as the matrix square root).
3. Compute the Jordan form of $RW_cR^* = U\Sigma^2 U^*$, selecting U to be unitary and ordering the eigenvalues as before. The transformation matrix that takes the system to its balanced form is $T = \Sigma^{-1/2}U^*R$, with $A_b = TAT^{-1}$, $B_b = TB$, $C_b = CT^{-1}$.

It is also easy to verify that this is an internally balanced realization and that the transformation matrix $T = T_2 T_1$ is nonsingular under certain assumptions (Problem 9).

In both cases, after a balanced realization is obtained, a truncation step needs to be performed to reduce the system's order.

4. Separate the balanced realization matrices A_b, B_b, and C_b as follows:

$$A_b = \begin{bmatrix} A_r & A_{12} \\ A_{21} & A_e \end{bmatrix}, \quad B_b = \begin{bmatrix} B_r \\ B_e \end{bmatrix}$$
$$C_b = \begin{bmatrix} C_r & C_e \end{bmatrix}$$

where $A_r \in \mathbb{R}^{r \times r}$. The reduced-order model is

$$G_r(s) \equiv \left[\begin{array}{c|c} A_r & B_r \\ \hline C_r & 0 \end{array} \right] \tag{9.128}$$

9.5.1 Approximation Error

Selecting the order r of the reduced model entails trading-off model size versus approximation error. It is desirable to have a small number of states due to computer speed and memory size constraints, but this may increase the approximation error. To this end it is important to consider the following bounds, taking into account that the r selected states correspond to the larger Hankel singular values $\{\sigma_i^H, i = 1, r\}$.

$$\|G(s) - G_r(s)\|_H \leq 2 \sum_{r+1}^{n} \sigma_i^H \tag{9.129}$$

$$\|G(s) - G_r(s)\|_\infty \leq 2 \sum_{r+1}^{n} \sigma_i^H \tag{9.130}$$

If the original system is not strictly proper, that is, $D \neq 0$, we select for the reduced-order system $D_r \triangleq D$. This choice does not modify the Hankel singular values or the approximation error bound and causes the approximation error at higher frequencies tend to zero, that is, $\lim_{s \to \infty} \|G(s) - G_r(s)\|_\alpha \to 0$.

If the system has a stable and an unstable part, we can reduce the order of the stable part and retain the unstable one.

$$G(s) = G_s(s) + G_u(s) \tag{9.131}$$

$$G_r(s) = G_{sr}(s) + G_u(s) \tag{9.132}$$

$$\|G(s) - G_r(s)\|_\alpha = \|G_s(s) - G_{sr}(s)\|_\alpha, \quad \alpha = H, \infty \tag{9.133}$$

Again, the approximation error remains unchanged as well as the number of unstable poles. This is an important assumption for robust stability analysis (Chapters 2 and 4). If we consider $G_r(s)$ as the nominal model from which a robust controller will be designed, we may interpret the error $[G(j\omega) - G_r(j\omega)]$ as frequency-weighted additive dynamic uncertainty. This family of models should "cover" the plant $G(s)$ by means of a nominal reduced-order model $G_r(s)$ and a frequency-weighted bound. Additionally, the number of unstable poles of each member of this family should remain the same, in order for the necessary and sufficient condition that guarantees robustness of this class of (additive) dynamic uncertainty sets to be valid.

Equivalently, if $G_r(s)$ is invertible, we may consider the frequency-weighted multiplicative uncertainty $[G(j\omega) - G_r(j\omega)] G_r^{-1}(j\omega)$. In most cases, the weight for multiplicative uncertainty is lower at low frequencies and increases at higher ones, up to a critical ω for which it increases above 100%. On the other hand, the behavior of the balanced-truncated approximation error is just the opposite: it is higher at lower frequencies and goes to zero as $\omega \to \infty$. Therefore, for robust control applications, it is important to *invert* this frequency characteristic of the approximation error.

This can be achieved by balancing the realization of $G(s)$, but truncating a new system $\tilde{G}(s) = G(1/s)$. The reduced model is given by $G_r(s) = \tilde{G}_r(1/s)$. This transforms lower into higher frequencies and preserves stability, Gramians, and, as a consequence, also balancing (see [246]). Furthermore, the error bounds remain the same, although the error as a function of frequency changes. Specifically, the algorithm proceeds as follows:

1. Transform the balanced system $G(s)$ to $\tilde{G}(s)$:

$$\left[\begin{array}{c|c} A & B \\ \hline C & D \end{array}\right] \xrightarrow{1/s} \left[\begin{array}{c|c} A^{-1} & A^{-1}B \\ \hline -CA^{-1} & D - CA^{-1}B \end{array}\right] \tag{9.134}$$

2. Define:

$$A^{-1} = \begin{bmatrix} A_{11} & A_{12} \\ A_{21} & A_{22} \end{bmatrix}^{-1} = \begin{bmatrix} K_{11} & K_{12} \\ K_{21} & K_{22} \end{bmatrix} \tag{9.135}$$

based on the partition of the original balanced system in (9.102). Then the truncated system $\tilde{G}(s)$ is

$$\tilde{G}_r(s) \equiv \left[\begin{array}{c|c} K_{11} & K_{11}B_1 + K_{12}B_2 \\ \hline -C_1 K_{11} - C_2 K_{21} & D - CA^{-1}B \end{array} \right] \quad (9.136)$$

3. Finally, invert the frequencies on the reduced-order model $\tilde{G}_r(s)$. Simplifying we obtain (prove)

$$G_r(s) \equiv \left[\begin{array}{c|c} A_{11} - A_{12}A_{22}^{-1}A_{21} & B_1 - A_{12}A_{22}^{-1}B_2 \\ \hline C_1 - C_2 A_{22}^{-1} A_{21} & D - C_2 A_{22}^{-1} B_2 \end{array} \right] \quad (9.137)$$

□

Note that the matrix A_{22} is always invertible, because it is stable. Thus $G_r(s)$ is well defined.

There are other algorithms that solve some of the numerical problems inherent in obtaining balanced realizations ([263]). It is also possible to perform balanced truncation over badly conditioned realizations or nonminimal ones. These algorithms are implemented in commercial CAD tools: Mu-Tools and Robust Toolbox from MathWorks ([20, 262]).

9.6 PROBLEMS

1. Prove that the Gramian W in (9.9) is symmetric and positive semidefinite, and therefore its eigenvalues are all real and nonnegative.

2. In the proof of Lemma 9.3, verify that signal $u(t)$ takes the initial state $z(0)$ to $z(\tau)$ with minimum energy.

3. Obtain the expressions in (9.68) and (9.69) as functions of the matrices of the equivalent discrete time system. Prove that substituting in (9.72) we obtain the controllability Gramian Lyapunov equation (9.61).

4. Show, by means of an example, that the Hankel singular values of the difference of a system and its balanced-truncated version are not equal to the Hankel singular values of the difference of their Hankel operators. In which case could they be equal?

5. Prove the relation between matrices T_{ib}, T_{in}, and T_{on}, as indicated in (9.88) and (9.89).

6. Consider a stable, strictly proper minimal realization represented by matrices A, B, and C. Prove that any balanced realization of this system can be transformed to another balanced realization by means of a unitary T satisfying $T\Sigma = \Sigma T$, with $W_c = W_o = \Sigma$. Thus there is not a unique balanced realization.

7. Prove that Moore's algorithm transforms a realization into its balanced form. Which is the assumption that guarantees T is a transformation matrix?

8. Prove that the Laub/Glover algorithm transforms a realization into its balanced form. Which is the assumption that guarantees T is a transformation matrix?

9. From the point of view of computational cost, what is the difference between both algorithms?

10. Consider the Jordan form of the product of Gramians of a system, $W_c W_o = T\Sigma^2 T^*$ with unitary T. Can T be used as the transformation matrix which takes the realization to its internally balanced form?

11. Verify the equations for the algorithm that provides zero error at $\omega = 0$ presented in Section 9.5.1.

10

ROBUST IDENTIFICATION

10.1 INTRODUCTION

In the first part of this book, the analysis and synthesis of robust controllers were considered. A basic point that has not yet been addressed is the method by which a set of models that represents a specific physical process can be obtained. Before the appearance of systematic methods, families of models were obtained by *ad hoc* procedures (see Example 11.2). At the end of the 1980s, the first algorithmic strategies were introduced, based on approximation techniques that provide a uniform error bound ([134, 139, 321]).

Classical identification procedures ([184, 293]) are not adequate for robust control, because usually they identify, using stochastic methods, a set of parameters of a fixed mathematical structure. These procedures are used to identify parametric models in adaptive control. Within this framework, a fixed model order is assumed, which is not the case when using robust design and analysis tools. Even if families of models with parametric uncertainty could be obtained in this way, there is a limited design machinery for robust analysis and synthesis of this class of uncertain plants. In general, the analysis of parametric sets of models is a NP-hard problem ([60, 248]).

Robust control analysis and design tools are based on a deterministic worst-case approach, with no previous assumption on the order of the system. For this reason intense research was initiated in the 1990s on deterministic identification procedures, which can be used as a first step in a robust control design methodology. This area is called *robust identification*.

The robust identification problem has been posed in [143, 144, 145, 146] and has attracted considerable attention since then. It considers model uncertainty due to two different sources: measurement noise and lack of knowledge of the system itself due to the limited information supplied by the experimental data. Therefore these new identification procedures are based not only on the experimental data (*a posteriori* information) but also on the *a priori* assumptions on the class of systems to be identified. The algorithms produce a nominal model based on the experimental information and a *worst-case* bound over the set of models defined by the *a priori* information.

If time response experiments are performed, an ℓ_1 error bound can be defined ([75, 193, 194, 197]), which is adequate for ℓ_1 optimal control. Furthermore, these types of experiments are useful in verifying the (approximate) stability, time invariance, and linearity of the systems ([272]), for example, before a frequency response experiment[1] is performed.

The uncertainty bounds can also be measured in terms of the $\mathcal{H}_\infty$ norm ([9, 137, 138, 146, 147, 191, 196]), which is more adequate for $\mathcal{H}_\infty$ optimal control or μ-synthesis procedures. In this case, the experimental data are obtained from the frequency response of the physical system, which is also assumed to be linear and time invariant.

A general abstract formulation of the problem will be presented next. We postpone the specific algorithms and convergence results using $\mathcal{H}_\infty$ and ℓ_1 metrics to later sections of this chapter.

10.2 GENERAL SETUP

In this chapter, we present algorithms for solving the problem of (robust) control-oriented identification of physical plants. The outcome of these procedures is a family of models, represented by a nominal one and an identification error. We start with an abstract and general formulation ([304]), which does not take into consideration specific metrics (used to evaluate identification errors and convergence) or model and noise sets. The objective of presenting a general formulation is to point out the differences with classical parametric identification methods ([184, 293]). It is also useful to understand the fundamental limitations that appear when working with corrupted and incomplete data and model *set* inclusion, from an information theoretic point of view. As a consequence, the structure of the information on the plant will be stressed.

Furthermore, this general setup is useful to indicate the similarities with *approximation* procedures of infinite-dimensional models. These will be included as particular cases of the identification ones. Recall that approximation is loosely defined as a procedure that represents a known large-order model (even infinite-dimensional) by a simplified smaller-order one. On the other hand, identification is meant as model computation from (possibly) *a priori* information and experimental data.

---

[1] Although all experiments are performed in the *time* domain, a sinusoidal sweep at the input of the plant to obtain an output magnitude and phase response is defined here as a frequency response experiment (see Section 10.3.2).

10.2.1 Input Data

The outcome of a robust identification procedure is a family of models that should include the real physical plant behavior. This family is specified by a nominal model and an uncertainty error measured in a certain norm. For the problem to be nontrivial, classes of candidate models and measurement noise should be assumed,[2] as will be explained later in greater detail. The uncertainty bound should be valid for any plant included in the class of models or noise inside the class of measurement noise. This naturally produces a *worst-case* approach. Therefore the input data to the problem are the class of candidate models and measurement noise assumed, called *a priori* information, and the experimental data, called *a posteriori* information.

The *a posteriori* information is a vector $\mathbf{y} \in \mathbb{C}^M$ of experimental data corrupted by noise η. The data can be frequency and/or time noisy samples of the system to be identified. For a model g and a given noise vector η, the experiment can be defined in terms of the operator $E(\cdot, \cdot)$ as follows:

$$\mathbf{y} = E(g, \eta) \tag{10.1}$$

For simplicity we assume $E(\cdot, \cdot)$ linear with respect to both variables, the noise being additive.

Note that this is not an injective operator because the same outcome $\mathbf{y}$ may be produced by different combinations of model and noise. This is a restatement of the fact that the information provided by $\mathbf{y}$ is incomplete (M samples) and corrupt (noise η). Therefore the operator is not invertible and no direct operation over $\mathbf{y}$ will provide the model g. Instead, a type of *set inversion* will be attempted. By this we mean the computation of a set of models, described by a (central) nominal one g_{id} and an uncertainty bound c_u. This set "covers" all possible sets of plants in the *a priori* class, which could have produced the *a posteriori* information $\mathbf{y}$. In order to make this a nontrivial problem, it is necessary to add extra information on the class of models and measurement noise. Otherwise the combinations of models g and measurement noises η, which could have produced $\mathbf{y}$, would not only be infinite but would also form unbounded sets. The set of candidate models would have an uncertainty bound $c_u \to \infty$.

This last point can be exemplified as follows. Assume we have an experiment that produces the vector of real numbers $\mathbf{y} = \begin{bmatrix} y_0 & \cdots & y_{M-1} \end{bmatrix}^T$. These represent the (truncated) impulse response of a certain (unknown) system $g_\star$ measured with (unknown) noise $\eta_\star$ at times $t_k = k\Delta t$, that is, $y_k = g_\star(t_k) + \eta_\star(t_k)$, $k = 0, \ldots, M - 1$. If the only available information is the vector $\mathbf{y} \in \mathbb{R}^M$, there is no way to identify plant $g_\star$. Note that we can always represent every measurement point by $y_k = g(t_k) + \eta(t_k)$, $k = 0, \ldots, M - 1$, where $g(t_k) = y_k + z$ and $\eta(t_k) = -z$, for any arbitrary real value z. There-

---

[2] Otherwise the uncertainty bound could be infinite and both the modeling and control problems would not make sense.

fore any (possibly unbounded) model or noise could generate this same set of data points.

Even without measurement noise, that is, $\eta_\star \equiv 0$, the candidate models $g(\cdot)$ could be almost anything, as long as $y_k = g(t_k)$, $k = 0, \ldots, M - 1$. We can interpret this result in terms of the more familiar *aliasing* concept. Specifically it is related to Shannon's theorem, which poses an upper bound on the class of plants that could be identified, using this experimental information $\mathbf{y}$. Only if $g_\star$ belongs to the FDLTI class of models with bandwidth less than $1/2\Delta t$ can it be computed from the available data. If the bandwidth is greater than $1/2\Delta t$ but bounded, a smaller (and bounded) set of candidate models (which include $g_\star$) could be determined.

The above argument justifies the need to define a class $\mathcal{S}$ of models and $\mathcal{N}$ of measurement noise. Specific classes will be described in further sections of this chapter, according to the norms used to bound the uncertainty error.

Hence the input data to a robust identification algorithm are composed of both *a priori* $(\mathcal{S}, \mathcal{N})$ and *a posteriori* ($\mathbf{y}$) information. Due to the fact that the assumed *a priori* information is a quantification of the engineering common sense or simply a "leap of faith," there is no guarantee that it will be coherent with the experimental *a posteriori* data. Therefore consistency of both types of information should be tested before an algorithmic procedure can be applied to compute the family of models.

10.2.2 Consistency

Consistency is a concept that can easily be understood if we first define the set of all possible models, which could have produced the *a posteriori* information $\mathbf{y}$, in accordance with the class of measurement noise:

$$\mathcal{S}(\mathbf{y}) \triangleq \{g \in \mathcal{S} \mid \mathbf{y} = E(g, \eta), \ \eta \in \mathcal{N}\} \tag{10.2}$$

Therefore $\mathcal{S}(\mathbf{y}) \subset \mathcal{S}$ is the smallest set of models, according to all the available input data (*a priori* and *a posteriori*), which are indistinguishable from the point of view of the input information. This means that with the knowledge of $(\mathbf{y}, \mathcal{S}, \mathcal{N})$ there is no way to select a smaller set of candidate models. Note that the above set is independent of the metric selected and depends on all the input information.

As we mentioned in the last section, for a robust identification problem to make sense, the identification error should be bounded. Since the "size" of the family of models is defined by this error bound and should include the set $\mathcal{S}(\mathbf{y})$, this set should be bounded as well. For this it is sufficient to have a bounded class $\mathcal{S}$ of models (see Figure 10.1). Furthermore, the "size" of the set $\mathcal{S}(\mathbf{y})$ places a lower bound on the identification error, which cannot be decreased unless we add some extra information to the problem. The latter can take the form of new experimental data or more detailed information on the sets $\mathcal{S}$ and/or $\mathcal{N}$. This lower bound on the uncertainty error holds for

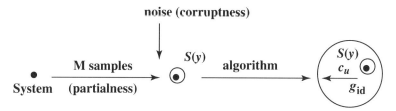

Figure 10.1. General statement of the problem.

any identification algorithm and represents a type of *uncertainty principle* of identification theory (see Lemma 10.4).

It is clear from the above arguments that consistency of *a priori* and *a posteriori* information can be defined as follows.

Definition 10.1 *The* a priori *information* $(S, \mathcal{N})$ *is consistent with the experimental* a posteriori *information* $\mathbf{y}$ *if and only if the set* $S(\mathbf{y})$ *is nonempty.*

It should be noted that once consistency between *a priori* and *a posteriori* information is established, both are equally reliable. Nevertheless, this does not imply that the "real" plant $g_\star$ belongs to S. This statement deserves a more detailed explanation, which will be presented next.

It could happen that, although $(S, \mathcal{N})$ and $\mathbf{y}$ are consistent, the set S does not include the real plant $g_\star$, that is, the *a priori* information is false. This could be the case of very noisy measurements ($\mathcal{N}$ "large"), which allows the set $S(\mathbf{y})$ not to be empty, although $g_\star \notin S(\mathbf{y})$, that is, there exists $g_1 \in S(\mathbf{y})$ such that $E(g_1, \eta_1) = E(g_\star, \eta_\star) = \mathbf{y}$ with $\eta_1, \eta_\star \in \mathcal{N}$.

From a practical point of view we can detect this problem by performing more experiments, say, $\{\mathbf{y}^1, \mathbf{y}^2, \ldots, \mathbf{y}^N\}$, $\mathbf{y}^i \in \mathbb{R}^M, i = 1, \ldots, N$. If we define this set of experiments as Λ, the new consistency set is

$$S(\Lambda) = \left\{ g \in S \mid \mathbf{y}^i = E(g, \eta^i), \ \eta^i \in \mathcal{N} \text{ and } \mathbf{y}^i \in \Lambda \right\} \qquad (10.3)$$

It is clear that if one of the experiments in Λ is $\mathbf{y}$ in equation (10.2), then $S(\Lambda) \subset S(\mathbf{y})$. In fact, we can equivalently define $S(\Lambda) = \cap_{i=1}^{N} S(\mathbf{y}^i)$. If the sets S and $\mathcal{N}$ meet certain conditions (detailed in further sections) and the experimental operator $E(\cdot, \cdot)$ makes sense, as we increase the number of experiments, the consistency set might get "smaller." Also, the *a posteriori* local error bound may decrease [see equation (10.5)].

Nevertheless, from a worst-case point of view, although N new experiments are performed, we could always obtain the same set of measurements $\mathbf{y}$. In this sense, the new consistency set would not "decrease," that is, $S(\Lambda) \equiv S(\mathbf{y})$. This can be interpreted in terms of the *a priori* lowest error bound (see Lemma 10.4). Because this *a priori* error is based on the worst-case experiment, it will not change with new experimental data. Therefore to decrease the consistency set (and the *a priori* lower error bound) new *types* of experiments need to be performed, for example, increasing the number M of measurements, decreasing the measurement noise bound, and incorporating a new type of experimental data (time and/or frequency).

However, and this concerns any empirical theory, no matter how many experiments we perform there is always the possibility of having the real plant excluded from the set $\mathcal{S}$, even for consistent input information. This is an epistemological problem that relates to the following facts:

- Any empirical theory (or model) can only be *falsified* (see [245]).
- The number of experiments is always finite.

At this point, the following practical remark can be made: To reduce the possibilities of having the real plant $g_\star$ excluded from $\mathcal{S}$, as many different *types* of experiments (time, frequency, changing sensors, and/or data points) as possible should be performed. The final verification is always through a new consistency test. This is the best that can be done in the framework of the scientific methodology.

In the case of approximation, the set $\mathcal{S}$ has only a single element (usually infinite-dimensional). Therefore the "experiment" that consists of a partial sample of this system (M data points), "measured" without noise, is always consistent. The uncertainty in this case arises only from the partialness (M data points) of the information. The only element of $\mathcal{S}$ (the "real" system) cannot be used as the desired model g_{id} because of computational reasons.

In later sections, the consistency problem will be stated as determining the existence of a model $g \in \mathcal{S}$, such that the difference between its output and $\mathbf{y}$ is included in set $\mathcal{N}$, that is, it is only due to measurement noise. This may be solved as an interpolation problem. The latter holds for additive measurement noise and linearity of the operator $E(\cdot,\cdot)$ with respect to model and noise, which we have assumed.

10.2.3 Identification Error

The *a priori* knowledge of the real system and measurement noise present in the experiment $\mathbf{y}$ is stated in terms of sets $\mathcal{S}$ and $\mathcal{N}$. The statement of the problem does not assign probabilities to particular models or noises; therefore it is deterministic in nature. In addition, the modeling error should be valid no matter which model $g \in \mathcal{S}$ is the real plant (or $\eta \in \mathcal{N}$ the noise vector) leading to a worst-case approach. In this deterministic worst-case framework, the identification error should "cover" all models $g \in \mathcal{S}$ that, combined with all possible noise vectors $\eta \in \mathcal{N}$, are consistent with the experiments. Ideally then, the error should only include the models $g \in \mathcal{S}(\mathbf{y})$. In practice, however, the family of models conservatively covers this "tight" uncertainty set. Hence it provides an upper bound for the distance (in a certain metric) from a model (consistent or not) to the real plant. In this framework, the worst-case error

GENERAL SETUP

is defined as follows:

$$e(\mathcal{A}) \triangleq \sup_{\eta \in \mathcal{N}, g \in \mathcal{S}} m\{g, \mathcal{A}[E(g, \eta), \mathcal{S}, \mathcal{N}]\} \tag{10.4}$$

where $E(\cdot, \cdot)$ is the experiment operator and $m(\cdot, \cdot)$ a specific metric.

The identification algorithm $\mathcal{A}$ maps both *a priori* and *a posteriori* information to a candidate nominal model. In this case the algorithm is said to be *tuned* to the *a priori* information. On the other hand, if it only depends on the experimental *a posteriori* data, it is called *untuned*.[3] Almost all classical parameter identification algorithms belong to the latter class ([184]). For simplicity, in the sequel, the identification algorithm will be denoted as $\mathcal{A}$, without specifying its arguments unless needed.

The identification error (10.4) can be considered as *a priori*, in the sense that it takes into account all possible experimental outcomes consistent with the classes $\mathcal{N}$ and $\mathcal{S}$, *before* the actual experiment is performed. Since it considers all possible combinations of models g and noises η, that is, all experimental data $\mathbf{y}$, it is a *global* identification error in terms of the experiments. A *local* error that applies only to a specific experiment $\mathbf{y}$ can be defined as follows:

$$e(\mathcal{A}, \mathbf{y}) \triangleq \sup_{g \in \mathcal{S}(\mathbf{y})} m[g, \mathcal{A}(\mathbf{y}, \mathcal{S}, \mathcal{N})] \tag{10.5}$$

From the above, it can be shown that the global worst-case error can be defined equivalently as follows.

Lemma 10.1

$$e(\mathcal{A}) = \sup_{\mathbf{y} \in \mathbf{Y}} e(\mathcal{A}, \mathbf{y}) \tag{10.6}$$

where $\mathbf{Y}$ is the set of all possible experimental data, consistent with sets $\mathcal{S}$ and $\mathcal{N}$.

Proof. By definition $\mathbf{Y} \triangleq \{E(g, \eta) | g \in \mathcal{S}, \eta \in \mathcal{N}\}$; therefore the set $\{g \in \mathcal{S}(\mathbf{y}), \mathbf{y} \in \mathbf{Y}\}$ is equivalent to $\mathcal{S}$. From the definition of $\mathbf{Y}$ and replacing $\mathbf{y}$ by $E(g, \eta)$ in (10.5), we obtain

$$\sup_{g \in \mathcal{S}} \sup_{\eta \in \mathcal{N}} m\{g, \mathcal{A}[E(g, \eta), \mathcal{S}, \mathcal{N}]\} = \sup_{\mathbf{y} \in \mathbf{Y}} \sup_{g \in \mathcal{S}(\mathbf{y})} m\{g, \mathcal{A}[\mathbf{y}, \mathcal{S}, \mathcal{N}]\}$$

□

---

[3] Obviously the algorithm always depends on the *a posteriori* information for the computation of an empirical model.

Clearly we always have $e(\mathcal{A}, \mathbf{y}) \leq e(\mathcal{A})$. To decrease the local error more experiments need to be performed, whereas to decrease the global error new *types* of experiments, compatible with new *a priori* classes, should be performed, for example, reducing the experimental noise and changing $\mathcal{N}$ accordingly.

As mentioned before, there is an *uncertainty principle* in identification, which produces a lower limit to the possible global error, closely related to the consistency set $\mathcal{S}(\mathbf{y})$. To this end we first define the radius and diameter of a set and prove several properties.

Definition 10.2 *The* radius *and* diameter *of a subset A of a metric space* (X, m), *are*

$$r(A) = \inf_{x \in X} \sup_{a \in A} m(x, a) \tag{10.7}$$

$$d(A) = \sup_{x, a \in A} m(x, a) \tag{10.8}$$

The radius can be interpreted as the maximum error, measured in a certain metric, when considering the set A as represented by a single "central" point (which may not belong to A). The diameter is the maximum distance between any two points in the set. From these definitions, the following properties can be proved.

Lemma 10.2

$$r(A) \leq d(A) \leq 2r(A) \tag{10.9}$$

Proof. From Definition 10.2 we have

$$r(A) \leq \sup_{a \in A} m(x, a) \leq d(A), \quad \forall x \in A \tag{10.10}$$

which proves the first inequality. From the definition of $r(A)$ it follows that:

$$\forall \epsilon > 0, \, \exists y \in X \mid \forall x \in A \; m(x, y) \leq r(A) + \epsilon/2 \tag{10.11}$$

Using the triangular inequality we obtain

$$d(A) = \sup_{x \in A} \sup_{a \in A} m(x, a) \leq \sup_{x \in A} \sup_{a \in A} [m(x, y) + m(a, y)]$$

$$\leq \sup_{x \in A} m(x, y) + \sup_{a \in A} m(a, y) \leq r(A) + r(A) + \epsilon = 2r(A) + \epsilon$$

which holds for all positive ϵ and proves the second inequality. □

GENERAL SETUP

Based on the concepts of radius and diameter of a set, we next quantify the "size" of the available information.

Definition 10.3 *The radius and diameter of information are defined as follows:*

$$\mathcal{R}(\mathcal{I}) \triangleq \sup_{\mathbf{y} \in \mathbf{Y}} r\left[\mathcal{S}(\mathbf{y})\right] \qquad (10.12)$$

$$\mathcal{D}(\mathcal{I}) \triangleq \sup_{\mathbf{y} \in \mathbf{Y}} d\left[\mathcal{S}(\mathbf{y})\right] \qquad (10.13)$$

In some sense, the radius and diameter of information represent the "size" in a certain metric, of the largest set of indistinguishable models according to all possible *a priori* and *a posteriori* information. In fact, it depends only on the *a priori* information because it does not correspond to a particular experiment, but to all possible experiments (holds for all $\mathbf{y} \in \mathbf{Y}$). Hence there always exists the possibility of performing the experiment $\mathbf{y}_\star$ that achieves the worst-case (largest) class of models $\mathcal{S}(\mathbf{y}_\star)$.

Moreover, this "worst case" (minimal information) experiment can sometimes be determined beforehand. To this end, define the following (desirable) properties of a set.

Definition 10.4 *A set A in a linear space X is called symmetric if and only if there exists an element $c \in X$ such that for any $a \in X$ for which $c + a \in A$ then $c - a \in A$. The element c is called the symmetry point of set A (note that it may not belong to A).*

Next, we prove that if the sets $\mathcal{S}$ and $\mathcal{N}$ are symmetric and convex, then the worst-case experiment, which defines the diameter of information, is achieved by both symmetry points of these sets. The classes of *a priori* information defined in the $\mathcal{H}_\infty$ and ℓ_1 identification sections are symmetric sets with the zero element as their symmetry points.

Lemma 10.3 *If the a priori sets $\mathcal{S}$ and $\mathcal{N}$ are symmetric and convex, with symmetry points $c_\mathcal{S}$ and $c_\mathcal{N}$, respectively, the diameter of information satisfies:*

$$\mathcal{D}(\mathcal{I}) = \sup_{\mathbf{y} \in \mathbf{Y}} d\left[\mathcal{S}(\mathbf{y})\right] = d\left[\mathcal{S}(\mathbf{y}_0)\right], \qquad \mathbf{y}_0 = E(c_\mathcal{S}, c_\mathcal{N}) \qquad (10.14)$$

Furthermore,

$$d\left[\mathcal{S}(\mathbf{y}_0)\right] = 2 \sup_{g \in \mathcal{S}(\mathbf{y}_0)} m(g, c_\mathcal{S}) \qquad (10.15)$$

Proof. Clearly $\mathcal{D}(\mathcal{I})$ is greater than or equal to $d[\mathcal{S}(\mathbf{y}_0)]$. Now assume

$$\sup_{\mathbf{y} \in \mathbf{Y}} d[\mathcal{S}(\mathbf{y})] > d[\mathcal{S}(\mathbf{y}_0)] \tag{10.16}$$

As a consequence, there exists an experiment $\mathbf{y}_\star$, a pair of models $g_1, g_2 \in \mathcal{S}(\mathbf{y}_\star)$, and noises $\eta_1, \eta_2 \in \mathcal{N}$, such that:

$$E(g_1, \eta_1) = E(g_2, \eta_2) = \mathbf{y}_\star \tag{10.17}$$

$$m(g_1, g_2) > d[\mathcal{S}(\mathbf{y}_0)] \tag{10.18}$$

By symmetry, the element $(2c_\mathcal{S} - g_1) \in \mathcal{S}$, and by convexity the element $h_1 \triangleq \frac{1}{2}[g_2 + (2c_\mathcal{S} - g_1)] \in \mathcal{S}$ as well. Hence the element $h_2 \triangleq c_\mathcal{S} + \frac{1}{2}(g_1 - g_2)$ belongs to the set $\mathcal{S}$, due to its symmetry. Replacing $\mathcal{S}$ by $\mathcal{N}$ we can prove in a similar way that the elements $\tilde{\eta}_1 \triangleq c_\mathcal{N} + \frac{1}{2}(\eta_1 - \eta_2)$ and $\tilde{\eta}_2 \triangleq c_\mathcal{N} + \frac{1}{2}(\eta_2 - \eta_1)$ belong to $\mathcal{N}$. Using the linearity of the experiment operator $E(\cdot, \cdot)$, we have

$$\mathbf{y}_0 = \frac{1}{2}[E(g_1, \eta_1) - E(g_2, \eta_2)] + E(c_\mathcal{S}, c_\mathcal{N}) \tag{10.19}$$

$$= E(h_1, \tilde{\eta}_1) \tag{10.20}$$

and similarly $E(h_2, \tilde{\eta}_2) = \mathbf{y}_0$. This proves that $h_1, h_2 \in \mathcal{S}(\mathbf{y}_0)$, with $m(h_1, h_2) = m(g_1, g_2)$, which contradicts (10.18).

It is easy to verify that $\mathcal{S}(\mathbf{y}_0)$ is a symmetric set, and therefore

$$\sup_{x,y} m(x, y) \leq m(x, c_\mathcal{S}) + m(y, c_\mathcal{S}) \tag{10.21}$$

$$\leq 2 \max[m(x, c_\mathcal{S}), m(y, c_\mathcal{S})] \tag{10.22}$$

which achieves equality when $y = -x$. The diameter is computed among the symmetric elements, which proves (10.15). $\square$

If the sets belong to normed linear spaces, we may use the metric $m(a, b) = \|a - b\|$.

The above proves that the experimental output achieved by the centers of both *a priori* sets provides the least amount of information (largest diameter of indistinguishable set). In the case where these centers are the null elements, it means that the experimental instance with null output provides minimal information, which in some sense is intuitively clear.

From the previous discussions, it seems important to have identification procedures that can compute models inside the consistency set. These are called *interpolation* algorithms and are defined as those that produce models included in the set $\mathcal{S}(\mathbf{y})$ for all possible experiments $\mathbf{y} \subset \mathbf{Y}$. The name comes from the fact that algorithms that interpolate the experimental data produce a model g that satisfies $\mathbf{y} = E(g, 0)$, and therefore $g \in \mathcal{S}(\mathbf{y})$ if $g \in \mathcal{S}$.

A lower bound for the worst-case error of any identification algorithm and an upper bound for any interpolation algorithm are presented next.

Lemma 10.4 *The worst-case identification error defined in (10.6) satisfies the following inequality:*

$$e(\mathcal{A}) \geq \mathcal{R}(\mathcal{I}) \geq \frac{1}{2}\mathcal{D}(\mathcal{I}) \qquad (10.23)$$

for any algorithm $\mathcal{A}$. The following upper bound holds for any interpolation algorithm $\mathcal{A}_I$:

$$\mathcal{D}(\mathcal{I}) \geq e(\mathcal{A}_I) \qquad (10.24)$$

Proof. It is clear that

$$\sup_{\mathcal{A}(\mathbf{y}) \in \mathcal{S}(\mathbf{y})} e(\mathcal{A}, \mathbf{y}) \geq e(\mathcal{A}, \mathbf{y}) \geq \inf_{\mathcal{A}(\mathbf{y}) \in \mathcal{S}} e(\mathcal{A}, \mathbf{y}) \qquad (10.25)$$

Next, take the supremum over all experiments $\mathbf{y} \in \mathbf{Y}$ on each member of the above inequality. Based on definitions for radius and diameter of information and equation (10.6), we obtain for the first inequality

$$\mathcal{D}(\mathcal{I}) \geq e(\mathcal{A}_I) \qquad (10.26)$$

The above holds only for interpolation algorithms because the supremum has been taken over all $\mathcal{A}_I(\mathbf{y}) \in \mathcal{S}(\mathbf{y})$. Using Lemma 10.2, we obtain for the last inequality

$$e(\mathcal{A}) \geq \mathcal{R}(\mathcal{I}) \geq \frac{1}{2}\mathcal{D}(\mathcal{I}) \qquad (10.27)$$

which holds for any algorithm $\mathcal{A}$. □

As mentioned before, $e(\mathcal{A})$, $\mathcal{D}(\mathcal{I})$, and $\mathcal{R}(\mathcal{I})$ are global in terms of the experimental outcome $\mathbf{y}$. Therefore the above *a priori* bounds hold for any experiment (consistent with $\mathcal{S}$ and $\mathcal{N}$) and may be used before any measurement is performed. As a consequence of the above lemma, interpolation algorithms can obtain the minimal error bound $\mathcal{R}(\mathcal{I})$ up to a factor of 2, that is, $\mathcal{D}(\mathcal{I}) \geq e(\mathcal{A}_I) \geq \frac{1}{2}\mathcal{D}(\mathcal{I})$, "almost" optimal.

In terms of local errors, an important subclass of interpolation procedures, called *central* algorithms, generate a model $g_\star$ that satisfies

$$e(\mathcal{A}, \mathbf{y}) = \sup_{g \in \mathcal{S}(\mathbf{y})} m(g, g_\star) = r\left[\mathcal{S}(\mathbf{y})\right] \qquad (10.28)$$

This "central" element may not belong to the set itself (see Definition 10.2), may be nonunique, or may not exist. If it does exist, then it is called *Chebyshev center* and represents the best approximation by a single element of the smallest indistinguishable set $S(\mathbf{y})$. From Lemma 10.4 it follows that these algorithms provide the smallest possible worst-case error. Therefore they may be considered optimal in that sense. The importance of symmetric sets is that the symmetry point is a *Chebyshev center* of the set.

Optimal Algorithms It is important to consider algorithms that make an efficient use of the available information and produce the smallest possible identification errors. These optimal algorithms ([75]) are defined next.

Definition 10.5 *The local optimal identification error of a particular experiment* $\mathbf{y} \in \mathbf{Y}$ *is*

$$e^\star(\mathbf{y}) \triangleq \inf_{\mathcal{A}} e(\mathcal{A}, \mathbf{y}) \qquad (10.29)$$

An algorithm $\mathcal{A}^\star$ *is called* locally optimal *if and only if* $e(\mathcal{A}^\star, \mathbf{y}) = e^\star(\mathbf{y})$.

Definition 10.6 *The global optimal identification error is defined as*

$$e^\star \triangleq \inf_{\mathcal{A}} e(\mathcal{A}) \qquad (10.30)$$

An algorithm $\mathcal{A}^\star$ *is called* globally optimal *if and only if* $e(\mathcal{A}^\star) = e^\star$. *Furthermore, an algorithm is called* strongly optimal *if it is locally optimal for all experiments* $\mathbf{y} \in \mathbf{Y}$.

Note that strong optimality is much stronger than just global optimality. It implies that, regardless of the set of available data, the identification procedure provides an estimation of the model, which minimizes the worst-case error. The central algorithms defined before are strongly optimal.

10.2.4 Convergence

A question that might be posed is the following: What happens with the family of models when the amount of information increases? It would be desirable to produce a "smaller" set of models as input data increase, that is, model uncertainty should decrease. This could be interpreted as having a monotonically decreasing uncertainty bound with increasing input information. Certainly, we would expect the set of models to tend to the real system (monotonically or not) when the uncertainty of the input information goes to zero. This latter property can be stated in terms of convergence as follows: An identification algorithm $\mathcal{A}$ is said to be convergent when the worst case global identification error $e(\mathcal{A})$ in (10.4) goes to zero as the input information tends to be "completed." The latter means that the "partialness" and "corruption" of the available information, both *a priori* and *a posteriori*, tend to zero simultaneously ([146]).

Input information is corrupted solely by measurement noise in this scheme. Therefore "corruption" tends to zero when the set $\mathcal{N}$ only contains a single known element, say, the zero element. On the other hand, partialness of information can disappear in two different ways: by *a priori* assumptions or *a posteriori* measurements. In the first case, partialness disappears when the set $\mathcal{S}$ tends to have only one element (the real system); in the second case, when the amount of experimental information is completed by the remaining (usually infinite) data points.

It is convenient to unify the above concepts as follows. The available information (*a priori* and *a posteriori*) is completed when the consistency set $\mathcal{S}(\mathbf{y})$ tends to only one element: the real system. When this happens there is no uncertainty, that is, the input information is complete.[4] Hence convergence can be defined as follows:

Definition 10.7 *An identification algorithm $\mathcal{A}$ converges if and only if*

$$\lim_{\mathcal{D}(\mathcal{I}) \to 0} e(\mathcal{A}) = 0 \qquad (10.31)$$

The diameter of information tends to zero when the noise set $\mathcal{N}$ reduces to a single element (usually zero noise) and either the set $\mathcal{S}$ tends to the real plant or the experiment operator provides all necessary data. Note that as the consistency set $\mathcal{S}(\mathbf{y})$ reduces to a single element, the experiment operator tends to be invertible. Since the identification error is defined in a worst-case sense, its convergence is uniform with respect to the *a priori* sets $\mathcal{N}$ and $\mathcal{S}$.

The definition of convergence that has been presented reflects the information structure of the problem and generalizes the ones that will be stated in the following sections. When considering convergence in terms of completing the *a posteriori* experimental information, care must be taken because infinite data points (time or frequency response experiments) do not in general guarantee complete information ([236]). Take, for example, a frequency response experiment with data only at integer frequencies $\{\omega_k = k, \ k = 0, \ldots, M-1\}$. By making $M \to \infty$ the *a posteriori* information will not determine the unique real system since $\mathcal{D}(\mathcal{I})$ does not vanish (see Definition 10.7), that is, information is still incomplete. For this reason, all necessary (infinite) data points to complete the information should be considered. For example, it takes an infinitely countable number of points in the time response or an infinitely uncountable number of points for the frequency response of a discrete IIR time system.

According to Lemma 10.4, all interpolation algorithms are convergent.

---

[4] Here we assume that the real plant always belongs to the consistency set. According to the comments at the end of Section 10.2.2, this will eventually be the case as long as the set $\mathcal{S}(\mathbf{y})$ gets smaller and the consistency tests are positive.

Moreover, since these procedures generate a nominal model in the consistency set, they are always tuned to the *a priori* information. In later sections, convergent noninterpolating algorithms, which may be untuned, will also be sought. Convergent untuned algorithms are said to be *robustly* convergent. This is an important property because algorithmic convergence is dissociated from *a priori* assumptions. Therefore convergence is guaranteed, even if the *a priori* information is false, although the resulting bounds are no longer valid. This has a practical implication because most of the *a priori* information is based on a "leap of faith" or a combination of engineering "intuition" and "common sense."

As mentioned before, model approximation can be considered as a special case of identification. In that case, the sets S and $\mathcal{N}$ already have single known elements: the real model and zero noise, respectively. Uncertainty is only due to the partialness of the *a posteriori* data. In model approximation the objective is not to exactly invert the experiment but to produce a lower-order model (due to computational tractability). Convergence is therefore defined in the limit as the number of data points tends to be completed ($M \to \infty$), rather than with the diameter of information going to zero.

Convergence and a priori class S The worst-case error defined by equation (10.4) depends on the classes of models and measurement noise. In turn, the convergence of this error to zero also depends on these classes. Thus there are necessary conditions that the sets S and $\mathcal{N}$ should meet in order to guarantee the *existence* of convergent algorithms. Next, we present a condition on S that precludes the existence of these types of algorithms ([194]).

Lemma 10.5 *If the a priori class S, which belongs to metric space $(\mathcal{V}, m)$, contains a ball of size γ, say, $\mathcal{B}_\gamma(g_0) = \{g \in \mathcal{V}, \ m(g, g_0) \leq \gamma\} \in S$, then $\mathcal{R}(\mathcal{I}) \geq \gamma$ for any number of measurements and any noise level.*

Proof. The experimental evaluation of the plant g_0 produces $\mathbf{y} = E(g_0, \eta) = \begin{bmatrix} g_0^0 + \eta_0^0 & \cdots & g_0^{M-1} + \eta_0^{M-1} \end{bmatrix}^T$, for $\eta_0 \in \mathcal{N}$. We can always find a model $g_\star$ such that $g_\star^k = g_0^k$, $k = 0, \ldots, M-1$ and $m(g_0, g_\star) = \gamma$, that is, $g_\star \in \mathcal{B}_\gamma(g_0)$. Therefore independently of the value of M, we have two indistinguishable elements g_0 and $g_\star$ inside the ball. Thus these elements should be included in the consistency set. Since $m(g_0, g_\star) = \gamma$ then $r[S(\mathbf{y})] \geq \gamma$. □

Several comments concerning the above result are presented next.

- If a set is defined only in terms of the distance among its elements, there is no relation between the number of measurements M and the "size" of the set. Therefore the *a priori* class should not contain this type of set if convergence is desired.
- The element $g_\star$ described above belongs to the kernel of the operator $[E(\cdot, \eta_0) - \mathbf{y}]$. Therefore the "size" of this kernel should decrease to zero

as $M \to \infty$ and $\eta \to 0$. This is related to the inversion of the experiment operator $E(\cdot, \cdot)$ when we have complete information, as mentioned in Section 10.2.1.
- In [194] and [229] specific plants g_* are provided for the metrics defined by the ℓ_1 and $\mathcal{H}_\infty$ norms.
- From Lemma 10.5 it follows that S should not have interior points if we expect to have convergent algorithms.

There are yet two other elements that should be limited by the *a priori* information: the intersample behavior and the "tail" of the experiment, that is, $\{y_k, \ k \geq M\}$. Both are a consequence of the finite number of experimental data points obtained from systems with infinite time or frequency responses. To this end define the following:

Definition 10.8 *Consider a metric space $(\mathcal{V}, m)$ and a family of functions $A : \mathcal{V} \to \mathcal{V}$. The family A is equicontinuous if and only if for any $\epsilon > 0$, there exists $\delta > 0$ such that if $x, y \in \mathcal{V}$, $m(x, y) < \delta$ implies $m[f(x), f(y)] < \epsilon$ for all $f \in A$.*

The above is a generalization of the uniform continuity of a function, extended to a whole family A. Such a family has a uniform bound over the intersample behavior of each of its functions. An equivalent characterization of bounded equicontinuous sets is given by the *Arzela–Ascoli* theorem ([201]).

Theorem 10.1 *The following statements are equivalent:*

(a) The family A is equicontinuous and uniformly bounded.
(b) The closure $\bar{A}$ is compact.

Therefore selecting compact sets for the *a priori* class of models will limit the intersample behavior.

To limit the "tail" of the experimental output, which is unknown due to the finite number of samples, additional restrictions should be added to the *a priori* set of models. When considering the last portion of the time response, this condition is related to the stability of the class of plants. On the other hand, when considering the frequency response the "tail" of the experiment is restricted by the "properness" of the models.

To recap, in order to have convergent algorithms, we need the *a priori* class of models S to have the following properties:

1. It should not contain interior points so that it converges as $M \to \infty$.
2. It should be an equicontinuous set to limit the intersample behavior. It should also be uniformly bounded in order to have a bounded consistency set, therefore it should be compact.
3. It should include only stable and proper models so that the (unknown) tail portions of the (time and/or frequency) experiments are bounded.

338 ROBUST IDENTIFICATION

For the sets considered above it can be proved that pointwise convergence is sufficient for uniform (worst-case) convergence ([196]).

Lemma 10.6 *For a set of M measurements and noise level ϵ, any untuned identification algorithm $\mathcal{A}$ satisfying*

$$\lim_{M \to \infty} \lim_{\epsilon \to 0} \|\mathcal{A}[E(g, \eta)] - g\| = 0 \tag{10.32}$$

for all $g \in \mathcal{V}$, also satisfies

$$\lim_{M \to \infty} \lim_{\epsilon \to 0} \sup_{g \in \mathcal{S}} \|\mathcal{A}[E(g, \eta)] - g\| = 0 \tag{10.33}$$

for any relatively compact set[5] $\mathcal{S} \subset \mathcal{V}$.

Proof. See [196]. □

10.2.5 Validation

The main objective of robust identification is to compute a set of models that can be used in a robust control setup. A general framework to describe robust control problems uses LFTs, as described in the initial chapters of this book. In this setup, there is a block that represents the nominal model (including robustness and performance weights), a controller block, a model uncertainty one, uncertain disturbance inputs, and error outputs. Both the model uncertainty block and the uncertain disturbance inputs are usually defined as bounded sets.

Similarly to the consistency test that verifies that *a priori* and *a posteriori* information are coherent, another procedure should establish whether or not the outcome of the identification procedure is coherent with the setup in which it will be used by being able to "explain" all available experimental information. Therefore, if a set of models is to be used in the context of a Robust Control framework, it seems reasonable to verify the existence of an element of the model uncertainty block and an uncertain input signal (both included in their respective sets) that reproduces the experimental data. This is a reasonable test irrespective of how the set of models has been obtained.

This same procedure is used in general to test an *empirical theory*. In a scientific framework, an empirical theory must be capable of being *falsified* ([245]). Similarly, an empirical model[6] should be tested against all available experimental data. If positive, this provides a verification that the model is *not false*. Obviously this test does not prove that the model is completely valid, because a new set of experimental data may invalidate it. For this reason, a model or set of models cannot be validated, only invalidated. The

---

[5] A set is said to be relatively compact if its closure is compact.
[6] It is not necessary that the model computation used the aforementioned experimental data. How the model has been obtained is not relevant at this point.

procedure called *invalidation* tests the model or set of models against all available experimental information. It provides a sufficient condition for the model to be invalid, if a particular experiment does not fit in the framework. Otherwise it only provides a necessary condition for the model to be valid when all experiments can be explained in this setup.

For different reasons, the problem of invalidation has usually been called the *validation* problem. It is not the purpose of this section to justify the use of this terminology (which can be explained probably better by psychological reasons) but only to introduce this concept, which will not be developed further. Nevertheless, care must be taken when using the word validation in this context.

If the analysis framework is identical to the one used for the experiment, a set of models computed using a robust identification algorithm will be always validated by the experimental data used to compute it. Therefore the only two reasons by which a robustly identified set of models may be invalidated are:

- A different structural setup.
- A new set of experimental data.

The usual framework in which the robust identification experiments are performed is coherent with the use of the LFTs as a general control framework. The consistency among both structures should include the norms used to bound disturbance inputs, model uncertainty, and output errors as well.

In practice, the invalidation problem involves computing the minimum norm bound on the disturbances and model uncertainty to validate the experiments. It is clear that with large enough disturbances or uncertainty, any experiment can be validated. Hence this minimization problem also measures the conservativeness of the set of models.

The problem of validation has been initiated in [292], for models given in a LFT form, using the $\mathcal{H}_\infty$ norm to bound uncertainty and disturbances. The problem has a solution in terms of the structured singular value and includes structured dynamic model uncertainty in the model description. The statement of the problem is given entirely in the frequency domain. For a practical application see [291]).

The use of a time-domain description and bounds in the ℓ_1 norm have been proposed in [242]. The solution involves a convex matrix optimization problem based on the experimental data and the proposed set of models.

10.3 FREQUENCY DOMAIN IDENTIFICATION

In this section we present Robust Identification procedures for *a priori* classes of models in $\mathcal{H}_\infty$ and experiments performed in the frequency domain. The scheme follows the general framework presented in the last section, although specific issues pertaining to this metric will be stressed.

10.3.1 Preliminaries

We consider causal, linear time invariant, stable, either continuous or discrete time models. For simplicity we consider only SISO systems, although all results throughout this section can be applied to MIMO systems as well, either element by element or as in [76]. To unify the treatment we denote the models as $H(z) = H_c(\lambda \frac{1-z}{1+z})$, $\lambda > 0$ in the continuous time case or $H(z) = H_d(1/z)$, for discrete time systems, with $z \in \mathbb{C}$. Note that the latter (known as the λ-transform in the bibliography [87]) is the inverse of the usual z transform. Therefore in the sequel, causal stable models $H(z)$ will be analytic inside the unit disk. In this context the identification problem reduces to that of computing a model, analytic in a certain region, based on partial and corrupt data measured at the boundary of this region.

In order to define the *a posteriori* information, we introduce the following sets:

$$\mathcal{H}_{\infty,\rho} \triangleq \left\{ H(z) \text{ analytic in } |z| < \rho, \sup_{|z|<\rho} |H(z)| < \infty \right\} \quad (10.34)$$

$$\mathcal{H}_\infty(\rho, K) \triangleq \left\{ H \in \mathcal{H}_{\infty,\rho}, \sup_{|z|<\rho} |H(z)| < K \right\} \quad (10.35)$$

$$A(D) \triangleq \{H \in \mathcal{H}_\infty, \text{ continuous in } |z| = 1\} \quad (10.36)$$

$$\mathcal{BH}_\infty \triangleq \{f \in \mathcal{H}_\infty \,;\, \|f\|_\infty < 1\} \quad (10.37)$$

$$\overline{\mathcal{BH}_\infty} \triangleq \{f \in \mathcal{H}_\infty \,;\, \|f\|_\infty \leq 1\} \quad (10.38)$$

where D denotes the unit disk. Note that the set $A(D)$ is a *subalgebra of the (unit) disk*, since addition and product are closed operations in the set.

Recall from Chapter 8 (Lemma 8.1 and Section 8.2.1) that all ℓ_∞-BIBO stable systems are included in the set $A(D)$ which in turn is included in the set of ℓ_2 stable systems. Note that not all models in $\mathcal{H}_\infty$ belong to $A(D)$ (see Example 8.1). Therefore a smaller subset of models should be considered. The following result can be proved based on the previous definitions (prove) and indicates the relations among these sets.

Lemma 10.7 *For any $\rho > 1$ and finite K, the following holds:*

$$\mathcal{H}_\infty(\rho, K) \subset \mathcal{H}_{\infty,\rho} \subset A(D) \subset \mathcal{H}_\infty \quad (10.39)$$

Note that $\mathcal{H}_\infty(\rho, K)$ is a strict subset of $A(D)$ because there are elements in the subalgebra that do not belong to $\mathcal{H}_\infty(\rho, K)$ (give an example).

The *a priori* class of models is defined as the set $\mathcal{S} = \mathcal{H}_\infty(\rho, K)$. This corresponds to exponentially stable systems, not necessarily finite dimensional.

From a practical viewpoint, the models defined by the *a priori* information have a stability margin of $(\rho - 1)$ and a peak response to complex exponential inputs of K. The following bound is satisfied in the time domain (prove):

$$|h(k)| \leq K\rho^{-k} \tag{10.40}$$

The *a posteriori* experimental information is the set of M samples of the frequency response of the system, measured with additive bounded noise:

$$y_k = H(e^{j\Omega_k}) + \eta_k, \quad k = 0, \ldots, M - 1 \tag{10.41}$$

which represent the components of vector $\mathbf{y} \in \mathbb{C}^M$. The measurement noise is bounded by ϵ in the ℓ_∞ norm.

To recap, the *a priori* and *a posteriori* input data are, respectively,

$$\mathcal{S} = \{H \in \mathcal{H}_\infty(\rho, K), \quad \rho > 1, \quad K < \infty\} \tag{10.42}$$

$$\mathcal{N} = \left\{\eta \in \mathbb{C}^M, \quad |\eta_k| \leq \epsilon, \quad k = 0, 1, \ldots, M - 1\right\} \tag{10.43}$$

$$\mathbf{y} = E(H, \eta) = H(e^{j\Omega_k}) + \eta_k, \quad k = 0, \ldots, M - 1 \tag{10.44}$$

Since the *a priori* class $\mathcal{S}$ is relatively compact, Lemma 10.6 applies. Also note that both sets, $\mathcal{S}$ and $\mathcal{N}$, are symmetric with centers $c_\mathcal{S}$ and $c_\mathcal{N}$ in the zero model and noise vector, respectively. Based on the general statement of the last section, the global error and convergence criteria for algorithm $\mathcal{A}$ are as follows:

$$e(\mathcal{A}) = \sup_{\eta \in \mathcal{N}, H \in \mathcal{S}} \|\mathcal{A}[E(H, \eta), \mathcal{S}, \mathcal{N}] - H\|_\infty \tag{10.45}$$

$$\lim_{\epsilon \to 0, M \to \infty} e(\mathcal{A}) = 0 \tag{10.46}$$

According to Lemma 10.6 if $\mathcal{A}$ is untuned, that is, $\mathcal{A}(\mathbf{y})$, the convergence criteria need only be applied to the error corresponding to plant H: $\|\mathcal{A}(\mathbf{y}) - H\|_\infty$ instead of to the above uniform error.

Using the results of Lemmas 10.2 and 10.3, a simple lower bound for the information radius can be found.

Lemma 10.8 *The radius of information cannot be smaller than* $\min(\epsilon, K)$.

Proof. Define $C = \min(\epsilon, K)$ and consider the constant models $H_1(z) = -H_2(z) = C$ and noises $\eta_1 = -\eta_2 = -C$, which belong to the *a priori* classes $\mathcal{S}$ and $\mathcal{N}$, respectively. Models H_1 and H_2 also belong to the consistency set $\mathcal{S}(\mathbf{y}_0)$, for the null experiment $\mathbf{y}_0 \equiv 0$, that is, $H_1 + \eta_1 = H_2 + \eta_2 = 0$. Hence, from Lemmas 10.2 and 10.3,

$$\mathcal{R}(\mathcal{I}) \geq \tfrac{1}{2}\mathcal{D}(\mathcal{I}) = \tfrac{1}{2}d[\mathcal{S}(0)] = \tfrac{1}{2}\|H_1 - H_2\|_\infty \geq \min(\epsilon, K) \qquad (10.47)$$

□

Again, this bound is global and holds for all possible experiments.

A feature of $\mathcal{H}_\infty$ identification is that it assumes that the experimental data is in the frequency domain. Real world experiments are always performed in the time domain. For this reason, the time-domain data needs to be pre-processed in order to obtain the required complex frequency-domain *a posteriori* information. In the next subsection, two convergent algorithms, which compute the data vector **y** and the equivalent frequency measurement error bound ϵ, are presented.

10.3.2 Sampling Procedure

To obtain the *a posteriori* information, an experiment should be performed to compute the frequency response of the system and the data vector $\mathbf{y} \in \mathbb{C}^M$. For practical reasons, this experiment involves bounded sequences of inputs and outputs measured in the time domain. The usual procedure is to input sinusoidal signals and measure the phase and magnitude of the output in relation to the input. The *a priori* information indicates the time at which the transient response is irrelevant and only the steady-state values are significant. Additional practical considerations involve verifying the *a priori* assumptions ([272]), that is, stability, linearity, and time invariance. The computation of the *a posteriori* data points can be performed using the same framework as the model identification itself, as shown next.

Consider the system initially at rest, a sequence of sinusoidal input signals bounded by α and measurement noise bounded in magnitude by γ.

Inputs. The complex sinusoidal sequence

$$u_n = \begin{cases} \alpha e^{-j\Omega_k n}, & n \geq 0 \\ 0, & n < 0 \end{cases} \qquad (10.48)$$

for the set of frequencies $\{\Omega_k;\ k = 0, \ldots, M-1\}$.

Outputs. The measured vector $\mathbf{x} \in \mathbb{C}^N$, with components

$$\begin{aligned} x_n &= (h * u)_n + v_n, & n = 0, \ldots, N-1 \\ |v_n| &\leq \gamma, & n = 0, \ldots, N-1 \end{aligned} \qquad (10.49)$$

Two possible algorithms ([146]) that identify the values y_k in (10.41) are the following:

FREQUENCY DOMAIN IDENTIFICATION

Algorithms. The data points at each frequency Ω_k are computed by algorithms $\mathcal{P}_1$ or $\mathcal{P}_2$ as follows:

$$\mathcal{P}_1(\mathbf{x})|_k = y_k = \frac{x_{N-1}}{\alpha e^{-j\Omega_k(N-1)}} \tag{10.50}$$

$$\mathcal{P}_2(\mathbf{x})|_k = y_k = \frac{1}{\alpha N} \sum_{n=0}^{N-1} x_n e^{j\Omega_k n} \tag{10.51}$$

Errors. The above values have the following error bounds:

$$\text{Method 1:} \quad |\eta_k| \le \frac{K\rho}{\rho - 1} \rho^{-N} + \frac{\gamma}{\alpha} \stackrel{\triangle}{=} \epsilon \tag{10.52}$$

$$\text{Method 2:} \quad |\eta_k| \le \frac{K\rho(1 - \rho^{-N})}{N(\rho - 1)^2} + \frac{\gamma}{\alpha} \stackrel{\triangle}{=} \epsilon \tag{10.53}$$

for all $k = 0, \ldots, M-1$ as shown in [146]. Note that v is the time-domain measurement noise bounded by γ from which a bound ϵ over the complex frequency-domain noise η_k, $k = 0, \ldots, M-1$, is calculated.

For the real system the inputs are replaced by sinusoids, although the bounds and outputs are similar. For simplicity the frequency points are equally spaced $\Omega_k = 2\pi/M$.

Note that the first term of both error bounds is related to the transient response, which vanishes as time increases ($N \to \infty$). The second term, on the other hand, is constant and represents a signal-to-noise ratio that can only be decreased with a better measurement procedure, that is, decreasing γ.

The above frequency point identification procedures can be recast in the same framework as model identification. The *a priori* information are the noise bound γ and the values (ρ, K) of the plant set $\mathcal{H}_\infty(\rho, K)$. The *a posteriori* data are the time measurements $\mathbf{x}$. The error is defined in a worst-case sense over the set of measurement noises v_k defined above and the class of models $\mathcal{H}_\infty(\rho, K)$. As in the general case, there is also a lower bound to the identification error, which can be specified in terms of the radius of information (see Section 10.2.3). Specifically, for all frequency point identification procedures, the following result holds.

Lemma 10.9 *The worst-case error for any frequency point identification algorithm $\mathcal{P}$ is always greater than $\min(K, \gamma/\alpha)$.*

Proof. The proof is similar to the one in Lemma 10.8 by defining the constant models as $\pm \min(K, \gamma/\alpha)$; therefore it will not be repeated here. □

It is clear from the bounds of Methods 1 and 2 that both are convergent. In fact, since they are untuned with respect to the *a priori* information, they are also robustly convergent:

$$\lim_{N \to \infty} \lim_{\gamma \to 0} e(\mathcal{P}) = 0 \tag{10.54}$$

Furthermore, according to the above lemma, both algorithms converge when $K \geq \gamma/\alpha$, to the (optimal) lower bound γ/α as $N \to \infty$. Therefore these procedures are robustly convergent and asymptotically optimal. We can always assume that $K \geq \gamma/\alpha$, since otherwise the input signal magnitude α can be scaled such that this condition is guaranteed.

For practical reasons, it is clear that the frequencies Ω_k cannot be generated exactly. Nevertheless, the frequency error δ_Ω can be included in the bounds as well ([229]).

10.3.3 Consistency

Recall from Section 10.2 that consistency of *a priori* and *a posteriori* information establishes the validity of the identification error bounds. As a consequence, the covergence of tuned algorithms depends on the solution to this problem. Although untuned identification procedures converge even for inconsistent input data, the use of the set of models in a robust control framework is based on the validity of these bounds and hence also depends on the consistency test.

Given the experimental data $\{y_k = H(e^{j\Omega_k}) + \eta_k, \ k = 0, \ldots, M-1\}$ and the *a priori* sets $\mathcal{S}$ and $\mathcal{N}$ defined before, the consistency set $\mathcal{S}(\mathbf{y})$ can be specified using (10.2). The consistency problem can be rephrased as the existence of a model $\hat{H} \in \mathcal{S}$ whose difference with $\mathbf{y}$ at the sampling points can be attributed solely to measurement noise in $\mathcal{N}$, that is,

$$\left| \hat{H}(e^{j\Omega_k}) - y_k \right| \leq \epsilon \quad \text{for all } k = 0, \ldots, M-1 \tag{10.55}$$

We refer to this problem as an approximate interpolation one because $\hat{H}(e^{j\Omega_k})$ matches y_k up to an error of size ϵ. For $\mathcal{H}_\infty$ identification, the consistency problem can be cast as an approximate interpolation problem and solved via the Nevanlinna–Pick procedure ([74]). A brief statement of this latter result follows ([21]).

Consider M values w_k measured in the interior of the unit disk $\bar{D}$ at points z_k. There exists a function $f \in \mathcal{BH}_\infty$ that satisfies $f(z_k) = w_k$ for all $k = 0, \ldots, M-1$ if and only if the Pick matrix P,

$$P_{ij} = \frac{1 - w_i \bar{w}_j}{1 - z_i \bar{z}_j} \ ; \quad i, j = 0, \ldots, M-1 \tag{10.56}$$

FREQUENCY DOMAIN IDENTIFICATION

is positive semidefinite. If $P > 0$ then there exist infinite interpolation functions f, which can be described as a linear fractional function of a free parameter in $\mathcal{BH}_\infty$. These solutions will be described in further detail in the context of interpolation algorithms (see Section 10.3.4 and Problem 6.5).

This result can be extended to $\mathcal{H}_\infty(\rho, k)$ using the fact that $f(z) \in \mathcal{BH}_\infty \iff Kf(z/\rho) \in \mathcal{H}_\infty(\rho, K)$ (verify). Hence the rescaled set of Pick matrices can be defined as follows:

$$P_{ij} = \frac{1 - \frac{1}{K^2}(y_i + \eta_i)(\bar{y}_j + \bar{\eta}_j)}{1 - \frac{1}{\rho^2}e^{j(\Omega_i - \Omega_j)}}; \quad i,j = 0,\ldots,M-1 \quad (10.57)$$

for any vector of measurement noise $\eta \in \mathcal{N}$. Therefore an equivalent characterization of the consistency problem can be stated as follows

Lemma 10.10 *A set of experimental data $\mathbf{y} \in \mathbb{C}^M$ is consistent with the a priori sets $\mathcal{S}$ and $\mathcal{N}$ defined in terms of constants (K, ρ, ϵ) if and only if there exists a vector $\eta \in \mathcal{N}$ such that the Pick matrix in (10.57) is positive definite*[7].

Proof. Follows from the fact that $f(z) \in \mathcal{BH}_\infty \iff Kf(z/\rho) \in \mathcal{S}$ and the classical Nevanlinna–Pick theory. □

From the above lemma it follows that a necessary condition for consistency is that all experimental samples remain below $K + \epsilon$. This is based on only part of the *a priori* information and stems from the fact that for all diagonal elements in P to be positive, it is necessary to have

$$|y_k + \eta_k| < K \Rightarrow |y_k| - |\eta_k| < K \Rightarrow |y_k| < K + \epsilon \quad (10.58)$$

for all $k = 0, \ldots, M-1$ and $\eta \in \mathcal{N}$, where a signal-to-noise ratio greater than 1 has been assumed, that is, $|y_k| \geq |\eta_k|$. Clearly the above is an intuitive upper bound for the experimental data based only on the magnitude of the signal and the measurement noise.

Finally, to test consistency, at least one noise vector $\eta_c \in \mathcal{N}$ should be found. A computational practical solution to this problem can be obtained by solving a minimization problem ([74]):

Lemma 10.11 *The following statements are equivalent:*

1. *The a posteriori data $\mathbf{y} \in \mathbb{C}^M$ and the a priori sets $(\mathcal{S}, \mathcal{N})$ defined by (K, ρ, ϵ) are consistent.*

---

[7] The nongeneric case of $P \geq 0$, which has a unique solution, will not be considered here but can be found in [21].

2.
$$\min_{\eta \in \mathcal{N}} \bar{\lambda}(-R) < 0 \qquad (10.59)$$

$$R = \begin{bmatrix} Q^{-1} & \frac{1}{K}[Y + X(\eta)]^* \\ \frac{1}{K}[Y + X(\eta)] & Q \end{bmatrix} \qquad (10.60)$$

where $Y = \text{diag}\begin{bmatrix} y_0 & \cdots & y_{M-1} \end{bmatrix}$, $X = \text{diag}\begin{bmatrix} \eta_0 & \cdots & \eta_{M-1} \end{bmatrix}$ and $Q_{ij} = \left[1 - e^{j(\Omega_i - \Omega_j)}/\rho^2\right]^{-1}$.

3. *The linear matrix inequality (LMI) problem*

$$R[X(\eta)] > 0 \qquad (10.61)$$

$$\begin{bmatrix} \epsilon I & X(\eta) \\ X^*(\eta) & \epsilon I \end{bmatrix} > 0 \qquad (10.62)$$

has a feasible solution for a vector $\eta \in \mathbb{C}^M$.

4.
$$\min_{\eta \in \mathcal{N}} \bar{\sigma}\left\{Q^{-1/2}[Y + X(\eta)]Q^{1/2}\right\} < K \qquad (10.63)$$

Proof. After some straightforward algebra, the Pick matrix can be written as

$$P = Q - \frac{(Y + X)Q(Y + X)^*}{K^2} \qquad (10.64)$$

Using the Schur complements, $P > 0 \iff R > 0$ and the equivalence of statements 1 and 2 follows. Matrix $R(X)$ is linear in X, which in turn is linear in η. The positiveness condition in equation (10.62) is equivalent to $|\eta_i| < \epsilon$, $i = 0, \ldots, M - 1$. Therefore the LMI feasibility problem 3 is equivalent to consistency. Finally the equivalence between statements 4 and 1 is derived in [74]. □

Statement 2 is a standard generalized eigenvalue minimization problem that can be solved via the methods developed in [214] or by semidefinite programming ([308]). Problem 3 is a standard LMI and can be solved using available commercial software ([126]). Finally, convex programming can solve problem 4.

An advantage of testing consistency through an approximate interpolation problem is that, from the parametrization of all interpolating functions, an interpolatory identification algorithm can be obtained directly. This will be detailed in the next subsection.

10.3.4 Identification Procedures

Once consistency has been established, the computation of a nominal model and a *valid* model error bound can be attempted. In this section two different types of algorithms will be presented, as well as their corresponding worst-case model error bounds.

The first type of procedure, initiated in [144] and extended to $\mathcal{H}_\infty$ identification in [137], consists of two steps, the second of which is an optimal Nehari approximation. In this approach most of the research effort has been concentrated on the first step, aimed towards improving the algorithm's performance. These procedures can be untuned to the *a priori* information; therefore they can be robustly convergent when the worst-case error vanishes as the input information is completed. This means that even if the model error bounds are invalid (if *a priori* and *a posteriori* information are inconsistent), the convergence of the algorithm is guaranteed. On the other hand, the nominal model generated by these *two-step* procedures is not guaranteed to be in the consistency set, which in principle may lead to a larger model error bound.

The second type of algorithm presented in this section is *interpolatory*, hence the nominal model is always in the consistency set. From the results presented in Section 10.2, it follows that the algorithm is always tuned and convergent. Furthermore, this procedure is also strongly optimal, in terms of the identification error, to within a factor of 2. This interpolatory algorithm is based on the consistency test performed in Section 10.3.3 and can be solved using the parametrization of all solutions of the same Nevanlinna–Pick problem. This *interpolatory* algorithm was first presented in [74]. Another type of interpolation algorithm has been reported in [233, 236], but will not be presented here. As a comparison, from a computational point of view, the first step of the *two-stage* algorithms can be solved via a FFT for equidistant experimental data and may take longer in the case of nonuniformly spaced data ([8, 127]). The second step (Nehari approximation) has a computational cost similar to the Nevanlinna–Pick problem.

Finally, note that the above identification algorithms produce a constant uncertainty bound as a function of frequency. Therefore the description of the set of models does not admit any frequency-dependent uncertainty weight. In the case of model approximation since the "real" system is available, the difference with the nominal model can be computed explicitly as a function of frequency. The uncertainty weight can be computed much in the same way as the nominal model. In this process the order of both the nominal model and the uncertainty weight that constitute the augmented plant will be the same as that of the nominal model itself. This will not be explained here but is detailed in [127].

In the sequel we illustrate the identification algorithms through the following example.

Example 10.1 *The physical system under consideration is a Euler–Bernoulli beam with viscous damping, which can be described using the following physical model that relates the vertical displacement y to time t and the longitudinal* *coordinate x:*

$$\rho a \frac{\partial^2 y}{\partial t^2} + E^* I \frac{\partial^4 y}{\partial x^4 \partial t} + EI \frac{\partial^4 y}{\partial x^4} = 0, \quad x \in [-1, 1] \tag{10.65}$$

with boundary conditions

$$\frac{\partial^2 y}{\partial x^2}(\pm 1, t) = 0; \qquad \frac{\partial^3 y}{\partial x^3}(\pm 1, t) = 0 \tag{10.66}$$

Here a is the cross-sectional area, ρ the mass density of the beam, I the moment of inertia, E the Young modulus of elasticity, and E^* the normal strain rate. The values adopted in this case are: $a\rho = 46$, $E^*I = 0.46$, and $EI = 55.2$, and the dynamics are evaluated at $x = 0$.

Since the PDE (10.65) is linear, by using the Laplace transform we obtain the following (infinite-dimensional) transfer function:

$$H(x, s) = \frac{1}{4\lambda^3(sE^*I + EI)} \cdot$$

$$\left[\sinh \lambda x - \sin \lambda x + \frac{(\sin \lambda \sinh \lambda - \cos \lambda \cosh \lambda - 1) \cos \lambda x}{\cos \lambda \sinh \lambda + \sin \lambda \cosh \lambda} \right.$$

$$\left. - \frac{(\sin \lambda \sinh \lambda + \cos \lambda \cosh \lambda + 1) \cosh \lambda x}{\cos \lambda \sinh \lambda + \sin \lambda \cosh \lambda} \right] \tag{10.67}$$

$$\lambda^4 = -\frac{s^2 \rho a}{sE^*I + EI} \tag{10.68}$$

Due to the rigid mode of the structure, this transfer function has an unstable part $1/2\rho s^2$. Although in practice it is not possible to physically separate the unstable and stable responses, the purpose of this example is only to test the different identification methods over the stable portion. Thus in the sequel we assume exact knowledge of the unstable portion of the model, and we subtract its response from the experimental data in order to identify only the stable part. At the end of this chapter some comments on identification of unstable systems in closed-loop experiments are presented.

In this case, we have generated three "experimental" data sets of 150 points each from the frequency response of the flexible structure using the software package ([173]). Figure 10.2 shows the total magnitude response of the system (stable+unstable), the stable part, and three sets of simulated "measurement" noise in $\ell_\infty(\epsilon)$ with $\epsilon = 8 \times 10^{-5}$.

The three sets of "experimental" measurements can be combined to produce an equivalent set of data points and a new measurement error bound. This can be achieved by considering each data point as the center of a ball of radius ϵ_i in $\mathbb{C}$ for each experiment $i = 1, 2, 3$. At each frequency, the center and radius of a new ball should be computed such that it "covers" the inter-

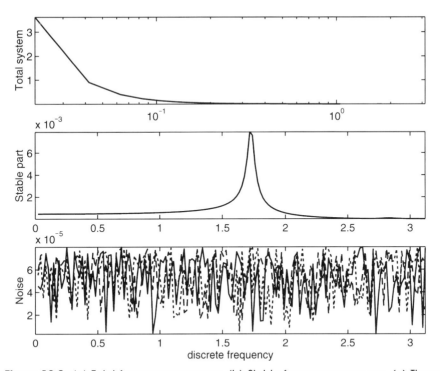

Figure 10.2. (a) Total frequency response. (b) Stable frequency response. (c) Three sets of "experimental" measurement noise.

section of all the balls corresponding to the experiments to be combined. This produces an equivalent set of data points and a frequency-dependent error bound that includes all previous experimental data. This frequency-dependent noise bound should be smaller than the individual bounds for each experiment and could provide a less conservative description of the noise, that is, $\epsilon(j\omega) \leq \min[\epsilon_1, \epsilon_2, \epsilon_3]$.

In this particular example, the equivalent set of measurements has been superimposed on the three initial data sets in the first illustration of Figure 10.3. The frequency-dependent error bound of this new measurement set appears in the second part of Figure 10.3. For simplicity throughout this chapter we will use the constant upper bound $\epsilon = 8 \times 10^{-5}$.

Two-Stage Algorithms A first intuitive algorithmic approach to obtain a model from the *a posteriori* measurements would use the discrete Fourier transform (DFT) and its inverse (IDFT), as follows:

$$h_{s1}(k) = \frac{1}{M} \sum_{n=0}^{M-1} y_n e^{-j\Omega_k n}, \quad \Omega_k = \frac{2\pi}{M}k \qquad (10.69)$$

350 ROBUST IDENTIFICATION

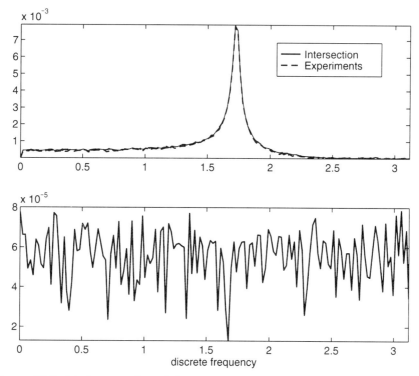

Figure 10.3. (a) Superposition of three measurement sets and their combination. (b) Equivalent measurement noise at each frequency.

$$H_{s1}(z) = \sum_{k=0}^{M-1} h_{s1}(k) z^k \qquad (10.70)$$

Here h_{s1} corresponds to an "equivalent" FIR impulse response obtained from the experimental samples **y** and H_{s1} is the identified model. Model $H_{s1}(z)$ interpolates the experimental data at M frequencies in the unit circle, that is, $y_k = H_{s1}(e^{j\Omega_k})$, $k = 0, \ldots, M-1$. If the real system has a finite impulse response (FIR) with less than or equal to M nonzero samples and in the absence of measurement noise, then $H_{s1}(z)$ coincides with the real system and only the M values $H_{s1}(e^{j\Omega_k})$, $k = 0, \ldots, M-1$, are necessary to represent it. If, instead, the real system has an infinite impulse response (IIR) and/or the samples are affected by noise, the following two problems appear:

- The equivalent impulse response h_{s1} suffers from *time aliasing*. Thus the identified model $H_{s1}(z)$ could be unstable, even if the real plant is stable. This becomes clear by shifting the sum in (10.70) by $(M-1)/2$ samples[8]

---

[8] Assume without loss of generality that M is odd.

to compute the (periodic) function $H_{s1}(z)$ as follows:

$$H_{s1}(z) = \sum_{k=-(M-1)/2}^{(M-1)/2} h_{s1}(k) z^k \tag{10.71}$$

Hence if the (periodic) function $h_{s1}(k) \neq 0$ for $-(M-1)/2 \leq k < 0$ (the same as for $(M+1)/2 \leq k \leq M-1$), it has an anticausal component that corresponds to unstable poles of $H_{s1}(z)$.

- Even if the number of samples M is large enough so that the values of the anticausal samples $\{h_{s1}(k), -(M-1)/2 \leq k < 0\}$ were negligible, measurement noise would contribute to make this portion of the response different from zero. The latter phenomenon does not happen in model approximation ([127]). Therefore the procedure is simpler in that case.

Furthermore, this procedure does not converge as noise level tends to zero and the number of samples to infinity, as illustrated by the following example.

Example 10.2 *Suppose the "real" plant is $H \equiv 0$, and the following (consistent) experimental set of measurements are available:*

$$y_k = \epsilon e^{j[(M-2k-1)/2M]\pi} \quad k = 0, \ldots, M-1 \tag{10.72}$$

Assume for simplicity and also for practical reasons that $\epsilon \leq K$. Using the above algorithm, the following interpolation model can be found:

$$H_{s1}(z) = \frac{1}{M} \sum_{n=0}^{M-1} K_M(e^{-j(2\pi/M)n} z) y_n \tag{10.73}$$

with

$$K_M(z) = \sum_{k=0}^{M-1} z^k = \frac{z^M - 1}{z - 1} \tag{10.74}$$

which satisfies

$$\left| K_M(e^{j\Omega}) \right| = \left| \frac{\sin(M\Omega/2)}{\sin(\Omega/2)} \right| \tag{10.75}$$

The phase of $K_M(e^{j\Omega})$ is piecewise continuous in Ω; nevertheless, by defining $\Omega_* = -\pi/M$ and $\Omega_k = (2\pi/M)k$, it can be calculated as $\angle K_M(\Omega_* - \Omega_k) = \angle K_M(\Omega_*) + \Omega_k/2$, where $\angle$ denotes phase. Finally, the identification error is *given by*

$$\|H_{s1}(e^{j\Omega})\|_\infty = \left\|\frac{1}{M}\sum_{k=0}^{M-1} K_M\left(e^{j(\Omega-\Omega_k)}\right) y_k\right\|_\infty \geq |H_{s1}(e^{j\Omega_*})|$$

$$= \frac{\epsilon}{M}\sum_{k=0}^{M-1}\left|K_M\left(e^{j(\Omega_*-\Omega_k)}\right)\right|$$

$$= \frac{\epsilon}{M}\sum_{k=0}^{M-1}\left|\frac{\sin\left[\frac{M}{2}(\Omega_*-\Omega_k)\right]}{\sin\frac{(\Omega_*-\Omega_k)}{2}}\right| = \frac{\epsilon}{M}\sum_{k=0}^{M-1}\frac{1}{\sin(\frac{2k+1}{2M}\pi)}$$

Since $\sin\theta \leq \theta$ *if* $\theta \geq 0$,

$$\|H_{s1}(e^{j\Omega})\|_\infty \geq \frac{2\epsilon}{\pi}\sum_{k=0}^{M-1}\frac{1}{2k+1} \longrightarrow \frac{\epsilon}{\pi}\log(M) + \mathcal{O}(1) \quad (10.76)$$

As a consequence, the worst-case interpolating function diverges as the number of samples increases.

The above problems motivate the *two-stage* algorithms ([146, 137]):

1. Using the IDFT, an equivalent impulse response is obtained from the experimental data **y**. This response, suitably weighted so that the convergence is uniform as $M \to \infty$ and $\epsilon \to 0$, is used to compute a model $H_{s1}(z)$. This part of the procedure is linear in the *a posteriori* data y_k, $k = 0, \ldots, M-1$.
2. The model $H_{s1}(z)$ may have an unstable part, although the original plant was stable. Therefore a projection of the model over the stable set $\mathcal{H}_\infty$ should be performed.

The details of each step are described next.

Stage 1. The input is the set of *a posteriori* information $\{y_n,\ n=0,\ldots, M-1\}$, from which h_{s1} is computed.

$$h_{s1}(k) = \frac{1}{M}\sum_{n=0}^{M-1} y_n e^{-j\Omega_k n}, \quad \Omega_k = \frac{2\pi}{M}k \quad (10.77)$$

The preidentified model H_{s1} is calculated using the following weighted sum:

$$H_{s1}(z) = \sum_{k=-(M-1)/2}^{(M-1)/2} w(k) h_{s1}(k) z^k \quad (10.78)$$

Here the weight values $w(k)$ are suitably selected to have the required convergence characteristics. Because of its reminiscence to signal processing procedures, it is often called "window" w. By replacing h_{s1} in the last equation it is clear that this part of the algorithm is linear in the input data $\mathbf{y}$,

$$H_{s1}(z) \triangleq I_M(\mathbf{y}) = \frac{1}{M} \sum_{k=0}^{M-1} y_k K_M(ze^{-j\Omega_k}) \tag{10.79}$$

where

$$K_M(z) = \sum_{k=-(M-1)/2}^{(M-1)/2} w(k) z^k \tag{10.80}$$

is the *kernel* of the interpolation operator $I_M : \mathbb{C}^M(\ell_\infty) \to \mathcal{L}_\infty$ (the input data are equipped with the ℓ_∞ norm). The design of the algorithm depends on the selection of the weight w or, equivalently, the kernel K_M or the interpolation operator I_M. Convergence is guaranteed for certain classes of weights w or operators I_M. This will be discussed at the end of this section in further detail.

Stage 2. The optimal projection from $H_{s1} \in \mathcal{L}_\infty$ to a stable causal model $H_{id} \in \mathcal{H}_\infty$ can be found by solving:

$$\min_{H \in \mathcal{H}_\infty} \|H_{s1} - H\|_\infty \tag{10.81}$$

This is precisely the *Nehari* problem solved in Section 6.6 (Theorem 6.5). The second step differentiates this identification procedure from the standard filtering problem, which arises in digital signal processing, where the noncausality of the filter can be solved via a delay.

The projection from $\mathcal{L}_\infty$ to $\mathcal{H}_\infty$ of H_{s1} should be performed only on its antistable (anticausal) part. To see this, consider $H_{s1}(z)$ in (10.78), which can be separated into its stable and antistable[9] parts $H_{s1}(z) = H_s(z) + H_a(z)$. It is clear that the optimal projection only depends on $H_a(z)$ because the stable part is already in $\mathcal{H}_\infty$ and can be subtracted. As in Section 6.6, a solution to the Nehari problem can be calculated in terms of the singular value decomposition of the Hankel matrix for the antistable part $H_a(z)$ as follows:

$$H_{id}(z) = H_{s1}(z) - \frac{\bar{\sigma} \sum_{k=0}^{N-1} \bar{u}_{N-k} z^k}{z^N \sum_{k=0}^{N-1} \bar{v}_{k+1} z^k} \tag{10.82}$$

---

[9] Recall that the antistable part has all its poles inside the unit circle, in this case at $z = 0$.

where $\bar{\sigma}$ is the maximum singular value of the Hankel matrix of $H_a(z)$, $(\bar{v}, \bar{u})$ its corresponding right and left singular vectors, and $N = (M - 1)/2$. This optimal approximation is rational and has degree less than or equal to N. Nevertheless, care must be taken when using the above equation, due to the fact that the second term should cancel *exactly* the antistable part of $H_{s1}(z)$. More efficient numerical solutions presented in state-space form can be found in [21] and [332] for both the continuous and discrete time cases.

The above procedures are untuned, since the only input data are the *a posteriori* information. The second stage is nonlinear in **y**, as opposed to the first stage, which is linear; hence the complete algorithm is nonlinear in the input data. In fact, it can be proved that there are no robustly convergent $\mathcal{H}_\infty$ identification procedures that map the input data **y** linearly to the nominal model ([234]).

When solving a model approximation problem ([127]), the second step is not necessary if the number of samples is high enough, so that the values of $\{h_{s1}(k),\ k = -1, \ldots, -(M - 1)/2\}$ are negligible. This is due to the fact that there is no measurement noise. In that case only the causal values $\{h_{s1}(k),\ k = 0, \ldots, (M - 1)/2\}$ need to be kept to compute the approximate model, which corresponds to a rectangular unitary one-sided window. Therefore the whole procedure is linear in the input data. Moreover, it can be shown that it is robustly convergent ([139, 321]).

In a similar way, in many practical identification applications when measurement noise is small enough only the first step of the identification algorithm needs to be applied ([9, 138]). In this latter case, the divergence of the identification error due to measurement noise is $\mathcal{O}[\epsilon \log(M)]$.

Convergence and Error Bounds It is possible to consider the identification errors at each stage in order to evaluate their convergence separately. It can be proved that the convergence of the whole procedure depends solely on the convergence of the first stage. Due to the optimality of the Nehari solution to problem (10.81) we have:

$$\|H_{s1} - H_{id}\|_\infty \leq \|H_{s1} - H\|_\infty \quad (10.83)$$

where $H \in \mathcal{S}$ is the system to be identified. Using the triangle inequality and the above equation, the total error can be bounded as follows:

$$\|H - H_{id}\|_\infty \leq \|H_{s1} - H\|_\infty + \|H_{s1} - H_{id}\|_\infty \leq 2 \|H - H_{s1}\|_\infty \quad (10.84)$$

Therefore the whole error converges if the identification error of the (linear) first step does, as $M \to \infty$ and $\epsilon \to 0$. Furthermore, the total worst-case error is bounded by twice the first-stage error. This indicates that the design of the algorithm and its convergence depend only on the selection of a convenient window function w for the first stage.

Necessary and sufficient conditions for stage 1 to be convergent are presented next ([137]).

Lemma 10.12 *The first step of the two-stage identification algorithm is robustly convergent if and only if the following two conditions are satisfied:*

$$\lim_{M \to \infty} \sup_{H \in \mathcal{S}} \|H_{pi} - H\|_\infty = 0 \tag{10.85}$$

$$\lim_{M, \frac{1}{\epsilon} \to \infty} \sup_{\eta \in \mathcal{N}} \|\eta_{pi}\|_\infty = 0 \tag{10.86}$$

where the preidentified model of step 1 (see (10.79)) has been separated into $H_{s1} = H_{pi} + \eta_{pi}$, using the fact that the noise is additive.

Proof. Both conditions are necessary by selecting $\eta = 0 \in \mathcal{N}$ and $H = 0 \in \mathcal{S}$, respectively. Sufficiency can be proved by applying the triangle inequality to $H_{s1} = H_{pi} + \eta_{pi}$ as follows:

$$\sup_{H \in \mathcal{S}, \eta \in \mathcal{N}} \|H_{s1} - H\|_\infty \leq \sup_{H \in \mathcal{S}} \|H_{pi} - H\|_\infty + \sup_{\eta \in \mathcal{N}} \|\eta_{pi}\|_\infty$$

The convergence of these two conditions guarantees the convergence of stage 1. □

Note that the above two conditions are met by any plant $H \in \mathcal{S}$ measured without noise and by the null plant $H = 0$ being robustly identified, respectively.

In the case of model approximation, only condition (10.85) needs to hold for convergence, where $H_{pi} = H_{s1} = H_{id}$. Therefore the worst-case identification noise-free error $\sup_{H \in \mathcal{S}} \|H_{pi} - H\|_\infty$ is called *approximation* error. Similarly, $\sup_{\eta \in \mathcal{N}} \|\eta_{pi}\|_\infty$ is termed the *noise* error. The uncertainty bound of the identification algorithms and their convergence will depend on both these errors.

There are many different windows that may guarantee convergence of the first stage, and hence of the algorithm. For example, the triangular window, known in digital signal processing as the Bartlett window, given by:

$$w(k) = \begin{cases} 1 - \frac{|k|}{N}, & |k| < N \\ 0, & |k| \geq N \end{cases} \tag{10.87}$$

where N has been defined in stage 2. In this case, the sum in (10.78) corresponds to the average of partial sums of the $h_{s1}(k)$, also known as their *Cesaro sum*, which guarantees uniform convergence ([139, 321]) as $N \to \infty$. It can be proved that this type of window produces a *noise* error bound of

ϵ. Hence it is optimal, because this is the lowest possible uncertainty bound, when $K \geq \epsilon$ (see Lemma 10.8). Another rapidly convergent procedure uses the De La Vallée Poussin sums ([137, 233]).

In [137], necessary and sufficient conditions are given, guaranteeing that a particular window will generate a (robust) convergent identification algorithm. These conditions therefore define a *class* of two-stage algorithms that includes, as special cases, all other two-stage algorithms cited in the bibliography. For each different algorithm, there is a compromise between the *approximation* and the *noise* identification errors. From the many possible convergent windows that may be generated, a subset that quantifies this compromise can be parametrized as a function of an integer q as follows:

$$w(k) = \begin{cases} 1, & -q \leq k \leq q \\ \dfrac{n-k}{n-q}, & q \leq k \leq n \\ \dfrac{n+k}{n-q}, & -n \leq k \leq -q \\ 0, & \text{otherwise} \end{cases} \quad (10.88)$$

with $0 < q < n < N$. Note that if $q \to n$ and the window is shifted by n to the right ($k \to k+n$), it approaches the one-sided rectangular window. This type of weight produces a robustly convergent approximation error. As mentioned before, this is the case in model approximation when N is large enough so that no significant time aliasing is produced. In the case of identification, the noise error diverges as $\mathcal{O}[\epsilon \log(n)]$.

On the other hand, when $q \to 0$, the window approaches the triangular one, which optimizes the noise error, as mentioned before. Hence the parameter q quantifies the trade-off between both components of the error.

For the window (10.88), the following worst-case identification error bound can be established exploiting the inequality (10.84):

$$2\left\{\frac{K}{\rho-1}\left[\rho^{-N+1} + \rho^{-(N-n+q-1)} + 2\rho^{-2q}\right] + \left[\frac{(n+q)}{n-q}\right]\epsilon\right\} \quad (10.89)$$

In the above equation the first term corresponds to the approximation error and the second to the noise error. Note that when $q \to 0$ (triangular window), the total noise bound due *only* to the first stage is ϵ, which corresponds to the minimal reachable value for the radius of information.

Example 10.3 *From the equivalent measurement set and noise bound determined in Example 10.1, we can compute a nominal model and error bound by using the two-stage algorithm. To this end we have used a trapezoidal window as shown in the first part of Figure 10.4. The linear identification part of the*

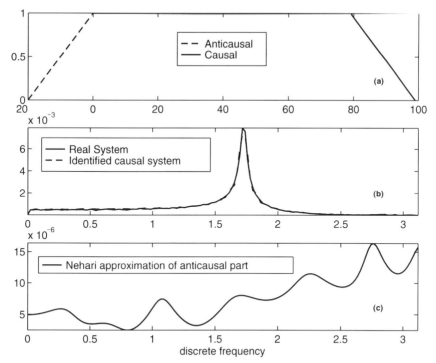

Figure 10.4. (a) Window. (b) Causal portion identification. (c) Optimal noncausal portion identification.

algorithm produces an identified system, depicted in the second portion of the same figure. Finally, the optimal approximation of the noncausal part of the model is shown in the last portion of Figure 10.4. Figure 10.5 compares the resulting nominal model and the combined measurements of all three samples.

Interpolatory Algorithms An interpolatory identification procedure, which produces a nominal model in the consistency set, can be obtained directly from the consistency solution. As mentioned in Section 10.2, interpolatory algorithms have certain advantages over two-stage ones. Since the identified model is in the set $\mathcal{S}(\mathbf{y})$, its distance to the Chebyshev center is within the diameter of information ([194]). As a consequence the algorithm is optimal up to a factor of 2 compared to central strongly optimal procedures. For the same reasons, it is also convergent. Therefore the identification error tends to zero as the information is completed.

The procedure ([74]) is based on the parametrization of all solutions to the Nevanlinna–Pick problem ([21]). For simplicity, here only the generic case

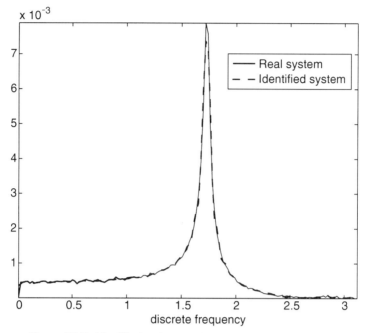

Figure 10.5. Identified system and combined measurements.

where the Pick matrix is strictly positive definite is considered; the degenerate case $(P \geq 0)$ can be found in [21, 74].

The algorithm depends on the noise sequence η_c computed in Section 10.3.3, to establish consistency. Recall that if the *a priori* and *a posteriori* information are consistent, a noise vector $\eta_c \in \mathcal{N}$ must be available such that the conditions in Lemma 10.11 hold. This is important not only to initiate the procedure, but also to guarantee that the identification error bounds are valid, since the algorithm is tuned to the *a priori* information. The algorithm proceeds as follows:

- Compute the Pick matrix P in (10.57) (which should be positive definite), replacing η by η_c.

- Define $z_k \triangleq e^{j\Omega_k}$ and $w_k \triangleq H(e^{j\Omega_k})$ for $k = 0, \ldots, M-1$ and compute

$$T(z) = R + \rho \begin{bmatrix} \dfrac{-1}{\rho - z\bar{z}_0} & \cdots & \dfrac{-1}{\rho - z\bar{z}_{M-1}} \\ \dfrac{\bar{w}_0}{K(\rho - z\bar{z}_0)} & \cdots & \dfrac{\bar{w}_{M-1}}{K(\rho - z\bar{z}_{M-1})} \end{bmatrix} P^{-1} W R$$

where

$$R = I + \rho^2 \begin{bmatrix} \frac{-1}{z_0} & \cdots & \frac{-1}{\bar{z}_{M-1}} \\ \frac{\bar{w}_0}{K\bar{z}_0} & \cdots & \frac{\bar{w}_{M-1}}{K\bar{z}_{M-1}} \end{bmatrix} P^{-1} Z$$

$$W = \begin{bmatrix} 1 & -\frac{w_0}{K} \\ \vdots & \vdots \\ 1 & -\frac{w_{M-1}}{K} \end{bmatrix}$$

$$Z = \begin{bmatrix} \frac{1}{\rho - z_0} & \frac{-w_0}{K(\rho - z_0)} \\ \vdots & \vdots \\ \frac{1}{\rho - z_{M-1}} & \frac{-w_{M-1}}{K(\rho - z_{M-1})} \end{bmatrix}$$

- The interpolating function is

$$H_{id}(z) = K \frac{T_{11}(\frac{z}{\rho}) q(z) + T_{12}(\frac{z}{\rho})}{T_{21}(\frac{z}{\rho}) q(z) + T_{22}(\frac{z}{\rho})} \tag{10.90}$$

where $q(z) \in \overline{B\mathcal{H}_\infty}$ is a free parameter. In particular, if $q(z)$ is constant, then $H_{id}(z)$ is of order M.

The degrees of freedom available suggest a further optimization step to select the optimal η_c and $q(z)$, for instance to minimize certain criteria. The computational size of the problem corresponds to the number of experimental data points.

As for the identification error, in [74] the following bound has been determined for the diameter of information:

$$\mathcal{D}(\mathcal{I}) \leq 2 \frac{\epsilon + K\psi}{1 + (\epsilon/K)\psi} \tag{10.91}$$

with

$$\psi \triangleq \frac{2\rho \sin(\pi/M)}{\sqrt{(\rho^2 - 1)^2 + 4\rho^2 \sin^2(\pi/M)}} \tag{10.92}$$

for the case of equidistant frequency sample points. Clearly this bound goes to zero as the information is completed.

Example 10.4 *From the equivalent measurement set and noise bound found in Example 10.1, we can compute a nominal model and error bound by using the interpolatory algorithm described above. It is clear that any interpolatory algorithm will fit the experiment within the measurement noise bound. Therefore it is unnecessary to compare this identification with the one in the previous example.*

On the other hand, the order of the resulting model will be at least the number of measurement data points. As a consequence, in this example we shall consider a smaller number of representative data points to reduce the order of the identified nominal model and the computational burden. To this end we have limited the number of measurements to 13, represented by asterisks in Figures 10.6 and 10.7, where the total number of data points (150) have been represented by a full line.

In Figure 10.6 the nominal model obtained by the interpolatory procedure described before is depicted. Note that for this particular type of system, the fitting is not very satisfactory in the intersample regions, due to the fact that the number of data points has been reduced an order of magnitude with respect to the total number of measurements in the previous example.

An interesting solution to this problem, especially in these types of applications, is to combine a parametric identification procedure with the (nonparametric) interpolatory robust identification one described in this subsection. The approach is to use a parametric identification procedure to determine the peak

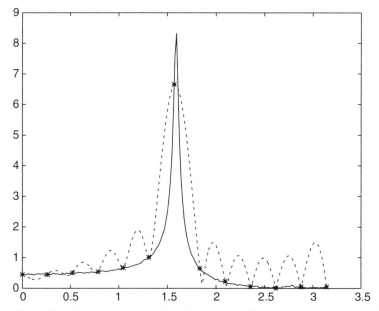

Figure 10.6. Complete set of experimental data (full), reduced set (asterisks) and nominal model (dots) identified by nonparametric interpolation.

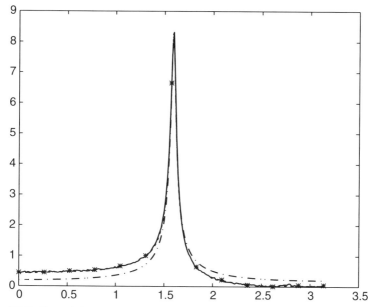

Figure 10.7. Complete set of experimental data (full), reduced set (asterisks), parametrically identified system (dash-dot) and nominal model (dots) identified by parametric/nonparametric interpolation.

of the response and the interpolatory (nonparametric) algorithm to compute a nominal model for the difference. For this example we have used the method in [232], which combines both parametric and nonparametric methods and produces a nominal model almost coincident with the experiment, represented in Figure 10.7 as a dotted line. In the same figure, the parametric identification produces the dash-dot line. Therefore the interpolatory algorithm is used only to identify the difference between the latter and the asterisks. Additional comments on these techniques will be presented at the end of this chapter.

10.4 TIME-DOMAIN IDENTIFICATION

In this section robust identification based on time series will be presented. In particular, the ℓ_1 norm as a measure to quantify the identification error will be considered. The general structure is similar to the $\mathcal{H}_\infty$ worst-case identification, as pointed out in the abstract setting of Section 10.2. The main differences are related to the experimental *a posteriori* information. In ℓ_1 identification the specific input signal to the system to be identified should be selected, whereas in $\mathcal{H}_\infty$ identification these are always sinusoids. The *a posteriori* information is obtained directly from the experiment with no intermediate procedure to "translate" the data points, as with the sampling procedure in $\mathcal{H}_\infty$ identification (see Section 10.3.2).

362 ROBUST IDENTIFICATION

These procedures fit the ℓ_1 robust control design methodology covered in Chapter 8. Therefore they can be used in combination with this technique when the experimental data originates from an adequate time-domain experiment performed over the system to be controlled.

An important point that was raised in Section 8.2.2 is that the ℓ_1 norm induces a stronger topology than the $\mathcal{H}_\infty$ norm. For this reason, all ℓ_1 identification procedures produce a set of models that not only may be used in an ℓ_1 optimal control setting but can also be used with an $\mathcal{H}_\infty$ optimal control procedure. As we will see in this section the algorithmic procedures are simpler than in $\mathcal{H}_\infty$ identification. However, not all input sequences can produce untuned algorithms.

A general framework for ℓ_1 identification was introduced in [194]. Further results appeared in [159, 160, 183, 205, 305, 306]. In [193] a specific input leading to an untuned identification algorithm was proposed. The untuned nature of the algorithms based on time series is strongly dependent on the input sequence. In fact, in the sequel it will be proved that there are no untuned algorithms that identify a system from impulse response measurements ([160]).

The tuned nature of the algorithms impose on the definition of convergence the assumption of the "correctness" of the *a priori* information, that is, a positive consistency test. The algorithms proposed in this section are all interpolatory. Recall (from Section 10.2.4) that this implies that they provide models in the consistency set. Hence the *a priori* and *a posteriori* information are automatically consistent as long as such models exist. This is similar to the dependence of the $\mathcal{H}_\infty$ interpolatory algorithm presented in the last section on the consistency test. For this reason, the consistency problem will not be repeated here.

10.4.1 Preliminaries

Next, the *a priori* and *a posteriori* information are presented. The systems considered, represented by their impulse response h, are discrete time, causal, linear, and ℓ_∞-BIBO stable. Thus the Markov parameters sequence $\{h\}$ satisfies:

$$\|h\|_1 = \sum_{k=0}^{\infty} |h(k)| < \infty \tag{10.93}$$

The class of models $\mathcal{S}$ is defined as follows:

$$\mathcal{S} \stackrel{\triangle}{=} \{h \,;\, |h(k)| \leq \Phi(k), \quad \Phi \in \ell_1 \text{ nonincreasing}\} \tag{10.94}$$

The above set includes, as a special case, the one used in $\mathcal{H}_\infty$ identification, that is, $\Phi(k) = K\rho^{-k}$.

The experimental data from the system are the set of its first M response samples corrupted by additive noise $\eta \in \ell_\infty(\epsilon)$, as follows:

$$\mathbf{y} = E(h, \eta) = h * u + \eta \tag{10.95}$$

$$y_k = \sum_{i=0}^{\infty} h(i)u(k-i) + \eta_k, \quad k = 0, \ldots, M-1 \tag{10.96}$$

where $*$ represents the convolution operator. The algorithm $\mathcal{A}$ maps the data vector $\mathbf{y} = \begin{bmatrix} y_0 & \cdots & y_{M-1} \end{bmatrix}^T$ to the model $H_{id} \in \ell_1$. For simplicity, the same notation for the experimental data $\mathbf{y}$ and the measurement noise η as in the $\mathcal{H}_\infty$ identification case will be used.

Without loss of generality (by scaling the noise and inputs if necessary), in the sequel we will assume that the input u is bounded by one in ℓ_∞. Furthermore, if the experimental data are collected starting at time $k = 0$, a known bound on the input prior to $k < 0$ will be assumed.

The *a priori* and *a posteriori* sets and the experiment function are therefore

$$\mathcal{S} = \{h \,;\, |h(k)| \leq \Phi(k), \Phi \in \ell_1 \text{ nonincreasing}\}$$

$$\mathcal{N} = \left\{\eta \in \mathbb{R}^M, \; |\eta_k| \leq \epsilon, \; k = 0, 1, \ldots, M-1\right\}$$

$$\mathbf{y} = E(h, \eta) = \sum_{i=0}^{\infty} h(i)u(k-i) + \eta_k \,;\quad k = 0, \ldots, M-1, \; u \in \mathcal{U}_\delta$$

$$\mathcal{U}_\delta = \left\{u \,\middle|\, \begin{array}{l} |u(n)| \leq \delta, n < 0 \\ |u(n)| \leq 1, n \geq 0 \end{array}\right\}$$

The global error is calculated as the worst-case over all models and measurement noises, as described in Section 10.2:

$$e(\mathcal{A}) = \sup_{\eta \in \mathcal{N}, h \in \mathcal{S}} \|\mathcal{A}[E(h, \eta), \mathcal{S}, \mathcal{N}] - h\|_1 \tag{10.97}$$

As before, the algorithms are said to be convergent when the identification error goes to zero as "partialness" and "corruptedness" vanish, that is

$$\lim_{\epsilon \to 0, M \to \infty} e(\mathcal{A}) = 0 \tag{10.98}$$

Restrictions and Bounds Recall from Section 10.2 that the consistency set definition allows the computation of global *a priori* bounds, valid for any possible identification algorithm that operates over the input data. The next lemma provides the minimal attainable bound on the identification error.

364 ROBUST IDENTIFICATION

Lemma 10.13 *For any input sequence $u \in \mathcal{U}_\delta$, the information diameter satisfies the following lower bound:*

$$\mathcal{D}(\mathcal{I}) \geq 2\min(\epsilon, K_0) \tag{10.99}$$

where $K_0 = \Phi(0)$.

Proof. Consider the models $h = \pm\min(\epsilon, K_0)$, consistent with the null output. Equation (10.99) follows immediately from the general results in Section 10.2. □

The above bound is independent of the duration of the experiment and the nature of the input sequence; therefore it represents an *absolute* lower limit on the error. Next, we obtain bounds for the impulse response experiment that depend on the number of samples taken ([230]).

Lemma 10.14 *The information diameter for the ℓ_1 identification procedure with an impulse response experiment is given by:*

$$\mathcal{D}_M(\mathcal{I}) = 2\left[\sum_{k=0}^{M-1} \min[\epsilon, \Phi(k)] + \sum_{k=M}^{\infty} \Phi(k)\right] \tag{10.100}$$

Proof. Define the following:

$$h(k) = \begin{cases} \min[\Phi(k), \epsilon], & k < M \\ \Phi(k), & k \geq M \end{cases} \quad \eta(k) = h(k)$$

Next, consider the pair of models and measurement noise vectors:

$$\begin{cases} h_1(k) = h(k) \\ \eta_1(k) = -\eta(k) \end{cases} \text{ and } \begin{cases} h_2(k) = -h(k) \\ \eta_2(k) = \eta(k) \end{cases}$$

$$\Longrightarrow h_1, h_2 \in \mathcal{S}, \quad \eta_1, \eta_2 \in \mathcal{N}$$

Hence the above produces zero outputs to an impulse excitation, that is, $E(h_1, \eta_1) = E(h_2, \eta_2) \equiv 0$.

Note that this is the worst-case model (maximum ℓ_1 norm) compatible with the experimental output and in the consistency set $\mathcal{S}(0)$. Equation (10.100) follows now from Lemma 10.3 and the fact that $\mathcal{S}$ and $\mathcal{N}$ are symmetric and convex by considering $d[\mathcal{S}(0)] = \|h_1 - h_2\|_1 = 2\|h\|_1$, $\Phi(k)$ and ϵ being positive. □

The following comments are in order:

- There is a significant quantitative difference between the bound in (10.100) and the global one in (10.99), the latter being a conservative lower bound for the identification error.

- The diameter of information tends to zero as $M \to \infty$ and $\epsilon \to 0$. This guarantees the existence of convergent algorithms, for example, interpolation procedures.
- For a given noise bound level, there exists a maximum number of experimental samples $M_\star$, which decreases the diameter of information. In other words, for $M \geq M_\star$, the diameter remains constant, which means that any extra experimental sample will not increase the amount of *useful* information. This fact will be used in the formulation of an interpolatory convergent algorithm ([230]) in the next subsection.

Under certain conditions ([163]) it can be proved that the impulse input sequence is optimal in terms of the diameter of information, for a given experiment duration M. These conditions are related to the fact that the *a priori* information does not exclude models in the consistency set of the null experimental outcome. The optimality stems from the fact that this input produces outputs directly related to each time coefficient. Therefore all the uncertainties are *decoupled*, that is, each measurement noise element is related to only one model coefficient.

For example, for the step response experiment, the reader may verify that the diameter of information satisfies

$$\mathcal{D}_M(\mathcal{I}) = \min[\epsilon, \Phi(0)] + 2\sum_{i=1}^{M-1} \min[\epsilon, \Phi(i)] + \sum_{i=M}^{\infty} \Phi(i) \qquad (10.101)$$

The proof is based on the same arguments as before, but using the following model:

$$h(k) = \begin{cases} \min[\Phi(0), \epsilon], & k = 0 \\ 2(-1)^k \min[\Phi(k), \epsilon], & 1 \leq k < M \\ \Phi(k), & k \geq M \end{cases} \qquad (10.102)$$

As noted, the above bound almost doubles the one obtained from the impulse experiment.

Next, the fact that there are no untuned algorithms based solely on the impulse response experiments ([160]) will be established.

Lemma 10.15 *There exists no untuned convergent algorithm based solely on the impulse response a posteriori experiment.*

Proof. Consider the untuned algorithm $\mathcal{A}(\mathbf{y})$ and two plants that produce the same experiment $\mathbf{y} = 0 \in \mathbb{R}^M$: $h_0 \equiv 0$ and $h_1(k) = \epsilon$, $k = 0, \ldots, M-1$, $\|h_1\|_1 > 2$. Since $\mathcal{A}(\cdot)$ is based only on the experimental information, it cannot distinguish between both plants, but should converge for all plants in $\mathcal{S}(0)$. Therefore if

$$\lim_{M \to \infty, \epsilon \to 0} \|\mathcal{A}(\mathbf{y}) - h_0\|_1 = 0$$

$$\Rightarrow \exists (M_0, \epsilon_0) \text{ such that } \|\mathcal{A}(\mathbf{y})\|_1 \leq 1, \quad M \geq M_0, \quad \epsilon \leq \epsilon_0$$

As a consequence for $M \geq M_0$ and $\epsilon \leq \epsilon_0$,

$$\|\mathcal{A}(\mathbf{y}) - h_1\|_1 \geq \|h_1\|_1 - \|\mathcal{A}(\mathbf{y})\|_1 \geq 1 \tag{10.103}$$

which contradicts the assumption that $\mathcal{A}$ converges to a plant in $\mathcal{S}(0)$ (see also [160]). □

In [237] the following necessary and sufficient condition for the existence of a robustly convergent ℓ_1 identification algorithm is established:

$$\exists \, \delta > 0 \text{ such that } \|h * u\|_\infty \geq \delta \|h\|_1, \quad \forall \, h \in \ell_1 \tag{10.104}$$

where $\|u\|_\infty \leq 1$ without loss of generality. The above condition excludes nonpersistent inputs for which $u(t) \to 0$ as $t \to \infty$. Specific untuned algorithms, based on experiments performed with inputs known as Galois sequences, have been proposed in [193]. Furthermore, in [238] it is proved that there are no linear robust convergent algorithms in ℓ_1, as was the case for $\mathcal{H}_\infty$ identification.

10.4.2 Identification Procedures

Next, several tuned convergent algorithms for ℓ_1 robust identification are presented. These are based on the impulse response experiment although, with minor changes, can be applied to the step response experiment as well. The fact that these algorithms produce a nominal model in the consistency set makes it unnecessary to test consistency, as long as such a model exists. Untuned convergent algorithms for additional input signals, can be found in [141, 193, 194, 306].

Based on the First $n \leq M$ Samples In general, when performing an identification experiment, there is good knowledge of the measurement errors of the sensors. Therefore, from the practical point of view, the consequences of tuning the algorithm with the measurement noise bound (ϵ) or with all the *a priori* information on the class of models and noise are quite different. Specifically, the possibility of assuming a wrong bound and producing a divergent algorithm, in the first case is much smaller than in the latter. Also, the fact that these methods are used as a first step of a robust control design procedure means that an uncertainty bound (or frequency-dependent weight, [127]) should be obtained. This depends on the identification error, which in turn depends on the *a priori* information. If this information is to be used anyway, why not take advantage in the algorithm of the piece of information for which there is a more certain knowledge: ϵ.

Three identification algorithms, two tuned to the value of ϵ and the third a globally asymptotically optimal one tuned to all the *a priori* information, will be presented in this subsection. All three algorithms generate a nominal model as well as an ℓ_1 error bound starting from impulse response data. The identified nominal model for these algorithms is based on the first $n \leq M$ noisy samples of the experiment, as follows:

$$H_{id}^n(z) = \sum_{k=0}^{n-1} y_k z^k \qquad (10.105)$$

where n depends on the *a priori* information. To assess the optimality of the algorithms, a bound on the identification error is obtained next.

Lemma 10.16 *The worst-case identification error $e_\star$ over all possible algorithms is bounded as follows:*

$$e_\star = \inf_{\mathcal{A}} e(\mathcal{A}) = \inf_{\mathcal{A}} \sup_{h \in \mathcal{S}, \eta \in \mathcal{N}} \|h - \mathcal{A}(\mathbf{y})\|_1 \geq \sum_{k=0}^{\infty} \min[\epsilon, \Phi(k)] \qquad (10.106)$$

Proof. This proof follows a similar one in [146]. For every $h \in \mathcal{S}$, $\eta \in \mathcal{N}$ and for any algorithm $\mathcal{A}$, the following holds:

$$\|h - \mathcal{A}(\mathbf{y})\|_1 \leq e(\mathcal{A}) \qquad (10.107)$$

where $e(\mathcal{A})$ is the worst-case local error defined in (10.106). Consider the following models and measurement noises:

$$\begin{cases} h_1(k) = \min[\Phi(k), \epsilon] \\ \eta_1(k) = -h_1(k) \end{cases} \qquad \begin{cases} h_2(k) = -h_1(k) \\ \eta_2(k) = -h_2(k) \end{cases}$$
$$\Longrightarrow h_1, h_2 \in \mathcal{S}, \quad \eta_1, \eta_2 \in \mathcal{N}, \quad E(h_1, \eta_1) = E(h_2, \eta_2) \equiv 0$$

Now, applying equation (10.107),

$$\|h_1 - \mathcal{A}(0)\|_1 \leq e(\mathcal{A}) \quad \text{and} \quad \|h_1 + \mathcal{A}(0)\|_1 \leq e(\mathcal{A})$$
$$\Longrightarrow \|h_1\|_1 \leq \tfrac{1}{2} \|h_1 - \mathcal{A}(0)\|_1 + \tfrac{1}{2} \|h_1 + \mathcal{A}(0)\|_1 \leq e(\mathcal{A})$$

To obtain (10.106), simply take the infimum over $\mathcal{A}$. □

Lemma 10.17 *For the identified model*

$$H_{id}^n(z) = \sum_{k=0}^{n-1} y_k z^k, \quad n \leq M$$

the following identification error bound holds:

$$\|h - H_{id}^n\|_1 \leq n\epsilon + \sum_{k=n}^{\infty} \Phi(k) \qquad (10.108)$$

The value of $n_\star$ that minimizes the above bound is the smallest integer n achieving $\Phi(n) \leq \epsilon$.

Proof. The first part follows from:

$$\|h - H_{id}^n\|_1 = \sum_{k=0}^{n-1} |h(k) - y_k| + \sum_{k=n}^{\infty} |h(k)|$$

$$\leq n\epsilon + \sum_{k=n}^{\infty} \Phi(k) \qquad (10.109)$$

For the second part, define

$$\alpha(n) = n\epsilon + \sum_{k=n}^{\infty} \Phi(k)$$

Therefore $\alpha(n+1) - \alpha(n) = \epsilon - \Phi(n)$ is nondecreasing because Φ is nonincreasing. As a consequence, the minimum of $\alpha(n)$ is achieved at the smallest $n_\star$ satisfying $\alpha(n+1) \geq \alpha(n)$. □

An intuitive interpretation of the above result is the following. Since $|h(k)| \leq \Phi(k)$ and Φ is nonincreasing, for values of k beyond the point for which $\Phi(k) \leq \epsilon$, there is no useful information on the system which can be obtained from the data.

It is clear from (10.109) that the value of n should depend not only on M but also on ϵ in order for the algorithm to converge. This is supported by the fact that there are no untuned identification algorithms based on impulse response data ([160, 193, 194]) as proved before. There are many values of n that may be selected for the algorithm to converge. In the sequel, three of them are presented.

Lemma 10.18 *The algorithms obtained by selecting n as either $n_1(M, \epsilon) = \min(M, c_0 - c_1 \ln \epsilon)$ or $n_2(M, \epsilon) = \min(M, c_0 + c_1 \epsilon^{-r})$, $r \in (0, 1)$ and c_0, c_1 real constants, are tuned to ϵ and convergent. A third, asymptotically globally optimal algorithm tuned to (Φ, ϵ) can be obtained using $n_3 = \min(M, n_\star)$, $n_\star$ defined in Lemma 10.17.*

Proof. For $n = n_1$, $\alpha(n_1) \leq (c_0 - c_1 \ln \epsilon + 1)\epsilon + \sum_{k=n_1+1}^{\infty} \Phi(k)$. The first term vanishes as $\epsilon \to 0$, and $n_1 \to \infty$ as $M \to \infty$ and $\epsilon \to 0$. Since $\Phi \in \ell_1$, this implies that the second term also goes to zero. Convergence for $n = n_2$ can be proved in a similar way. For $n = n_3$, when $M \to \infty$ then $n_3 \to n_\star$ and therefore the upper bound in (10.108) achieves the optimal value in (10.106). Also, for any given $\delta > 0$, there is always a q such that $\sum_{k=q+1}^{\infty} \Phi(k) < \delta/2$, and $\epsilon > 0$ such that $(q+1)\epsilon < \delta/2$. Because the lower bound in (10.106) is global, the last procedure is asymptotically globally optimal. □

As a special case of the above results, consider again the set of models $\mathcal{H}_\infty(\rho, K)$, which includes the exponentially stable discrete time linear systems considered in the $\mathcal{H}_\infty$ identification setup ([146]). For $\rho > 1$ and $K \geq 0$, $\mathcal{H}_\infty(\rho, K) \subset A(D)$, the disk algebra of functions analytic in the open unit disk and continuous on the unit circle. It is well known that the impulse response of this class of systems is bounded above by the sequence $\{K\rho^{-k},\ k \geq 0\}$. This leads to the following corollary.

Corollary 10.1 *For the identified model H_{id}^n defined in (10.105), the identification error can be bounded as follows:*

$$\|h - H_{id}^n\|_1 \leq n\epsilon + \frac{K\rho^{-(n-1)}}{\rho - 1} \tag{10.110}$$

This bound is minimized by selecting the smallest integer $n \geq n_\star$, with

$$n_\star = \left\lfloor \frac{\ln(K/\epsilon)}{\ln \rho} - 1 \right\rfloor \tag{10.111}$$

Proof. Follows immediately from Lemma 10.17 by noting that $\mathcal{H}_\infty(\rho, K) \subset \mathcal{S}$ when selecting $\Phi(k) = K\rho^{-k}$. □

Notes

1. Depending on the resulting value of n, it may be necessary to balance and truncate the identified model H_{id}^n, to obtain a smaller-order model. This is a standard procedure, which may be simplified, avoiding the computation of Gramians, using the algorithm in [45, 127]. The new error bounds are given now in terms of the $\|\cdot\|_\infty$ or the Hankel norm. Similarly, although computationally more expensive, an optimal Hankel model reduction may be performed.
2. For the case of systems that have all positive impulse response values, the ℓ_1 bound coincides with the $\mathcal{H}_\infty$ bound. Therefore the asymptotically ℓ_1 optimal algorithm is also asymptotically $\mathcal{H}_\infty$ optimal.
3. When the value of $n_\star$ is very high, for example in cases where M is high and ϵ is very small, care must be taken when balancing, in particular, when the first singular vector of the Hankel matrix is computed ([45, 127]), to avoid numerical errors that may increase the approximation error.
4. The identified system can be represented by a family of models, in terms of the nominal H_{id}^n and the identification error bound. This family could be unnecessarily conservative if the convergence of the algorithm

is not monotonically decreasing with $\epsilon \to 0$. In this case there might exist $\epsilon_* \geq \epsilon$ for which the identification error is smaller than the one considered and still "covers" the class of models $\mathcal{S}$ and noise $\mathcal{N} = \ell_\infty(\epsilon) \subset \ell_\infty(\epsilon_*)$. All the algorithms presented here have monotonically decreasing convergence.

Based on all Samples Next we present, a strongly optimal convergent algorithm, that is, locally optimal for any set of experimental data. This algorithm is based on the one in [160].

Consider the vector of experimental data $\mathbf{y}$ of length M and define

$$h_U(k) = \min[y_k + \epsilon, \Phi(k)] \tag{10.112}$$

$$h_L(k) = \max[y_k - \epsilon, -\Phi(k)] \tag{10.113}$$

Note that $h_L(k)$ is the lowest value of $h(k)$ compatible with the *a priori* information, therefore consistent. Similarly, $h_U(k)$ is the upper bound on $h(k)$ for each k, so that model h belongs to the consistency set. In addition, the *consistency interval* $[h_L(k), h_U(k)]$ has length $2\min[\epsilon, \Phi(k)]$.

The algorithm is as follows:

$$\mathcal{A}(\mathbf{y}, \Phi, \epsilon)|_k = \begin{cases} \frac{1}{2}[h_L(k) + h_U(k)], & k < M \\ 0, & k \geq M \end{cases} \tag{10.114}$$

The above algorithm is called central because it selects the model in the center of the consistency set; therefore it is also strongly optimal. This last property can be proved as follows.

Lemma 10.19 *The worst-case error for algorithm (10.114) is*

$$e[\mathcal{A}(\mathbf{y}, \Phi, \epsilon)] = \sum_{k=0}^{M-1} \min[\epsilon, \Phi(k)] + \|\Phi\|_1 - \sum_{k=0}^{M-1} |\Phi(k)| \tag{10.115}$$

and tends to zero in the limit $M \to \infty$ and $\epsilon \to 0$. The algorithm is strongly optimal.

Proof. The bound can be obtained as in Lemma 10.17, using the fact that

$$|\mathcal{A}(\mathbf{y}, \Phi, \epsilon)_k - h(k)| \leq \min[\epsilon, \Phi(k)], \quad k = 0, \ldots, M-1$$

since the model is in the center of the interval $[h_L(k), h_U(k)]$. The above bound can be achieved with a suitable choice of system and measurement noise. It is clear that the consistency set is symmetric with respect to the model $H_{id} = \mathcal{A}(\mathbf{y}, \Phi, \epsilon)$; therefore the algorithm is also central.

Note that the first term of the error is equivalent to

$$n_*\epsilon + \sum_{k=n_*}^{M-1} |\Phi(k)| \tag{10.116}$$

with n_* defined in Lemma 10.17. This error converges to zero as $(1/\epsilon, M) \to \infty$. Using the above equation, it is easily seen that the total error can be rewritten as the global lower bound in equation (10.106). It follows that the algorithm is globally optimal. $\square$

In the special case where $\Phi(k) = K\rho^{-k}$, the bound (10.115) reduces to ([160]):

$$e[\mathcal{A}(\mathbf{y}, \Phi, \epsilon)] = \sum_{k=0}^{M-1} \min[\epsilon, \Phi(k)] + \frac{K}{\rho^{M-1}(\rho - 1)} \tag{10.117}$$

The above algorithm is clearly tuned to all *a priori* information. Next, another interpolatory algorithm, which is tuned only to the model information, that is, Φ, will be presented. This algorithm is given by:

$$\mathcal{A}(\mathbf{y}, \Phi)|_k = \begin{cases} \text{sign}(y_k) \min[|y_k|, \Phi(k)] & 0 \le k \le M-1, \\ 0, & k \ge M \end{cases} \tag{10.118}$$

Lemma 10.20 *The worst-case error for algorithm (10.118) is:*

$$e[\mathcal{A}(\mathbf{y}, \Phi)] = \sum_{k=0}^{M-1} \min[\epsilon, 2\Phi(k)] + \|\Phi\|_1 - \sum_{k=0}^{M-1} |\Phi(k)| \tag{10.119}$$

which tends to zero in the limit $M \to \infty$ and $\epsilon \to 0$.

Proof. The proof is similar to the one above, considering the fact that

$$|\mathcal{A}(\mathbf{y}, \Phi)_k - h(k)| \le \min[\epsilon, 2\Phi(k)]$$

when $|y_k| \le \Phi(k)$. If, however, $|y_k| \ge \Phi(k)$, the following holds:

$$|\mathcal{A}(\mathbf{y}, \Phi)_k - h(k)| \le 2\Phi(k)$$

Since the experiment and the noise have the same sign, in this case

$$|\mathcal{A}(\mathbf{y}, \Phi)_k - h(k)| \le |h(k) - y_k| \le \epsilon$$

In both cases this leads to the desired bound. $\square$

For the particular case when $\Phi(k) = K\rho^{-k}$, the bound (10.119) simplifies to ([160]):

$$e[\mathcal{A}(\mathbf{y}, \Phi)] = \sum_{k=0}^{M-1} \min[\epsilon, 2\Phi(k)] + \frac{K}{\rho^{M-1}(\rho - 1)}$$

10.5 FURTHER RESEARCH TOPICS

Due to the fact that robust identification is a currently active research area, there are yet many theoretical and computational aspects that have not been developed fully. In this section, several items that have not been included in the main part of this chapter will be commented on and referenced. An excellent recent survey of the area of robust identification can be found in [199], which includes an extensive list of references.

10.5.1 Unstable Systems

The methods developed in this chapter apply to stable LTI systems. For slowly divergent systems, that is, with slow unstable poles, care must be taken when carrying out the experiment, especially when considering its duration ([127]). In general, neither time nor frequency response experiments can be performed over open-loop unstable systems. Rather, they should be carried out after a stabilizing controller has been applied to the plant. In that case, the closed-loop input/output data should be used to identify the open-loop plant. The identification error should also take into account the fact that both model and plant are unstable. Thus the $\mathcal{H}_\infty$ norm should be replaced by a suitable distance function.

In [196] a complete framework for the robust identification of strongly[10] stabilizable plants is presented. The measure of identification error is computed in terms of the gap, graph, or chordal metrics.

10.5.2 Nonuniformly Spaced Experimental Points

In the $\mathcal{H}_\infty$ identification two-stage procedures, it is assumed that the frequency experimental data points are equally spaced. Since this may not be the case in practical situations, the standard algorithm should be modified accordingly. In [8] and [235] interpolation is used to perform identification with nonuniformly spaced experimental data points. In [127] a procedure that uses explicitly the unequally separated data points is presented, in the context of model approximation.

---

[10] A *strongly* stabilizable plant is the one that can be stabilized by a stable controller.

10.5.3 Model Reduction

In many of the algorithms presented in this chapter, the model order remains high. As a consequence, a further model reduction step needs to be performed and the model error bounds should be modified accordingly. This has been attempted in [142] for ℓ_1 identification and in [198, 253] for $\mathcal{H}_\infty$ identification.

10.5.4 Continuous Time Plants

The robust identification framework presented so far considers both discrete and continuous time systems represented by $H(z)$ in Section 10.3. Nevertheless, in this last case, some extra considerations related to the frequency aliasing should be taken into account. Specifically, strictly proper stable LTI plants with an *a priori* upper bound on their roll-off frequencies should be considered. This problem has been treated in [9, 147].

10.5.5 Sample Complexity

For application of the identification techniques to practical problems, it is important to know whether the experiments can be performed in a reasonable amount of time. Specifically, the following question may be asked. *Which is the minimum time length of an experiment in order to arrive at a certain degree of accuracy, measured by the worst-case identification error?* A general framework for this problem has been considered in [183] and specific results have been obtained in [163, 85, 243, 303]. These results are presented next.

Consider the FIR systems of length M, bounded experimental inputs $\|u\|_\infty \leq 1$, and a (optimal over all algorithms and inputs) worst-case identification error bounded by

$$\inf_{\mathcal{A}} \inf_{\|u\|_\infty \leq 1} e(\mathcal{A}, u) \leq 2Q\epsilon \qquad (10.120)$$

where ϵ is the measurement noise uniform upper bound and the parameter $Q \geq 1$ defines the level of uncertainty ($Q = 1$ corresponds to the optimal error bound 2ϵ). For this case, the minimum length of the experiment duration to achieve this bound is $\mathcal{O}(2^{Mf(1/Q)})$, with $f(\cdot)$ a strictly monotonically increasing function in $(0, 1)$, with $f(0) = 0$ and $\lim f(x) = 1$ as $x \to 1$. It can be shown that this is a tight bound.

A consequence of this result is that the worst case class of measurement noise (uniformly bounded by ϵ) considered in time-domain experiments may be very conservative. Therefore, as suggested in [243], the class of measurement noise should be relaxed or noise should be considered in the average. On the other hand, the worst-case class of models does not need to be modified.

In the case of $\mathcal{H}_\infty$ identification, the sample complexity is only quadratic in M ([199]).

374 ROBUST IDENTIFICATION

10.5.6 Mixed Time/Frequency Experiments

When applying robust identification techniques to either frequency- or time-domain experiments separately, it may happen that a model that efficiently fits the data in one domain may have poor performance in the other domain. For example, with only frequency experimental data points, a "good" frequency response fitting (small $\mathcal{H}_\infty$ error norm) may lead to a "poor" fitting in the impulse response. Additionally, from an information theoretic viewpoint, more experiments produce a smaller consistency set of indistinguishable models, and as a consequence a smaller worst-case error. From a practical standpoint, in many cases robust identification algorithms are applied to systems that may not be exactly LTI. Under these circumstances it is desirable to perform both time and frequency response experiments ([272]) to assess the validity of these assumptions. Thus, in these cases, an algorithm that combines both types of experimental data could take advantage of the additional data available.

Recent papers [77, 333] have proposed interpolatory algorithms that use data obtained from time-domain experiments and generate a nominal model together with an $\mathcal{H}_\infty$ bound on the identification error. Therefore, although they combine time domain measurements with frequency domain error bounds, they do not use both types of experimental information simultaneously.

Alternatively, some recently proposed algorithms exploit both time and frequency domain experiments at the same time. In [273] necessary and sufficient conditions for the consistency of mixed time/frequency experimental data for FIR systems are presented and an interpolatory identification algorithm is proposed. In [231] the above result was extended to IIR systems, based on a general interpolation theorem ([21]), which allows the simultaneous interpolation of both the time and frequency experiments. The solution to this problem leads to a convex LMIs optimization.

10.5.7 Mixed Parametric/Nonparametric Models

All of the results in this chapter address nonparametric identification of models with a worst-case global bound. In many cases, part of the model has a clear parametric structure. In these cases disregarding this information may lead to conservative results (see Example 10.4). This is usually the case in mechanical flexible structures, which have a well defined parametric model for the lower-frequency modes and an unknown higher-frequency behavior that naturally leads to a nonparametric identification ([121]).

In general, the parametric information does not appear explicitly in the *a priori* knowledge (K, ρ) usually considered. Therefore it is important to include it, if available, so that less conservative *a priori* estimates of (K, ρ) for the nonparametric portion can be derived. This is the case of systems exhibiting large peaks in the frequency response (for instance, a poorly damped flexible structure), which must be "covered" by large values of K ([121]). This is a consequence of the fact that the usual *a priori* information characterizes only the smoothness and magnitude of the whole class of models but cannot distinguish among other properties, for example, low frequency model structure. Instead, the *a priori* parametric knowledge provides more "structured" information.

There are several results that address this problem, for example, in [121, 175, 205, 232, 333]. In particular, the identification of affine parametric and nonparametric models using simultaneous time and frequency response experimental data can be solved as a convex optimization problem ([232]). This includes cases of practical interest such as flexible mechanical structures, as well as models that can be described concisely in terms of a set of Laguerre or Kautz functions or, more generally, any other basis of $\mathcal{H}_2$ ([148]).

11

APPLICATION EXAMPLES

11.1 SAC-C ATTITUDE CONTROL ANALYSIS

11.1.1 Introduction

The SAC-C satellite is a joint project between the National Commission of Space Activities (Argentina) and NASA (United States). It is essentially an Earth observation satellite that carries an Argentine CCD camera for Earth resources observation, a Global Positioning System (GPS) experiment from JPL (NASA), and a magnetic sensor from Denmark. The latter has a Star Imager and two Magnetometers mounted on an 8 meter boom, which will be deployed in orbit. The purpose of the boom is to magnetically isolate the magnetometers from the spacecraft. The length of this boom makes it very flexible; therefore the model of the spacecraft cannot be considered as a rigid body (see Figure 11.1).

The attitude objective is to point the camera toward the vertical with a precision of $\pm 1°$ around the roll and pitch axis; that is, the line of sight of the camera should be within a $1°$ square with respect to the vertical direction. Based on optical considerations, the jitter of the control should be on the order of $0.05°/s$.

Due to several practical limitations, it is important to have the simplest possible controller and the least amount of hardware. These limitations stem from the fact that the microprocessor onboard should not only implement the mission control mode but also several alternative safe-hold modes, as well as failure logics. Therefore a low-order controller is desirable. In addition, sensor and actuator hardware adds weight to the spacecraft, which in turn decreases the available payload. For this reason, only the minimum required rotations along the axes are sensed. Similarly only the minimum number

378 APPLICATION EXAMPLES

of torques are applied by the actuators. Thus only the roll and pitch axes are sensed by means of a scan wheel horizon sensor (SW) and the control torques are applied only in the pitch and yaw axes. These control torques act through magnetic torque coils (MTCs) and momentum wheels, the latter built into the SW. In the design, the torque bounds of both actuators should be considered, due to limitations in the maximum magnetic moment of the MTCs and the maximum size and acceleration of the momentum wheels.

The desired attitude of the satellite in Mission Mode (MM) is the following: roll along the velocity vector, yaw opposite to the vertical, and pitch perpendicular to the orbital plane (see Figure 11.1). Our goal is to design a minimal order controller to achieve this nominal attitude. This controller feeds back the roll and pitch angles and rates as well as the momentum wheel velocities, using static LQR feedback. The only dynamics involved in the controller arise from a yaw angle and rate observer (the only axis without sensor). Therefore the complete controller has only two states, is applied to only two of the three axes, and senses two out of three axes: as simple as it can be. The possibility of obtaining such a low-order controller and not having to sense or actuate over all axes is due to the fact that the spacecraft and the orbital dynamics are coupled.

The attitude for the Earth Pointing Safe Hold Mode (EPSHM) is the same as for MM, except that the momentum wheels are kept at nominal constant speed. Here the control is applied using only the Earth magnetic field, through the MTCs.

The controller design is carried-out using a very simple procedure, that takes into consideration the limitations in the controller order, sensors, and actuators. This design is validated in Section 11.1.4 through an analysis involving all the performance objectives as well as model uncertainties. Using the structured singular value, we assess the robustness of the controller against the unknown high-order dynamics of the boom, considered as global uncertainty. Robust performance is guaranteed using μ for both modes: MM and EPSHM. Several time simulations of the complete system are presented involving random measurement noise and torque perturbations. Further details of this analysis are reported in [274].

11.1.2 Linear Model

The general nonlinear dynamic equation of the system is:

$$\underbrace{T_w + T_m}_{\text{control}} + T_g + T_p = I_T \dot{W} + W \times I_T W + W \times h$$

$$T_w^s \triangleq \begin{bmatrix} 0 \\ T_{wy} \\ T_{wz} \end{bmatrix} = - \begin{bmatrix} 0 \\ \dot{h}_y \\ \dot{h}_z \end{bmatrix}, \quad T_m^s \triangleq \begin{bmatrix} 0 \\ T_{my} \\ T_{mz} \end{bmatrix}$$

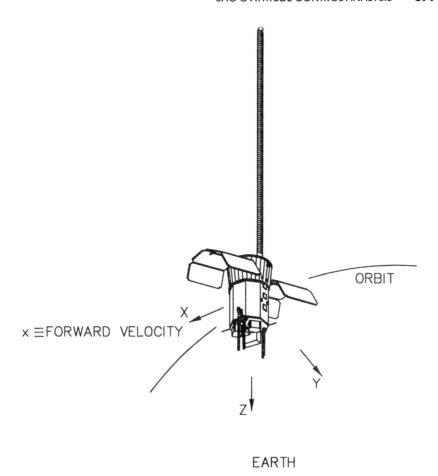

Figure 11.1. SAC-C Earth observation satellite.

Here W is the angular rate of the spacecraft with respect to inertial space, represented as the rotation of a frame attached to the spacecraft with respect to the inertial reference frame. The inertia of the rigid part of the spacecraft is I_T and h represents the wheels momentum in the spacecraft frame. This frame is the usual roll (x), pitch (y), and yaw (z) axes mentioned before (see Figure 11.1).

The external torques consist of the perturbation (T_p), gravitational (T_g), and magnetic control (T_m), that is, $T = T_p + T_g + T_m$. The gravitational torque is due to the effect of the rigid body model of the boom and the magnetic control is produced by the MTCs. The perturbation torques are due to the aerodynamic drag (almost constant in pitch and yaw along the orbit) and the magnetic and gravitational residual torques, both periodic and negligible as compared with the former.

The wheels and magnetic torque coil torques are used for control purposes: the first one depends on the velocity change of the momentum exchange wheels and therefore does not change the total momentum of the spacecraft, the latter is an external torque that does change the total momentum. Note that these torques are applied only in the pitch and yaw axes. The reason is that a control torque in the roll axis is unnecessary due to the orbital coupling between roll and yaw, as will be seen next. This reduces the number of actuators, therefore simplifying the design and minimizing costs. The external perturbation torques are bounded but otherwise unknown.

Next, a linearization is computed around the equilibrium conditions, corresponding to the nominal attitude of the satellite: the roll and yaw axes colinear with the velocity and local vertical vectors, respectively; and pitch perpendicular to the orbital plane, which completes the spacecraft frame (boom opposite to the local vertical). The roll, pitch, and yaw angles are denoted as θ_R, θ_P, and θ_Y, respectively. The incremental values of the wheels momenta are δh_y in pitch and δh_z in the yaw direction. For simplicity we can assume that $I_T = \mathrm{diag}\begin{bmatrix} I_x & I_y & I_z \end{bmatrix}$.

Since the roll and yaw dynamics are coupled but independent from the pitch, we define the system state as

$$x^T = \begin{bmatrix} x_{ry}^T & x_p^T \end{bmatrix} \qquad (11.1)$$

$$x_{ry} = \begin{bmatrix} \dot\theta_R & \theta_R & \dot\theta_Y & \theta_Y & \delta h_z \end{bmatrix}^T, \qquad x_p = \begin{bmatrix} \dot\theta_P & \theta_P & \delta h_y \end{bmatrix}^T$$

Adding the equations for the wheels torques $T_{wy} = -\delta \dot h_y$ and $T_{wz} = -\delta \dot h_z$, we obtain the state equations for both the [roll–yaw] and [pitch] dynamics:

$$\dot x_{ry} = A_{ry} x_{ry} + B_{ry} \begin{bmatrix} T_{wz} \\ T_{mz} \end{bmatrix} + B_z T_{pz} \qquad (11.2)$$

$$\dot x_p = A_p x_p + B_{\mathrm{pitch}} \begin{bmatrix} T_{wy} \\ T_{my} \end{bmatrix} + B_y T_{py} \qquad (11.3)$$

$$\implies \begin{cases} \dot x(t) = A x(t) + B u(t) + B_p T_p \\ y = \begin{bmatrix} I_{2\times 2} & 0_{2\times 6} \\ 0_{4\times 4} & I_{4\times 4} \end{bmatrix} x(t) + 0_{6\times 4} u(t) \stackrel{\triangle}{=} C x(t) + D u(t) \end{cases}$$

where $x(t)$ has been defined in (11.1) and where:

$$B_p \stackrel{\triangle}{=} \begin{bmatrix} 0_{2\times 2} \\ \frac{1}{I_z} & 0 \\ 0_{2\times 2} \\ 0 & \frac{1}{I_y} \\ 0_{2\times 2} \end{bmatrix}, \qquad u(t) \stackrel{\triangle}{=} \begin{bmatrix} T_{wz} \\ T_{mz} \\ T_{wy} \\ T_{my} \end{bmatrix}, \qquad T_p \stackrel{\triangle}{=} \begin{bmatrix} T_{pz} \\ T_{py} \end{bmatrix} \qquad (11.4)$$

Note that the system output consists only of the measured states, that is roll and pitch. Matrices $A_{ry}, B_{ry}, A_p, B_{\text{pitch}}, B_y$, and B_z have been defined in [274].

11.1.3 Design Constraints

To simplify the controller as much as possible, a state feedback has been designed. The only controller dynamics will be the ones from the yaw angle and rate observer, due to the fact that only roll and pitch sensors are available (scan wheels).

A static LQR controller has been designed, taking into account the desired bounds for the different states and actuators when computing the weights to solve the minimization problem. These constraints are as follows:

$$\begin{bmatrix} T_{wz} \\ T_{mz} \\ T_{wy} \\ T_{my} \end{bmatrix} \leq \begin{bmatrix} 0.012 \sin \alpha \text{ N} \cdot \text{m} \\ \frac{\sqrt{2}}{2} 6\, 10^{-4} \text{ N} \cdot \text{m} \\ 0.024 \cos \alpha \text{ N} \cdot \text{m} \\ \frac{\sqrt{2}}{2} 6\, 10^{-4} \text{ N} \cdot \text{m} \end{bmatrix} \quad (11.5)$$

This is due to the fact that the torque bound in each wheel is 12 mNm and that in pitch both add their torques projected through $\cos \alpha$ ($\alpha \leq 10°$). In yaw, on the other hand, they subtract both torques projected by $\sin \alpha$. The bound on the magnetic torque coils is computed as the average torque along the orbit. Each MTC has 15 Am^2 and the average magnetic field along the orbit is $(\sqrt{2}/2)\, 0.4 \times 10^{-4}$ Tesla. Therefore the actuator weight in the LQR problem is a diagonal matrix that has the inverse of these bounds as each element (Bryson's Rule, see Chapter 5).

The state weighting matrix is also diagonal with the inverses of the maximum allowable bound in each of the states. These are given by:

$$\begin{bmatrix} \dot{\theta}_R \\ \theta_R \\ \dot{\theta}_Y \\ \theta_Y \\ \delta h_z \\ \dot{\theta}_P \\ \theta_P \\ \delta h_y \end{bmatrix} \leq \begin{bmatrix} 0.1°/\text{s} \\ 1° \\ 0.1°/\text{s} \\ 1° \\ 0.06 \text{ N} \cdot \text{m} \cdot \text{s} \\ 0.1°/\text{s} \\ 1° \\ 0.06 \text{ N} \cdot \text{m} \cdot \text{s} \end{bmatrix} \quad (11.6)$$

The bounds in both δh_z and δh_y correspond to a maximum deviation from the nominal wheel speed of 400 rpm. For a total momentum of 3.2 N·m·s (2000 rpm, $I_w = 0.0077$) along the pitch axis, these bounds are 0.11 and 0.64 N·m·s, respectively.

The observer involves only the [roll–yaw] dynamics, due to the fact that pitch is uncoupled. Furthermore, a reduced order observer will be used so that the controller has the minimum possible order. The measured, estimated, and input variables to this observer are, respectively,

$$x_m \triangleq \begin{bmatrix} \dot{\theta}_R \\ \theta_R \\ \delta h_z \end{bmatrix}, \quad x_Y \triangleq \begin{bmatrix} \dot{\theta}_Y \\ \theta_Y \end{bmatrix}, \quad T_{cz} \triangleq \begin{bmatrix} T_{wz} \\ T_{mz} \end{bmatrix}$$

Closed-Loop System The observer system is given by:

$$\left[\begin{array}{c|c} A_o & B_o \\ \hline C_o & D_o \end{array} \right] = \left[\begin{array}{c|cc} A_o & B_{o1} & B_{o2} \\ \hline C_o & D_{o1} & 0_{2\times 2} \end{array} \right] \tag{11.7}$$

with output $\hat{x}_Y$ and input $\begin{bmatrix} x_m^T & T_{cz}^T \end{bmatrix}^T$. In this case the observer gain can be computed analytically ([274]).

The control signals, LQR gain matrix, and state feedback are

$$T_c \triangleq \begin{bmatrix} T_{wz} \\ T_{mz} \\ T_{wy} \\ T_{my} \end{bmatrix} = - \begin{bmatrix} F_R & F_Y & F_{hz} & 0_{2\times 3} \\ 0_{2\times 2} & 0_{2\times 2} & 0_{2\times 1} & F_P \end{bmatrix} \begin{bmatrix} \dot{\theta}_R \\ \theta_R \\ \hat{x}_Y \\ \delta h_z \\ \dot{\theta}_P \\ \theta_P \\ \delta h_y \end{bmatrix} \tag{11.8}$$

where the feedback matrix has been partitioned according to the roll, yaw, δh_z, and pitch variables. From the above equation and the observer dynamics we compute the controller as follows:

$$\dot{z} = \underbrace{(A_o - B_{o2} F_Y C_o)}_{A_k} z + \overbrace{\{B_{o1} - B_{o2}(F_Y D_{o1} + \begin{bmatrix} F_R & F_{hz} \end{bmatrix})\}}^{B_{k1}} x_m$$

$$T_{cz} = \underbrace{-F_Y C_o}_{C_{k1}} z \overbrace{- (F_Y D_{o1} + \begin{bmatrix} F_R & F_{hz} \end{bmatrix})}^{D_{k1}} x_m \quad \begin{bmatrix} T_{wy} \\ T_{my} \end{bmatrix} = \underbrace{-F_P}_{D_{k2}} \begin{bmatrix} \dot{\theta}_P \\ \theta_P \\ \delta h_y \end{bmatrix}$$

or equivalently,

$$K(s) \equiv \left[\begin{array}{c|cc} A_k & B_{k1} & 0_{2\times 3} \\ \hline C_{k1} & D_{k1} & 0_{2\times 3} \\ 0_{2\times 2} & 0_{2\times 3} & D_{k2} \end{array} \right]$$

$$= \left[\begin{array}{c|c} A_k & B_k \\ \hline C_k & D_k \end{array} \right] \tag{11.9}$$

Note that the controller has the minumum number of states, given the available sensors (roll and pitch) and the minimum number of actuator outputs (yaw and pitch) given the the coupled dynamics.

Using the above definitions for the controller, input, state, and output variables, the closed-loop system is given by:

$$\dot{x} = (A + BD_kC)x + BC_kz + B_pT_p + BD_kn$$
$$\dot{z} = B_kCx + A_kz + B_kn$$
$$e = \begin{bmatrix} x \\ u \end{bmatrix}$$

Thus the state–space realization of the generalized plant $G_{cl}(s)$ with states $\begin{bmatrix} x^T & z^T \end{bmatrix}^T$, disturbance inputs $p \triangleq \begin{bmatrix} T_p^T & n^T \end{bmatrix}^T$, and e the output to be minimized is:

$$\left[\begin{array}{cc|cc} A + BD_kC & BC_k & B_p & BD_k \\ B_kC & A_k & 0_{2\times 2} & B_k \\ \hline I_{8\times 8} & 0_{8\times 2} & 0_{8\times 4} & 0_{8\times 4} \\ D_kC & C_k & 0_{4\times 2} & D_k \end{array} \right] = \left[\begin{array}{c|c} A_c & B_c \\ \hline C_c & D_c \end{array} \right]$$

11.1.4 Robustness Analysis

The closed-loop dynamics computed in the last section correspond to the nominal system, which takes into account only the rigid body modes of both satellite and boom. Nevertheless, the flexible modes of the boom, should be taken into account explicitly in the robustness analysis.

These modes can be modeled as uncertainty in the high-order dynamics, using experimental data provided by the boom manufacturer ([295]), to compute the frequency distribution of the uncertainty. The boom dynamics is considered in this setup as follows:

$$u_\Delta = G_{\text{boom}}(s)y_\Delta, \quad u_\Delta \triangleq \begin{bmatrix} T_{px} \\ T_{py} \end{bmatrix}, \quad y_\Delta \triangleq \begin{bmatrix} \theta_R \\ \theta_P \end{bmatrix}$$

The reason is that the driving input, which initiates the flexible dynamics oscillations, is the angular rotation of the spacecraft in roll and pitch (the torsion around the yaw axis is negligible). On the other hand, the perturbation of the boom on the satellite structure is represented by the (internal) torques applied in the roll and pitch axes due to these oscillations. The experiments in [295] determined a lower frequency dynamics of 1 Hz. Therefore a convenient weight that "covers" the uncertain high-frequency dynamics of the boom and assumes a 20% modeling error at lower frequencies is

$$W_\Delta(s) = 0.4 \frac{5s + \pi}{s + 2\pi} I_{2\times 2}$$

384 APPLICATION EXAMPLES

The performance specifications call for minimizing the states and bounding the control torques of MWs and MTCs. Therefore a suitable performance weight is a constant diagonal matrix with the inverse of the upper bounds in (11.5) and (11.6) as its elements.

To perform the robustness analysis of the satellite the nominal and uncertain portions, and weights for robustness and performance are connected in the LFT form shown in Figure 11.2.

The closed-loop nominal system computed in Section 11.1.3 corresponds to the lower portion of the general system $R(s) = F_e(P, K)$, connecting the inputs p to the outputs e, that is, $G_{cl}(s) = R_{22}(s)$. The total system, without weights, connects the inputs (u_Δ, p) to the outputs (y_Δ, e) and has the following state–space realization:

$$R(s) \equiv \left[\begin{array}{c|c} A_c & \begin{matrix} 1 & 0 \\ \mathbf{0}_{4\times2} \\ 0 & 1 \\ \mathbf{0}_{4\times2} \end{matrix} \quad B_c \\ \hline \begin{matrix} 0 & 1 & 0 & 0 & 0 & 0 & 0 & 0 & 0 \\ 0 & 0 & 0 & 0 & 0 & 0 & 1 & 0 & 0 \\ & & & & C_c & & & & \end{matrix} & \begin{matrix} \mathbf{0}_{2\times2} & \mathbf{0}_{2\times8} \\ \mathbf{0}_{12\times2} & D_c \end{matrix} \end{array} \right]$$

$$= \left[\begin{array}{c|c} A_r & B_r \\ \hline C_r & D_r \end{array} \right] \tag{11.10}$$

To perform the robustness analysis the output weights must be added to the system as indicated in Figure 11.2. Furthermore, to normalize the analysis, we scale the inputs in vector p so that the magnitude of the worst-case perturbations has unity norm. To this end we use a constant input weighting matrix having the maximum perturbation torques and measurement noise as its elements. These perturbation torques are due mainly to the atmospheric

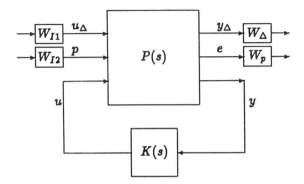

Figure 11.2. Linear fractional transformation setup for robustness analysis.

drag effect in yaw and pitch with a maximum of 10^{-5} N·m, and to the residual magnetic torques in roll with a maximum of 10^{-6} N·m. The angular measurement error due to the scan wheels is $0.3°$ in roll and pitch, the only measured axes (recall yaw is estimated with an observer). The angular rate measurement error in the same axes stems from the error in the LSB of the word that spans the total angle, this is $0.01°$. The rate is computed as the difference between two correlative angle samples, divided by the sampling time T. For this reason, the rate "measurement" noise weight is similar to a notch filter centered at $\omega = 2\pi/T$ and scaled by $0.01°/s$:

$$W_{\text{rate}}(s) = \frac{0.01s}{(s + \pi/5)(s/20\pi + 1)} \qquad (11.11)$$

Finally, the rate measurement errors of the wheel velocity, which contribute to the error in their angular momentum in yaw (δh_z) and pitch (δh_y), amount to a maximum of 2% of h_{yo}.

The necessary and sufficient conditions to guarantee robust (internal) stability, nominal performance, and robust performance are

$$\text{RS} \iff \|W_\Delta(s)R_{11}(s)W_{I1}(s)\|_\infty \leq 1 \qquad (11.12)$$

$$\text{NP} \iff \|W_p(s)R_{22}(s)W_{I2}(s)\|_\infty \leq 1 \qquad (11.13)$$

$$\text{RP} \iff \sup_{s=j\omega} \mu_\Delta [W_O(s)R(s)W_I(s)] \leq 1 \qquad (11.14)$$

where $W_O(s) = \text{diag}\,[W_\Delta(s) \quad W_p(s)]$, $W_I(s) = \text{diag}\,[W_{I1}(s) \quad W_{I2}(s)]$, and the uncertainty structure Δ contains two full blocks, one for the boom high-order dynamics, the second for the performance objectives.

11.1.5 Simulations

The following are simulated outputs of the main variables to be minimized, in Mission Mode (MM). Figures 11.3 to 11.6 show the angles, angular rates, and momentum wheel and magnetic control torques, all below the desired bounds. Figure 11.7 shows the robust stability and nominal and robust performance conditions, just below the value 1.

A different controller has been designed with the same methodology to control the spacecraft only with the MTCs, called Earth Pointing Safe-Hold Mode (EPSHM). In this case the wheels are kept at a constant speed and provide the necessary angular mmomentum to the system. In particular, the weight on the MW control torques for the design is large enough so that the wheel torques are below 6×10^{-6} Nm, that is, almost constant speed. The robust analysis curves based on the infinity norm and the structured singular values are presented in Figure 11.8.

386 APPLICATION EXAMPLES

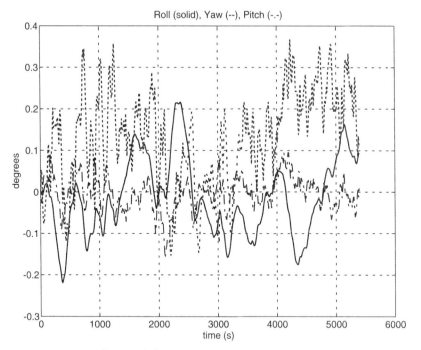

Figure 11.3. Rotation angular errors (MM).

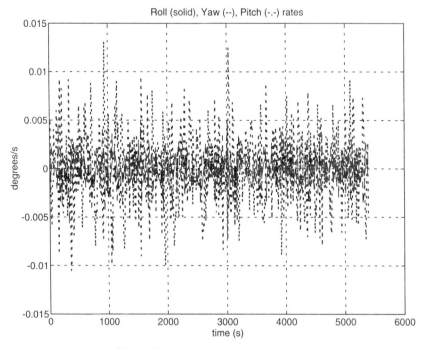

Figure 11.4. Rotation rate errors (MM).

SAC-C ATTITUDE CONTROL ANALYSIS 387

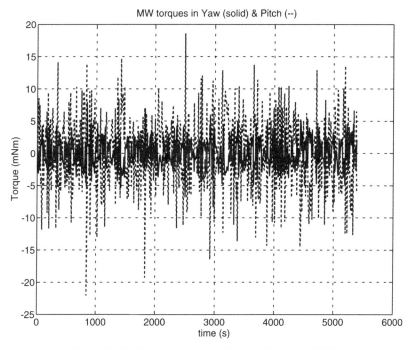

Figure 11.5. Momentum wheel control torques (MM).

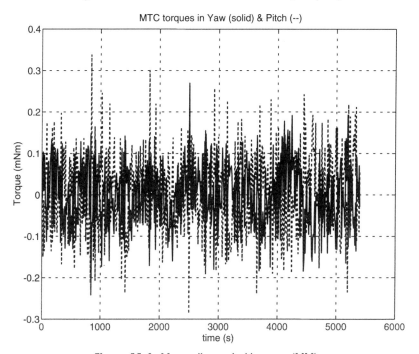

Figure 11.6. Magnetic control torques (MM).

388 APPLICATION EXAMPLES

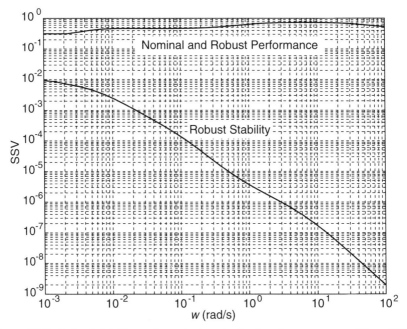

Figure 11.7. Robust stability and nominal and robust performance conditions (MM).

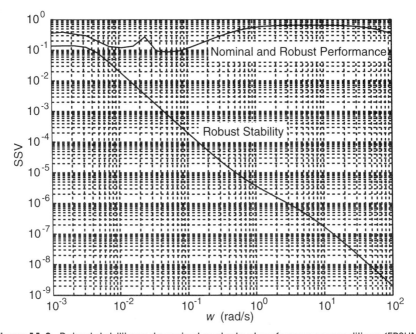

Figure 11.8. Robust stability and nominal and robust performance conditions (EPSHM).

11.2 CONTROLLER DESIGN FOR A D_2O PLANT

In this section we illustrate by step, the robust control methodology through an application to a D_2O plant. The mathematical model of the plant is a linearized and simplified version of a nonlinear system with a delay. This particular application is a standard problem in the process control area. The purpose of this example is to assess the incidence of model and disturbance uncertainty in this typical application. The model uncertainty arises from the time delay and the gains and poles of the transfer function, which depend on the particular linearization point.

The usual control structure to handle systems with time delays is the Smith predictor ([289]). This classical method in process control has the advantage that, when the delay is known *exactly*, the controller designed to achieve (nominal) stability needs to take into account only the rational (nondelayed) part of the plant. In this example, based on the Smith predictor, we perform the robust analysis and design of a controller for this particular uncertain system, which exhibits all the relevant issues of process control problems. The following example is based on [268], and the interpretation of Smith predictors with uncertainty is described in [211].

11.2.1 Model of the Plant

The plant to be considered is an experimental process that obtains D_2O by isotopic exchange between H_2O and H_2S (see [128, 206]). The plant has already been constructed but is not yet operational.

A general diagram of the process can be seen in Figure 11.9. This diagram is a simplification of the process flow sheet (there are more cold and hot towers in the actual plant) but retains the main issues of interest. The input water flow and liquid flows are marked with full lines and the gas flows with dashed ones. Two compressors have been included, as well as three heat exchangers that control the temperature in the process. The greatest D_2O concentration in each stage is obtained between the hot and cold towers. A fraction of the partially enriched water of the first stage is sent to the second stage for further enrichment. The product is extracted from the second stage after the H_2S stripper eliminates the gas. The isotopic exchange process that takes place in the cold and hot columns, respectively, is the following:

$$H_2O + HDS \xrightarrow{30°C} HDO + H_2S$$

$$HDO + H_2S \xrightarrow{130°C} H_2O + HDS$$

The controlled variable is the ratio L/G, with L and G representing the liquid and gas operational flows, respectively. The value of G remains almost constant due to the fact that it circulates on a closed circuit, as seen in the same figure; therefore the control is performed by changing the value of L.

390 APPLICATION EXAMPLES

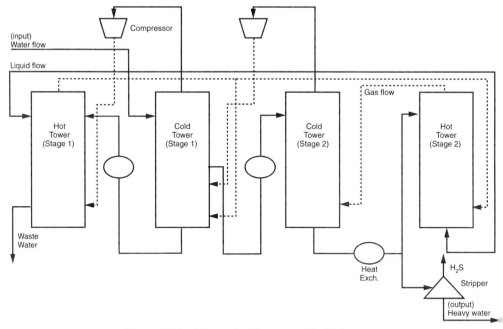

Figure 11.9. Schematic diagram of the D_2O plant.

A general nonlinear model of the whole process has been implemented ([206, 228]) for simulation purposes. This general model has been obtained from physical considerations and is too complicated for both analysis and control design. Therefore the controller analysis and synthesis are based on a linear model that takes differential H_2O flow as an input and differential D_2O outlet concentration as the output. The linearization has been made at 0.01%, 0.3%, and 1% disturbances in the input water flow, under the restriction of 15% outlet concentration of D_2O. The structure of the linear model remains the same for all three linearization points, but the parameters (gains and poles) differ from one point to another. Assuming continuity between the linearization points (water flow disturbance) and the model parameters, the linear process may be described by a *set* of models with the same linear structure but with parameters lying in real bounded intervals.

The delay is a consequence of the measurement procedure for the output variable. It consists of the collection of samples at each column, which takes around 15 minutes per column, with a transportation time between columns of another 15 minutes. All these samples are measured with a mass spectrometer, to obtain, after approximately a 4 hour test, the D_2O concentration. Depending on particular conditions, the elapsed time between the sample extraction and the final output value may differ significantly. Therefore for analysis purposes, we consider a nominal delay of 5 hours, obtained as an average value, with an uncertainty of ±1 hour.

In this linear model we also consider additive disturbances in the output. These disturbances include several liquid and gas flows, which go through part of the process and which, for operational purposes, are bounded. If these bounds are exceeded, the required level of D_2O extraction will not be met (or we will have no extraction at all). Due to the fact that these flows circulate through most of the columns, we may assume that the dynamics of the disturbances is very slow (see Figure 11.9). For the above reasons, we model the additive disturbances at the output as a *set* of bounded signals with unknown and slow dynamics.

The set of linear models representing the process is the following:

$$G(s) = e^{-s\tau} \left[\frac{k_1}{s+r_1} + \frac{k_2}{s+r_2} + \frac{k_3}{s+r_3} + \frac{k_4}{s+r_4} \right] = e^{-s\tau} G_M(s)$$

$$k_i \in [l_i, u_i], \qquad i = 1, \ldots, 4$$
$$r_j \in [v_j, w_j], \quad v_j > 0, \quad j = 1, \ldots, 4 \qquad (11.15)$$
$$\tau = \tau_o + \delta, \quad |\delta| < 1, \quad \delta \in \mathbb{R},$$

where the nominal delay is $\tau_o = 5$ hours. The uncertainty intervals for poles and gains are obtained from their values at the three linearization points mentioned before ([228]).

The disturbances are represented as a weighted and bounded set, that is, $\{W_d(s)d(s) \mid \|d\|_2 \leq 1\}$. Here $W_d(s)$ takes care of the fact that the disturbances have a higher energy content in the lower frequencies (slow dynamics).

For the controller design step, a rational approximation of the nominal time delay needs to be made. Many different methods exists, but a standard, simple procedure is the use of Padé approximants. Nevertheless, we will verify the robust performance condition with the exact values of the delay, without approximation errors.

11.2.2 Robustness Analysis

In this section we obtain conditions on the controller such that stability, performance, and robustness are achieved. Due to the fact that the system has a delay, we work with the Smith predictor control structure, but including the uncertainty in the delay, poles, and gains.

The classical Smith predictor structure ([289]) is presented in Figure 11.10. The stable plant $G(s)$ consists of a rational term $G_M(s, \mathbf{p})$ with a vector of uncertain parameters $\mathbf{p}$, $p_i \in [a_i, b_i]$ (in this case the poles and gains in (11.15)) and an uncertain delay $\tau = \tau_o + \delta$, $|\delta| < 1$. Define the nominal model as $G_o(s) = e^{-s\tau_o} G_M(s, \mathbf{p}_o)$ which will be used for the design. Next, we analyze step by step nominal and robust stability and performance.

Nominal Stability The Smith predictor operates by substracting from the feedback signal the difference between the nominal model and the corresponding minimal phase part (Figure 11.10). This is equivalent to the inclusion of a prediction in the system since the feedback signal $f(t)$ is exactly the output at a future time $(t + \tau_o)$. This procedure is effective only when there is complete knowledge of the system (the parameters take their nominal values $\tau = \tau_o$ and $\mathbf{p} = \mathbf{p}_o$), otherwise we will not be able to have a correct prediction of y.

From the same figure, we observe that the nominal transfer function between u and the feedback f, equals $G_{Mo} \triangleq G_M(s, \mathbf{p}_o)$. Therefore in terms of nominal stability, only a controller $K(s)$ that stabilizes the nominal minimum phase part $G_{Mo}(s)$, regardless of the time delay τ_o, needs to be designed. This is the main advantage of the classical Smith predictor, but this result is no longer valid when considering performance or robustness issues, as will be seen next.

Nominal Performance Based on the results in previous chapters, nominal performance depends on the transfer function between the disturbance and the output error signal. Figure 11.11 yields the following transfer function in the nominal case, that is, $G(s) = G_o(s)$:

$$y_d(s) = \left[1 - e^{-s\tau_o} T_M(s)\right] W_d(s) d(s) \qquad (11.16)$$

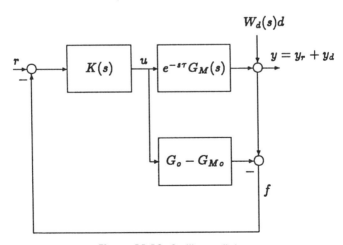

Figure 11.10. Smith predictor.

where we have defined the complementary sensitivity as

$$T_M(s) \triangleq G_{Mo}(s)K(s)\left[1 + G_{Mo}(s)K(s)\right]^{-1}$$

The necessary and sufficient condition for nominal performance is

$$\text{Nominal performance} \iff \|y_d(s)\|_2 \leq 1, \quad \forall \, \|d(s)\|_2 \leq 1$$

$$\iff \left\|\left[1 - e^{-s\tau_o}T_M(s)\right]W_d(s)\right\|_\infty \leq 1 \quad (11.17)$$

This leads to the following $\mathcal{H}_\infty$ control problem:

$$\min_{\text{stabilizing } K(s)} \left\|\left[1 - e^{-s\tau_o}T_M(s)\right]W_d(s)\right\|_\infty \quad (11.18)$$

An important point is that the nominal performance condition **includes** the nominal delay τ_o. As we mentioned before, the classical Smith predictor does not simplify the analysis step for performance, as it did with nominal stability. A controller that solves equation (11.18) can be obtained by a standard $\mathcal{H}_\infty$ optimal control algorithm, with a previous approximation of the time delay. The optimal solution without approximating the delay can be obtained from [116], but this approach cannot solve the robust performance problem.

Robust Stability In this section we analyze the closed-loop stability for all possible delays between 4 and 6 hours and all possible parameters **p** [poles and gains in (11.15)] of the minimum phase part of the plant.

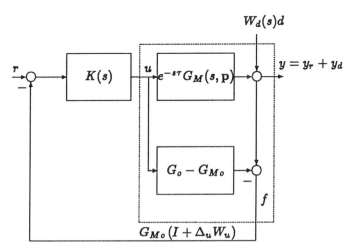

Figure 11.11. Smith predictor with uncertainty.

A practical[1] way to analyze models with parametric uncertainty is to "cover" them with dynamic uncertainty. In this case we have used a multiplicative dynamic description for the set of plants, as follows:

$$\mathcal{M} \triangleq \{[1 + \Delta_u W_\Delta(s)]\, G_{M_o}(s)\,,\ \Delta_u \in \mathbb{C}\,,\ |\Delta_u| < 1\} \quad (11.19)$$

In this application, there are two different sources of parameter uncertainty, which in principle could be covered separately. This will produce a "smaller" set of plants than in the case where the coverage is performed over both parametric uncertainties together. As a consequence, the analysis results for the "smaller" set of models will always be less conservative than in the case where both parametric uncertainties are *globally* covered.

For the delay, we transform the dashed block of Figure 11.11 into a nominal system with global multiplicative uncertainty, with $\mathbf{p} = \mathbf{p}_o$. To this end, consider the family of models

$$\mathcal{G} \triangleq G(s) - G_o(s) + G_{M_o}(s) \quad (11.20)$$

$$= \left[1 + \left(e^{-s\delta} - 1\right)e^{-s\tau_o}\right] G_{M_o}(s),\quad \delta \in \mathbb{R}\,,\ |\delta| < 1$$

which should be *covered* by a family of models with multiplicative dynamic uncertainty as in (11.19). The latter set of models does not take into account any phase information because Δ_u is bounded in magnitude; therefore at each frequency ω in a Nyquist plot, it represents the interior of a disk of radius $W_\Delta(j\omega)$. To cover the parametric uncertainty in the delay, this circle should include the set $\mathcal{G}$, for which it is sufficient to achieve the following inequality at each $s = j\omega$:

$$\sup\{|\Delta_u W_\Delta(j\omega)|\,,\ \Delta_u \in \mathbb{C}\,,\ |\Delta_u| < 1\}$$
$$\geq \sup\left\{\left|\left(e^{-j\omega\delta} - 1\right)e^{-j\omega\tau_o}\right|\,,\ \delta \in \mathbb{R},\ |\delta| < 1\right\}$$

which in turn is equivalent to

$$|W_\Delta(j\omega)| \geq \max_{|\delta|<1}\left|e^{-j\omega\delta} - 1\right| \quad (11.21)$$

Figure 11.12 shows the above condition in a magnitude Bode plot. A rational

---

[1] An *ad hoc* analysis considering directly parametric uncertainty could be made, but it is not the purpose of this example. The results on analysis of parametric uncertainty structures are either analytical methods that solve particular cases (and do not fit in this application) or algorithmic methods that solve general cases but are computationally involved, in fact a NP-complete problem ([60, 248]). There is not yet a synthesis method that supports a general parametric analysis. An excellent survey can be found in [35].

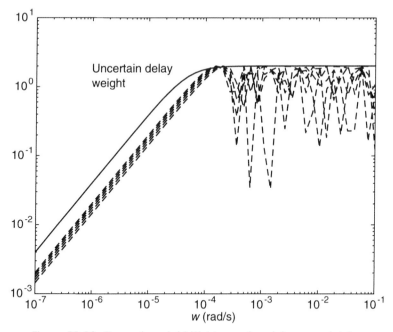

Figure 11.12. Dynamic weight $W_\Delta(s)$ covering delay uncertainty.

approximation $W_\Delta(s)$ for the weight can be computed from the same figure. Note that the set of models that covers the delay uncertainty is independent of the nominal delay τ_o.

Following a similar procedure, a set of models with dynamic multiplicative uncertainty, which covers the parametric uncertainty in gains and poles can be obtained. For briefness it will not be repeated here, but the corresponding weighting function may be seen in Figure 11.13.

The necessary and sufficient condition to guarantee the stability of *all* members of a family of models with multiplicative uncertainty[2] is

$$\|W_u(s)T_M(s)\|_\infty < 1 \qquad (11.22)$$

From the above equation, we observe that the critical point in terms of robust stability is the frequency ω_c for which $|W_u(j\omega_c)| \geq 1$. From (11.22) it is clear that, for $\omega > \omega_c$, the values of $|T_M(j\omega)|$ should be less than one to achieve the robust stability condition. In some sense this frequency defines an upper bound for the closed-loop bandwidth, the latter interpreted in terms of $T_M(s)$.

---

[2] For the family $\mathcal{M}$ the condition is necessary and sufficient when replacing W_Δ with W_u, but since $\mathcal{G} \subset \mathcal{M}$, for the actual set of models it is only sufficient.

APPLICATION EXAMPLES

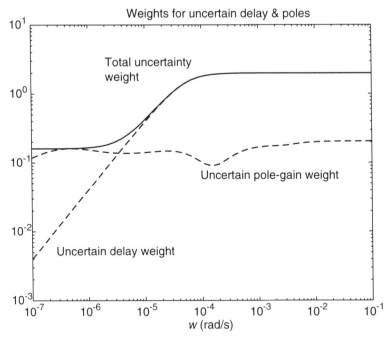

Figure 11.13. Dynamic weights covering pole gain and delay uncertainties. Total uncertainty weight is $W_u(s)$.

The above condition could be useful in determining the relative importance of both types of uncertainties for the robust stability of the plant. By comparing the weights in Figure 11.13 we observe that, for the frequency range of interest, the curve due to the pole and gain uncertainty remains below one. Therefore the limiting factor in the robust stability condition is the delay uncertainty. Nevertheless, in the lower-frequency range, the uncertainty in poles and gains is larger than the one due to the delay. Therefore, from a practical point of view, we should consider the total uncertainty weight $W_u(s)$, which covers both types of parametric uncertainty (see Figure 11.13).

Finally, it should be noted that the Smith predictor structure allows this analysis to be independent of the *nominal* delay τ_o. Only the uncertainty is taken into account, which appears indirectly in the weights $W_\Delta(s)$ (see equation (11.21)) and $W_u(s)$.

Robust Performance For robust performance we need to achieve the performance objective for all models in the set $\{[1 + \Delta_u W_u(s)] G_{Mo}(s), |\Delta_u| < 1\}$. From Figure 11.11 we obtain the following transfer function between d and the output signal:

$$y_d(s) = \left[1 - e^{-s\tau_o} T_M(s)\right] \left[1 + \Delta_u W_u(s) T_M(s)\right]^{-1} W_d(s) d(s) \qquad (11.23)$$

where we have replaced $G(s) = G_o(s) + \Delta_u W_u(s) G_{Mo}(s)$. A necessary and sufficient condition for robust performance can be obtained from the above equation in the following way:

$$\left|(1 - e^{-sT_o} T_M)(1 + \Delta_u W_u T_M)^{-1} W_d\right| \leq 1, \quad \forall s = j\omega, \ |\Delta_u| < 1$$

$$\iff \left|(1 - e^{-sT_o} T_M) W_d\right| \leq \inf_{|\Delta_u|<1} |(1 + \Delta_u W_u T_M)|, \quad \forall s = j\omega$$

$$\iff \left|(1 - e^{-sT_o} T_M) W_d\right| + |W_u T_M| < 1, \quad \forall s = j\omega \tag{11.24}$$

where we have selected a worst-case $|\Delta_u| < 1$ at each frequency.

Equation (11.24) is a combination of both the robust stability and nominal performance conditions of previous sections. The main difference with a standard *mixed sensitivity* problem is that both terms of this equation are applied to different plants—one to $G_{Mo}(s)$ and the other to a combination of $G_{Mo}(s)$ and $G_o(s)$. To recast it as a standard mixed sensitivity problem we apply the following transformation:

$$1 - e^{-sT_o} T_M(s) = \left(1 - e^{-sT_o}\right) + e^{-sT_o} S_M(s) \tag{11.25}$$

where $S_M(s) \triangleq [1 + G_{Mo}(s) K(s)]^{-1}$ is the sensitivity function. The goal is now to design a controller that solves:

$$\min_{\text{stabilizing } K(s)} \gamma \left|S_M(s) \tilde{W}_d(s)\right| + |W_u(s) T_M(s)| \tag{11.26}$$

with $\tilde{W}_d = e^{-sT_o} W_d(s)$. The factor γ not only provides relative weight between the nominal performance and robust stability conditions, but also compensates for the extra term $(1 - e^{-sT_o})$ in equation (11.25). At the design stage in the next section, the value of γ is computed so that the robust performance condition (11.24) is achieved in the least conservative way.

11.2.3 Controller Design

The mixed sensitivity problem of equation (11.26) can be solved in two ways. $\mathcal{H}_\infty$ optimal control theory can be used to design a controller that solves:

$$\min_{\text{stabilizing } K(s)} \left\| \begin{bmatrix} W_u(s) T_M(s) \\ \gamma \tilde{W}_d(s) S_M(s) \end{bmatrix} \right\|_\infty \tag{11.27}$$

that is, it minimizes the norm of the transfer matrix from the input w_2 to the outputs (z_1, z_2) in Figure 11.14. Recall from Section 4.8 that the optimal value of (11.27) is at most twice the value of (11.26).

398 APPLICATION EXAMPLES

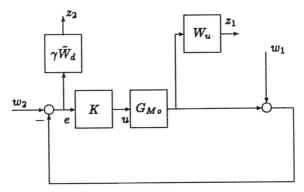

Figure 11.14. Mixed sensitivity problem.

Alternatively μ-synthesis can be used to design a controller that minimizes directly the value in (11.26). This is equivalent to

$$\min_{\text{stabilizing } K(s)} \sup_{s=j\omega} \mu \left\{ \begin{bmatrix} -W_u(s)T_M(s) & W_u(s)T_M(s) \\ -\gamma \tilde{W}_d(s)S_M(s) & \gamma \tilde{W}_d(s)S_M(s) \end{bmatrix} \right\} \quad (11.28)$$

the argument being the transfer function matrix between inputs (w_1, w_2) and outputs (z_1, z_2) of Figure 11.14.

The sensitivity $S_M(s)$ and its complement $T_M(s)$ are both functions of the nominal minimum phase part of the plant $G_{Mo}(s)$, which has the following state–space description:

$$\dot{x}(t) = \begin{bmatrix} -\frac{1}{395.6} & 0 & 0 & 0 \\ 0 & -\frac{1}{16183.6} & 0 & 0 \\ 0 & 0 & -\frac{1}{67522.7} & 0 \\ 0 & 0 & 0 & -\frac{1}{5748589.8} \end{bmatrix} x(t) + \begin{bmatrix} 0.395 \times 10^{-3} \\ 0.336 \times 10^{-4} \\ 0.17 \times 10^{-4} \\ 0.166 \times 10^{-5} \end{bmatrix} u(t)$$

$$y(t) = \begin{bmatrix} 1 & 1 & 1 & 1 \end{bmatrix} x(t)$$

The uncertainty weight, obtained from Figure 11.13, and the performance weight from design considerations are, respectively,

$$W_u(s) = 2 \frac{s + 4 \times 10^{-6}}{s + 5 \times 10^{-5}}, \quad W_d(s) = 10^3 \frac{1.66 \times 10^4 s + 1}{3 \times 10^8 s + 1} \quad (11.29)$$

When equation (11.26) is solved using $\mathcal{H}_\infty$ optimal control, the resulting optimal solution does not necessarily achieve the optimal robust performance in (11.24), where we have a combination of minimum and nonminimum phase models. Therefore we have iterated over γ until the design achieves the robust performance condition (11.24) in the least conservative way. In fact, for this case, the best performance has been obtained for the first iteration of a μ-synthesis procedure, that is, the $\mathcal{H}_\infty$ optimal design,[3] and a value of $\gamma_{max} = 1.55$.

For the design, we used a third order Padé approximation of the nominal delay. Nevertheless, to verify robust performance in (11.24), the exact values of the delay have been taken into account. The plots of the robust and nominal performance and robust stability can be seen in Figure 11.15. In the same figure we can see the structured singular value μ of the closed-loop system. We remark that, since the design has been made using the minimum phase part of the system $G_{Mo}(s)$, the value of μ is below the actual robust performance condition (11.24), which includes the delay. For the value γ_{max}, the latter is very close to unity at $\omega_o = 2 \cdot 10^{-4}$ rad/s. In Figure 11.16 we can see the response of the system to an 11 hour sinusoidal disturbance of unit magnitude for 4, 5, and 6 hour time delays. The frequency of the disturbance has been selected as a worst-case value according to the peak on robust performance in Figure 11.15 ($\omega_o = 2 \cdot 10^{-4}$ rad/s, worst-case disturbance).

Due to the fact that the present design optimizes condition (11.26), a

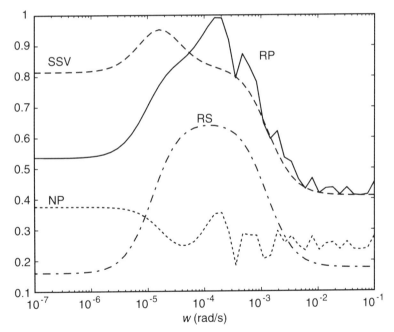

Figure 11.15. Robust and nominal performance, robust stability, and μ (for the design structure).

---

[3] In a standard problem, where the sensitivity and its complement correspond to the same model, the performance produced by a μ-synthesis design should have been better than or at least equal to the $\mathcal{H}_\infty$ optimal one.

400 APPLICATION EXAMPLES

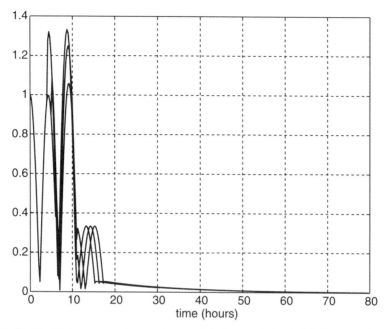

Figure 11.16. Time response for worst-case disturbance (robust disturbance rejection).

possible improvement could be obtained by the use of methods that consider directly parametric uncertainty and the fact that the system is infinite-dimensional. This is beyond the scope of this example, only intended to illustrate a complete step-by-step robust control analysis and design procedure on a typical process control application.

11.3 X-29 PARAMETRIC ANALYSIS

In this section we perform a robust stability analysis of the *lateral-directional* controlled dynamics of NASA's X-29 aircraft. This is a true multivariable control problem due to the coupling of the roll and yaw axes. The rigid body model of the aircraft with a controller in place has been taken into consideration and the analysis consists of computing the multivariable stability margin (or equivalently μ) for the uncertainty in the aerodynamic coefficients. Further details of this example can be found in [265].

11.3.1 Linear Model

First we introduce the notation used throughout this section ([189]) in Table 11.1, corresponding to Figure 11.17. An illustration of the aircraft is shown in Figure 11.18.

Table 11.1. Aircraft parameters

a_Y	Lateral acceleration (g)
b	Reference span (ft)
C_l, C_n	Roll and yaw moment coefficients
C_Y	Lateral force coefficient
g	Gravity acceleration (32.2 ft/s^2)
I_x, I_y, I_z	Roll, pitch, yaw axes moments of inertia (slug-ft^2)
I_{xz}	Roll and yaw axis cross inertia (slug-ft^2)
H	Altitude (ft)
m	Mass (slugs)
p, q, r	Roll, pitch, yaw rates (rad/s)
$\bar{q}$	Dynamic pressure (lb/ft^2)
S	Reference area (ft^2)
T	Sampling period (0.025 s)
V	Velocity (ft/s)
α	Angle of attack (rad)
β	Sideslip angle (rad)
δ_a, δ_r	Aileron and rudder deflection (rad)
θ, ϕ	Pitch and roll attitude (rad)

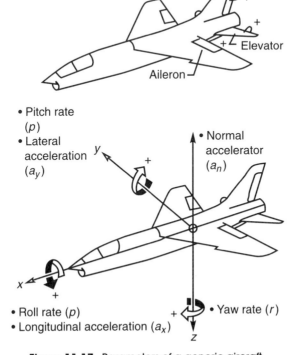

Figure 11.17. Parameters of a generic aircraft.

402 APPLICATION EXAMPLES

Figure 11.18. NASA X-29 aircraft.

The lateral-directional linear model of the aircraft ([189]) is presented next.

$$\dot{\beta} = \frac{\bar{q}S}{mV}C_Y + p\sin\alpha - r\cos\alpha + \frac{g}{V}\sin\phi\cos\theta \qquad (11.30)$$

$$\dot{p}I_x - \dot{r}I_{xz} = \bar{q}SbC_l + rq(I_y - I_z) + pqI_{xz} \qquad (11.31)$$

$$\dot{r}I_z - \dot{p}I_{xz} = \bar{q}SbC_n + pq(I_x - I_y) - rqI_{xz} \qquad (11.32)$$

$$\dot{\phi} = p + q\tan\theta\sin\phi + r\tan\theta\cos\phi \tag{11.33}$$

$$a_y = \frac{\bar{q}S}{mg}C_Y - \frac{Z_{a_y}}{g}\dot{p} + \frac{X_{a_y}}{g}\dot{r} \tag{11.34}$$

with the aerodynamic coefficients given by

$$C_Y = C_{Y_\beta}\beta + C_{Y_{\delta_a}}\delta_a + C_{Y_{\delta_r}}\delta_r \tag{11.35}$$

$$C_l = C_{l_\beta}\beta + C_{l_{\delta_a}}\delta_a + C_{l_{\delta_r}}\delta_r + C_{l_p}\frac{pb}{2V} + C_{l_r}\frac{rb}{2V} \tag{11.36}$$

$$C_n = C_{n_\beta}\beta + C_{n_{\delta_a}}\delta_a + C_{n_{\delta_r}}\delta_r + C_{n_p}\frac{pb}{2V} + C_{n_r}\frac{rb}{2V} \tag{11.37}$$

The subscripts β, δ_a, and δ_r indicate the variables with respect to which the derivatives C_Y, C_l, and C_n are taken; for example, C_{Y_β} represents the derivative of the lateral force with respect to the sideslip angle. These derivatives are computed by parameter identification procedures applied to the experimental data. In addition, error bounds on these parameters are obtained as well.

The linearized state–space equations, considering the input from the differential aileron and flap $\mathbf{u} = [\delta_a \ \ \delta_r]^T$ and the output $\mathbf{y} = [p \ \ r \ \ \phi \ \ a_Y]^T$, are:

$$\dot{x} = E^{-1}\tilde{A}x + E^{-1}\tilde{B}u \tag{11.38}$$

$$y = Cx + Du \tag{11.39}$$

where the state is $x = [\beta \ \ p \ \ r \ \ \phi]^T$. Matrices $(\tilde{A}, \tilde{B})$ explicitly contain the uncertain aerodynamic coefficients C_{l_β}, C_{n_β}, $C_{l_{\delta_a}}$, $C_{n_{\delta_a}}$, $C_{l_{\delta_r}}$, and $C_{n_{\delta_r}}$:

$$\tilde{A} = \begin{bmatrix} q_m C_{Y_\beta} & \sin\alpha & -\cos\alpha & (q/V)\cos\theta \\ q_s C_{l_\beta} & -\tilde{a}_{22} & \tilde{a}_{23} & 0 \\ q_s C_{n_\beta} & \tilde{a}_{32} & -\tilde{a}_{33} & 0 \\ 0 & 1 & \tan\theta & \tilde{a}_{44} \end{bmatrix} \tag{11.40}$$

$$\tilde{B} = \begin{bmatrix} q_m C_{Y_{\delta_a}} & q_m C_{Y_{\delta_r}} \\ q_s C_{l_{\delta_a}} & q_s C_{l_{\delta_r}} \\ q_s C_{n_{\delta_a}} & q_s C_{n_{\delta_r}} \\ 0 & 0 \end{bmatrix} \tag{11.41}$$

the additional matrices are:

$$E = \begin{bmatrix} 1 & 0 & 0 & 0 \\ 0 & I_x & -I_{xz} & 0 \\ 0 & -I_{xz} & I_z & 0 \\ 0 & 0 & 0 & 1 \end{bmatrix} \tag{11.42}$$

APPLICATION EXAMPLES

$$C = \begin{bmatrix} 0 & 1 & 0 & 0 \\ 0 & 0 & 1 & 0 \\ 0 & 0 & 0 & 1 \\ q_g C_{Y_\beta} & -Z_{a_Y}/g & X_{a_Y}/g & 0 \end{bmatrix} \tag{11.43}$$

$$D = \begin{bmatrix} 0 & 0 \\ 0 & 0 \\ 0 & 0 \\ q_g C_{Y_{\delta a}} & q_g C_{Y_{\delta r}} \end{bmatrix} \tag{11.44}$$

Here we have defined

$$q_m = \bar{q}S/mV, \qquad q_g = \bar{q}S/mg, \qquad q_s = \bar{q}Sb \tag{11.45}$$

$$q_v = \bar{q}Sb^2/2V, \qquad \tilde{a}_{22} = -(q_v C_{l_p} + qI_{xz}) \tag{11.46}$$

$$\tilde{a}_{23} = q_v C_{l_r} + q(I_y - I_z), \qquad \tilde{a}_{32} = q_v C_{n_p} + q(I_x - I_y) \tag{11.47}$$

$$\tilde{a}_{33} = -(q_v C_{n_r} - qI_{xz}), \qquad \tilde{a}_{44} = q\tan\theta \tag{11.48}$$

This model has four open-loop poles: the *dutch roll*, a complex conjugate pair that applies to the roll and yaw dynamics; the *roll mode*, a real stable pole; and the *spiral mode*, a low frequency unstable pole.

The controller structure for all flight conditions is the following:

$$K(s) = K_a(s) \cdot K_b(s) \tag{11.49}$$

where $K_a(s)$ is:

$$\begin{bmatrix} A_f(s)Z(s)\left[X_{kp3} + X_{ki3}T/2 + X_{ki3}/s\right] & 0 \\ 0 & A_r(s)Z(s)X_{kp4} \end{bmatrix} \tag{11.50}$$

and the elements of $K_b(s)$ are given by

$$(K_b)_{11} = P_1(s) \cdot S_p(s) \cdot [K_2 - \alpha \cdot \text{Blend} \cdot L(s) \cdot K_3] \tag{11.51}$$

$$(K_b)_{12} = K_3 \cdot S_r(s) \cdot F_r(s) \cdot P_1(s) \cdot L(s) \tag{11.52}$$

$$(K_b)_{13} = -K_3 \cdot P_2(s) \cdot L(s) \cdot g \cdot \text{Blend}/V \tag{11.53}$$

$$(K_b)_{14} = K_4 \cdot F_a(s) \cdot P_1(s) \tag{11.54}$$

$$(K_b)_{21} = P_1(s) \cdot S_p(s) \cdot [K_{16} - \alpha \cdot \text{Blend} \cdot L(s)] \tag{11.55}$$

$$(K_b)_{22} = K_{17} \cdot S_r(s) \cdot F_r(s) \cdot P_1(s) \cdot L(s) \tag{11.56}$$

$$(K_b)_{23} = -K_{17} \cdot P_2(s) \cdot L(s) \cdot g \cdot \text{Blend}/V \tag{11.57}$$

$$(K_b)_{24} = K_{18} \cdot F_a(s) \cdot P_1(s) \tag{11.58}$$

where Blend is a parameter.

The elements of both matrices are shown in Table 11.2.

Table 11.2. Matrices $K_a(s)$ and $K_b(s)$

$A_f(s)$	Fourth-order differential flap actuator model
$A_r(s)$	Fourth-order rudder actuator model
$Z(s)$	Padé approximation due to Sample & Hold delay
$P_1(s)$	Roll, yaw, and lateral acceleration prefilters
$P_1(s)$	Roll attitude prefilter
$S_p(s)$	Second-order Roll rate sensor model
$S_r(s)$	Second-order yaw rate sensor model
$F_r(s)$	Second-order yaw notch filter
$F_a(s)$	Second-order lateral acceleration notch filter
$L(s)$	Analog equivalence of digital filter $L(z)$

11.3.2 Results

The robust stability analysis was performed over two different linearization points. The first one corresponds to the nominal design flight conditions, the second one to a critical condition at sea level and with large angle of attack. These conditions are shown in Table 11.3.

The mathematical structure of the controller and nominal model of the plant remain unchanged, although the parameters have different values according to the flight condition. In the controller these parameters are K_2 to $K_{18}, X_{kp3}, X_{ki3}, X_{kp4}$, and Blend; in the nominal plant model the parameters are the aerodynamic coefficient derivatives. In this last case not only do the nominal values of the parameters change but also their uncertainty bounds.

Before a branch-and-bound analysis algorithm is applied to this example ([266, 281]), we proceed to compute the structure of the characteristic polynomial $f(s, \Delta)$ as a function of the uncertain parameters mentioned before; this is

$$G(s, \Delta) = D + C \left[sE - \tilde{A}(\Delta) \right]^{-1} \tilde{B}(\Delta) \qquad (11.59)$$

$$f(s, \Delta) = \det [I + K(s)G(s, \Delta)] \qquad (11.60)$$

Table 11.3. Linearization conditions

	No. 1	No. 2
Mach No.	0.9	0.4
Altitude	30,000 ft	sea level
α	3.78°	30°
V	895 ft/s	447 ft/s
θ	3.81°	−60°
q	0°/s	24.62°/s
$\bar{q}$	357 lb/ft^2	237 lb/ft^2

406 APPLICATION EXAMPLES

$$\Delta = \text{diag}\,[\delta_1 \quad \cdots \quad \delta_n] \quad (11.61)$$

with δ_i the bounded uncertainty of parameter p_i around its nominal value p_{oi}, that is, $p_i = p_{oi} + \delta_i$, $|\delta_i| \leq \bar{\delta}_i$. A symbolic manipulation solver has been used to determine the mathematical structure of this equation, in particular the exponents of the uncertain parameters, to determine which are *repeated* ones. The mapping theorem can be applied to the parameters that appear linearly. On the other hand the ones raised to powers larger than one appear in a polynomial fashion in the characteristic equation and therefore should be restricted by the equality contraints mentioned in Chapter 7, reducing them to a multilinear dependence. In this case, the characteristic polynomial $f(\cdot)$ is multilinear in all parameters but two, in this latter case with a quadratic dependence. The new set of parameters generated by the addition of equality restrictions (see Chapter 7) is:

$$p_1 = p_2 = C_{l_\beta} = p_{o1} + \delta_1 = p_{o2} + \delta_2 \quad (11.62)$$

$$p_3 = p_4 = C_{n_\beta} = p_{o3} + \delta_3 = p_{o4} + \delta_4 \quad (11.63)$$

$$p_5 = C_{l_{\delta a}} = p_{o5} + \delta_5 \quad (11.64)$$

$$p_6 = C_{n_{\delta a}} = p_{o6} + \delta_6 \quad (11.65)$$

$$p_7 = C_{l_{\delta r}} = p_{o7} + \delta_7 \quad (11.66)$$

$$p_8 = C_{n_{\delta r}} = p_{o8} + \delta_8 \quad (11.67)$$

The bounds on the uncertainties are

$$\begin{array}{lll} \bar{\delta}_1 = \bar{\delta}_2 = 33.3\%, & \bar{\delta}_3 = \bar{\delta}_4 = 13.7,\% & \bar{\delta}_5 = 12.8\% \\ \bar{\delta}_6 = 58\%, & \bar{\delta}_7 = 18.9\%, & \bar{\delta}_8 = 14.3\% \end{array} \quad (11.68)$$

for the first linearization point and

$$\begin{array}{lll} \bar{\delta}_1 = \bar{\delta}_2 = 33.3\%, & \bar{\delta}_3 = \bar{\delta}_4 = 11.1\%, & \bar{\delta}_5 = 32.0\% \\ \bar{\delta}_6 = 63.6\%, & \bar{\delta}_7 = 685\%, & \bar{\delta}_8 = 22.6\% \end{array} \quad (11.69)$$

for the second.

The multivariable stability margin $k_m(j\omega)$ $[\mu_{\Delta_p}(j\omega)]$ was computed for a certain range of frequencies. The minimum values for each flight condition were $k_m(j\omega) = 3.15$ $[\mu_{\Delta_p}(j\omega) = 0.317]$ at $\omega = 3.5$ rad/s in the first case, and $k_m(j\omega) = 2.89$ $[\mu_{\Delta_p}(j\omega) = 0.346]$ at $\omega = 10$ rad/s for the second case. Thus the design is robustly stable.

The worst-case parameter combination, from the stability viewpoint, can also be obtained from the algorithm. This is the vector p^* whose image first reaches the origin of the complex plane for the smallest scaling factor k. This is a very important element in the analysis and redesign (if necessary) of the controller.

This worst-case combination of parameters for flight conditions 1 and 2 were, in this case,

$$\Delta_1 = \text{diag}\begin{bmatrix} \bar{\delta}_1 & \bar{\delta}_2 & \bar{\delta}_3 & \bar{\delta}_4 & -(0.86 \cdot \bar{\delta}_5) & -\bar{\delta}_6 & -\bar{\delta}_7 & \bar{\delta}_8 \end{bmatrix} \quad (11.70)$$

$$\Delta_2 = \text{diag}\begin{bmatrix} \bar{\delta}_1 & \bar{\delta}_2 & -\bar{\delta}_3 & -\bar{\delta}_4 & -\bar{\delta}_5 & (0.18 \cdot \bar{\delta}_6) & \bar{\delta}_7 & \bar{\delta}_8 \end{bmatrix} \quad (11.71)$$

The design of $K(s)$ was made at both linearization points for the set of nominal parameters $p_N = p_o + \Delta_N$, with Δ_N as follows:

$$\Delta_N = \text{diag}\begin{bmatrix} -\bar{\delta}_1 & -\bar{\delta}_2 & \bar{\delta}_3 & \bar{\delta}_4 & -\bar{\delta}_5 & -\bar{\delta}_6 & \bar{\delta}_7 & -\bar{\delta}_8 \end{bmatrix} \quad (11.72)$$

As observed, this disagrees with the values obtained in (11.70) and (11.71) for both flight conditions. This is a consequence of the fact that it is difficult to develop a physical *intuition* on how the parameter variations will influence the stability of the system, especially in MIMO plants with a large number of uncertain parameters.

As stated previously (see Chapter 7), the exact computation of $k_m(j\omega)$ or $\mu_{\Delta_p}(j\omega)$ in the cases of parameter uncertainty has an exponential time computation with the number of parameters. Nevertheless, in many cases, as the one presented here, it can be computed in a practical way by branch-and-bound procedures. Further details of the algorithm used in this case can be found in [266, 281], and this particular example has been developed in [265] under a NASA grant.

11.4 CONTROL OF A DC-TO-DC RESONANT CONVERTER

11.4.1 Introduction

DC-to-DC resonant converters have been the object of much attention lately, since they have the potential to provide high-performance conversion, allowing for smaller, lighter power supplies. However, they require using a control circuit capable of maintaining the desired output voltage under different operating conditions. In this section (based on the recent paper [65]) we illustrate the μ-synthesis technique by designing a controller for a parallel resonant converter. In addition to guaranteeing stability for a wide range of load conditions, this controller rejects disturbances at the converter input while keeping the control input and the settling time within values compatible with a practical implementation.

11.4.2 The Conventional Parallel Resonant Converter

Figure 11.19 shows a diagram of a conventional second-order parallel resonant converter. The combinations of the diodes and transistors form bidirectional switches operating at 50% duty ratio. Thus, in each switching period, the resonant circuit L-C is alternatively excited by $+V_g$ and $-V_g$.

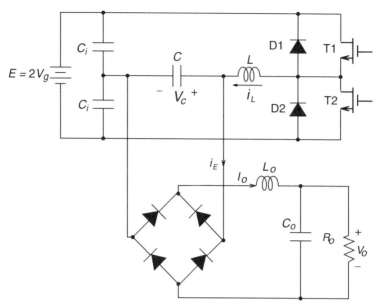

Figure 11.19. Conventional second-order PRC circuit diagram.

The large output inductor L_o and capacitor C_o are used to minimize the load effect on the resonant capacitor voltage and to ensure the constant output voltage through the output circuit [37]. As for notation, the resistor R_o and the voltages V_g and V_o represent the load, the line (input), and the output, respectively.

Throughout this section we use as nominal parameters the following values, taken from the design example in Chapter 2 of [37]:

$$L = 41.18\,\mu\text{H}$$

$$C = 11.30\,\text{nF}$$

$$R_o = 208.33\,\Omega$$

$$L_o = 10\,\text{mH}$$

$$C_o = 47\,\mu\text{F}$$

$$V_g = 100\,\text{V}$$

$$V_o = 250\,\text{V}$$

$$f_s = 200\,\text{kHz}$$

$$Z_o = \sqrt{L/C} = 60.39\,\Omega$$

For convenience, we introduce the following normalized variables:

$$V_{ng} = \frac{V_g}{V_g} = 1$$

$$Q_p = \frac{R_o}{Z_o} = 3.45$$

$$V_{no} = \frac{V_o}{V_g} = 2.5$$

$$F_{ns} = \frac{f_s}{f_o} = 0.86$$

where the resonant frequency $f_o = 1/2\pi\sqrt{LC}$.

11.4.3 Small Signal Model

Under steady-state conditions it can be shown that, for a PRC operating in the continuous conduction mode [37], there are four circuit modes in each switching period. Thus the converter is a nonlinear, variable structure system, with its steady-state trajectory uniquely determined by the normalized switching frequency F_{ns} and the load condition Q_p. For a given operating point, a discrete time, small signal model of the converter can be obtained by using a perturbation method. The sampling time for this discrete time model is equal to $T_s/2$, where $T_s = 1/f_s$ is the switching period. Therefore it follows that this model is correct under small-signal perturbations with frequencies up to the operating switching frequency $\omega_s = 2\pi f_s = 1.26 \times 10^6 \text{rad/s}$.

The discrete time model from the normalized switching frequency F_{ns} and the normalized line V_{ng} to the normalized output V_{no} (to simplify the notation, we use the same variables for both the steady state and its perturbation) at the nominal operating point is given by the following state–space realization [37]:

$$X(k+1) = AX(k) + BU(k)$$
$$V_{no}(k) = CX(k)$$

where

$$A = \begin{bmatrix} 0.8219 & 0.5504 & -2.1402 \\ -0.2767 & 0.6108 & -0.6644 \\ 0.0053 & 0.0075 & 0.9387 \end{bmatrix}$$

$$B = \begin{bmatrix} -6.4684 & 0.4834 \\ 10.6774 & 1.9499 \\ -0.0002 & 0.0162 \end{bmatrix}$$

$$C = [0 \quad 0 \quad 3.45]$$

The state variables and inputs are defined as

$$X(k) = \begin{bmatrix} i_{nl}(k) & v_{nc}(k) & I_{no}(k) \end{bmatrix}^T, U(k) = \begin{bmatrix} F_{ns}(k) & V_{ng}(k) \end{bmatrix}^T$$

where i_{nl}, v_{nc}, and I_{no} are the normalized resonant inductor current, capacitor voltage, and output current, respectively.

11.4.4 Control Objectives

Figure 11.20 illustrates the diagram used for control design. In the small-signal model of the converter there are two inputs: line voltage V_{ng} and switching frequency F_{ns}. The objective is to synthesize a controller having as input the error signal (obtained by comparing the output voltage versus the reference input r) and as output the switching frequency, F_{ns}, such that the output voltage is kept at a prescribed level (in our case $V_o = 250$ V, i.e., $V_{no} = 2.5$) at all operating points. This problem can further be divided into four parts:

- *Line Regulation* (*Nominal Performance*). The line voltage is often unregulated and could have a substantial range of variation, typically around $\pm 20\%$. This variation will be modeled as an external disturbance, thus leading to a disturbance rejection problem. Performance specifications for this type of problem are usually given in terms of time-domain quantities, such as:
    1. Zero steady-state error.
    2. Small overshoot at output (usually less than 10% for reference input step response).
    3. Appropriate settling times for both line and reference input step responses (5 ms at most in our case).

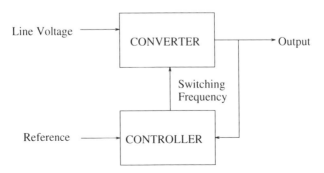

Figure 11.20. The diagram for control design.

4. A closed-loop bandwidth of at least 360 Hz in order to successfully suppress line ripple.

- *Load Regulation (Robust Stability).* On the other hand, the load condition could also vary over a wide range. Since the load R_o enters the dynamics of the model, load variations will appear as model uncertainty and could possibly lead to stability problems. Normally the load changes from 10% at low load to 90% at full load condition. Other model uncertainties, such as unmodeled high-frequency dynamics and uncertainties in the resonant inductor L and capacitor C, will also be considered.
- *Robust Performance.* Since the converter operates over a wide range of load conditions, the performance requirements must be satisfied at all operating points. This is equivalent to requiring satisfactory response under both line and load variations.

11.4.5 Analysis of the Plant

Control Characteristics For a PRC converter operating under steady-state conditions, the input–output relationships can be represented by the control characteristics curves, relating the output voltage to the load and switching frequency. Given any two variables among the normalized output V_{no}, switching frequency ratio F_{ns}, and output load Q_p, the third variable can be determined from the curves. Thus these curves allow one to easily visualize the effects of the switching frequency and load on the converter output. From a control point of view, the control characteristics curves allow us to make an initial estimate of the load change that can be tolerated and to see some of the difficulties inherent in the load regulation problem.

Figure 11.21 shows the control characteristics curves for various output loads Q_p, obtained analytically from the steady-state analysis. To maintain the output voltage constant in the presence of perturbations, the controller should adjust the switching frequency to keep the converter operating along the dashed line indicated in Figure 11.21. As the converter is perturbed away from the nominal operating point (marked with an asterisk in the figure) the plant dynamics may vary significantly, resulting in a difficult control problem.

Remark Note that from Figure 11.21 it follows that at lighter loads (higher R_o, larger Q_p, and lower I_o), a small frequency change will result in larger output changes. Thus we should expect that the control problem will become more difficult at larger Q_p values. In the next section we will show, through a frequency-domain analysis, that this is exactly the case.

Frequency Responses From the discrete time state–space model, we can easily get the z-transfer functions from the normalized switching frequency

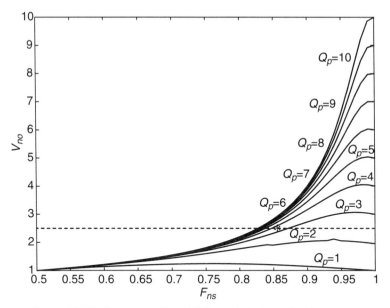

Figure 11.21. The conventional PRC control characteristics curves.

F_{ns} and the normalized line input V_{ng} to the normalized output V_{no}:

$$[G(z) \quad G_g(z)] = \left[\frac{V_{no}(z)}{F_{ns}(z)} \quad \frac{V_{no}(z)}{V_{ng}(z)}\right] = C[zI - A]^{-1}B \quad (11.73)$$

Following a common approach, we will carry out the analysis of the plant and the synthesis of a digital controller using a w-plane approach. To this effect, the bilinear transformation

$$z = \frac{1 + sT_s/4}{1 - sT_s/4} \quad (11.74)$$

is used to obtain the transfer functions in the frequency domain s. These transfer functions, still denoted as $G(s)$ and $G_g(s)$, are given by

$$G(s) = 2.652 \times 10^{-2} \frac{(s + 795041)(s - 792431)(s - 800003)}{(s + 29167)(s + 83363 \pm 202487i)} \quad (11.75)$$

$$G_g(s) = -1.367 \times 10^{-2} \frac{(s - 800003)(s + 484930 \pm 290253i)}{(s + 29167)(s + 83363 \pm 202487i)} \quad (11.76)$$

The above transfer functions correspond to the nominal load $R_o = 208.33\,\Omega$. As stated before, since the load enters the dynamics of the converter, load variations result in different transfer functions. Figures 11.22 and 11.23 show

CONTROL OF A DC-TO-DC RESONANT CONVERTER

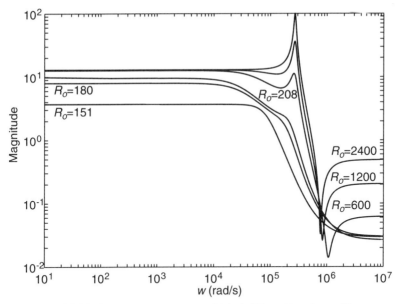

Figure 11.22. Frequency responses $G(s)$ at different load conditions.

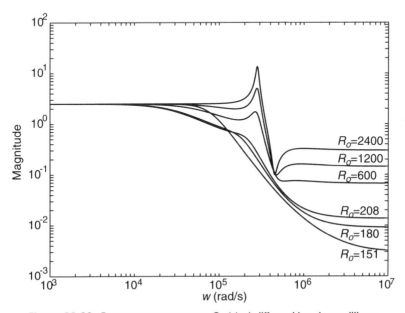

Figure 11.23. Frequency responses $G_g(s)$ at different load conditions.

414 APPLICATION EXAMPLES

the frequency responses of $G(s)$ and $G_g(s)$ corresponding to several different load conditions, respectively.

These figures show that as the load becomes lighter (larger R_o), the overshoot increases, leading to a more difficult control problem. This conclusion is consistent with the conclusion drawn in the last section from the study of the control characteristics. On the other hand, the control characteristics require that Q_p be greater than V_{no} in order to get the prescribed output voltage. Since in our design example the value of V_{no} is chosen to be 2.5, it follows that R_o should be greater than 151 Ω. Therefore, in the sequel, we will assume that R_o varies within the range 151 to 1200 Ω.

11.4.6 Control Design

As mentioned before, our goal is to design a controller that satisfies the performance specifications listed in Section 11.4.4 for all load conditions in the range 151 Ω $\leq R_o \leq$ 1200 Ω, assuming that the values of the components of the resonant tank are known within a 10% tolerance. In the sequel we solve this problem by recasting it into a robust performance synthesis form and using μ-synthesis. To this effect we need first to describe the family of plants corresponding to different values of the load as a nominal plant subject to uncertainty.

Plant Description and Uncertainty Weight Selection In this example we choose to model the uncertainty caused by load variations by using a single, norm-bounded, multiplicative uncertainty that covers all possible plants. Let $G^{R_o}(s)$ and $G_g^{R_o}(s)$ denote the transfer functions from the control input and line input to the output at operating points other than the nominal point ($R_o \neq 208.33$ Ω), respectively. In the sequel we represent these transfer functions as

$$G^{R_o}(s) = G(s)\left[1 + \Delta_I(s)W_I(s)\right] \tag{11.77}$$

$$G_g^{R_o}(s) = G_g(s)\left[1 + \Delta_g(s)W_g(s)\right] \tag{11.78}$$

where $G(s)$ and $G_g(s)$ are the nominal transfer functions given in equations (11.75) and (11.76), respectively, $W_I(s)$ and $W_g(s)$ are fixed weighting functions containing all the information available about the frequency distribution of the uncertainty, and $\Delta_I(s)$ and $\Delta_g(s)$ are stable transfer functions representing model uncertainty. Furthermore, without loss of generality (by absorbing any scaling factor into $W_I(s)$ and $W_g(s)$ if necessary), it can be assumed that $\|\Delta_I(s)\|_\infty \leq 1$ and $\|\Delta_g(s)\|_\infty \leq 1$. Thus $W_I(s)$ and $W_g(s)$ are such that their respective magnitude Bode plots cover the Bode plots of all possible plants. Some sample uncertainties corresponding to different values of the load R_o are shown in Figures 11.24 and 11.25. We can see that in both figures the multiplicative uncertainties have a peak around the resonant frequency. This peak becomes larger and steeper as the load resistance R_o increases.

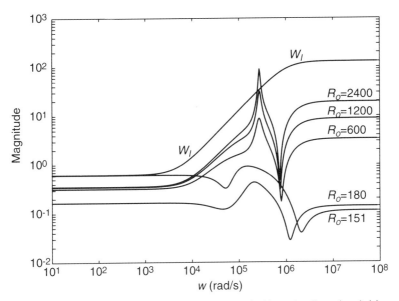

Figure 11.24. Multiplicative uncertainties (control to output) and weight.

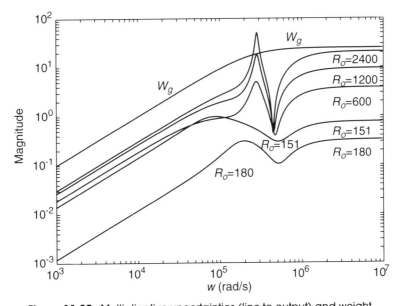

Figure 11.25. Multiplicative uncertainties (line to output) and weight.

Based on these plots, the following multiplicative uncertainty weights were chosen for control design:

$$W_I(s) = \frac{1.4 \times 10^{-4}s + 0.65}{10^{-6}s + 1} \quad (11.79)$$

$$W_g(s) = \frac{10^{-4}s}{4 \times 10^{-6}s + 1} \quad (11.80)$$

The magnitude frequency responses of $W_I(s)$ and $W_g(s)$ are also shown in Figures 11.24 and 11.25, respectively. These figures clearly show that attempting to cover the sharp peak around the resonant frequency will result in large gaps between the weight and the uncertainty at high frequencies, introducing conservatism at that frequency range. On the other hand, a tighter fit at high frequencies using higher-order functions will result in high-order controllers. The weights (11.79) and (11.80) used in our design provide a good trade-off between robustness and controller complexity.

We turn our attention now to the effects of changes in the values of L and C, the resonant tank components. Since these changes affect the location of the resonant peak, they could conceivably destabilize any controller design based on its nominal location. Figure 11.26 shows the changes in the transfer functions due to $\pm 10\%$ changes in the values of L and/or C. It is worth noticing that our choice of weighting functions W_I and W_g will also cover this family of plants, even at the extreme load conditions $R_o = 1200\,\Omega$ and $151\,\Omega$. Thus a robust controller designed using these weighting functions will be able to accommodate both changes in the load condition and uncertainty in L and C.

Figure 11.27 shows a block diagram of the converter, taking into account the uncertainty. Here Δ_I and Δ_g are scalar blocks, representing the model uncertainty perturbations from the control and line inputs, respectively, and W_I and W_g are the corresponding uncertainty weights.

Performance Weight Selection As discussed in Chapter 7, in order to guarantee robust performance we need to add to the structure shown in Figure 11.27 an additional (fictitious) uncertainty block Δ_p, along with the corresponding performance weights W_e and W_u, associated with the tracking/regulation error and the control effort, respectively, resulting in the block diagram shown in Figure 11.28.

The selection of the performance weights W_e and W_u entails a trade-off among good regulation versus peak control action. The weight on the control error $W_e(s)$ is usually selected to be very large at low frequencies in order to get good tracking and regulation. Additionally, the order of the weights should be kept low in order to keep the overall controller complexity low. A good compromise between performance and complexity is given by weighting

CONTROL OF A DC-TO-DC RESONANT CONVERTER 417

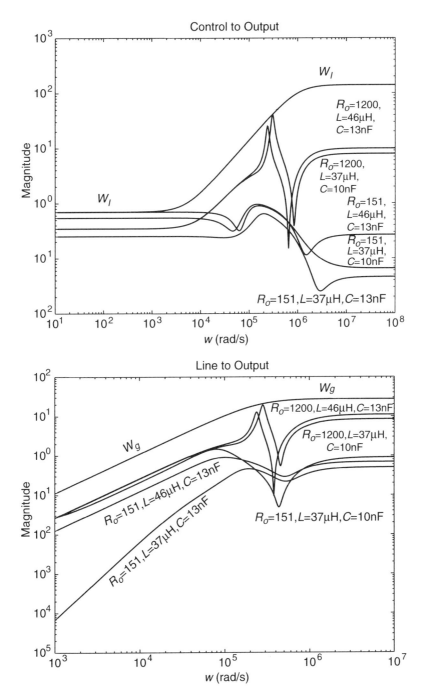

Figure 11.26. Uncertainties due to ±10% changes of L and/or C at extreme load conditions $R_o = 1200\,\Omega$ and $155\,\Omega$.

418 APPLICATION EXAMPLES

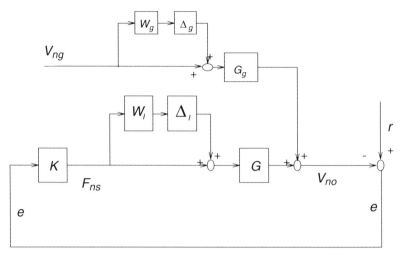

Figure 11.27. The block diagram of the converter including the uncertainty due to load and component variations.

functions of the form

$$W_e(s) = \frac{T_1 s + 1}{T_2 s + A} \tag{11.81}$$

where A is the desired steady-state error (A will be zero if zero steady state is required); T_1 approximately determines the bandwidth ($\omega_b \approx 1/T$) and

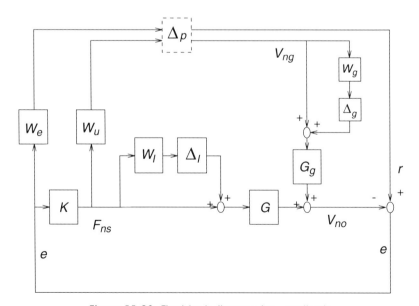

Figure 11.28. The block diagram for μ-synthesis.

hence the rising time and settling time; where the ratio T_2/T_1 is associated with performance requirements against high-frequency noise ([188]). Note that there is no exact relationship between the parameters T_1 and T_2 and time-domain performance specifications given in terms of rise-time, settling-time, and overshoot. The design of multiobjective robust controllers subject to both time- and frequency-domain specifications is, to a large extent, an open problem, although some progress has been made recently [301].

In this particular design example we selected the following weights:

$$W_e(s) = \frac{0.0006s + 1}{0.004s} \quad (11.82)$$

$$W_u(s) = \frac{10^{-4}s}{10^{-7}s + 1} \quad (11.83)$$

The weight on the control input $W_u(s)$ was chosen close to a differentiator to penalize fast changes and large overshoot in the control input. The weight $W_e(s)$ guarantees zero steady-state error. The frequency responses of $W_e(s)$ and $W_u(s)$ are shown in Figure 11.29. The closed-loop bandwidth corresponding to this choice of weights is approximately $1/0.0006 \approx 1700\,\text{rad/s}$.

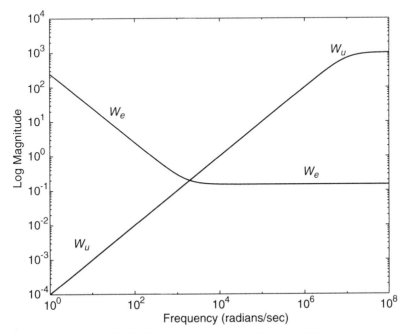

Figure 11.29. Performance weights $W_e(s)$ and $W_u(s)$.

11.4.7 Controller Synthesis

Combining the uncertainty description with the performance weights yields an uncertainty structure Δ consisting of two scalar blocks (corresponding to the robust stability requirements) and a 2×2 block (corresponding to the robust performance requirements). Since the Δ structure has only three blocks, the upper bound of μ, $\inf_{D \in \mathcal{D}} \|DMD^{-1}\|_\infty$, coincides with its exact value. The robust controller was synthesized using the μ Analysis and Synthesis Toolbox [20], applied to the block diagram shown in Figure 11.28. After four D-K iterations with third-order D-scalings, we obtained a 18th-order controller yielding $\mu_{RP} = 0.9721$. Finally, Hankel norm model reduction yielded a sixth-order controller with virtually no performance degradation ($\mu_{RP} = 0.9753 < 1$). The state–space description of this reduced-order controller is given by

$$K = C_k(sI - A_k)^{-1}B_k + D_k \qquad (11.84)$$

where

$$A_k = \begin{pmatrix} -0.25 & 1.708 & -1.144 & 1.414 & -0.1161 & 1.296 \\ 1.708 & -1.320e+5 & 9.980e+4 & -3.190e+5 & 1.460e+4 & -2.213e+5 \\ -1.144 & 9.980e+4 & -7.670e+4 & 3.208e+5 & -1.200e+4 & 1.983e+5 \\ -1.414 & 3.190e+5 & -3.208e+5 & -2.874e+5 & 1.729e+5 & -4.053e+5 \\ -1.161 & 1.460e+4 & -1.200e+4 & -1.729e+5 & -2.664e+3 & 1.013e+5 \\ -1.296 & 2.213e+5 & -1.983e+5 & -4.053e+5 & -1.013e+5 & -8.045e+5 \end{pmatrix}$$

$$B_k^T = (-2.338 \quad 7.983 \quad -5.345 \quad -6.610 \quad -0.543 \quad -6.060) \times 10^{-2}$$

$$C_k = (-0.935 \quad 3.193 \quad -2.138 \quad 2.644 \quad 0.217 \quad 2.424) \times 10^4$$

$$D_k = 0$$

In order to benchmark the performance of the robust controller, we also designed a phase-lag controller using classical design tools, based on the plant frequency responses at the various operating points shown in Figure 11.22. To improve performance, this controller was further tuned by trial and error. The transfer function of the final controller is given by

$$K_{pl} = \frac{0.02s + 200}{s + 0.2} \qquad (11.85)$$

The frequency responses of both the μ and the phase-lag controllers are shown in Figure 11.30. Both controllers have similar responses at low frequencies, while at high frequencies the gain of the μ controller decays faster in order to accommodate the model uncertainties at high frequencies.

Figure 11.31 shows the closed-loop frequency responses for the nominal plant and for the lightest load considered in the design. Note that in both cases the μ controller provides lower gain and better roll-off at high frequencies. Moreover, while the response corresponding to the phase-lag controller is acceptable for the nominal plant, it exhibits a large peak at the resonant frequency at light loads. As we show next, this peak results in significant performance deterioration at these loads.

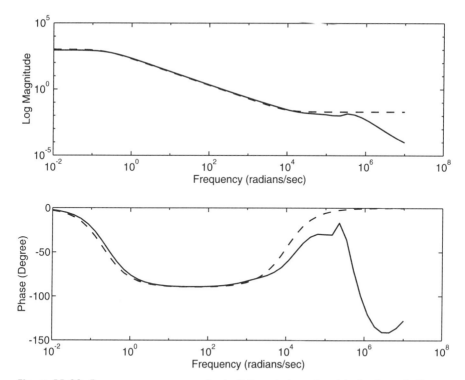

Figure 11.30. Frequency responses of μ (solid) and phase-lag (dashed) controllers.

11.4.8 Simulation Results

Linear Simulations The closed-loop system corresponding to the μ controller was simulated at the nominal operating point $R_o = 208\,\Omega$ and at two extreme cases $R_o = 151\,\Omega$ and $1200\,\Omega$, using the corresponding linear model of the plant. The time responses to 20% step change in line voltage V_{ng} and reference input r are shown in Figure 11.32.

For the nominal case $R_o = 208\,\Omega$, the settling time is about 2.5 ms for both line voltage change and reference input change. The output responses are satisfactory since the settling time is smaller than the required 5 ms, with no overshoot. The control action in these responses is also adequate, without overshoot or abrupt change. This is due to the choice of the weight W_u, penalizing fast changes and overshoots in the control action.

When the operating point moves to $R_o = 151\,\Omega$, the settling times for both step changes are about 4 ms. This increase is due mainly to the significant decrease in plant static gain (see Figure 11.22). The μ controller is undertuned at this operating point in order to achieve robust performance. When the operating point moves toward lighter loads, the responses are almost the same as for the nominal plant, except that for the case $R_o = 1200\,\Omega$ (note

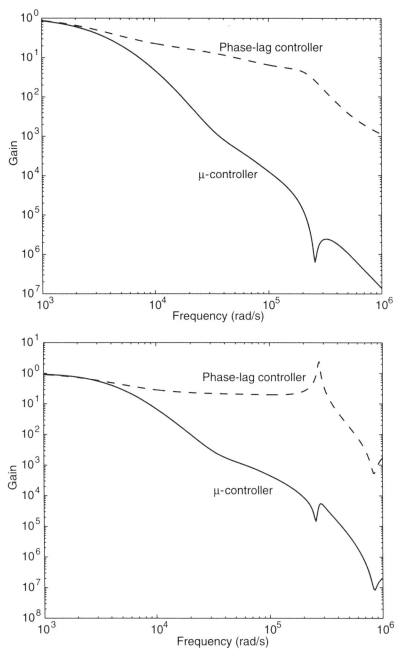

Figure 11.31. Closed-loop frequency responses for μ (solid) and phase-lag (dashed) controllers: (a) $R_o = 208\,\Omega$ and (b) $R_o = 1200\,\Omega$.

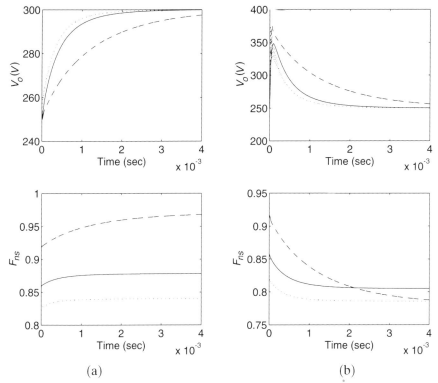

Figure 11.32. Linear simulation results with μ controller at different operating points $R_o = 208$ Ω (solid), 151 Ω (dashed), and 1200 Ω (dotted): (a) reference input step change (20%) and (b) line voltage step change (20%).

that this load is the lightest load considered in our design), some chattering in both the output and control input starts to show up at the beginning of the responses. The occurrence of the chattering is linked to the large peak in the plant frequency response at lighter loads, barely covered by the uncertainty weights W_l and W_g.

From the simulation results it follows that the controller achieves robust performance, since all performance specifications are satisfied at all operating points of interest. However, significant variation of performance is also observed. This is a direct result of the large variation in the plant dynamics, and any fixed linear controller can do very little in this respect. To reduce this variation will require using a nonlinear, gain scheduling controller.

The same simulation was performed for the closed-loop system corresponding to the phase-lag controller. The time responses to 20% step change in line voltage V_{ng} and reference input r at three different operating points — $R_o = 151$ Ω, 208 Ω, and 1200 Ω—are shown in Figure 11.33. They are

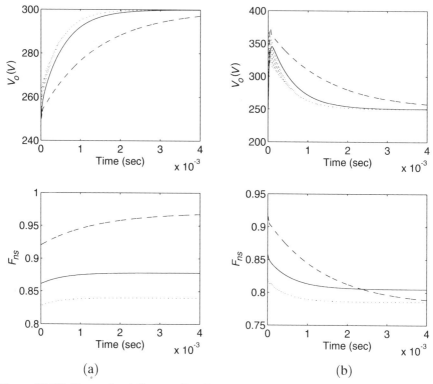

Figure 11.33. Linear simulation results with phase-lag controller at different operating points $R_o = 208\,\Omega$ (solid), $151\,\Omega$ (dashed), and $1200\,\Omega$ (dotted): (a) reference input step change (20%) and (b) line voltage step change (20%).

similar to the responses with the μ controller except that the performance is far worse for $R_o = 1200\,\Omega$. This is due to the phase-lag controller's inability to provide enough attenuation to counteract the increment in the magnitude of the resonant peak of the plant at heavy loads, as shown in Figure 11.31. Furthermore, as shown in Figure 11.34, at $R_o = 2400\,\Omega$ the phase-lag controller fails to stabilize the system, while the μ controller can still produce acceptable performance.

Nonlinear Simulations and Validation While simulations using the linearized model of the plant corresponding to different load conditions can usually provide an approximate evaluation of load regulation performance, this is usually insufficient to assess the performance of a highly nonlinear system such as the converter. Thus to further validate the controller, a nonlinear simulation of the PRC circuit was performed using P-Spice. Figure 11.35 shows the responses due to reference input and line voltage step changes. Note that these results are similar to those obtained using a linear simulation

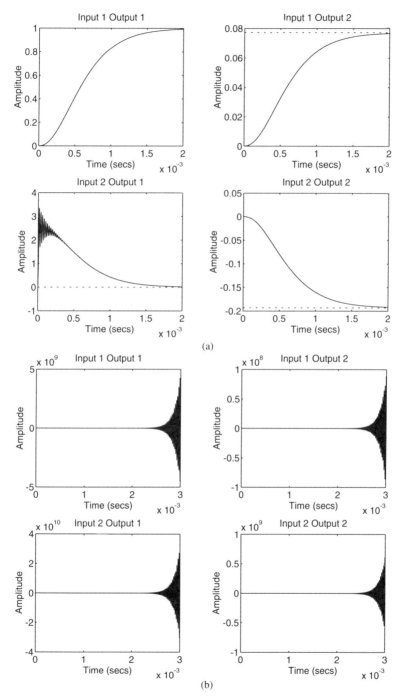

Figure 11.34. Step responses of the closed-loop system at $R_o = 2400\,\Omega$: (a) μ controller (stable) and (b) phase-lag controller (unstable).

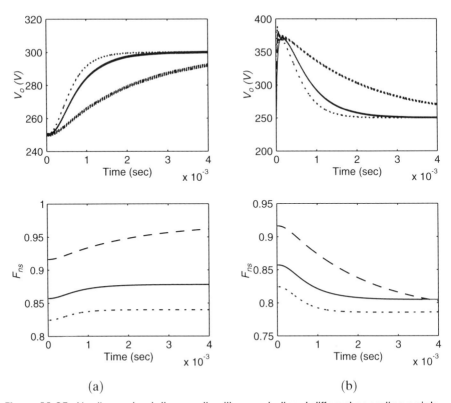

Figure 11.35. Nonlinear simulation results with μ controller at different operating points $R_o = 208\ \Omega$ (solid), 151 Ω (dashed) and 1200 Ω (dotted) Ω: (a) reference input step change (20%) and (b) line voltage step change (20%).

as shown in Figure 11.32. In the responses to line voltage step change and reference step change, settling times are slightly larger than those in the linear simulation. The chattering observed in the output voltage is substantial, due to the periodic switching behavior of the converter (periodic charge and discharge of capacitor).

BIBLIOGRAPHY

[1] Abdallah C., "On the robust control of uncertain nonlinear systems," *Proceedings American Control Conference*, San Diego, 1990.

[2] Abdallah C., Dawson D., Dorato P., Jamshidi M., "Survey of robust control for rigid robots," *IEEE Control Systems Magazine*, Vol. 11, No. 2, 1991.

[3] Abedor J., Nagpal K., Poola K., "A linear matrix inequality approach to peak to peak gain minimization," *International Journal of Robust and Nonlinear Control*, Vol. 6, pp. 899–927, 1996.

[4] Ackermann J., "Parameter space design of robust control systems," *IEEE Transactions on Automatic Control*, Vol. 25, pp. 1058–1072, 1980.

[5] Ackermann J., *Robust Control: Systems with Uncertain Physical Parameters*, Springer-Verlag, New York, 1993.

[6] Adamjan V.M., Arov D.Z., Krein M.G., "Analytic properties of Schmidt pairs for a Hankel operator and the generalized Schur–Takagi problem," *Math USSR Sbornik*, Vol. 15, pp. 31–73, 1971.

[7] Adamjan V.M., Arov D.Z., Krein M.G., "Infinite Hankel block matrices and related extension problems," *AMS Translations*, Vol. 111, No. 133, 1978.

[8] Akcay H., Gu G., Khargonekar P., "Identification in $\mathcal{H}_\infty$ with nonuniformly spaced frequency response measurements," *Proceedings American Control Conference*, Chicago, 1992.

[9] Akcay H., Gu G., Khargonekar P., "A class of algorithms for identification in $\mathcal{H}_\infty$: continuous-time case," *IEEE Transactions on Automatic Control*, Vol. 38, No. 2, 1993.

[10] Anderson B.D.O., Vongpanitlerd S., *Network Analysis and Synthesis: A Modern Systems Theory Approach*, Prentice-Hall, Englewood Cliffs, NJ, 1973.

[11] Anderson B., Kraus F., Mansour M., Dasgupta S., "Easily testable sufficient conditions for the robust stability of systems with multiaffine parameter dependence," in *Robustness of Dynamic Systems with Parameter Uncertainties*, M. Mansour, S. Balemi, W. Truöl, Eds. Birkhäuser, Berlin, 1992.

[12] Anderson B.D.O., Moore, J.B., *Optimal Control: Linear Quadratic Methods*, Prentice-Hall, Englewood Cliffs, NJ, 1990.

[13] Åstrom K., Eykhoff P., "System identification: a survey," *Automatica*, Vol. 7, pp. 123–162, 1971.

[14] Åstrom K., Wittenmark B., *Computer Controlled Systems: Theory and Design*, Prentice-Hall, Englewood Cliffs, NJ, 1984.

[15] Athans M., Falb P., *Optimal Control: An Introduction to the Theory and Its Applications*, McGraw-Hill, New York, 1966.

[16] Athans M., Kapasouris P., Kappos E., Spang H.A. III, "Linear quadratic gaussian with loop transfer recovery methodology for the F-100 engine," *AIAA Journal of Guidance, Control and Dynamics*, Vol. 9, No. 1, 1986.

[17] Balakrishnan V.S., Boyd S., Balemi S., "Branch and bound algorithm for computing the minimum stability degree of parameter-dependent linear systems," Technical Report. Information Systems Laboratory, Stanford University, 1991.

[18] Balakrishnan V.S., Boyd S.P., "On computing the worst case peak gain of linear systems," *Systems and Controls Letters*, Vol. 19, 1992.

[19] Balas G., Doyle J.C., "Identification of flexible structures for robust control," *IEEE Control Systems Magazine*, Vol. 10, No. 4, 1990.

[20] Balas G., Doyle J.C., Glover K., Packard A., Smith R., μ-*Analysis and Synthesis Toolbox*, The MathWorks Inc., Natick, MA, 1991.

[21] Ball J.A., Gohberg I., Rodman L., *Interpolation of Rational Matrix Functions*, Operator Theory: Advances and Applications, Vol. 45, Birkhäuser, Berlin, 1990.

[22] Ball J.A., Helton J.W., Walker M.L., "$\mathcal{H}_\infty$ control for nonlinear systems with output feedback," *IEEE Transactions on Automatic Control*, Vol. 38, No. 4, 1993.

[23] Bancroft A.R., "Heavy water G-S process research and development achievements," *Atomic Energy Canada Limited (AECL)* Report No. 6215, 1979.

[24] Banks H.T., Burns J.A., "Hereditary control problems: numerical methods based on averaging approximations," *SIAM Journal of Control and Optimization*, No. 16, pp. 169–208, 1978.

[25] Banks H.T., Kappel F., "Spline approximations for functional differential equations," *Journal of Differential Equations*, Vol. 34, pp. 496–522, 1979.

[26] Barabanov A.E., Granichin O.N., "Optimal controller for a linear plant with bounded noise," *Automation and Remote Control*, Vol. 5, pp. 39–46, 1984.

[27] Barmish B.R., "Invariance of the strict Hurwitz property for polynomials with perturbed coefficients," *IEEE Transactions on Automatic Control*, Vol. 29, 1984. (First presented at the 1983 *Conference on Decision and Control*, Texas.)

[28] Barmish B., Khargonekar P.P., "Robust stability of feedback control systems with uncertain parameters and unmodelled dynamics." *Proceedings American Control Conference*, Georgia, 1988.

[29] Barmish B.R., "New tools for robustness analysis," Plenary Session, *Proceedings Conference on Decision and Control*, Austin, Texas, 1988.

[30] Barmish B., Ackermann J.E., Hu H.Z. "The tree structured decomposition: a new approach to robust stability analysis," *Proceedings Conference on Information Sciences and Systems*, Johns Hopkins University, Baltimore, 1989.

[31] Barmish B.R., "A generalization of Kharitonov's four polynomial concept for robust stability problems with linearly dependent coefficient perturbations," *IEEE Transactions on Automatic Control*, Vol. 34, No. 2, 1989.

[32] Barmish B., Khargonekar P.P., Shi Z.C., Tempo R., "Robustness margin need not be a continuous function of the problem data," *System and Control Letters*, Vol. 15, 1990.

[33] Barmish B.R., Shi Z., "Robust stability of a class of polynomials with coefficient depending multilinearly on perturbations," *IEEE Transactions on Automatic Control*, Vol. 35, No. 9, 1990.

[34] Barmish B., Kang H., "A survey of extreme point results for robustness of control systems," *Automatica*, Vol. 29, No. 1, 1993.

[35] Barmish B.R., *New Tools for Robustness Analysis*, Mcmillan, London, 1994.

[36] Bartlett A.C., Hollot C., Lin H., "Root locations of an entire polytope of polynomials: it suffices to check the edges," *Mathematics of Control, Signals and Systems*, Vol. 1, No. 1, 1988.

[37] Batarseh I., *Steady State Analysis and Design of High Order Resonant Converters*, Ph.D. Thesis, University of Illinois, Chicago, June 1990.

[38] Beck C., Doyle J.C., "Mixed μ upper bound computation," *Proceedings Conference on Decision and Control*, Tucson, 1992.

[39] Bertsekas D.P., Rhodes I.B., "On the minmax reachability of target set and target tubes," *Automatica*, Vol. 7, pp. 233–247, 1971.

[40] Bertsekas D.P., "Infinite-time reachability of state-space regions by using feedback control," *IEEE Transactions on Automatic Control*, Vol. AC-17, No. 4, pp. 604–613, 1972.

[41] Blanchini F., Sznaier M., "Rational $\mathcal{L}^1$ suboptimal compensators for continuous–time systems," *IEEE Transactions on Automatic Control*, Vol. 39, No. 7, pp. 1487–1492, 1994.

[42] Blanchini, F., "Non-quadratic Lyapunov function for robust control," *Automatica,* Vol. 31, No. 3, 451–461, 1994.

[43] Blanchini F., Sznaier M., "Persistent disturbance rejection via static state feedback," *IEEE Transactions on Automatic Control*, Vol. 40, No. 6, pp. 1127–1131, June 1995.

[44] Becker N., Grimm W.M., "On $\mathcal{L}_2$ and $\mathcal{L}_\infty$ stability approaches for the robust control of robot manipulators," *IEEE Transactions on Automatic Control*, Vol. 33, No. 1, 1988.

[45] Beliczynski B., Kale I., Cain G.D., "Approximation of FIR by IIR digital filters: an algorithm based on balanced model reduction," *IEEE Transactions on Signal Processing*, Vol. 40, No. 3, 1992.

[46] Bernstein, D., Wie, B., Organizers, Invited Session on "A Benchmark Problem for Robust $\mathcal{H}_2 / \mathcal{H}_\infty$ Control Design," Session WP-15, *Proceedings of the American Control Conference*, San Diego, pp. 961–973, 1990.

[47] Bhattacharyya S.P., Chapellat H., Keel L.H., *Robust Control: The Parametric Approach*, Prentice-Hall, Englewood Cliffs, NJ, 1995.

[48] Bialas S., "A necessary and sufficient condition for stability of interval matrices," *International Journal of Control*, Vol. 37, pp. 717–722, 1983.

[49] Bialas S., Garloff J., "Stability of polynomials under coefficient perturbations," *IEEE Transactions on Automatic Control*, Vol. 30, 1985.

[50] Biernacki R.M., Hwang H., Bhattacharyya S.P., "Robust stability with structured real parameter perturbations," *IEEE Transactions on Automatic Control*, Vol. 32, 1987.

[51] Blondel V. and Gevers, M., "Simultaneous stabilization of three linear systems is rationally undecidable," *Mathematics of Control, Signals and Systems*, Vol. 6, pp. 135–145, 1993.

[52] Bode H.W., *Network Analysis and Feedback Amplifier Design*, Van Nostrand, Princeton, NJ, 1945.

[53] Bodson M., Chiasson J.N., Novotnak R.T., Rekowski R.B., "High-performance nonlinear feedback control of a permanent magnet stepper motor," *IEEE Transactions on Control Systems Technology*, Vol. 1, No. 1, 1993.

[54] Bose N.K., "A system theoretic approach to stability of sets of polynomials," *Contemporary Mathematics*, Vol. 47, pp. 25–34, 1985.

[55] Boyd S., Desoer C.A., "Subharmonic functions and performance bounds in linear time invariant feedback systems," *IMA Journal of Mathematics, Control and Information*, Vol. 2, pp. 153–170, 1985.

[56] Boyd S.P., Doyle J.C., "Comparison of peak and RMS gains for discrete–time systems," *Systems and Controls Letters*, Vol. 21, 1993.

[57] Boyd S., Balakrishnan V., Kabamba P., "A bisection method for computing the $\mathcal{H}_\infty$ norm of a transfer matrix and related problems," *Mathematics of Control, Signals and Systems*, No. 2, 1989.

[58] Boyd S., Barrat C.H., *Linear Controller Design Limits of Performance*, Prentice-Hall Information and System Sciences Series, Prentice-Hall, Englewood Cliffs, NJ, 1991.

[59] Boyd S., El Ghaoui L., Feron E., Balakrishnan V., *Linear Matrix Inequalities in Systems and Control Theory*, SIAM Studies in Applied Mathematics, Vol. 15, SIAM, Philadelphia, 1994.

[60] Braatz R.D., Young P.M., Doyle J.C., Morari M., "Computational complexity of μ calculation," *IEEE Transactions on Automatic Control*, Vol. 39, No. 5, 1994.

[61] Brockett R.W., *Finite Dimensional Linear Systems*, Wiley, New York, 1970.

[62] Bruck J., Blaum M., "Neural networks, error-correcting codes and polynomials over the binary n-cube," *IEEE Transactions on Information Theory*, No. 9, 1989.

[63] Bryson A.E., Ho Y.C., *Applied Optimal Control*, Hemisphere Publishing, New York, 1975.

[64] Bu J., Sznaier M., Holmes M., "A linear matrix inequality approach for synthesizing low order l^1 controllers," *Proceedings 35th IEEE Conference on Decision and Control*, Kobe, Japan, pp. 1875–1880, 1996.

[65] Bu J., Sznaier M., Batarseh I., Wang Z.Q., "Robust controller design for a parallel resonant converter using μ-synthesis," *IEEE Transactions on Power Electronics*, Vol. 12, No. 5, pp. 837–853, September 1997.

[66] Bu J., *Multi-Objective Robust Control: Design of Mixed $\ell^1/\mathcal{H}_\infty$ Controllers*, Ph.D. Dissertation, The Pennsylvania State University, 1997.

[67] Burns J., Peichl G., "On robustness of controllability for finite dimensional approximations of distributed parameter systems," Technical Report, Dept. of

Mathematics, Virginia Polytechnic Institute and State University, Blacksburg, 1985.

[68] Chapellat H., Bhattacharyya S.P., "A generalization of Kharitonov's theorem: robust stability of interval plants," *IEEE Transactions on Automatic Control*, Vol. 34, No. 3, 1989.

[69] Chapellat H., Bhattacharyya S.P., "An alternative proof of Kharitonov's theorem," *IEEE Transactions on Automatic Control*, Vol. 34, No. 4, 1989.

[70] Chapellat H., Dahleh M., Bhattacharyya S.P., "Robust stability under structured and unstructured perturbations," *IEEE Transactions on Automatic Control*, Vol. 35, No. 10, 1990.

[71] Chapellat H., Dahleh M., Bhattacharyya S.P., "Robust stability manifolds for multilinear interval systems," *IEEE Transactions on Automatic Control*, Vol. 38, No. 2, 1993.

[72] Chen C.T., *Linear System Theory and Design*, Holt, Rinehart & Winston, New York, 1984.

[73] Chen J., Fan M., Nett C., "The structured singular value and stability of uncertain polynomials: a missing link," *ASME Annual Winter Meeting*, Atlanta, 1991.

[74] Chen J., Nett C., Fan M., "Worst-case system identification in $\mathcal{H}_\infty$: validation of a priori information, essentially optimal algorithms, and error bounds," *Proceedings American Control Conference*, Chicago, 1992; also in *IEEE Transactions on Automatic Control*, Vol. 40, No. 7, 1995.

[75] Chen J., Nett C., Fan M., "Optimal non-parametric system identification from arbitrary corrupt finite time series: a control-oriented approach," *Proceedings American Control Conference*, Chicago, 1992.

[76] Chen J., Farrell J., Nett C., Zhou K., "$\mathcal{H}_\infty$ identification of multivariable systems by tangential interpolation methods," *Proceedings Conference on Decision and Control*, Florida, 1994.

[77] Chen J., Nett C., "The Carathéodory–Fejér problem and the $\mathcal{H}_\infty/\ell^1$ identification: a time domain approach," *IEEE Transactions on Automatic Control*, Vol. 40, No. 4, 1995.

[78] Chiang, R. and Safonov, M.G., "$\mathcal{H}_\infty$ Robust control synthesis for an undamped, non-colocated spring-mass system," *Proceedings American Control Conference*, pp. 966–967, 1990.

[79] Curtain R., Glover K., "Balanced realisations for infinite dimensional systems," *Workshop on Operator Theory and Its Applications*, Amsterdam, 1985.

[80] Dahleh M.A., Pearson J., "ℓ^1-Optimal controllers for discrete time systems," *Proceedings American Control Conference*, Seattle, pp. 1964–1968, 1986.

[81] Dahleh M.A., Pearson J., "ℓ^1-Optimal controllers for MIMO discrete time systems," *IEEE Transactions on Automatic Control*, Vol. 32, No. 4, 1987.

[82] Dahleh M.A., Boyd Pearson J., "$\mathcal{L}^1$–Optimal compensators for continuous time systems," *IEEE Transactions on Automatic Control*, Vol. 32, No. 10, pp. 889–895, 1987.

[83] Dahleh M.A., Pearson J., "Optimal rejection of persistent disturbances, robust stability and mixed sensitivity minimization," *IEEE Transactions on Automatic Control*, Vol. 33, No. 8, 1988.

[84] Dahleh M.A., Ohta Y., "A necessary and sufficient condition for robust BIBO stability," *Systems and Control Letters*, Vol. 11, pp. 271–275, 1988.

[85] Dahleh M.A., Theodosopoulos T., Tsitsiklis J., "The sample complexity of worst-case Identification of F.I.R. linear systems," *Systems and Control Letters*, Vol. 20, pp. 157–166, 1993.

[86] Dahleh M.A., Khammash M., "Controller design for plants with structured uncertainty," *Automatica*, Vol. 29, No. 1, pp. 37–56, 1993.

[87] Dahleh M.A., Díaz-Bobillo I.J., *Control of Uncertain Systems: A Linear Programming Approach*, Prentice-Hall, Englewood Cliffs, NJ, 1995.

[88] Fu M., Dasgupta S., Blondel V., "Robust stability under a class of nonlinear parametric perturbations," *IEEE Transactions on Automatic Control*, Vol. 40, No. 2, 1995.

[89] D'Attellis C.E., *Introducción a los Sistemas Nolineales de Control y sus Aplicaciones*, Editorial Control S.R.L. (AADECA), Buenos Aires, 1992.

[90] D'azzo J., Houpis C., *Feedback Control System Analysis and Synthesis*, McGraw-Hill, New York, 1966.

[91] deGaston R.R., *Nonconservative Calculation of the Multiloop Stability Margin*, Ph.D. Dissertation, University of Southern California, 1985.

[92] de Gaston R., Safonov M., "Exact calculation of the multiloop stability margin," *IEEE Transactions on Automatic Control*, Vol. 33, 1988.

[93] Deodhare G., Vidyasagar M., "ℓ^1 Optimality of feedback control systems, the SISO discrete time case," *IEEE Transactions on Automatic Control*, Vol. 35, No. 9, pp. 1082–1085, 1990.

[94] Desoer C., Liu R., Murray J., Saeks R., "Feedback design: the fractional representation approach to analysis and synthesis," *IEEE Transactions on Automatic Control*, Vol. 25, No. 3, 1980.

[95] Díaz-Bobillo, I. J., Dahleh, M. A.,"Minimization of the maximum peak-to-peak gain: The general multiblock problem," *IEEE Transactions on Automatic Control*, Vol. AC-38, pp. 1459–1482, October 1993.

[96] Dorato P. (Ed.), *Robust Control*, IEEE Press, New York, 1987.

[97] Dorato P., Yedavalli R. (Eds.), *Recent Advances in Robust Control*, IEEE Press, New York, 1990.

[98] Dorato, P., Abdallah, C., Cerone V., *Linear-Quadratic Control: An Introduction*, Prentice-Hall, Englewood Cliffs, NJ, 1995.

[99] Dorf R.C., *Modern Control Systems*, Addison-Wesley, Reading, MA, 1983.

[100] Doyle J.C., "Robustness of multiloop feedback systems," *Proceedings Conference on Decision and Control*, San Diego, 1978.

[101] Doyle J.C., Stein G., "Multivariable feedback design: concepts for a classical/modern synthesis," *IEEE Transactions on Automatic Control*, Vol. 26, No. 1, pp. 4–16, 1981.

[102] Doyle J.C., "Analysis of feedback systems with structured uncertainty," *IEE Proceedings*, Vol. 129, Pt. D, No. 6, 1982.

[103] Doyle J.C., "Lecture notes on advances in multivariable control," *ONR Honeywell Workshop*, Minneapolis, 1984.

[104] Doyle J.C., Glover K., Khargonekar P., Francis B., "State-space solutions to standard $\mathcal{H}_2$ and $\mathcal{H}_\infty$ control problems," *IEEE Transactions on Automatic Control*, Vol. 34, 1989.

[105] Doyle J.C., Francis B., Tannenbaum A., *Feedback Control Theory*, Maxwell/Macmillan, New York, 1992.

[106] Dwyer T.A.W. III, "Exact nonlinear control of large angle rotational maneuvers," *IEEE Transactions on Automatic Control*, Vol. 29, No. 9, 1984.

[107] Dwyer T.A.W. III, "Exact nonlinear control of spacecraft slewing maneuvers with internal momentum transfer," *AIAA Journal of Guidance, Control and Dynamics*, Vol. 9, No. 2, 1986.

[108] Eising R., "The distance between a system and the set of uncontrollable systems," *Proceedings MTNS*, Beer-Sheva, Springer-Verlag, Berlin, 1984.

[109] Enns D., Bugajski D., Klepl M., "Flight control for the F8 oblique wing research aircraft," *IEEE Control Systems Magazine*, Vol. 8, No. 2, 1988.

[110] Enns D. "Multivariable flight control for an Attack Helicopter," *IEEE Control Systems Magazine*, Vol. 7, No. 2, 1988.

[111] Eszter E., Sánchez-Peña R., "Value set boundary computation of uncertainty structures," *Proceedings American Control Conference*, Chicago, 1992; also presented at the *IFAC International Workshop on Robustness of Control Systems*, Kappel Am Albis, Switzerland, 1991.

[112] Eszter E., Sánchez-Peña R., "Computation of algebraic combinations of uncertainty value sets," *IEEE Transactions on Automatic Control*, Vol. 39, No. 11, 1994.

[113] Fan M., Tits A., "Characterization and efficient computation of the structured singular value," *IEEE Transactions on Automatic Control*, Vol. 31, 1986.

[114] Fan M.K.H., Tits A.L.,"m-Form numerical range and the computation of the structured singular value," *IEEE Transactions on Automatic Control*, Vol. AC-33, 1988.

[115] Fan M., Tits A., Doyle J.C., "On robustness under parametric and dynamic uncertainties," *Proceedings American Control Conference*, Georgia, 1988.

[116] Foias C., Tannenbaum A., Zames G., "Weighted sensitivity minimization," *IEEE Transactions on Automatic Control*, Vol. 31, No. 8, pp. 763–766, 1986.

[117] Francis B. and Doyle, J. C., "Linear control with an $\mathcal{H}_\infty$ optimality criterion," *Siam Journal of Control and Optimization*, Vol. 25, No. 4, pp. 815–844, 1987.

[118] Francis B., *A Course in $\mathcal{H}_\infty$ Control Theory*, Springer-Verlag, Berlin, 1987.

[119] Franklin G., Powell J., *Digital Control of Dynamic Systems*, Addison-Wesley, Reading, MA, 1980.

[120] Franklin G., Powell J., Emami-Naeini A., *Feedback Control of Dynamic Systems*, Addison-Wesley, Reading, MA, 1986.

[121] Friedman J., Khargonekar P., "Application of identification in $\mathcal{H}_\infty$ to lightly damped systems: two case studies," *IEEE Transactions on Control Systems Technology*, Vol. 3, No. 3, 1995.

[122] Fu M., Barmish B.R., "Maximal unidirectional perturbation bounds for stability of polynomials and matrices," *System and Control Letters*, No. 11, pp. 173–179, 1988.

[123] Gahinet P., "Reliable computation of $\mathcal{H}_\infty$ central controllers near the optimum," *Proceedings American Control Conference*, pp. 738–742, 1992.

[124] Gahinet P., Apkarian P., "A linear matrix inequality approach to $\mathcal{H}_\infty$ control," *International Journal of Robust and Nonlinear Control*, Vol. 4, pp. 421–448, 1994.

[125] Gahinet P., "On the game Riccati equations arising in $\mathcal{H}_\infty$ optimization problems," *SIAM Journal of Control and Optimization*, May 1994.

[126] Gahinet P., Nemirovski A., Laub A., Chilali M., *LMI Control Toolbox*, The MathWorks Inc., Natick, MA, 1995.

[127] Galarza C., Sánchez-Peña R., "Robust approximation and control," *IEEE Control Systems Magazine*, Vol. 16, No. 5, 1996.

[128] Galley M.R., Bancroft A.R., "Canadian heavy water production 1970 to 1980," *Second World Congress of Chemical Engineering*, Montreal, Canada, 1981.

[129] Gantmacher F.R., *The Theory of Matrices*, Chelsea Publishing Company, New York, 1977.

[130] Gel'fand I.M., *Lecture Notes on Linear Algebra*, Interscience Publishers, New York, 1961.

[131] Glover J.D., Schweppe F.C., "Control of linear dynamic systems with set constrained disturbances," *IEEE Transactions on Automatic Control*, Vol. AC-16, pp. 411–423, 1971.

[132] Glover K., "All optimal Hankel norm approximations of linear multivariable systems and their L^∞ error bounds," *International Journal of Control*, Vol. 39, 1984.

[133] Glover K., Doyle J.C., "State space formulae for all stabilizing controllers that satisfy an $\mathcal{H}_\infty$ norm bound and relations to risk sensitivity," *System and Control Letters*, No. 11, 1988.

[134] Glover K., Curtain R., Partington J., "Realisation and approximation of linear infinite dimensional systems with error bounds," *SIAM Journal on Control and Optimization*, Vol. 26, No. 4, 1988.

[135] Glover K., Limebeer D.J.N., Doyle J.C., Kasenally, E.M., Safonov M.G., "A characterization of all solutions to the four block general distance problem," *SIAM Journal of Control and Optimization*, Vol. 29, No. 2, pp. 283–324, Mar. 1991.

[136] Green M., Limebeer D., *Linear Robust Control*, Prentice-Hall, Englewood Cliffs, NJ, 1995.

[137] Gu G., Khargonekar P., "A class of algorithms for identification in $\mathcal{H}_\infty$," *Automatica*, Vol. 28, No. 2, 1992.

[138] Gu G., Khargonekar P., "Linear and nonlinear algorithms for identification in $\mathcal{H}_\infty$ with error bounds," *IEEE Transactions on Automatic Control*, Vol. 37, No. 7, 1992.

[139] Gu G., Khargonekar P., Lee E., "Approximation of infinite dimensional systems," *IEEE Transactions on Automatic Control*, Vol. 34, No. 6, 1989.

[140] Guiver J.P., Bose N.K., "Causal and weakly causal 2-D filters with applications in stabilization," in *Multidimensional Systems Theory*, N.K. Bose, Ed., Reidel Publishing, Amsterdam, Holland, 1985.

[141] Gustafsson T.K., Mäkilä P., "Modelling of uncertain systems via linear programming," *IFAC World Congress*, Vol. 5, pp. 293–298, Sydney, 1993.

[142] Hakvoort R., "Worst-case system identification in ℓ^1: error bounds, optimal models and model reduction," *Proceedings Conference on Decision and Control*, Tucson, pp. 499–504, 1992.

[143] Helmicki J., Jacobson C., Nett C., "$\mathcal{H}_\infty$ identification of stable LSI systems: a scheme with direct application to controller design," *Proceedings American Control Conference*, Pittsburgh, pp. 1428–1434, 1989.

[144] Helmicki J., Jacobson C., Nett C., "Identification in $\mathcal{H}_\infty$: a robustly convergent, nonlinear algorithm," *Proceedings American Control Conference*, San Diego, 1990.

[145] Helmicki J., Jacobson C., Nett C., "Identification in $\mathcal{H}_\infty$: linear algorithms," *Proceedings American Control Conference*, San Diego, 1990.

[146] Helmicki J., Jacobson C., Nett C., "Control oriented system identification: a worst case/deterministic approach in $\mathcal{H}_\infty$," *IEEE Transactions on Automatic Control*, Vol. 36, No. 10, 1991.

[147] Helmicki J., Jacobson C., Nett C., "Worst-case/deterministic identification in $\mathcal{H}_\infty$: the continuous-time case," *IEEE Transactions on Automatic Control*, Vol. 37, No. 5, 1992.

[148] Heuberger P., Van den Hof P., Bosgra O., "A generalized orthonormal basis for linear dynamical systems," *IEEE Transactions on Automatic Control*, Vol. 40, No. 3, 1995.

[149] Hollot C.V., Looze D.P., Bartlett A.C., "Unmodelled dynamics: performance and stability via parameter space methods," *Proceedings Conference on Decision and Control*, Los Angeles, 1987.

[150] Hollot C.V., Xu Z.L., "When is the image of a multilinear function a polytope? A conjecture," *Proceedings Conference on Decision and Control*, Florida, 1989.

[151] Hollot C.V., Tempo R., "$\mathcal{H}_\infty$ performance of weighted interval plants: complete characterization of vertex results," *Proceedings American Control Conference*, San Francisco, 1993.

[152] Horowitz I.M., *Synthesis of Feedback Systems*, Academic Press, New York, 1963.

[153] Hunt L., Su R., Meyer G., "Global transformations of nonlinear systems," *IEEE Transactions on Automatic Control*, Vol. 28, No. 1, 1983.

[154] Hurwitz A., "On the conditions under which an equation has only roots with negative real parts," *Matematische Annalen*, Vol. 46, 1895.

[155] Special Issue on System Identification for Robust Control Design, *IEEE Transactions on Automatic Control*, Vol. 37, No. 7, 1992.

[156] Isidori A., *Nonlinear Control Systems: An Introduction*, Comm. and Control Engineering Series, Springer-Verlag, New York, 1989.

[157] Isidori A., Astolfi A., "Disturbance attenuation and $\mathcal{H}_\infty$ control via measurement feedback in nonlinear systems," *IEEE Transactions on Automatic Control*, Vol. 37, No. 9, pp. 1283–1293, 1992.

[158] Iwasaki T., Skelton R., "A complete solution to the general $\mathcal{H}_\infty$ control problem: LMI existence conditions and state-space formulas," *Automatica*, 1994.

[159] Jacobson C., Nett C., "Worst-case system identification in ℓ^1: optimal algorithms and error bounds," *Proceedings American Control Conference*, Boston, 1991.

[160] Jacobson C., Nett C., Partington J.R., "Worst-case system identification in ℓ^1: optimal algorithms and error bounds," *System and Control Letters*, No. 19, pp. 419–424, 1992.

[161] Jahn J., Sachs E., "Generalized quasiconvex mappings and vector optimization," *SIAM Journal of Control and Optimization*, Vol. 24, 1986.

[162] Jury E., Blanchard J., "A stability test for linear discrete time systems in table form," *Proceedings IRE*, Vol. 49, 1961.

[163] Kacewicz B., Milanese M., "Optimal finite-sample experiment design in worst-case system identification," *Proceedings Conference on Decision and Control*, Tucson, 1992.

[164] Kalman R.E., "Contributions to the theory of optimal control," *Boletín de la Sociedad Matemática Mejicana*, No. 5, 1960.

[165] Kalman R.E., "Irreducible realizations and the degree of a rational matrix," *SIAM Journal of Applied Mathematics*, Vol. 13, No. 2, 1965.

[166] Kalman R.E., "Algebraic structure of finite dimensional dynamical systems," *Proceedings National Academy of Science USA*, Vol. 54, 1965.

[167] Kalman R.E., Bucy R.S., "New results in linear filtering and prediction theory," *Journal on Basic Engineering, Transactions ASME, Ser. D*, No. 83, 1961.

[168] Kalman R.E., "When is a linear control system optimal?" *Journal of Basic Engineering*, Vol. 86, 1964.

[169] Kappel F., Salamon D., "Spline approximations for retarded systems and the Ricatti equation," Mathematics Research Center, University of Wisconsin, Technical Summary Report No. 2680, Apr. 1984.

[170] Kato T., *A Short Introduction to Perturbation Theory for Linear Operators*, Springer-Verlag, New York, 1982.

[171] Kenney C., Laub A.J., "Controllability and stability radii for companion form systems," *Mathematics of Control, Signals and Systems*, Vol. 1, No. 3, 1988.

[172] Kharitonov V., "Asymptotic stability of an equilibrium position of a family of linear differential equations," *Differencialnye Uravneniya*, Vol. 14, 1978.

[173] Klompstra M.H., *SIMSAT: Simulation Package for Flexible Systems*, Master's Thesis, Department of Mathematics, University of Groningen, 1987.

[174] Kolmogorov A.N., Fomin S.V., *Elements of the Theory of Functions and Functional Analysis*, Vols. I & II, Graylock Press, Rochester, NY, 1957.

[175] Kosut R.L., Lau M., Boyd S., "Identification of systems with parametric and nonparametric uncertainty," *Proceedings American Control Conference*, San Diego, pp. 2412–2417, 1990.

[176] Krause J., Khargonekar P., "A comparison of classical stochastic estimation and deterministic robust estimation," *IEEE Transactions on Automatic Control*, Vol. 37, No. 7, 1992.

[177] Kuraoka H., Ohka N., Ohba M., Hosoe S., Zhang F., "Application of H-infinity design to automotive fuel control," *IEEE Control Systems Magazine*, Vol. 10, No. 3, 1990.

[178] Kucera V., *Discrete Linear Control: The Polynomial Equation Approach*, Wiley, New York, 1972.

[179] Kwakernaak H., Sivan R., *Linear Optimal Control Systems*, Wiley-Interscience, New York, 1972.

[180] Laub A., *Proceedings Joint Automatic Control Conference*, San Francisco, 1980.

[181] Lehtomaki N.A., Sandell N.R., Athans M., "Robustness results in linear quadratic gaussian based multivariable control," *IEEE Transactions on Automatic Control*, Vol. AC-26, No. 1, pp. 75–92, 1981.

[182] Limebeer D., Kasenally E., Jaimoukha I., Safonov M., "All solutions to the four blocks general distance problem," *Proceedings 27th IEEE CDC*, pp. 875–880, 1988.

[183] Lin L., Wang L.Y., Zames G., "Uncertainty principles and identification n-widths for LTI and slowly varying systems," *Proceedings American Control Conference*, Chicago, 1992.

[184] Ljung L., *System Identification: Theory for the User*, Prentice-Hall, Englewood Cliffs, NJ, 1987.

[185] Lu W.M., Zhou K., Doyle J.C., "Stabilization of uncertain linear systems: an LFT approach," *IEEE Transactions on Automatic Control*, Vol. AC-41, No. 1, pp. 50–65, 1996.

[186] Luenberger D.G., "Observing the state of a linear system," *IEEE Transactions on Military Electronics*, No. 8, 1964.

[187] Luenberger D.G., *Optimization by Vector Space Methods*, Wiley, New York, 1969.

[188] Lundstrom P., Skogestad S., Wang Z.Q., "Performance weight selection for $\mathcal{H}_\infty$ and μ control methods," *Transactions of the Institute of Measurement and Control*, Vol. 13, No. 5, pp. 241–252, 1991.

[189] Maine R.E., Iliff K.W., "Application of parameter estimation to aircraft stability and control: the output error approach," *NASA Reference Publication 1168*, 1986.

[190] Mäkilä P., "On $\mathcal{H}_\infty$ identification of stable systems and optimal approximation," Report 89-7, Department of Chemical Engineering, Åbo Akademi, Finlandia, 1989.

[191] Mäkilä P., "Identification of stabilizable systems: closed-loop approximation," *International Journal of Control*, Vol. 54, No. 3, 1991.

[192] Mäkilä P., "Laguerre methods and $\mathcal{H}_\infty$ identification of continuous-time systems," *International Journal of Control*, Vol. 53, No. 3, 1991.

[193] Mäkilä P., "Robust identification and Galois sequences," *International Journal of Control*, Vol. 54, No. 5, 1991.

[194] Mäkilä P.M., "Worst-case input-output identification," *International Journal of Control*, Vol. 56, No. 3, pp. 673–689, 1992.

[195] Mäkilä P.,"Robust approximate modelling from noisy point evaluations," *Proceedings American Control Conference*, San Francisco, CA, 1993.

[196] Mäkilä P., Partington J., "Robust identification of strongly stabilizable systems," *IEEE Transactions on Automatic Control*, Vol. 37, No. 11, 1992.

[197] Mäkilä P., Partington J., "Worst-case identification from closed-loop time series," *Proceedings American Control Conference*, pp. 301–306, Chicago, 1992.

[198] Mäkilä P., Partington J., "Robust approximate modelling of stable linear systems," *International Journal of Control*, Vol. 58, No. 3, pp. 665–683, 1993.

[199] Mäkilä P.M., Partington J.R., Gustafsson T.K., "Worst-case control-relevant identification," *Automatica*, Vol. 31, No. 12, pp. 1799–1819, 1995.

[200] Mangasarian O.L., *Nonlinear Programming*, McGraw-Hill, New York, 1969.

[201] Marsden J., *Elementary Classical Analysis*, W.H. Freeman and Co., San Francisco, 1974.

[202] Martin J., "State space measures for stability robustness," *IEEE Transactions on Automatic Control*, Vol. 32, 1987.

[203] Meyer D.G., "Two properties of ℓ^1 optimal controllers," *IEEE Transactions on Automatic Control*, Vol. 33, No. 9, pp. 876–878, 1988.

[204] Milanese M., Vicino A., "Optimal Estimation theory for dynamic systems with set membership uncertainty: an overview," *Automatica*, Vol. 27, No. 6, pp. 997–1009, 1991.

[205] Milanese M., "Worst-case ℓ_1 identification," in *Bounding Approaches to System Identification*, M. Milanese, X. Norton, X. Piet-Lahanier, X. Walter, Eds. Plenum Press, New York, 1994.

[206] Miller M., "Process simulation of heavy water plants," *Atomic Energy Canada Limited (AECL)*, Report No. 6178, 1978.

[207] Minnichelli R.J., Anagnost J.J., Desoer C., "An elementary proof of Kharitonov's stability theorem with extensions," *IEEE Transactions on Automatic Control*, Vol. 34, No. 9, pp. 995–998, 1989.

[208] Mirsky L.Q., *Quarterly Journal of Mathematics*, Vol. 2, No. 11, 1960.

[209] Moore B.C., "Principal component analysis in linear systems: controllability, observability and model reduction," *IEEE Transactions on Automatic Control*, Vol. 26, No. 1, 1981.

[210] Morari M., Ray H., *CONSYD*, based on Holt et al. "CONSYD: Control system design," *Computers and Chemical Engineering*, 1986.

[211] Morari M., Zafirou E., *Robust Process Control*, Prentice-Hall, Englewood Cliffs, NJ, 1989.

[212] Nehari Z., "On bounded bilinear forms," *Annals of Mathematics*, No. 65, pp. 153–162, 1957.

[213] Neimark Y.I., "On the problem of the distribution of the roots of polynomials," *Doklady Akademii Nauk*, Vol. 58, 1947.

[214] Nesterov Y., Nemirovski A., *Optimization Over Positive Definite Matrices: Mathematical Background and User's Manual*, USSR Academy of Sciences, Central Economics Mathematical Institute, 1990.

[215] Nett C., Jacobson C., Balas M., "A connection between state-space and doubly coprime fractional representation," *IEEE Transactions on Automatic Control*, Vol. 29, No. 9, 1984.

[216] Newlin M., Young P.M., "Mixed μ problems and branch and bound techniques," *Proceedings Conference on Decision and Control*, Tucson, 1992.

[217] Norman S., Boyd S.P., "Numerical solution of a two-disk problem," *Proceedings American Control Conference*, June 21–23, 1989, Pittsburgh, PA, pp 1745–1747, 1989.

[218] Nyquist H., "Regeneration theory," *Bell Systems Technical Journal*, Vol. 11, 1932.

[219] Packard A., *What's New with μ? Structured Uncertainty in Multivariable Control*. Ph.D. Dissertation, University of California, Berkeley, 1988.

[220] Packard A., Doyle J.C., "The complex structured singular value," *Automatica*, Vol. 29, No. 1, 1993.

[221] Packard A., Pandey P., "Continuity properties of the real/complex structured singular value," *IEEE Transactions on Automatic Control*, Vol. 38, No. 3, 1993.

[222] Packard A., "Gain scheduling via linear fractional transformations," *System and Control Letters*, Vol. 22, pp. 79–92, 1994.

[223] Paige C., "Properties of numerical algorithms related to computing controllability," *IEEE Transactions on Automatic Control*, Vol. 26, 1981.

[224] Palka B.P., *An Introduction to Complex Function Theory*, Springer-Verlag Undergraduate Texts in Mathematics Series, Springer-Verlag, New York, 1991.

[225] Parrot S., "On a quotient norm and the Sz–Nagy Foais lifting theorem," *Journal of Functional Analysis*, Vol. 30, pp. 311–328, 1978.

[226] Papadimitrou C.H., Steiglitz K., *Combinatorial Optimization: Algorithms and Complexity*, Prentice-Hall, Englewood Cliffs, NJ, 1982.

[227] Papoulis A., *Probability, Random Variables and Stochastic Processes*, McGraw-Hill, New York, 1984.

[228] Paredes S.O., "Dinámica y Control de una planta de Agua Pesada," *Trabajo especial*, Instituto Balseiro, Centro Atómico Bariloche, CNEA, 1986.

[229] Parrilo P., *Identificación Robusta de Sistemas Dinámicos*, Electrical Engineering Thesis, University of Buenos Aires, 1994.

[230] Parrilo P., Sánchez-Peña R., Galarza C., "Tuned ℓ^1 identification from impulse response data: application to a fluid dynamics problem," *IEEE Transactions on Control Systems Technology*, Vol. 4, No. 3, 1996.

[231] Parrilo P., Sznaier M., Sánchez-Peña R., "Mixed time/frequency based robust identification," *Proceedings Conference on Decision and Control*, Kobe, Japan, 1996.

[232] Parrilo P., Sánchez-Peña R., Sznaier M., "A parametric extension of mixed time/frequency robust identification," *IEEE Transactions on Automatic Control*, to appear.

[233] Partington J., "Robust identification and interpolation in $\mathcal{H}_\infty$," *International Journal of Control*, Vol. 54, No. 5, pp. 1281–1290, 1991.

[234] Partington J., "Robust identification in $\mathcal{H}_\infty$," *Journal of Mathematical Analysis and Applications*, Vol. 166, pp. 428–441, 1992.

[235] Partington J., "Algorithms for identification in $\mathcal{H}_\infty$ with unequally spaced functions measurements," *International Journal of Control*, Vol. 58, pp. 21–31, 1993.

[236] Partington J.R., "Interpolation in normed spaces from the values of linear functionals," *Bulletin of the London Mathematical Society*, Vol. 26, pp. 165–170, 1994.

[237] Partington J.R., Mäkilä P., "Worst case analysis of identification — BIBO robustness for closed loop data," *IEEE Transactions on Automatic Control*, Vol. 39, pp. 2171–2176, 1994.

[238] Partington J.R., Mäkilä P., "Analysis of linear methods for robust identification in ℓ^1," *Automatica*, Vol. 31, No. 5, pp. 755–758, 1995.

[239] Pernebo L., "An algebraic theory for the design of controllers for linear multivariable systems: Parts I and II," *IEEE Transactions on Automatic Control*, Vol. 26, No. 1, 1981.
[240] Pernebo L., Silverman L.M., "Model reduction via balanced state space representations," *IEEE Transactions on Automatic Control*, Vol. 27, No. 2, 1982.
[241] Polyak B.T., "Robustness analysis for multilinear perturbations," in *Robustness of Dynamic Systems with Parameter Uncertainties*, M. Mansour, S. Balemi, W. Truöl, Eds. Birkhäuser, Berlin, 1992.
[242] Poolla K., Khargonekar P., Tikku A., Krause J., Nagpal K., "A time-domain approach to model validation," *Proceedings American Control Conference*, Chicago, 1992.
[243] Poolla K., Tikku A., "On the time complexity of worst-case system identification," *IEEE Transactions on Automatic Control*, Vol. AC-39, pp. 944–950, 1994.
[244] Pontryagin L., Boltyanskii V., Gamkrelidze R., Mishchenko E., *The Mathematical Theory of Optimal Processes*, Interscience Publishers, New York, 1962.
[245] Popper K., *The Logic of Scientific Discovery*, Hutchinson & Co., London, 1958.
[246] Prakash R., "Properties of a low-frequency approximation balancing method of model reduction," *IEEE Transactions on Automatic Control*, Vol. 39, No. 5, 1994.
[247] Robel, "On computing the ∞-norm," *IEEE Transactions on Automatic Control*, Vol. 34, No. 8, 1989.
[248] Rohn J., Poljak R., "Checking robust nonsingularity is NP hard," *Mathematics of Control, Signals and Systems*, 1992.
[249] Rotea M., "The generalized $\mathcal{H}_2$ control problem," *Automatica*, Vol. 29, No. 2, pp. 373–385, 1993.
[250] Rotstein H., Desages A., Romagnoli J., "Calculation of highly structured stability margins," *International Journal of Control*, Vol. 49, 1989.
[251] Rotstein H., Sánchez-Peña R., Desages A.; Romagnoli J., "Robust characteristic polynomial assignment," *Automatica*, Vol. 27, No. 4, 1991.
[252] Routh E.J., *Dynamics of a System of Rigid Bodies*, Macmillan, New York, 1892.
[253] Rubin N.P., Limebeer D.J.N., "$\mathcal{H}_\infty$ identification for robust control," *Proceedings American Control Conference*, Baltimore, pp. 2040–2044, 1994.
[254] Saeki M., "Method of robust stability analysis with highly structured uncertainties," *IEEE Transactions on Automatic Control*, Vol. 31, pp. 935–940, 1986.
[255] Safonov M., Athans M., "Gain and phase margins for multiloop LQG regulators," *IEEE Transactions on Automatic Control*, Vol. 22, 1977.
[256] Safonov M., "Tight bounds on the response of multivariable systems with component uncertainty," *Proceedings of the 16th Allerton Conference*, Illinois, 1978.
[257] Safonov M., "Stability margins of diagonally perturbed multivariable feedback systems," *Proceedings Conference on Decision and Control*, 1981.
[258] Safonov M., Laub A., Hartmann G., "Feedback properties of multivariable systems: the role and use of the return difference matrix," *IEEE Transactions on Automatic Control*, Vol. 26, 1981.
[259] Safonov M., Doyle J.C., "Minimizing conservativeness of robustness singular

values," in *Multivariable Control*, S.G. Tzefestas, Ed. D. Reidel Pub. Co., Dordrecht, The Netherlands, 1984.

[260] Safonov M., Verma M., "Multivariable L^∞ sensitivity optimization and Hankel approximation," *Proceedings American Control Conference*, 1984.

[261] Safonov M., Chiang R., "CACSD using the state space $\mathcal{L}_\infty$ theory — a design example," *IEEE Transactions on Automatic Control*, Vol. 33, No. 5, 1988.

[262] Safonov M., Chiang R., *Robust Control Toolbox*, The MathWorks Inc., Natick, MA, 1988.

[263] Safonov M., Chiang R., "A Schur method for balanced truncation model reduction," *IEEE Transactions on Automatic Control*, Vol. 34, No. 7, 1989.

[264] Safonov M., Limebeer D.J.N., Chiang R., "Simplifying the $\mathcal{H}_\infty$ theory via loop–shifting, matrix pencil and descriptor concepts," *International Journal of Control*, Vol. 50, No. 6, pp. 2467–2488, 1989.

[265] Sánchez-Peña R., *Robust Analysis of Feedback Systems with Parametric and Dynamic Structured Uncertainty*, Ph.D. Dissertation, California Institute of Technology, Pasadena, 1988.

[266] Sánchez-Peña R., Sideris A., "A general program to compute the multivariable stability margin for systems with real parametric uncertainty," *Proceedings American Control Conference*, Georgia, 1988.

[267] Sánchez-Peña R., Sideris A., "Robustness with real parametric and structured complex uncertainty," *International Journal of Control*, Vol. 52, No. 3, 1990.

[268] Sánchez-Peña R., "Robust analysis and control of a D_2O plant," *Latin American Applied Research*, No. 24, pp. 125–136, 1994.

[269] Sánchez-Peña R., García R., Fernández Berdaguer E., "Margin computation of state space system properties," *Latin American Applied Research*, Vol. 26, pp. 167–176, 1996.

[270] Sánchez-Peña R., Eszter E., Aguilera N., "Part I: Robustness of uncertain zero, pole and delay systems," presented at the *IFAC International Workshop on Robustness of Control Systems*, Kappel Am Albis, Switzerland, 1991, Technical Report University of Buenos Aires.

[271] Sánchez-Peña R., *Introducción a la Teoría de Control Robusto*, Editorial Control S.R.L. (AADECA), Buenos Aires, 1992.

[272] Sánchez-Peña R., Galarza C., "Practical issues in robust identification," *IEEE Transactions on Control Systems Technology*, Vol. 2, No. 1, 1994.

[273] Sánchez-Peña R., Sznaier M., "Robust identification with mixed time/frequency experiments: consistency and interpolation algorithms," *Proceedings Conference on Decision and Control*, New Orleans, 1995.

[274] Sánchez-Peña R., Alonso R., "Structured singular value analysis of the SAC-C satellite attitude control," *Proceedings 7th Latin American Congress of Automatic Control*, Buenos Aires, 1996.

[275] Scherer C., *The Riccati Inequality and State–Space $\mathcal{H}_\infty$ Optimal Control*," Ph.D. Dissertation, Universitat Wurzburg, Germany, 1990.

[276] F. Schweppe, *Uncertain Dynamic Systems*, Prentice-Hall, Englewood Cliffs, NJ, 1973.

[277] Shaked U., "A general transfer function approach to linear stationary filtering

and steady state optimal control problems," *International Journal of Control*, Vol. 24, 1976.

[278] Shaked U., "A general transfer function approach to the steady state linear quadratic Gaussian stochastic control problem," *International Journal of Control*, Vol. 24, 1976.

[279] Sideris A., *Robust Feedback Synthesis via Conformal Mappings and $\mathcal{H}_\infty$ Optimization*, Ph.D. Dissertation, University of Southern California, Los Angeles, 1985.

[280] Sideris A., de Gaston R., "Multivariable stability margin calculation with uncertain correlated parameters," *Proceedings Conference on Decision and Control*, Athens, 1986.

[281] Sideris A., Sánchez-Peña R., "Fast computation of the multivariable stability margin for real interrelated uncertain parameters," *IEEE Transactions on Automatic Control*, Vol. 34, No. 12, 1989.

[282] Sideris A., Sánchez-Peña R., "Robustness margin calculation with dynamic and real parametric uncertainty," *IEEE Transactions on Automatic Control*, Vol. 35, No. 8, 1990.

[283] Sideris A., "An efficient procedure to check the robust stability of polynomials with coefficients in a polytope," *Mathematics of Control, Signals and Systems*, Vol. 4, 1991.

[284] Sideris A., "Elimination of frequency search from robustness tests," *IEEE Transactions on Automatic Control*, Vol. 37, No. 10, 1992.

[285] Siljak D.D., *Nonlinear Systems*, Wiley, New York, 1969.

[286] Silverman L., Bettayeb M., "Optimal approximation of linear systems," *Proceedings Joint Automatic Control Conference*, San Francisco, 1980.

[287] Skogestad S., Morari M., Doyle J.C., "Robust control of ill-conditioned plants: high purity distillation," *IEEE Transactions on Automatic Control*, Vol. 33, 1988.

[288] Skogestad S., Morari M., "L-V control of a high purity distillation column," *Chemical Engineering Science*, Vol. 43, 1988.

[289] Smith O., "Closer control of loops with dead time," *Chemical Engineering Progress*, No. 53, 1957.

[290] Smith R., *Model Validation for Uncertain Systems*, Ph.D. Dissertation, California Institute of Technology, Pasadena, 1990.

[291] Smith R., "Model validation for robust control: an experimental process control application," *Automatica*, Vol. 31, No. 11, pp. 1637–1647, 1995.

[292] Smith R., Doyle J.C., "Model validation: a connection between robust control and identification," *IEEE Transactions on Automatic Control*, Vol. 37, No. 7, 1992.

[293] Söderström T., Stoica P., *System Identification*, Prentice-Hall International, Englewood Cliffs, NJ, 1989.

[294] Stein G., Athans M., "The LQG/LTR procedure for multivariable feedback control design," *IEEE Transactions on Automatic Control*, Vol. 32, No. 2, 1987.

[295] Steen Larsen H., *Boom Test: Experimental Values for Eigenfrequencies, Torsion and Flexure*, Technical Note Ørsted Satellite Project, Per Udsen Co., December 1995.

[296] Stein G., Doyle J.C., "Beyond singular values and loop shapes," *AIAA Journal of Guidance, Control and Dynamics*, Vol. 14, No. 1, 1991.

[297] Strang, *Linear Algebra and Its Applications*, Academic Press, New York, 1976.

[298] Stoorvogel A.A., "The singular $\mathcal{H}_2$ control problem," *Automatica*, Vol. 28, No. 3, pp. 627–631, 1992.

[299] Sznaier M., "Mixed $\ell^1/\mathcal{H}_\infty$ controllers for SISO discrete-time systems," *Systems and Control Letters*, Vol. 23, No. 3, pp. 179–186, 1994.

[300] Sznaier M., "An exact solution to mixed $\mathcal{H}_2/\mathcal{H}_\infty$ problems via convex optimization," *IEEE Transactions on Automatic Control*, Vol. AC-39, No. 12, pp. 2511–2517, 1994.

[301] Sznaier M., Dorato P., organizers of: "Multiobjective robust control," *Proceedings 33 IEEE CDC*, Lake Buena Vista, FL, pp. 2672–2707, 1994.

[302] Tierno J.E., Young P.M., "An improved μ lower bound via adaptive power iteration," *Proceedings Conference on Decision and Control*, Tucson, 1992.

[303] Tikku A., Poolla K., "Sample complexity for worst-case system identification problems," *Proceedings American Control Conference*, San Francisco, 1993.

[304] Traub J.F., Wasilkowski G.W., Wozniakowski H., *Information-Based Complexity*, Academic Press, New York, 1988.

[305] Tse D.N.C, Dahleh M.A., Tsitsiklis J.N., "Optimal and robust identification in ℓ^1," *Proceedings American Control Conference*, Boston, 1991.

[306] Tse D.N.C, Dahleh M.A., Tsitsiklis J.N., "Optimal asymptotic identification under bounded disturbances," *IEEE Transactions on Automatic Control*, Vol. 38, pp. 1176–1190, 1993.

[307] Tsypkin Y.Z., Polyak B.T., "Frequency domain criteria for ℓ^p robust stability of continuous linear systems," *IEEE Transactions on Automatic Control*, Vol. 36, 1991.

[308] Vanderberghe L., Boyd S., "SP-Software for semidefinite programming," *SIAM Review*, Vol. 38, pp. 49–95, 1996.

[309] Van der Schaft A.J., "On a state-space approach to nonlinear $\mathcal{H}_\infty$ control," *Systems and Control Letters*, Vol. 16, pp. 1–8, 1991.

[310] Van der Schaft A.J., "$\mathcal{L}_2$ gain analysis of nonlinear systems and $\mathcal{H}_\infty$ nonlinear control," *IEEE Transactions on Automatic Control*, Vol. 37, pp. 770–784, 1992.

[311] Van Loan C.F., "How near is a stable matrix to an unstable one?" *Contemporary Mathematics*, No. 47, 1985.

[312] Vicino A., Tesi A., Milanese M., "Computation of nonconservative stability perturbation bounds for systems with nonlinearly correlated uncertainties," *IEEE Transactions on Automatic Control*, Vol. 35, No. 7, 1990.

[313] Vidyasagar M., *Nonlinear Systems Analysis*, Prentice-Hall, Englewood Cliffs, NJ, 1978.

[314] Vidyasagar M., *Control System Synthesis: A Factorization Approach*, The MIT Press, Cambridge, MA, 1985.

[315] Vidyasagar M. "Optimal rejection of persistent bounded disturbances," *IEEE Transactions on Automatic Control*, Vol. 31, pp. 527–535, 1986.

[316] Wang Z.Q., Sznaier M., Blanchini F. "Further results on rational approximations

of $\mathcal{L}^1$-optimal controllers," *IEEE Transactions on Automatic Control*, Vol. 40, No. 3, pp. 552–557, Mar. 1995.

[317] Wang Z.Q., Sznaier M., Blanchini F., "Continuity properties of $\mathcal{L}^1/l^1$ optimal controllers for plants with stability boundary zeros," *Proceedings of the 1995 American Control Conference*, Seattle, pp. 961–963, 1995.

[318] Wei K.H., Yedavalli R.K., "Robust stabilizability for linear systems with both parameter variation and unstructured uncertainty," *Proceedings Conference on Decision and Control*, Los Angeles, 1987.

[319] Tin-Yung Wen J., Kreutz-Delgado K., "The attitude control problem," *IEEE Transactions on Automatic Control*, Vol. 36, No. 10, 1991.

[320] Wilson D.A., "Convolution and Hankel operator norms for linear systems," *IEEE Transactions on Automatic Control*, Vol. AC-34, pp. 94–97, 1989.

[321] Wu N., Gu G., "Discrete Fourier transform and $\mathcal{H}_\infty$ approximations," *IEEE Transactions on Automatic Control*, Vol. 35, No. 9, 1990.

[322] Yeung K.S., Wang S.S., "A simple proof of Kharitonov's theorem," *IEEE Transactions on Automatic Control*, Vol. AC-32, 1987.

[323] Youla D.C., Jabr H., Bongiorno J., "Modern Wiener-Hopf design of optimal controllers: the SISO case," *IEEE Transactions on Automatic Control*, Vol. 21, 1976.

[324] Youla D.C., Jabr H., Bongiorno J., "Modern Wiener-Hopf design of optimal controllers: the multivariable case," *IEEE Transactions on Automatic Control*, Vol. 21, 1976.

[325] Young P.M., Doyle J.C., "Computation of μ with real and complex uncertainty," *Proceedings Conference on Decision and Control*, Hawaii, 1990.

[326] Young P.M., Newlin M., Doyle J.C., "μ analysis with real parametric uncertainty," *Proceedings Conference on Decision and Control*, England, 1991.

[327] Young P.M., Newlin M., Doyle J.C., "Computing bounds for the mixed μ problem," *International Journal of Robust and Nonlinear Control*, Vol. 5, pp. 573–590, 1995.

[328] Zadeh L., Desoer C., *Linear System Theory: The State Space Approach*, McGraw-Hill, New York, 1963.

[329] Zames G., "Feedback and optimal sensitivity: model reference transformations, multiplicative seminorms and approximate inverses," *IEEE Transactions on Automatic Control*, Vol. 26, pp. 301–320, 1981.

[330] Zames G., Francis, B.A., "Feedback, minimax sensitivity, and optimal robustness," *IEEE Transactions on Automatic Control*, Vol. 28, pp. 585–601, 1983.

[331] Zhang Z., Freudenberg J.S., "LTR for nonminimum phase plants," *IEEE Transactions on Automatic Control*, Vol. 35, No. 5, 1990.

[332] Zhou K., Doyle J.C., Glover K., *Robust and Optimal Control*, Prentice-Hall, Englewood Cliffs, NJ, 1996.

[333] Zhou T., Kimura H., "Time domain identification for robust control," *System and Control Letters*, Vol. 20, pp. 167–178, 1993.

APPENDIX A

MATHEMATICAL BACKGROUND

Here we present the basic necessary mathematical background on which the main part of the book is based. The objective is not to deliver an exhaustive treatment for each of the subjects. The material in this appendix can be found in further detail in [118, 130, 170, 174, 297].

A.1 ALGEBRAIC STRUCTURES

The algebraic structures are instrumental in understanding the problems in control theory from an abstract point of view. For example, controllability can be interpreted as the range space of an operator, and observability as the kernel of another one. Also, certain optimal filtering and regulation problems can be viewed as projections over a certain vector space.

A.1.1 Field

Definition A.1 *A field as $(\mathcal{F}, \&, \star)$ is an algebraic structure composed of a set $\mathcal{F}$ and two operations $\&$ and $\star$, with the following properties:*

1. Set $\mathcal{F}$ is closed with respect to $\&$, that is, $a, b \in \mathcal{F} \implies (a \& b) \in \mathcal{F}$.
2. Operation $\&$ is associative, that is, $(a \& b) \& c = a \& (b \& c) = a \& b \& c$ for $a, b, c \in \mathcal{F}$.
3. Operation $\&$ is commutative, that is, $a \& b = b \& a$ for $a, b \in \mathcal{F}$.
4. Set $\mathcal{F}$ contains the neutral element $n_\&$ with respect to $\&$, that is, there exists $n_\& \in \mathcal{F}$ such that $(a \& n_\&) = a$ for all $a \in \mathcal{F}$.

5. Set $\mathcal{F}$ contains the inverse element $a_\&^I$ with respect to &, that is, for all $a \in \mathcal{F}$, there exists $a_\&^I \in \mathcal{F}$ such that $(a \,\&\, a_\&^I) = n_\&$.
6. Set $\mathcal{F}$ is closed with respect to $\star$.
7. Operation $\star$ is associative.
8. Set $\mathcal{F}$ contains the neutral element $n_\star$ with respect to $\star$.
9. Set $\mathcal{F}$ contains the inverse element $a_\star^I$ with respect to $\star$.
10. Operation $\star$ is distributive with respect to &, that is, $(a \,\&\, b) \star c = (a \star c) \,\&\, (b \star c)$ for $a, b, c \in \mathcal{F}$.

Take, for example, set $\mathbb{R}$ equipped with operations $(+, \times)$ as $(\&, \star)$ respectively, then $n_\& = 0$, $n_\star = 1$, $a_\&^I = -a$, and $a_\star^I = a^{-1}$ ($a \neq 0$).

A.1.2 Linear Vector Space

Definition A.2 *A set $\mathcal{V}$ is a* linear vector space *over the field $(\mathcal{F}, +, \times)$ if and only if the following properties are satisfied: (In the sequel the elements of $\mathcal{F}$ and $\mathcal{V}$ will be called scalars and vectors, respectively).*

1. Set $\mathcal{V}$ is closed with respect to $+$.
2. Operation $+$ is associative in $\mathcal{V}$.
3. Operation $+$ is commutative in $\mathcal{V}$.
4. Set $\mathcal{V}$ contains the neutral element with respect to $+$.
5. Set $\mathcal{V}$ contains the inverse element with respect to $+$.
6. $\mathcal{V}$ is closed with respect to operation $\times$ between scalars and vectors.
7. Operation $\times$ among scalars and vectors is associative in the scalars, that is, $(a \times b) \times v = a \times (b \times v) = a \times b \times v$ for $a, b \in \mathcal{F}$ and $v \in \mathcal{V}$.
8. Distributive 1: $(a + b) \times v = (a \times v) + (b \times v)$ for $a, b \in \mathcal{F}$ and $v \in \mathcal{V}$.
9. Distributive 2: $(u + v) \times a = (u \times a) + (v \times a)$ for $a \in \mathcal{F}$ and $u, v \in \mathcal{V}$.
10. Field $\mathcal{F}$ contains the neutral element of operation $\times$ between vectors and scalars, that is, $n_\times \times v = v$ for $n_\times \in \mathcal{F}$ and $v \in \mathcal{V}$.

In the sequel, linear vector space, linear space, or vector space will be used equivalently.

Definition A.3 *The set of vectors $\{v_1, \ldots, v_n\}$ in the linear space $\mathcal{V}$, defined over the field $\mathcal{F}$, is* linearly independent *if and only if the linear combination $\alpha_1 v_1 + \cdots + \alpha_n v_n = 0 \iff \alpha_1 = \cdots = \alpha_n = 0$, with $\alpha_i \in \mathcal{F}$, $i = 1, \ldots, n$. Otherwise they are called* linearly dependent.

Definition A.4 *A set of linearly independent vectors $\{u_1, \ldots, u_n\}$ of a linear space $\mathcal{V}$ constitute a* base *of $\mathcal{V}$ if and only if any $v \in \mathcal{V}$ can be represented as a linear combination of these vectors. The* dimension *of $\mathcal{V}$ is the minimum number of vectors that constitute a base, in this case n.*

ALGEBRAIC STRUCTURES

Linear spaces may have finite or infinite dimension, as will be shown next by means of several examples.

Example A.1 *The matrices in $\mathbb{R}^{n\times 1}$ or $\mathbb{C}^{n\times 1}$, constitute finite-dimensional linear spaces. The representations of position in physical space with respect to a certain reference frame also constitute a finite-dimensional linear space.*

The sequences of real numbers $S_\mathbb{R}$ or the set of continuous functions defined in a real interval $C_{[0,1]}$ form infinite-dimensional linear spaces. These last two cases may represent the pulse sequences in digital systems or certain classes of signals in analog circuits, respectively.

A.1.3 Metric, Norm, and Inner Products

Definition A.5 *A metric space $(\mathcal{V}, d(\cdot,\cdot))$ is defined in terms of a linear vector space $\mathcal{V}$ and a real function $d(\cdot,\cdot): \mathcal{V}\times\mathcal{V} \to \mathbb{R}_+$, satisfying the following conditions:*

1. $d(x,y) \geq 0, \quad \forall\, x,y \in \mathcal{V}.$
2. $d(x,y) = 0 \iff x = y.$
3. $d(x,y) = d(y,x), \quad \forall\, x,y \in \mathcal{V}.$
4. $d(x,z) \leq d(x,y) + d(y,z), \quad \forall\, x,y,z \in \mathcal{V}.$

Here $\mathbb{R}_+ \triangleq \{x \in \mathbb{R},\ x \geq 0\}$.

Definition A.6 *A normed space $(\mathcal{V}, \|\cdot\|)$ is defined in terms of a linear vector space $\mathcal{V}$ and a real function $\|\cdot\|: \mathcal{V} \to \mathbb{R}_+$ that satisfies the following conditions:*

1. $\|x\| \geq 0, \quad \forall\, x \in \mathcal{V}.$
2. $\|x\| = 0 \iff x = \underline{0}.$
3. $\|\alpha x\| = |\alpha| \cdot \|x\|, \quad \forall\, x \in \mathcal{V}, \alpha \in \mathcal{F}.$
4. $\|x + y\| \leq \|x\| + \|y\|, \quad \forall\, x,y \in \mathcal{V}.$

Here $|\cdot|$ represents the magnitude of a scalar and $\underline{0}$ is the neutral element of $\mathcal{V}$ with respect to addition.

Example A.2 *The following are examples of normed spaces:*

1. *Take the linear space $\mathbb{C}^n$ and define*

$$\|x\|_p \triangleq \sqrt[p]{\sum_{i=1}^{n} |x_i|^p}, \quad p \geq 1 \tag{A.1}$$

$$\|x\|_\infty \triangleq \max_{1\leq i \leq n} |x_i| \tag{A.2}$$

2. *Take the linear space of sequences $S_\mathbb{R}$ and define*

$$\|x\|_p \triangleq \sqrt[p]{\sum_{i=1}^{\infty} |x_i|^p}, \quad p \geq 1 \tag{A.3}$$

$$\|x\|_\infty \triangleq \max_{i \geq 1} |x_i| \tag{A.4}$$

3. *Take the linear space of continuous functions $C_{[a,b]}$ and define*

$$\|f\|_p \triangleq \sqrt[p]{\int_a^b |f(t)|^p \, dt}, \quad p \geq 1 \tag{A.5}$$

$$\|f\|_\infty \triangleq \max_{t \in [a,b]} |f(t)| \tag{A.6}$$

Definition A.7 *An* inner product space $(\mathcal{V}, <\cdot,\cdot>)$ *is defined in terms of a linear vector space $\mathcal{V}$ and a function $<\cdot,\cdot>: \mathcal{V} \times \mathcal{V} \to \mathcal{F}$ that satisfies the following conditions $\forall\, x, y, z \in \mathcal{V}$ and $\alpha \in \mathcal{F}$:*

1. $<x, y> = \overline{<y, x>}$.
2. $<x + y, z> = <x, z> + <y, z>$.
3. $<\alpha x, y> = \alpha <x, y>$.
4. $<x, x> \geq 0, \quad \forall x \neq 0$.
5. $<x, x> = 0 \iff x = 0$.

Here $\bar{a}$ represents the complex conjugate of element a.

Example A.3 *The three normed spaces from the last example can be made inner product spaces with the following definitions:*

1. *Take vectors in $\mathbb{C}^n$ and define*

$$<x, y> \triangleq \sum_{i=1}^{n} \overline{x_i} y_i \tag{A.7}$$

2. *Take sequences in $S_\mathbb{R}$ and define*

$$<x, y> \triangleq \sum_{i=1}^{\infty} x_i y_i \tag{A.8}$$

3. Take continuous functions $f \in C_{[a,b]}$ such that $f : [a,b] \to \mathbb{C}$, and define

$$<f,g> \triangleq \int_a^b \overline{f(\omega)} g(\omega) \, d\omega \tag{A.9}$$

An inner product space generates a normed vector space when using its inner product to define a norm, as follows: $\|x\| = \sqrt{<x,x>}$. In addition, a normed vector space generates a metric space when using its norm to define the distance function, as follows: $d(x,y) = \|x-y\|$.

Definition A.8 *Two vectors x,y in $(\mathcal{V}, <\cdot,\cdot>)$ are said to be* orthogonal *if and only if $<x,y> = 0$. If, in addition, $<x,x> = 1$ and $<y,y> = 1$, they are called* orthonormal.

A.2 FUNCTION SPACES

A.2.1 Introduction

Next, some important properties of infinite-dimensional vector spaces are briefly reviewed.

Definition A.9 *A sequence $\{x_n\}$ in a metric space $(\mathcal{V},d)$ is said to be a* Cauchy *sequence if and only if $\forall \epsilon > 0 \; \exists N$ such that $n,m \geq N \implies d(x_n, x_m) < \epsilon$.*

Definition A.10 *A sequence $\{x_n\}$ in a metric space $(\mathcal{V},d)$ converges to $x_\star \in \mathcal{V}$ if and only if $\forall \epsilon > 0 \; \exists N$ such that $n \geq N \implies d(x_n, x_\star) < \epsilon$.*

It is easy to prove that any convergent sequence is a Cauchy sequence. On the other hand, the converse, is not necessarily true.

Example A.4 *Consider among the rational numbers $\mathcal{Q}$ the sequence $\{3 \quad 3.14 \quad 3.1415 \quad \cdots\}$. This is a Cauchy sequence, although it does not converge because $\pi \notin \mathcal{Q}$.*

Definition A.11 *A metric space $(\mathcal{V},d)$ is said to be* complete *if and only if all Cauchy sequences are convergent in $\mathcal{V}$.*

A.2.2 Banach and Hilbert spaces

Definition A.12 *A complete normed vector space $(\mathcal{V}, \|\cdot\|)$ is called a* Banach *space.*

Definition A.13 *A complete inner product space $(\mathcal{V}, <\cdot,\cdot>)$ is called a* Hilbert *space.*

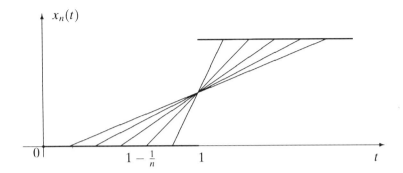

Figure A.1. Nonconvergent sequence in $C_{[0,1]}$.

Example A.5 *The infinite-dimensional vector space of continuous functions defined in the unit interval $C_{[0,1]}$ is not complete when using the metrics $d_1(x, y) = \|x - y\|_1$ or $d_2(x, y) = \|x - y\|_2$, where these norms have been defined in (A.5).*

Note in Figure A.1 that the sequence $x_n(t)$ of continuous functions converges to a discontinuous one. Therefore the space of continuous functions is not complete, under these metrics.

A.2.3 Operator and Signal Spaces

Many of the useful function spaces, which will be described in this subsection, include other than continuous functions. In these cases integration should be understood in the *Lebesgue* integral sense (*Riemann* integral is valid only for continuous functions). In addition, the concepts of minimum and maximum of a function (min, max) are replaced by their generalizations: *infimum* and *supremum* (inf, sup). Finally, *essential supremum*, denoted as ess sup, refers to the supremum of a function for all but *sets of measure zero* in its domain. These functional analysis concepts can be found in greater detail in [174].

Time Domain

- $\mathcal{L}_1(\mathbb{R}, \mathbb{R})$ is the space of absolutely integrable scalar functions, equipped with the norm

$$\|f\|_1 \triangleq \int_{-\infty}^{\infty} |f(t)|\, dt \tag{A.10}$$

FUNCTION SPACES 451

- $\mathcal{L}_1(\mathbb{R}, \mathbb{R}^{n\times m})$ is the space of absolutely integrable matrix functions, equipped with the norm

$$\|f\|_1 \stackrel{\Delta}{=} \sup_{1 \leq i \leq n} \left\{ \sum_{j=1}^{m} \|f_{ij}\|_1 \right\} \tag{A.11}$$

- $\mathcal{L}_2(\mathbb{R}, \mathbb{R}^{n\times m})$ is the space of square integrable matrix functions of $t \in \mathbb{R}$, with inner product

$$<f, g> = \int_{-\infty}^{\infty} \operatorname{trace}\left[f(t)g(t)^T\right] dt \tag{A.12}$$

and norm $\|f\|_2 \stackrel{\Delta}{=} \sqrt{<f,f>}$.

- $\mathcal{H}_2(\mathbb{R}, \mathbb{R}^{n\times m})$ is the subspace of $\mathcal{L}_2(\mathbb{R}, \mathbb{R}^{n\times m})$ formed by *causal* functions, that is, zero for $t < 0$. The notation $\mathcal{L}_2([0, \infty), \mathbb{R}^{n\times m})$ is used equivalently.
- $\mathcal{H}_2^\perp(\mathbb{R}, \mathbb{R}^{n\times m})$ is the subspace of $\mathcal{L}_2(\mathbb{R}, \mathbb{R}^{n\times m})$ consisting of *anticausal* functions, that is, zero for $t > 0$. The notation $\mathcal{L}_2((-\infty, 0], \mathbb{R}^{n\times m})$ is used equivalently.

Next, we show that for a causal stable system G, $\|G\|_2$ can be interpreted as the RMS value of its output $z = G * u$ when the input u is a white noise signal with unit covariance. To this effect, given a wide sense stationary random vector signal $u(t)$ define its autocorrelation as

$$R_{uu}(\tau) = \mathcal{E}\left[u(t+\tau)u(t)^T\right] \tag{A.13}$$

where $\mathcal{E}$ denotes expectation. The Fourier transform of $R_{uu}(\tau)$ is called the spectral density $S_{uu}(j\omega)$, that is,

$$S_{uu}(j\omega) = \int_{-\infty}^{\infty} R_{uu}(\tau) e^{-j\omega\tau} d\tau \tag{A.14}$$

It follows that $\mathcal{E}(u^T u) = \operatorname{trace}\left[\mathcal{E}(uu^T)\right] = \operatorname{trace}\left[R_{uu}(0)\right]$. It is a standard result that if $z = G * u$ then $S_{zz}(j\omega) = G(j\omega) S_{uu}(j\omega) G^*(j\omega)$. Thus

$$\begin{aligned} \mathcal{E}(z^T z) &= \operatorname{trace}\left[R_{zz}(0)\right] = \frac{1}{2\pi} \int_{-\infty}^{\infty} \operatorname{trace}\left[S_{zz}(j\omega)\right] d\omega \\ &= \frac{1}{2\pi} \int_{-\infty}^{\infty} \operatorname{Trace}\left[G(j\omega) G^*(j\omega)\right] = \|G\|_2^2 \end{aligned} \tag{A.15}$$

where we used the facts that $S_{uu}(j\omega) = I$ and that $\operatorname{trace}(GG^*) = \operatorname{trace}(G^*G)$.

Frequency Domain

- $\mathcal{L}_2(\jmath\mathbb{R}, \mathbb{C}^{n\times m})$ is the space of square integrable matrix functions of $\jmath\omega$, $\omega \in \mathbb{R}$, with inner product

$$<F, G> = \frac{1}{2\pi} \int_{-\infty}^{\infty} \text{trace}\,[F(\jmath\omega)G(\jmath\omega)^*]\,d\omega \qquad (A.16)$$

and norm $\|F\|_2 \triangleq \sqrt{<F, F>}$. Here we have defined $G^*(s) \triangleq G^T(-s)$.

- $\mathcal{H}_2(\jmath\mathbb{R}, \mathbb{C}^{n\times m})$ is the subspace of $\mathcal{L}_2(\jmath\mathbb{R}, \mathbb{C}^{n\times m})$ composed of analytic functions in $\text{Re}(s) > 0$ such that

$$\sup_{\sigma>0} \int_{-\infty}^{\infty} \text{trace}\,[F(\jmath\omega + \sigma)F(\jmath\omega + \sigma)^*]\,d\omega < \infty \qquad (A.17)$$

- $\mathcal{H}_2^{\perp}(\jmath\mathbb{R}, \mathbb{C}^{n\times m})$ is the subspace of $\mathcal{L}_2(\jmath\mathbb{R}, \mathbb{C}^{n\times m})$ composed of analytical functions in $\text{Re}(s) < 0$.
- $\mathcal{L}_\infty(\jmath\mathbb{R}, \mathbb{C}^{n\times m})$ is the (essentially) bounded matrix function space equipped with the following norm:

$$\|F\|_\infty = \text{ess}\,\sup_{\omega}\,\bar{\sigma}\,[F(\jmath\omega)] \qquad (A.18)$$

- $\mathcal{H}_\infty(\jmath\mathbb{R}, \mathbb{C}^{n\times m})$ is the subspace of $\mathcal{L}_\infty(\jmath\mathbb{R}, \mathbb{C}^{n\times m})$ of analytic and (essentially) bounded functions in $\text{Re}(s) > 0$.
- $\mathcal{H}_\infty^{\perp}(\jmath\mathbb{R}, \mathbb{C}^{n\times m})$ is the subspace of $\mathcal{L}_\infty(\jmath\mathbb{R}, \mathbb{C}^{n\times m})$ of analytic and (essentially) bounded functions in $\text{Re}(s) < 0$.

$\mathcal{L}_2(\mathbb{R}, \cdot)$ and $\mathcal{L}_2(\jmath\mathbb{R}, \cdot)$ are Hilbert spaces; $\mathcal{L}_\infty(\jmath\mathbb{R}, \cdot)$ is a Banach space. In many cases, when these spaces are restricted to the set of rational transfer functions or matrices, an $\mathcal{R}$ is added to the notation, that is, $\mathcal{RL}_2$, $\mathcal{RL}_\infty$, $\mathcal{RH}_\infty$, $\mathcal{RH}_\infty^{\perp}$. In the main part of the book, the arguments of these spaces will be dropped for simplicity, these being clear from the context.

A.2.4 Isomorphism

Definition A.14 *Two vector spaces V and W are said to be* isomorphic *if and only if there exists a linear transformation $L : V \to W$, satisfying the following conditions:*

1. $L(x) = L(y) \Rightarrow x = y \in V$, *that is,* $\mathbf{N}(L) = \{\underline{0}\}$. *This property is defined as* monomorphism *or one to one.*
2. $w = L(x)$ *for all* $w \in W$ *with* $x \in V$, *that is,* $\mathbf{R}(L) \equiv W$. *This property is called* epimorphism *or onto.*

Here $\mathbf{N}(\cdot)$ and $\mathbf{R}(\cdot)$ represent the kernel and range spaces of an operator, respectively.

Definition A.15 *Two normed vector spaces $(V, \|\cdot\|_V)$ and $(W, \|\cdot\|_W)$ are said to be* isometric *if and only if they are isomorphic and, in addition, the a linear transformation L satisfies $\|v\|_V = \|L(v)\|_W$ for any vector $v \in V$.*

Example A.6 *The normed linear vector spaces $\mathcal{L}_2(\mathbb{R}, \mathbb{R})$ and $\mathcal{L}_2(\mathbb{R}, \mathbb{C})$ are isomorphic and isometric under the linear transformation $\mathcal{F}: g(t) \to G(\jmath\omega)$ defined as the Fourier transform:*

$$\|g\|_2 \stackrel{\triangle}{=} \sqrt{\int_0^\infty |g(t)|^2\, dt} = \sqrt{\frac{1}{2\pi}\int_{-\infty}^\infty |G(\jmath\omega)|^2\, d\omega} \stackrel{\triangle}{=} \|G\|_2 \quad (A.19)$$

This is also known as Parseval's theorem. Furthermore, the inner product among vectors in both spaces is also preserved,

$$<f, g>_t \stackrel{\triangle}{=} \int_0^\infty f(t) g(t)\, dt \quad (A.20)$$

$$= \frac{1}{2\pi}\int_{-\infty}^\infty F^*(\jmath\omega) G(\jmath\omega)\, d\omega \stackrel{\triangle}{=} <F, G>_\omega \quad (A.21)$$

As a consequence, the "algebraic structure" of both spaces is similar. In addition, orthonormal vectors in one space will remain orthonormal in the transformed space, that is, "geometry" and "proportion" are preserved from one vector space to the other.

Due to the isometry that the Fourier transform establishes between $\mathcal{L}_2(\mathbb{R})$ and $\mathcal{L}_2(\jmath\mathbb{R})$, their subspaces can be related in the same way. Therefore $\mathcal{H}_2(\mathbb{R})$ and $\mathcal{H}_2(\jmath\mathbb{R})$ are isometric, as well as $\mathcal{H}_2^\perp(\mathbb{R})$ and $\mathcal{H}_2^\perp(\jmath\mathbb{R})$. Hence causal signals (or systems) are isometric with stable transfer matrices and anticausal signals (or systems) with antistable ones.

A.2.5 Induced Norms

Definition A.16 *Consider two normed linear vector spaces $(V, \|\cdot\|_V)$ and $(W, \|\cdot\|_W)$ and a linear transformation $L: V \to W$. The* induced norm *of the transformation is defined as follows:*

$$\|L\|_{V \to W} \stackrel{\triangle}{=} \sup_{\|v\|_V \neq 0} \frac{\|L(v)\|_W}{\|v\|_V} \quad (A.22)$$

It can be proved that equivalently the induced norm can be computed as $\sup_{\|v\|_V = 1} \|L(v)\|_W$.

Example A.7

1. Consider the operator $A \in \mathbb{R}^{n \times m}$ and use the Euclidean vector norm $\|\cdot\|_2$ in both $\mathbb{R}^n$ and $\mathbb{R}^m$:

$$\|A\|_{2 \to 2} \triangleq \sup_{\|x\|_2 \neq 0} \frac{\|Ax\|_2}{\|x\|_2} = \sup_{\|x\|_2 \neq 0} \sqrt{\frac{x^T A^T A x}{x^T x}}$$

$$= \sqrt{\frac{x^T x \bar{\lambda}(A^T A)}{x^T x}} = \sqrt{\bar{\lambda}(A^T A)} \quad (A.23)$$

The latter will be defined in the next section as the maximum singular value of A.

2. If the norm $\|\cdot\|_1$ is used instead in $\mathbb{R}^n$ and $\mathbb{R}^m$, the following induced norm is obtained:

$$\|A\|_{1 \to 1} = \max_j \sum_{i=1}^{m} |a_{ij}| \quad (A.24)$$

3. When $\|\cdot\|_\infty$ is used in $\mathbb{R}^n$ and $\mathbb{R}^m$ the following induced norm is obtained:

$$\|A\|_{\infty \to \infty} = \max_i \sum_{j=1}^{n} |a_{ij}| \quad (A.25)$$

The induced norm quantifies the maximum "amplification" of a linear operator, a generalization of the concept of *gain* in SISO transfer functions. Take, for example, the following matrix:

$$A = \begin{bmatrix} 0.001 & 10^9 \\ 0 & 0.001 \end{bmatrix} \quad (A.26)$$

which can be interpreted as a linear operator $A : \mathbb{R}^2 \to \mathbb{R}^2$. Its eigenvalues $\lambda_1 = \lambda_2 = 0.001$ do not measure the maximum amplification of vectors in $\mathbb{R}^2$ when multiplied by the matrix. On the other hand, its induced norm $\|A\|_{2 \to 2} = \bar{\sigma}(A) \approx 10^9$ does. For example the input $u = [0 \quad 1]^T$, produces an output having $\|Au\|_2 = \bar{\sigma}(A) \approx 10^9$.

An important property of induced norms is the fact that they are sub-multiplicative, that is, $\|L_2 \cdot L_1\|_{w \to w} \leq \|L_2\|_{w \to w} \|L_1\|_{w \to w}$.

Proof. From (A.22) we have that $\|Ly\|_w \leq \|L\|_{w \to w} \|y\|_w$. It follows that

$$\|L_2 L_1 y\|_w \leq \|L_2\|_{w \to w} \|L_1 y\|_w \leq \|L_2\|_{w \to w} \|L_1\|_{w \to w} \|y\|_w \quad (A.27)$$

Equivalently,

$$\frac{\|L_2 L_1 y\|_w}{\|y\|_w} \leq \|L_2\|_{w \to w} \|L_1\|_{w \to w} \quad (A.28)$$

The proof follows now directly from the definition of induced norms. □

FUNCTION SPACES 455

A.2.6 Some Important Induced System Norms

In this section we consider the $\mathcal{L}^\infty$ and $\mathcal{L}^2$ induced system norms and we show that they coincide with the $\mathcal{L}_1$ and $\mathcal{H}_\infty$ norms introduced in Section A.2.3.

The $\mathcal{L}^\infty \to \mathcal{L}^\infty$ Case For simplicity consider the SISO case. For any input $u(t) \in \mathcal{L}^\infty$, $\|u\|_\infty = 1$, we have that the corresponding output $y = g * u$ satisfies

$$\begin{aligned} |y(t)| &\leq \int_0^t |g(t-\tau)||u(\tau)|\, d\tau \\ &\leq \int_0^t |g(t-\tau)|\, d\tau \leq \int_0^\infty |g(\lambda)|\, d\lambda \\ &= \|g\|_1 \end{aligned} \qquad (A.29)$$

This shows that $\|g\|_{\mathcal{L}_\infty \to \mathcal{L}_\infty} \leq \|g\|_1$. To show that the equality holds, assume that $\|g\|_1 = \mu_o$. Then, given $\epsilon > 0$, there exist T such that

$$\mu_o - \epsilon \leq \int_0^T |g(t)|\, dt \leq \mu_o \qquad (A.30)$$

Consider the following signal

$$u(t) = \begin{cases} \text{sign}\,[g(T-t)], & 0 \leq t \leq T \\ 0, & \text{otherwise} \end{cases} \qquad (A.31)$$

It is easily seen that the corresponding output $y(t)$ satisfies $\mu_o \geq |y(T)| \geq \mu_o - \epsilon$. Thus

$$\|g\|_{\mathcal{L}_\infty \to \mathcal{L}_\infty} = \sup_{\|u\|_\infty = 1} \|g * u\|_\infty \geq \mu_o - \epsilon \qquad (A.32)$$

The proof follows now from the fact that ϵ is arbitrary.

The $\mathcal{L}^2 \to \mathcal{L}^2$ Case Consider any input $u \in \mathcal{L}_2, ; \|u\|_2 = 1$ and let $z = Gu$. Using the frequency-domain definition of $\|\cdot\|_2$, we have that

$$\|z\|_2^2 = \|Gu(s)\|_2^2 \qquad (A.33)$$

$$= \frac{1}{2\pi} \int_{-\infty}^\infty u^*(j\omega) G^*(j\omega) G(j\omega) u(j\omega)\, d\omega$$

this is not a multiplication (highest s.v. of G(ω))

$$\leq \frac{1}{2\pi} \int_{-\infty}^\infty \bar\sigma\,[G(j\omega)]\, u^*(j\omega) u(j\omega)\, d\omega$$

$$\leq \sup_\omega \bar\sigma\,[G(j\omega)] \frac{1}{2\pi} \int_{-\infty}^\infty u^*(j\omega) u(j\omega)\, d\omega$$

$$= \|G(s)\|_\infty \qquad (A.34)$$

This shows that the $\mathcal{H}_\infty$ norm is an upper bound of the $\mathcal{L}_2$ induced norm. To show equality we need to find some $u_\star \in \mathcal{L}_2, \|u_\star\|_2 = 1$, such that $\|z\|_2 = \|G\|_\infty$. Assume that $\sup_\omega \bar{\sigma}[G(j\omega)]$ is achieved at the frequency $\omega_\star$. Consider the input $u_\star(t) = k(\epsilon)e^{-\epsilon t^2}\cos(\omega_\star t)v_{\max}$, where $v_{\max}$ is the unitary right singular vector that corresponds to $\bar{\sigma}[G(j\omega_\star)]$ and $k(\epsilon)$ is such that $\|u_\star(s)\|_2 = 1$ for $\epsilon > 0$. As $\epsilon \to 0$ we have that

$$\begin{aligned}
\|z_\star\|_2^2 &= \lim_{\epsilon \to 0} \|G(s)u_\star(s)\|_2^2 \\
&\longrightarrow \frac{1}{2\pi}\int_{-\infty}^{\infty} \|G(j\omega)v_{\max}k(\epsilon)[\delta(\omega+\omega_\star)+\delta(\omega-\omega_\star)]\|^2 \, d\omega \\
&\longrightarrow \|G(j\omega_\star)v_{\max}\|^2 \\
&= \bar{\sigma}^2[G(j\omega_\star)] = \|G(s)\|_\infty^2
\end{aligned} \qquad (A.35)$$

The $\mathcal{H}_2$ Norm The $\mathcal{H}_2$ norm does not admit an interpretation as a norm induced by a single space (and thus it does not have the submultiplicative property). Nevertheless, it is instructive to show that it can be considered as an induced system norm, if the input and ouput spaces are allowed to have different norms. Given a signal $u(t)$ we can define its autocorrelation and spectral density proceeding exactly as in Section A.2.3. If $R_{uu}(\tau)$ and $S_{uu}(j\omega)$ exist and R_{uu} is bounded, then u is said to be a power signal, with power (semi)norm given by $\|u\|_\mathcal{P}^2 = \text{trace}[R_{uu}(0)]$. A power signal v is said to have bounded spectral density if $\|S_{vv}(j\omega)\|_\infty < \infty$. In this case we can define its spectral density norm $\|v\|_\mathcal{S} = \|S_{vv}(j\omega)\|_\infty$. Proceeding as in Section A.2.3 it can easily be shown (by considering a signal $u(t)$ having $S_{uu} = I$) that $\|G\|_2 = \|G\|_{\mathcal{S} \to \mathcal{P}}$.

Finally, in the SISO case, $\|.\|_2$ can be interpreted as the $\mathcal{L}_2 \to \mathcal{L}_\infty$ norm. To establish this fact consider $u \in \mathcal{L}_2, \|u\|_2 = 1$. The corresponding output satisfies

$$\begin{aligned}
|y(t)| &= \left|\int_{-\infty}^{\infty} g(t-\tau)u(\tau)\,d\tau\right| \\
&\leq \left(\int_{-\infty}^{\infty} g(t-\tau)^2 \, d\tau\right)^{1/2} \left(\int_{-\infty}^{\infty} u(\tau)^2 \, d\tau\right)^{1/2} \\
&= \|g\|_2
\end{aligned} \qquad (A.36)$$

where the second line follows from the Cauchy–Schwartz inequality. To complete the proof we need to show that there exists at least one signal $u(t) \in \mathcal{L}_2, ; \|u\|_2 = 1$ such that $\|y\|_\infty \geq \|g\|_2$. To this effect take

$$u(t) = \frac{g(-t)}{\|g\|_2} \qquad (A.37)$$

Obviously $\|u\|_2 = 1$ and

$$\|y\|_\infty \geq |y(0)| = \frac{1}{\|g\|_2} \int_{-\infty}^{\infty} |g(-\tau)|^2 \, d\tau = \|g\|_2 \tag{A.38}$$

A.3 DUALITY AND DUAL SPACES

A.3.1 The Dual Space

Consider a linear vector space X. A functional on X is a function $f: X \to \mathbb{R}$.

Definition A.17 *Consider a normed linear space X. Its dual space X^* is defined as the space of all bounded linear functionals on X, equipped with the norm $\|f\| = \sup_{\|x\| \leq 1} |f(x)|$.*

It can be shown [187] that X^* is a Banach space. In the sequel, given $x \in X$ and $r \in X^*$, $\langle x, r \rangle$ denotes the value of the linear functional r at the point x.

Example A.7: The Dual of ℓ^p, $1 \leq p < \infty$ Consider the space ℓ^p of all real sequences $h = \{h_i\}$ for which $\sum_{i=1}^{\infty} |h_i|^p < \infty$, equipped with the norm $\|h\|_p = (\sum_{i=1}^{\infty} |h_i|^p)^{1/p}$ (if $p = \infty$ then $\|h\|_\infty = \sup_k |h_k|$). Given $1 \leq p < \infty$, define the conjugate index q by the equation[1]

$$\frac{1}{p} + \frac{1}{q} = 1 \tag{A.39}$$

Next, we show that every linear functional f on ℓ^p has a unique representation of the form

$$f(h) = \sum_{i=1}^{\infty} h_i \eta_i \tag{A.40}$$

where $\eta = \{\eta_i\} \in \ell^q$. Assume first that $1 < p < \infty$. Consider the following sequences

$$e_i \in \ell^p \triangleq \{\overbrace{0, \ldots, 0}^{i-1}, 1, 0 \ldots\}$$

Given a bounded linear functional f on ℓ^p, define the sequence $\eta = \{\eta_i = f(e_i)\}$. As we show next $\eta \in \ell^q$. Consider the following element of ℓ^p:

$$h^N = \{h_i^N\}, \quad h_i^N = \begin{cases} |\eta_i|^{q/p} \, sign(\eta_i), & i \leq N \\ 0, & \text{otherwise} \end{cases} \tag{A.41}$$

---
[1] If $p = 1$ then $q = \infty$.

Then $\|h^N\| = (\sum_{i=1}^{\infty} |\eta_i|^q)^{1/p}$ and $f(h^N) = \sum_{i=1}^{N} |\eta_i|^{(q/p)+1} = \sum_{i=1}^{N} |\eta_i|^q$. Since f is bounded, it follows that

$$\sum_{i=1}^{N} |\eta_i|^q = |f(h^N)| \leq \|f\|_p \|h^N\|_p \Rightarrow \left(\sum_{i=1}^{N} |\eta_i|^q\right)^{1/q} \leq \|f\|_p \quad (A.42)$$

Thus $\sum_{i=1}^{N} |\eta_i|^q$ is bounded for all N, which implies that $\eta \in \ell^q$. Finally, if $p = 1$, define

$$h^N = \{h_i^N\}, \quad h_i^N = \begin{cases} 0, & i \neq N \\ \text{sign}(\eta_i), & i = N \end{cases} \quad (A.43)$$

Then $|\eta_N| = f(h^N) \leq \|f\| \|h^N\| = \|f\|$. Thus $\eta \in \ell^\infty$. Conversely, every element $\eta \in \ell^q$ defines a member of ℓ_p^* via (A.40). Moreover, it is easily shown (using Holder's inequality) that

$$\|f\| = \|\eta\|_q = \begin{cases} \left(\sum_{i=1}^{\infty} |\eta_i|^q\right)^{1/q}, & 1 < p < \infty \\ \sup_i |\eta_i|, & p = 1 \end{cases} \quad (A.44)$$

This establishes the fact that $\ell_p^* = \ell_q$. However, the dual of ℓ^∞ is not ℓ^1. This can easily be established by noting that while ℓ^1 is separable, ℓ^∞ is not. Indeed, $\ell^1 = c_o^*$, where c_o denotes the subspace of ℓ^∞ formed by sequences converging to zero.

A.3.2 Minimum Norm Problems

In this section we consider the solution of the following class of problems. Given a normed linear space X, a subspace $M \subset X$ and $x \in X$, find

$$\mu_o = \inf_{m \in M} \|x - m\| \quad (A.45)$$

and, if possible, $m_o \in M$ such that $\|x - m_o\| = \mu_o$. If X is a Hilbert space then the solution to this problem is given by the well known *projection theorem* ([187], Theorem 3.3.2): the unique solution m_o is the orthogonal projection of x onto M, that is, a vector m_o such that $x - m_o \perp M$. In the case of general Banach spaces, the solution to this problem is more complex and requires introducing several preliminary results.

Definition A.18 *Consider a normed vector space X, a subspace $M \subset X$, and a linear functional $f : M \to \mathbb{R}$. A linear functional $F : X \to \mathbb{R}$ is called an extension of f if $f(x) = F(x)$ for all $x \in M$.*

Definition A.19 *Consider a real normed space X and its dual X^*. An element $x \in X$ is aligned with $x^* \in X^*$ if $\langle x, x^* \rangle = \|x\| \|x^*\|$.*

Definition A.20 *Consider a subset $S \subset X$. The annihilator of S is defined as*

$$S^\perp = \{x^* \in X^* : \langle x, x* \rangle = 0 \text{ for all } x \in S\} \quad (A.46)$$

Theorem A.1 (Hahn–Banach, [187]) *Given a real normed space X and a bounded linear functional f defined in a subspace $M \subset X$, there exists a bounded linear functional $F: X \to \mathbb{R}$ which is an extension of f and such that*

$$\|F\| = \|f\|_M = \sup_{m \in M} \frac{|f(m)|}{\|m\|} \quad (A.47)$$

Using these tools we can now furnish a solution to the minimum norm problem (A.45).

Theorem A.2 [187] *Consider a real normed linear space X, its dual X^*, a subspace $M \subseteq X$, and its annihilator $M^\perp \subseteq X^*$. Let $x \in X$ be at a distance μ from M. Then*

$$\mu = \inf_{m \in M} \|x - m\| = \max_{x \in \overline{BM^\perp}} \langle x, x^* \rangle$$

where the maximum on the right is achieved for some $x_o^* \in M^\perp$. Moreover, if the infimum on the left is achieved for some $m_o \in M$ then $\langle x - m_o, x_o^* \rangle = \|x - m_o\| \cdot \|x_o^*\|$ (i.e., $x - m_o$ is aligned with x_o^*).

Proof. From the definition of μ, given any $\epsilon > 0$, there exists $m_\epsilon \in M$ such that

$$\|x - m_\epsilon\| \leq \mu + \epsilon \quad (A.48)$$

Thus for any $x^* \in \overline{BM^\perp}$ we have that

$$\langle x, x^* \rangle = \langle x - m_\epsilon, x^* \rangle \leq \|x^*\| \|x - m_\epsilon\|$$
$$\leq \mu + \epsilon \quad (A.49)$$

Since ϵ is arbitrary this shows that

$$\max_{x \in \overline{BM^\perp}} \langle x, x^* \rangle \leq \mu$$

To complete the first part of the proof we need to find $x_o^* \in \overline{BM^\perp}$ such that $\langle x, x_o^* \rangle = \mu$. Define the set $N = \{x + M\} \subset X$. Any element $n \in N$ can be written in a unique way as $n = \alpha x + m$, $m \in M$. Define the linear functional

$f: N \to \mathbb{R}$ as $f(n) = \alpha\mu$. Then

$$\|f\| = \sup \frac{|f(n)|}{\|n\|} = \sup \frac{|\alpha|\mu}{\|\alpha x + m\|}$$
$$= \frac{\mu}{\inf \|x + \frac{m}{\alpha}\|} = 1 \qquad (A.50)$$

From the Hahn–Banach theorem there exists a bounded linear functional $F: X \to \mathbb{R}$ with $\|F\| = 1$ and such that $F(n) = f(n)$ for all $n \in N$. Since $M \subset N$ and $f(m) = 0$ for $m \in M$ it follows that $F \in M^\perp$. Moreover, $F(x) = f(x) = \mu$. The proof follows now by selecting $x_o^* = F$. To complete the proof assume that there exist $m_o \in M$ such that $\|x - m_o\| = \mu$. Let $x_o^* \in \overline{BM^\perp}$ be such that $\langle x, x_o^* \rangle = \mu$. Then

$$\langle x - m_o, x_o^* \rangle = \langle x, x_o^* \rangle = \mu = \|x - m_o\| \cdot \|x_o\| \qquad (A.51)$$

□

Theorem A.2 has the following dual, which can be proved using similar techniques.

Theorem A.3 [187] *Consider a real normed linear space X, its dual X^*, a subspace $M \subseteq X$, and its annihilator $M^\perp \subseteq X^*$. Let $x^* \in X^*$ be at a distance μ from $M^\perp$. Then*

$$\mu = \min_{r^* \in M^\perp} \|x^* - r^*\| = \sup_{x \in \overline{BM}} \langle x, x^* \rangle$$

where the minimum is achieved for some $r_0^* \in M^\perp$. If the supremum on the right is achieved for some $x_0 \in M$, then $\langle x^* - r_0^*, x_0 \rangle = \|x^* - r_0^*\| \cdot \|x_0\|$ (i.e., $x^* - r_0^*$ is aligned with x_0).

A special case of the above theorems is the case when either $M^\perp$ (in Theorem A.2) or M (Theorem A.3) is finite-dimensional. In this case both problems admit a solution. This is precisely the case in ℓ^1 optimal control discussed in Chapter 8.

A.4 SINGULAR VALUES

A.4.1 Definition

Theorem A.4 *Any matrix $A \in \mathbb{C}^{n \times m}$ of rank r has the following singular value decomposition:*

$$A = U \begin{bmatrix} \Sigma_r & 0 \\ 0 & 0 \end{bmatrix} V^* \qquad (A.52)$$

with $\Sigma_r = \text{diag}[\sigma_1 \quad \cdots \quad \sigma_r]$, $\sigma_1 \geq \ldots \geq \sigma_r > 0$, and $U \in \mathbb{C}^{n \times n}$, $V \in \mathbb{C}^{m \times m}$ are unitary matrices.

Proof. Due to the fact that A^*A is symmetric and positive semidefinite, its eigenvalues are all real and nonnegative, and the corresponding set of eigenvectors span $\mathbb{R}^m$. Thus it is possible to select a set of orthonormal eigenvectors for the following Jordan decomposition:

$$V^*(A^*A)V = \begin{bmatrix} \Sigma_r^2 & 0 \\ 0 & 0 \end{bmatrix} \tag{A.53}$$

Let $\sigma_i = \sqrt{\lambda_i(A^*A)}$. Without loss of generality we assume that the eigenvalues of A^*A are ordered, that is, $\sigma_1 \geq \cdots \geq \sigma_r$. Define V_1 and V_2 as the first r and $(m-r)$ columns of V, respectively. Replacing $V = [V_1 \quad V_2]$ in the last equation yields:

$$V_1^* A^* A V_1 = \Sigma_r^2 \tag{A.54}$$

$$\implies \Sigma_r^{-1} V_1^* A^* A V_1 \Sigma_r^{-1} = I_r \tag{A.55}$$

$$V_2^* A^* A V_2 = 0 \tag{A.56}$$

where I_r is the r-dimensional identity matrix. Define also $U_1 \triangleq AV_1\Sigma_r^{-1} \in \mathbb{C}^{n \times r}$, thus $U_1^* U_1 = I_r$. The columns of U_1 can always be completed by an orthonormal set of vectors so that $\mathbb{C}^n$ is spanned; that is $U \triangleq [U_1 \quad U_2]$ with U unitary. Finally,

$$\begin{bmatrix} U_1^* \\ U_2^* \end{bmatrix} A \begin{bmatrix} V_1 & V_2 \end{bmatrix} = \begin{bmatrix} U_1^*AV_1 & U_1^*AV_2 \\ U_2^*AV_1 & U_2^*AV_2 \end{bmatrix} = \begin{bmatrix} \Sigma_r & 0 \\ 0 & 0 \end{bmatrix} \tag{A.57}$$

Here $U_1^*AV_1 = \Sigma_r$ and $AV_2 = 0$ were obtained from equations (A.54) and (A.56). By definition, the rows of U_2^* are orthogonal to the columns of $U_1 = AV_1\Sigma_r^{-1}$, therefore $U_2^* U_1 \Sigma_r = U_2^* AV_1 = 0$, which completes the proof. □

A.4.2 Properties and Applications

Bounds Define $\bar{\sigma}(A)$ and $\underline{\sigma}(A)$ as the maximum and minimum singular values of matrix A; then

$$\bar{\sigma}(A) = \frac{1}{\underline{\sigma}(A^{-1})}, \quad \text{if } A^{-1} \text{ exists} \tag{A.58}$$

$$\sigma_i(A) - \bar{\sigma}(B) \leq \sigma_i(A+B) \leq \sigma_i(A) + \bar{\sigma}(B) \tag{A.59}$$

Pseudoinverse The least mean square and minimum norm matrix problems are, respectively,

$$\min_x \|Ax - b\|_2, \ A \in \mathbb{C}^{n \times m}, \ n \geq m \quad (A.60)$$

$$\min \{\|x\|_2, \ Ax = b, \ A \in \mathbb{C}^{n \times m}, \ n \leq m\} \quad (A.61)$$

The solution to these problems can be obtained by means of the matrix *pseudoinverse* or generalized inverse A^+. The latter can efficiently be computed with the use of the singular value decomposition as follows:

$$A = U \begin{bmatrix} \Sigma_r & 0 \\ 0 & 0 \end{bmatrix} V^*, \qquad A^+ = V \begin{bmatrix} \Sigma_r^{-1} & 0 \\ 0 & 0 \end{bmatrix} U^* \quad (A.62)$$

Matrix Singularity To verify the singularity of a matrix, theoretically either the determinant, eigenvalues, or singular values may be used. Nevertheless, from the computational point of view, it is more sensible to test the singularity using the latter. This can be exemplified as follows. Consider the matrix

$$A = \begin{bmatrix} -1 & 1 & \cdots & \cdots & 1 \\ 0 & -1 & 1 & \cdots & \vdots \\ \vdots & \ddots & \ddots & \ddots & 1 \\ 0 & 0 & \cdots & 0 & -1 \end{bmatrix} \in \mathbb{R}^{n \times n} \quad (A.63)$$

The input vector $x = [1 \ \ 1/2 \ \ \cdots \ \ 1/2^{n-1}]^T$ yields $Ax = -2^{1-n}[1 \ \cdots \ 1]^T$ that tends to the zero vector as $n \to \infty$. Hence for a large enough n this matrix is *computationally* singular, although it is *theoretically* not.

The eigenvalues are $\lambda_1 = \cdots = \lambda_n = -1$ and the determinant is $\det(A) = (-1)^n$, none of which provide a clear measure of the computational singularity of this matrix. In this last case, the reason is that $\det(A) = \prod_1^n \lambda_i$, which therefore depends on all eigenvalues, not only on the smallest one. In the former case, the measure provided by the smallest magnitude eigenvalue only quantifies the effect of the matrix on the eigenvectors, but not on any arbitrary vector in the input space.

On the other hand, by considering the minimum value $\|Ax\|_2$ can take over all possible input vectors x (normalized by $\|x\|_2 \neq 0$ to avoid trivial solutions), the measure of singularity can be computed as

$$\inf_{\|x\|_2 \neq 0} \frac{\|Ax\|_2}{\|x\|_2} = \inf_{\|x\|_2 \neq 0} \sqrt{\frac{x^T A^T A x}{x^T x}}$$

$$= \sqrt{\frac{(x^T x)\underline{\lambda}(A^T A)}{x^T x}} = \underline{\sigma}(A) \quad (A.64)$$

which is directly the minimum singular value $\underline{\sigma}(A) \approx 2^{1-n}$.

Condition Number Consider the linear matrix equation $Ax = b$. The maximum relative error δb due to an error δx, can be found as follows:

$$\sup_{x,\delta x} \frac{\|\delta b\|_2}{\|b\|_2} = \sup_{x,\delta x} \frac{\|A\delta x\|_2}{\|Ax\|_2} \quad \text{with} \quad \|\delta x\|_2 = \|x\|_2 = 1$$

$$= \frac{\sup_{\delta x} \|A\delta x\|_2}{\inf_x \|Ax\|_2} \quad \text{with} \quad \|\delta x\|_2 = \|x\|_2 = 1$$

$$= \frac{\bar{\sigma}(A)}{\underline{\sigma}(A)} \triangleq \kappa(A) \tag{A.65}$$

where δx and x are independent vectors. The problem is equivalent to the one of finding the maximum relative error δx of the solution for a given error δb in the data vector b. The *condition number* $\kappa(\cdot)$ provides this measure and quantifies the relation between the maximum and minimum "gains" of A (according to different input vector "directions"). For example, orthogonal (rotation) or unitary matrices are "spherical" because $U^*U = I \implies \sigma_1 = \cdots = \sigma_n = 1$; hence in this case $\kappa(U) = 1$. On the other hand, for singular matrices $\kappa \longrightarrow \infty$.

APPENDIX B

SYSTEM COMPUTATIONS

In the following sections we present the state-space representations of the most usual interconnections between linear systems. These representations are used to manipulate systems in a computer code. There is standard software that implements these interconnections in state-space form, as in the case of the μ-Toolbox and the Robust Toolbox of Matlab.

B.1 SERIES

Consider the following linear systems:

$$G_1(s) \equiv \left[\begin{array}{c|c} A_1 & B_1 \\ \hline C_1 & D_1 \end{array} \right] \tag{B.1}$$

$$G_2(s) \equiv \left[\begin{array}{c|c} A_2 & B_2 \\ \hline C_2 & D_2 \end{array} \right] \tag{B.2}$$

with states, inputs, and outputs (x_1, u_1, y_1) and (x_2, u_2, y_2), respectively. The series connection can be seen in Figure B.1, where $u = u_1$, $y = y_2$, and $y_1 = u_2$.

The objective is to compute a state-space realization (not necessarily minimal) of the product system. To this end, consider the state of $G_2(s)G_1(s)$ as $x = \begin{bmatrix} x_1^T & x_2^T \end{bmatrix}^T$, and proceed as follows:

$$\dot{x}_1(t) = A_1 x_1(t) + B_1 u_1(t) \tag{B.3}$$

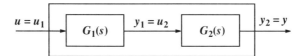

Figure B.1. Series connection.

$$\begin{aligned}\dot{x}_2(t) &= A_2 x_2(t) + B_2 u_2(t) = A_2 x_2(t) + B_2 y_1(t)\\ &= A_2 x_2(t) + B_2 [C_1 x_1(t) + D_1 u_1(t)] \end{aligned} \quad (B.4)$$

$$\begin{aligned} y(t) &= y_2(t) = C_2 x_2(t) + D_2 u_2(t)\\ &= C_2 x_2(t) + D_2 [C_1 x_1(t) + D_1 u_1(t)] \end{aligned} \quad (B.5)$$

Reordering the above equations, we have that the state-space realization for $G_2(s)G_1(s)$ is the following:

$$\left[\begin{array}{cc|c} A_1 & 0 & B_1 \\ B_2 C_1 & A_2 & B_2 D_1 \\ \hline D_2 C_1 & C_2 & D_2 D_1 \end{array}\right] \quad (B.6)$$

B.2 CHANGE OF VARIABLES

Consider the following realization of a linear system:

$$G(s) \equiv \left[\begin{array}{c|c} A & B \\ \hline C & D \end{array}\right] \quad (B.7)$$

with states, inputs, and outputs $x(t)$, $u(t)$, and $y(t)$, respectively. If we change the state, input, and output as follows—$\tilde{x}(t) = Tx(t)$, $\tilde{u}(t) = Pu(t)$, $\tilde{y}(t) = Ry(t)$—the new state-space realization is (Problem 1)

$$G(s) \equiv \left[\begin{array}{c|c} TAT^{-1} & TBP^{-1} \\ \hline RCT^{-1} & RDP^{-1} \end{array}\right] \quad (B.8)$$

where T and P are nonsingular. Since only the definition of the states, inputs, and outputs have been changed, it seems reasonable to expect that the poles remain unchanged. This is clear from the fact that the new dynamic matrix TAT^{-1} is similar to A.

B.3 STATE FEEDBACK

If we apply to the system $G(s)$ in (B.7), the feedback $u(t) = \tilde{u}(t) + Fx(t)$, the new system with input $\tilde{u}(t)$ is (Problem 1)

$$G_f(s) \equiv \left[\begin{array}{c|c} A + BF & B \\ \hline C + DF & D \end{array}\right] \tag{B.9}$$

B.4 STATE ESTIMATION

The state of system $G(s)$ can be estimated by observation of its input $u(t)$ and output $y(t)$ as follows:

$$\dot{\hat{x}}(t) = A\hat{x}(t) + Bu(t) + H\left[y(t) - C\hat{x}(t) - Du(t)\right] \tag{B.10}$$

where $G(s)$ has the state space representation (B.7). The realization of the observer with inputs $\begin{bmatrix} u^T & y^T \end{bmatrix}^T$ and output $\hat{x}$ is (Problem 1)

$$G_o(s) \equiv \left[\begin{array}{c|cc} A - HC & B - HD & H \\ \hline I & 0 & 0 \end{array}\right] \tag{B.11}$$

B.5 TRANSPOSE SYSTEM

The realization of the transpose of system $G(s)$ is

$$G^T(s) \equiv \left[\begin{array}{c|c} A^T & C^T \\ \hline B^T & D^T \end{array}\right] \tag{B.12}$$

This new system exchanges inputs and outputs.

B.6 CONJUGATE SYSTEM

The realization of the conjugate $\widetilde{G}(s) \triangleq G^T(-s)$ is the following:

$$\widetilde{G}(s) \equiv \left[\begin{array}{c|c} -A^T & -C^T \\ \hline B^T & D^T \end{array}\right] \tag{B.13}$$

In this case, the inputs and outputs are exchanged and the dynamics is the mirror image of $G(s)$ with respect to $s = j\omega$.

468 SYSTEM COMPUTATIONS

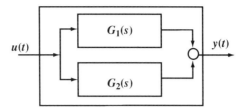

Figure B.2. Addition.

B.7 ADDITION

The addition of $G_1(s)$ and $G_2(s)$ can be seen in Figure B.2. As in Section B.1, the state is the combination of the states of both $G_1(s)$ and $G_2(s)$, which have been defined in (B.1) and (B.2). Therefore

$$\dot{x}_1(t) = A_1 x_1(t) + B_1 u(t) \tag{B.14}$$

$$\dot{x}_2(t) = A_2 x_2(t) + B_2 u(t) \tag{B.15}$$

$$y(t) = C_1 x_1(t) + C_2 x_2(t) + [D_1 + D_2] u(t) \tag{B.16}$$

From the above, the following realization of $G_1(s) + G_2(s)$ is obtained:

$$G_1(s) + G_2(s) \equiv \left[\begin{array}{cc|c} A_1 & 0 & B_1 \\ 0 & A_2 & B_2 \\ \hline C_1 & C_2 & D_1 + D_2 \end{array} \right] \tag{B.17}$$

B.8 OUTPUT FEEDBACK

The connection can be seen in Figure B.3, where the realizations of $G_1(s)$ and $G_2(s)$ are given in (B.1) and (B.2). It can be proved (Problem 2) that

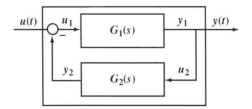

Figure B.3. Output feedback.

the realization of the closed-loop system $G_c(s)$ with input u and output y is:

$$A_c = \begin{bmatrix} A_1 - B_1 D_2 \Phi C_1 & B_1(-I + D_2 \Phi D_1) C_2 \\ B_2 \Phi C_1 & A_2 - B_2 \Phi D_1 C_2 \end{bmatrix} \quad \text{(B.18)}$$

$$B_c = \begin{bmatrix} B_1 - B_1 D_2 \Phi D_1 \\ B_2 \Phi D_1 \end{bmatrix} \quad \text{(B.19)}$$

$$C_c = \begin{bmatrix} \Phi C_1 & -\Phi D_1 C_2 \end{bmatrix} \quad \text{(B.20)}$$

$$D_c = \Phi D_1 \quad \text{(B.21)}$$

where we have defined $\Phi \triangleq [I + D_1 D_2]^{-1}$. For this to make sense, Φ should exist, which is equivalent to well posedness of the loop (see Section 2.2). In practice, well posedness is guaranteed if either one of the systems is strictly proper, that is, $D_1 = 0$ or $D_2 = 0$.

B.9 INVERSE

The right (left) inverse $G_r^{-1}(s)$ ($G_\ell^{-1}(s)$) is defined such that the serial connection with the input (output) of the system $G(s)$ produces the identity matrix. If the system $G(s)$ is square and invertible, the right and left inverses coincide. The realization of the inverse, in both cases, is

$$G^{-1}(s) \equiv \left[\begin{array}{c|c} A - BD^{-1}C & -BD^{-1} \\ \hline D^{-1}C & D^{-1} \end{array} \right] \quad \text{(B.22)}$$

where D^{-1} denotes the right or left inverse, respectively. In all cases, D must have full rank for this to make sense. When this is not the case, D^{-1} should be replaced by its pseudoinverse $D^\dagger$.

Without loss of generality, we can verify the square and right inverse cases as follows:

$$G(s)G^{-1}(s) \equiv \left[\begin{array}{cc|c} A & BD^{-1}C & BD^{-1} \\ 0 & A - BD^{-1}C & -BD^{-1} \\ \hline C & C & I \end{array} \right] \quad \text{(B.23)}$$

The change of variables $\tilde{x} = Tx$, with

$$T = \begin{bmatrix} I & I \\ 0 & I \end{bmatrix} \quad \text{(B.24)}$$

yields

$$G(s)G^{-1}(s) \equiv \left[\begin{array}{cc|c} A & 0 & 0 \\ 0 & A - BD^{-1}C & -BD^{-1} \\ \hline C & 0 & I \end{array} \right] = I \qquad (\text{B.25})$$

by eliminating the unobservable and uncontrollable states.

B.10 LINEAR FRACTIONAL TRANSFORMATIONS

Definition B.1 *A function $\mathcal{F}: Z \longrightarrow S$ of the form $S = (\alpha + \beta Z)(\gamma + \delta Z)^{-1}$ is called a linear fractional transformation.*

The following is an example of a linear fractional transformation (LFT), which is used in digital control and in many proofs of $\mathcal{H}_\infty$ control, model reduction, and robust identification.

Example B.1 *The bilinear transform between the Laplace variable $s \in \mathbb{C}$ and the z-transform variable $z \in \mathbb{C}$ is a special case of LFT. It can be defined as follows:*

$$s = \frac{2(z-1)}{T(z+1)} \qquad (\text{B.26})$$

where T is a parameter of the transformation.

It can be used to transform continuous LTI systems described by their Laplace transform, to discrete time LSI systems in the z variable [14, 119]. In this case, T represents the time interval between samples of the continuous system. It is used to compute a digital controller $D(z)$ from a previously designed continuous time controller $K(s)$. The advantage is that the stability regions of the continuous and discrete systems are mapped one to one, that is, the left-half plane is mapped to the unit disk (see Figure B.4).

An important application of these types of transformations is in robust control theory. A general structure for analysis and synthesis is shown in Figure B.5. This structure includes the nominal system, model uncertainty, the controller, exogenous signals (disturbances, measurement noise, references), errors to be minimized, and performance/robustness design weights. In this case, the parameters $\alpha, \beta, \gamma, \delta$ and the variables S and Z belong to the set of transfer matrices.

The usual interconnections can be seen in Figures B.6 and B.7. In the first case, the nominal model and the controller are separated, which will be used for controller *synthesis*. In the second case, the closed-loop nominal system is

LINEAR FRACTIONAL TRANSFORMATIONS

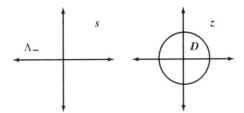

Figure B.4. Bilinear transformation, left-half s plane Λ_- is mapped to the unit disk $\mathcal{D}$.

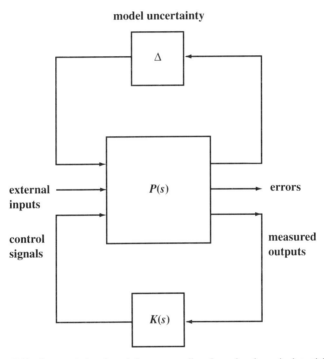

Figure B.5. General structure interconnection for robust control problems.

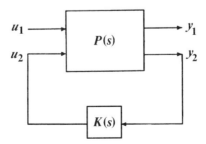

Figure B.6. Lower fractional interconnection $F_\ell\,[P(s), K(s)]$.

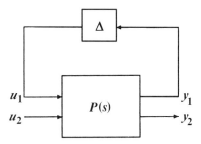

Figure B.7. Upper fractional interconnection $F_u[P(s),\Delta]$.

separated from the uncertainty. This is used for robustness and performance *analysis*. The notation is as follows:

$$y_1 = F_\ell[P(s), K(s)]\, u_1 \tag{B.27}$$

$$y_2 = F_u[P(s), \Delta]\, u_2 \tag{B.28}$$

$$F_\ell[P(s), K(s)] = P_{11} + P_{12}K\,[I - P_{22}K]^{-1} P_{21} \tag{B.29}$$

$$F_u[P(s), \Delta] = P_{22} + P_{21}\Delta\,[I - P_{11}\Delta]^{-1} P_{12} \tag{B.30}$$

These equations correspond to the interconnections of Figures B.6 and B.7. Note that $F_\ell(\cdot, \cdot)$ and $F_u(\cdot, \cdot)$ are special cases of the general LFT defined above. Next we illustrate with an example how to recast a control problem into a LFT form.

Example B.2 *Consider the control system of Figure B.8. There, a family of models is described by means of a nominal model $G_0(s)$, and multiplicative*

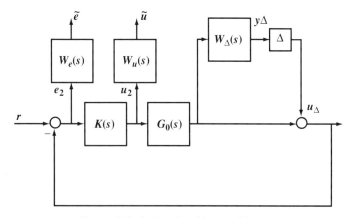

Figure B.8. Robust tracking problem.

dynamic uncertainty Δ weighted by $W_\Delta(s)$. A reference set of signals $r(s)$ needs to be followed with minimal weighted tracking error $W_e(s)e(s)$ and control signal $W_u(s)u(s)$.

To this end we put the problem in the standard form of Figure B.5. The controller $K(s)$, uncertainty Δ, nominal model $P(s)$ (including all weights), external reference $r(s)$, and signals to be minimized $\tilde{e}(s)$ and $\tilde{u}(s)$ are considered.

To compute $P(s)$, by linearity, we obtain all transfer functions between each input and output, considering all other inputs equal to zero. The resulting nominal system in this case is

$$P(s) = \begin{bmatrix} \begin{bmatrix} 0 & 0 \\ -W_e & W_e \\ 0 & 0 \\ [-I & I] \end{bmatrix} & \begin{bmatrix} W_\Delta G_0 \\ -W_e G_0 \\ W_u \\ [-G_0] \end{bmatrix} \end{bmatrix}$$

$$= \begin{bmatrix} P_{11}(s) & P_{12}(s) \\ P_{21}(s) & P_{22}(s) \end{bmatrix} \quad (B.31)$$

The above matrix has been divided into four blocks, according to the dimensions of the inputs

$$u_1 = \begin{bmatrix} u_\Delta \\ r \end{bmatrix} \quad \text{and} \quad u_2,$$

and the outputs

$$e_1 = \begin{bmatrix} y_\Delta \\ \tilde{e} \\ \tilde{u} \end{bmatrix} \quad \text{and} \quad e_2$$

Once the controller $K(s)$ is, designed, robust analysis can be performed by connecting it with the nominal $P(s)$:

$$F_\ell[P(s), K(s)] = P_{11}(s) + P_{12}(s)K(s)[I - P_{22}(s)K(s)]^{-1} P_{21}(s)$$

$$= T(s) = \begin{bmatrix} T_{11}(s) & T_{12}(s) \\ T_{21}(s) & T_{22}(s) \end{bmatrix} \quad (B.32)$$

The block separation in the analysis matrix $T(s)$ is made according to the inputs u_Δ and r and the outputs y_Δ and $\begin{bmatrix} \tilde{e} \\ \tilde{u} \end{bmatrix}$. Therefore, as has been explained in Chapters 2 and 3, we can assess nominal performance by computing $\|T_{22}(s)\|_\infty$ and robust stability by computing $\|T_{11}(s)\|_\infty$.

An important property of these types of interconnections is that the connection of two LFTs is another LFT. This is very useful when we need to parametrize a family of controllers, as in the next example.

Example B.3 *It can be shown (see Chapter 3) that for a given nominal model $P(s)$, all possible stabilizing controllers can be parametrized in terms of a free parameter $Q(s)$, which should be stable and proper. This set of stabilizing controllers can be represented by a LFT connection between a nominal closed-loop system $J(s)$ and the parameter $Q(s)$ (see Figure B.9), as follows:*

$$F_\ell [P(s), K(s)] = F_\ell \{P(s), F_\ell [J(s), Q(s)]\}$$
$$= F_\ell [T(s), Q(s)] \quad (B.33)$$

where $T(s)$ is computed by connecting $P(s)$ and $J(s)$. State-space realizations for J and T are given in (3.39) and (3.41) respectively.

To obtain the state-space realization of the LFTs in equations (B.29) and (B.30), we could use the formulas for series, addition, and output feedback of transfer functions. In that case, most probably, the order of the realization will be unnecessarily high. Instead, we will obtain the state-space realization directly from the realizations of each of the transfer matrices of the LFT. In the sequel we obtain a realization of the LFT in (B.29). It is left to the reader as an exercise to compute the realization of the LFT in (B.30) (Problem 3).

Consider the following realizations of $P(s)$ and $K(s)$:

$$P(s) \equiv \left[\begin{array}{c|cc} A_p & B_1 & B_2 \\ \hline C_1 & D_{11} & D_{12} \\ C_2 & D_{21} & D_{22} \end{array} \right] \quad (B.34)$$

$$K(s) \equiv \left[\begin{array}{c|c} A_k & B_k \\ \hline C_k & D_k \end{array} \right] \quad (B.35)$$

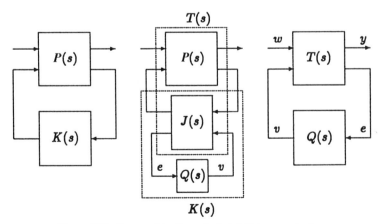

Figure B.9. Parametrization of all stabilizing controllers.

For the lower fractional transformation in Figure B.6, we define the state, input, and output of $P(s)$ as x_p, $\begin{bmatrix} u_1^T & u_2^T \end{bmatrix}^T$, and $\begin{bmatrix} y_1^T & y_2^T \end{bmatrix}^T$, respectively. The state, input, and output of $K(s)$ are x_k, y_2, and u_2, respectively. The state-space equations for the interconnection are the following (for simplicity we drop the dependency with t):

$$P(s) \begin{cases} \dot{x}_p = A_p x_p + B_1 u_1 + B_2 u_2 \\ y_1 = C_1 x_p + D_{11} u_1 + D_{12} u_2 \\ y_2 = C_2 x_p + D_{21} u_1 + D_{22} u_2 \end{cases} \quad (B.36)$$

$$K(s) \begin{cases} \dot{x}_k = A_k x_k + B_k y_2 \\ u_2 = C_k x_k + D_k y_2 \end{cases} \quad (B.37)$$

Replacing the expression for u_2 in y_2, we obtain

$$y_2 = (I - D_{22} D_k)^{-1} (C_2 x_p + D_{22} C_k x_k + D_{21} u_1) \quad (B.38)$$

Define $\Phi_\ell \triangleq (I - D_{22} D_k)^{-1}$ and $\Phi_r \triangleq (I - D_k D_{22})^{-1}$, which exist if the feedback loop is well posed (why?). Now replacing the above equation in the expression for u_2, we obtain

$$u_2 = \Phi_r (D_k C_2 x_p + C_k x_k + D_k D_{21} u_1) \quad (B.39)$$

where we have used the fact that $D_k \Phi_\ell = \Phi_r D_k$. Finally, we obtain the state-space equations for $F_\ell[P(s), K(s)]$, replacing both y_2 and u_2 in equations (B.36) and (B.37):

$$\dot{x}_p = (A_p + B_2 \Phi_r D_k C_2) x_p + (B_2 \Phi_r C_k) x_k + (B_1 + B_2 \Phi_r D_k D_{21}) u_1 \quad (B.40)$$

$$\dot{x}_k = (B_k \Phi_\ell C_2) x_p + (A_k + B_k \Phi_\ell D_{22} C_k) x_k + (B_k \Phi_\ell D_{21}) u_1 \quad (B.41)$$

$$y_1 = (C_1 + D_{12} \Phi_r D_k C_2) x_p + (D_{12} \Phi_r C_k) x_k \\ + (D_{11} + D_{12} \Phi_r D_k D_{21}) u_1 \quad (B.42)$$

Here we used the facts that $\Phi_\ell D_{22} = D_{22} \Phi_r$, $\Phi_\ell = I + D_{22} \Phi_r D_k$, and $\Phi_r = I + D_k \Phi_\ell D_{22}$.

B.11 NORM COMPUTATIONS

B.11.1 $\mathcal{H}_2$ Norm Computation

Define the controllability and observability Gramians as

$$W_c \triangleq \int_0^\infty e^{At} B B^T e^{A^T t} dt \in \mathbb{R}^{n \times n} \quad (B.43)$$

$$W_o \triangleq \int_0^\infty e^{A^T t} C^T C e^{At} dt \in \mathbb{R}^{n \times n} \quad (B.44)$$

476 SYSTEM COMPUTATIONS

It is simple to show that both are symmetric and positive semidefinite. In addition, they satisfy the following Lyapunov equations (Problem 5):

$$AW_c + W_c A^T + BB^T = 0 \tag{B.45}$$

$$A^T W_o + W_o A + C^T C = 0 \tag{B.46}$$

which can be proved by using the properties of the matrix exponential.

Applying the definitions above to the expression for $\|G(s)\|_2$, it is easy to see that (Problem 6)

$$\|G(s)\|_2^2 = \text{trace}\left[CW_c C^T\right] = \text{trace}\left[B^T W_o B\right] \tag{B.47}$$

B.11.2 $\mathcal{H}_\infty$ Norm Computation

Consider the stable, proper dynamical system in $\mathcal{L}_\infty$:

$$G(s) \equiv \left[\begin{array}{c|c} A & B \\ \hline C & D \end{array}\right] \tag{B.48}$$

and define its corresponding Hamiltonian,

$$H = \begin{bmatrix} A + BR^{-1}D^T C & BR^{-1}B^T \\ -C^T(I + DR^{-1}D^T)C & -(A + BR^{-1}D^T C)^T \end{bmatrix} \tag{B.49}$$

where $R \triangleq I - D^T D$ then the following result holds:

Lemma B.1 *The Hamiltonian H has no imaginary eigenvalues if and only if* $\|G(s)\|_\infty < 1$.

Proof. This result was obtained as an intermediate step in the proof of the Bounded Real Lemma (Lemma 6.1) □

Using the above result, it is possible to compute iteratively $\|G(s)\|_\infty$, scaling $G(s)$ by a parameter $\gamma \geq 0$, that is, $G_\gamma(s) = \gamma G(s)$, and calculating the Hamiltonian H_γ from $G_\gamma(s)$. If it has no imaginary eigenvalues then $\|G_\gamma(s)\|_\infty < 1$. Otherwise $\|G_\gamma(s)\|_\infty \geq 1$. Using a bisection method, this procedure can be iterated until the value of γ is such that $\|G_\gamma(s)\|_\infty \to 1$, which is equivalent to $\|G(s)\|_\infty \to 1/\gamma$, with any required level of precision. This procedure can be found in greater detail in [57, 104, 247].

B.11.3 ℓ^1 Norm Computation

Consider a SISO stable discrete time system

$$G = \left[\begin{array}{c|c} A & B \\ \hline C & D \end{array}\right]$$

Let $\rho(A) = \rho_o < 1$ and denote by g_i the Markov parameters of G. It follows that

$$G(z) = \sum_{i=0}^{\infty} g_i z^{-i} \tag{B.50}$$

is analytic outside any open disk $|z| < \delta < 1$ with $\delta > \rho_o$. Using Cauchy's residue formula to compute g_i yields

$$g_i = \frac{1}{2\pi} \oint_{|z|=\delta} G(z) z^{i-1} dz \tag{B.51}$$

It follows that

$$|g_i| \leq \frac{1}{2\pi} \|G\|_{\infty,\delta} \delta^i \int_0^{2\pi} d\theta = \|G\|_{\infty,\delta} \delta^i \tag{B.52}$$

where we have defined $\|G\|_{\infty,\delta} \triangleq \sup_{|z|=\delta} |G(z)|$. Thus $\|(I - P_N)G\|_1 = \sum_{i=N}^{\infty} |g_i| \leq \|G\|_{\infty,\delta} \delta^N (1-\delta)^{-1}$. It follows that given $\epsilon > 0$, there exists N such that $\|(I - P_N)G\|_1 \leq \epsilon$. Thus $\|G\|_1$ can be approximated arbitrarily close by the finite series $\sum_{i=0}^{N-1} |g_i|$. However, it turns out that in most cases the estimate of N obtained in this way is rather conservative, leading to unnecessary computations. In this section we indicate an alternative, more efficient, procedure for computing $\|G\|_1$. To this effect we need to introduce some preliminary results relating the ℓ^1 norm of a stable system with its Hankel singular values [56]. For simplicity, all the derivations are carried out in the SISO case (the MIMO version can be found in [87]).

Lemma B.2 *Consider a finite-dimensional SISO stable discrete time system G with McMillan degree n. Then:*

1. $\|G\|_{\mathcal{H}_\infty} \geq \sigma_1$
2. $\|G\|_1 \leq |d| + 2\sum_{i=1}^n \sigma_i$
3. $\|G\|_1 \leq (2n+1)\|G\|_{\mathcal{H}_\infty}$

where σ_i denote the Hankel singular values of G, in decreasing order.

Proof. The proof of 1 follows from Nehari's theorem (see Chapter 6). To prove 2 and 3 start by writing $\|G\|_1$ explicitly as

$$\begin{aligned} \|G\|_1 &= |d| + \sum_{i=0}^{\infty} |cA^i b| \\ &= |d| + \sum_{i=0}^{\infty} |cA^{2i} b| + \sum_{i=0}^{\infty} |cA^{2i+1} b| \end{aligned} \tag{B.53}$$

From the Cauchy–Schwartz inequality we have that

$$\sum_{i=0}^{\infty} |cA^{2i}b| \leq \sum_{i=0}^{\infty} \|(A^T)^i c^T\|_2 \|A^i b\|_2$$
$$\leq \left(\sum_{i=0}^{\infty} \|(A^T)^i c^T\|_2^2\right)^{1/2} \left(\sum_{i=0}^{\infty} \|A^i b\|_2^2\right)^{1/2} \quad (B.54)$$
$$= [\text{trace}(Q)]^{1/2} [\text{trace}(P)]^{1/2}$$

where P and Q denote the controllability and observability Gramians of G, respectively. Proceeding in the same way to obtain an upper bound of the second sum in (B.53) yields

$$\|G\|_1 \leq |d| + 2[\text{trace}(Q)]^{1/2}[\text{trace}(P)]^{1/2} \quad (B.55)$$

The proof of 2 follows now by selecting a balanced realization of G so that $\text{trace}(P) = \text{trace}(Q) = \sum_{i=1}^{n} \sigma_i$. Finally, part 3 follows from combining 1 and 2. $\square$

Next, we exploit this result to compute $\|G\|_1$ with arbitrary precision [18]. Note that for any N, $\|G\|_1 = \|P_N G\|_1 + \|(I - P_N)G\|_1$ and that $\|(I - P_N)G\|_1 = \|S_L^{(N-1)}(I - P_N)G\|_1$, where S_L denotes the shift-left operator. Thus, in order to bound the error entailed in approximating $\|G\|_1$ by $\|P_N G\|_1$, we need to bound $\|S_L^{(N-1)}(I - P_N)G\|_1$. Note that $G_{tail} \triangleq S_L^{(N-1)}(I - P_N)G$ has the following state-space realization:

$$G_{tail} = \left[\begin{array}{c|c} A & A^N B \\ \hline C & 0 \end{array}\right] \quad (B.56)$$

From Lemma B.2 we have that

$$\sigma_1 \leq \|G_{tail}\|_1 \leq 2\sum_{i=1}^{n} \sigma_i \quad (B.57)$$

where σ_i denote the Hankel singular values of G_{tail}. Hence in order to compute $\|G\|_1$ with an error $\leq \epsilon$, we need to find N such that $2\sum_{i=1}^{n} \sigma_i - \sigma_1 \leq \epsilon$. Let P_{tail} and Q_{tail} denote the controllability and observability Gramians of G_{tail}. Clearly $Q_{tail} = Q$. It can easily be shown (by considering its explicit expression as an infinite sum) that $P_{tail} = A^N P (A^T)^N$. Thus $\sigma_i = \sqrt{\lambda_i [QA^N P(A^T)^N]}$ and N is given by

$$N = \min\left\{N : 2\sum_{i=1}^{n} \sigma_i - \sigma_1 \leq \epsilon\right\} \quad (B.58)$$

B.12 PROBLEMS

1. Prove the state-space realizations for the change of variables, state feedback, and state estimation.

2. Prove that the state-space realization for the case of output feedback coincides with equations (B.18)–(B.21).

3. Obtain the state-space realization for the interconnection $F_u[P(s), \Delta]$, where Δ is a constant complex matrix.

4. Verify the formulas for series, addition, and output feedback as special cases of a LFT.

5. Verify that the Gramians W_c and W_o are symmetric, positive semidefinite and satisfy the Lyapunov equations (B.45) and (B.46), respectively.

6. Verify the computation of the 2-norm using the Gramians, as in (B.47).

APPENDIX C

RICCATI EQUATIONS

Recall that the Riccati equation played a key role in solving the $\mathcal{H}_2$ and $\mathcal{H}_\infty$ problems in Chapters 5 and 6. In this appendix we briefly examine some of the properties of the solutions to this equation.

Given $A, R, Q \in \mathbb{R}^{n \times n}$, with R and Q symmetric, an algebraic Riccati equation (ARE) is an equation of the form

$$A^T X + XA + XRX + Q = 0 \tag{C.1}$$

To this equation we can associate a Hamiltonian matrix $H \in \mathbb{R}^{2n \times 2n}$ of the form

$$H = \begin{bmatrix} A & R \\ -Q & -A^T \end{bmatrix} \tag{C.2}$$

By using the similarity transformation

$$T = \begin{bmatrix} 0 & -I \\ I & 0 \end{bmatrix} \tag{C.3}$$

it can be shown that H and $-H^T$ are similar. It follows that the eigenvalues of H are symmetric with respect to the imaginary axis. Assume that H has no eigenvalues on the $j\omega$ axis. Then it must have n eigenvalues in $\text{Re}(s) < 0$ and the remaining n in $\text{Re}(s) > 0$. Denote by $\mathcal{X}_-(H)$ and $\mathcal{X}_+(H)$ the n-dimensional invariant subspaces corresponding to the stable and antistable eigenvalues of H, respectively. Let

$$\mathcal{X} \triangleq \begin{bmatrix} X_1 \\ X_2 \end{bmatrix}, \quad X_1, X_2 \in \mathbb{R}^{n \times n} \tag{C.4}$$

be a matrix whose columns span $\mathcal{X}_-(H)$. A solution to the Riccati equation (C.1) can be obtained in terms of $\mathcal{X}$ as follows.

Lemma C.1 *Assume that H does not have eigenvalues on the $j\omega$ axis. If X_1 is invertible then $X = X_2 X_1^{-1}$ solves (C.1). Moreover, X is symmetric and stabilizing, that is, $A + RX$ is stable.*

Proof. Since $\mathcal{X}$ spans the stable invariant subspace of H there exists a stable matrix Λ such that
$$H\mathcal{X} = \mathcal{X}\Lambda \tag{C.5}$$
Postmultiplying by X_1^{-1} and using the explicit expressions for H and $\mathcal{X}$ yields
$$\begin{bmatrix} A & R \\ -Q & -A^T \end{bmatrix} \begin{bmatrix} I \\ X \end{bmatrix} = \begin{bmatrix} I \\ X \end{bmatrix} X_1 \Lambda X_1^{-1} \tag{C.6}$$
Thus $A + RX = X_1 \Lambda X_1^{-1}$. Stability of $A + RX$ follows now from the fact that Λ spans the stable invariant subspace of H. Premultiplying (C.6) by $[-X \quad I]$ yields
$$-XA - Q - XRX - A^T X = 0 \tag{C.7}$$
thus X solves (C.1). To show that X is symmetric, premultiply (C.5) by $\mathcal{X}^* T$ to obtain
$$\mathcal{X}^* T H \mathcal{X} = \mathcal{X}^* T \mathcal{X} \Lambda \tag{C.8}$$
Since $TH = -H^T T = H^T T^T$ it follows that the left-hand side (and hence also the right-hand side) of (C.8) is Hermitian. Thus
$$\mathcal{X}^* T \mathcal{X} \Lambda = \Lambda^* \mathcal{X}^* T^* \mathcal{X} = -\Lambda^* \mathcal{X}^* T \mathcal{X} \tag{C.9}$$
or, equivalently,
$$\mathcal{X}^* T \mathcal{X} \Lambda + \Lambda^* \mathcal{X}^* T \mathcal{X} = 0 \tag{C.10}$$
The (unique) solution to this Lyapunov equation is $\mathcal{X}^* T \mathcal{X} = 0$, which implies that $X_1^* X_2$ is Hermitian. Rewriting X as $X = (X_1^{-1})^* X_1^* X_2 X_1^{-1}$ shows that X is also Hermitian. Finally, from the fact that the subspace $\mathcal{X}_-(H)$ has conjugate symmetry it can be shown that X is real. $\square$

From Lemma C.1 it follows that when H does not have eigenvalues on the imaginary axis and X_1 is invertible then $X = X_2 X_1^{-1}$ is uniquely determined by H. We will denote the function $H \to X$ by $X = Ric(H)$. The domain of $Ric(\cdot)$ is the set of Hamiltonian matrices with the two properties mentioned above. Thus we will use the notation $H \in dom(Ric)$ to indicate Hamiltonian matrices with no eigenvalues on the $j\omega$ axis and where X_1 is invertible (equivalently, where the subspaces $\mathcal{X}_-(H)$ and $\mathrm{Span} \begin{bmatrix} 0 \\ I \end{bmatrix}$ are complementary). In the sequel we provide some sufficient conditions for $H \in dom(Ric)$.

Lemma C.2 *Assume that H has no eigenvalues on the $j\omega$ axis and R is either positive or negative semidefinite. Then $H \in dom(Ric)$ if and only if the pair (A, R) is stabilizable.*

Proof. ($\Leftarrow$) Let $\mathcal{X} = \begin{bmatrix} X_1 \\ X_2 \end{bmatrix}$. To show that $H \in dom(Ric)$ we need to show that X_1 is invertible, or equivalently that its null space $\mathcal{N}(X_1) = \{0\}$. Consider any $x \in \mathcal{N}(X_1)$. From (C.5) we have that

$$AX_1 + RX_2 = X_1 \Lambda \tag{C.11}$$

Premultiplying by $x^* X_2^*$, postmultiplying by x, and using the fact that $X_2^* X_1$ is Hermitian yields

$$x^* X_2^* A X_1 x + x^* X_2^* R X_2 x = x^* X_2^* X_1 \Lambda x = 0 \Rightarrow x^* X_2^* R X_2 x = 0 \tag{C.12}$$

Since R is semidefinite, it follows that $RX_2 x = 0$. Postmultiplying (C.11) by x yields

$$0 = AX_1 x + RX_2 x = X_1 \Lambda x \tag{C.13}$$

Thus if $x \in \mathcal{N}(X_1)$, then $\Lambda x \in \mathcal{N}(X_1)$. To show that $\mathcal{N}(X_1) = 0$, assume to the contrary that $\mathcal{N}(X_1) \neq 0$. Then $\Lambda|_{\mathcal{N}(X_1)}$ has at least an eigenvalue λ, $\text{Re}(\lambda) < 0$. Denote by x a corresponding eigenvector. Premultiplying (C.5) by $[0 \ I]$ and postmultiplying by x yields

$$-QX_1 x - A^T X_2 x = X_2 \Lambda x = \lambda X_2 x \Rightarrow x^* X_2^* (A + \lambda^* I) = 0 \tag{C.14}$$

Since $RX_2 x = 0$ this is equivalent to

$$x^* X_2^* [A + \lambda^* I \quad R] = 0 \tag{C.15}$$

Since (A, R) is stabilizable by assumption, it follows from the PBH test that $X_2 x = 0$. Thus $\mathcal{X} x = 0$, which implies (since $\mathcal{X}$ has full column rank) that $x = 0$.

($\Rightarrow$) If $H \in dom(Ric)$ then from Lemma C.1 $A + RX$ is stable. It follows that (A, R) is stabilizable. $\square$

Lemma C.3 *Assume that H has the form*

$$H = \begin{bmatrix} A & -BB^T \\ -C^T C & -A^T \end{bmatrix} \tag{C.16}$$

Then $H \in dom(Ric)$ and $X = Ric(H) \geq 0$ if and only if the pair (A, B) is stabilizable and (A, C) does not have any unobservable modes on the $j\omega$ axis. Moreover, if (A, C) is observable then $X > 0$.

Proof. If H does not have any eigenvalues on the $j\omega$ axis the first part of the proof follows immediately from Lemma C.2 by noting that stabilizability of (A, B) implies stabilizability of (A, BB^T). Thus, to complete the first part of the proof, we only need to show that H does not have eigenvalues on the imaginary axis. Assume to the contrary that there exist $x = [x_1^T \ x_2^T]^T \neq 0$ such that

$$\begin{bmatrix} A & -BB^T \\ -C^TC & -A^T \end{bmatrix} \begin{bmatrix} x_1 \\ x_2 \end{bmatrix} = j\omega \begin{bmatrix} x_1 \\ x_2 \end{bmatrix} \tag{C.17}$$

Premultiplying by $[x_2^* \ x_1^*]$ and rearranging we obtain

$$x_2^*(A - j\omega I)x_1 - x_2^*BB^Tx_2 - x_1^*C^TCx_1 - x_1^*(A - j\omega I)^*x_2 = 0 \tag{C.18}$$

or, equivalently,

$$2j\mathbb{I}\mathrm{m}\,[x_2^*(A - j\omega I)x_1] - x_2^*BB^Tx_2 - x_1^*C^TCx_1 = 0 \tag{C.19}$$

It follows that $x_2^*(A - j\omega I)x_1$ is real and that $B^Tx_2 = 0$ and $Cx_1 = 0$. Combining these equations with (C.17) yields

$$x_2^*[A - j\omega I \quad B] = 0 \tag{C.20}$$

$$\begin{bmatrix} A - j\omega I \\ C \end{bmatrix} x_1 = 0 \tag{C.21}$$

Since (A, B) is stabilizable, from (C.20) it follows that $x_2 = 0$. Similarly from (C.21) and the assumption that (A, C) does not have unobservable modes on the $j\omega$ axis, it follows that $x_1 = 0$. Thus $x = 0$ against the assumption. This establishes that H does not have $j\omega$-axis eigenvalues and thus $H \in dom(Ric)$ and $X = Ric(H)$ is well defined. To show that $X \geq 0$ consider the Riccati equation associated with H:

$$\begin{aligned} A^TX + XA - XBB^TX + C^TC &= 0 \\ &\iff \tag{C.22} \\ (A - BB^TX)^TX + X(A - BB^TX) + XBB^TX + C^TC &= 0 \end{aligned}$$

From Lemma C.1 we have that $A - BB^TX$ is stable. Thus the solution to (C.22) is given by

$$X = \int_0^\infty e^{(A-BB^TX)^Tt}(XBB^TX + C^TC)e^{(A-BB^TX)t}\,dt \geq 0 \tag{C.23}$$

Finally, we show that if (A, C) is observable then $X > 0$. Assume to the contrary that $\mathcal{N}(X) \neq 0$ and consider $x \in \mathcal{N}(X), x \neq 0$. Pre- and postmultiplying (C.22) by x^T and x, respectively, we have that $Cx = 0$. Using this fact and postmultiplying (C.22) by x we have that $XAx = 0$. Continuing along these lines we obtain $CAx = 0$, $CA^2x = 0, \ldots, CA^{n-1}x = 0$, which contradicts the observability of (A, C). □

Lemma C.4 *Assume that D has full column rank and let $R = D^T D > 0$. The matrix*

$$\begin{bmatrix} A - j\omega I & B \\ C & D \end{bmatrix} \tag{C.24}$$

has full column rank for all ω if and only if the pair $(A - BR^{-1}D^T C, C - DR^{-1}D^T C)$ has no unobservable modes on the $j\omega$ axis.

Proof. ($\Rightarrow$) Suppose that there exist $\omega, x \neq 0$ such that

$$\begin{bmatrix} A - BR^{-1}D^T C - j\omega I \\ C - DR^{-1}D^T C \end{bmatrix} x = 0 \tag{C.25}$$

This is equivalent to

$$\begin{bmatrix} A - j\omega I & B \\ C & D \end{bmatrix} \begin{bmatrix} I & 0 \\ -R^{-1}D^T C & I \end{bmatrix} \begin{bmatrix} x \\ 0 \end{bmatrix} = 0 \tag{C.26}$$

which contradicts the hypothesis.

($\Leftarrow$) Assume that there exist ω, $x = [x_1^T \ x_2^T]^T$ such that

$$\begin{bmatrix} A - j\omega I & B \\ C & D \end{bmatrix} \begin{bmatrix} x_1 \\ x_2 \end{bmatrix} = 0 \tag{C.27}$$

Define

$$\begin{bmatrix} y \\ z \end{bmatrix} \triangleq \begin{bmatrix} I & 0 \\ R^{-1}D^T C & I \end{bmatrix} \begin{bmatrix} x_1 \\ x_2 \end{bmatrix} \tag{C.28}$$

Then

$$\begin{bmatrix} A - BR^{-1}D^T C & B \\ C - DR^{-1}D^T C & D \end{bmatrix} \begin{bmatrix} y \\ z \end{bmatrix} = j\omega \begin{bmatrix} y \\ 0 \end{bmatrix} \tag{C.29}$$

Premultiplying this last equation by $[0 \ D^T]$ we get $z = 0$. Thus (C.29) is equivalent to

$$\begin{bmatrix} A - j\omega - BR^{-1}D^T C \\ C - DR^{-1}D^T C \end{bmatrix} y = 0 \tag{C.30}$$

that is, the pair $(A - BR^{-1}D^T C, C - DR^{-1}D^T C)$ has an unobservable mode on the $j\omega$ axis. □

Corollary C.1 *Assume that H has the form*

$$H = \begin{bmatrix} A - BR^{-1}D^TC & -BR^{-1}B^T \\ -C^T(I - DR^{-1}D^T)C & -(A - BR^{-1}D^TC)^T \end{bmatrix} \quad (C.31)$$

where D has full column rank and $R = D^TD$. Then $H \in dom(Ric)$ and $X = Ric(H) \geq 0$ if and only if (A, B) is stabilizable and the matrix (C.24) has full column rank for all ω. If, in addition, $(A - BR^{-1}D^TC, C - DR^{-1}D^TC)$ is observable then $X > 0$.

INDEX

a posteriori information, 325
a priori information, 324, 327
ACC Benchmark Problem, 153
achievable closed loop mappings, 79
actuator uncertainty, 55, 101, 102, 104
adjoint operator, 163, 298
affine dependence, 228
algebraic Riccati equation, 481
algebraic Riccati inequality (ARI), 181
aliasing, 326, 350, 356, 373
all pass embedding, 197
all pass system, 88, 197
μ-analysis, 53
annihilator, 256, 459
anticausal, 351, 451, 453
antistable, 453
Arzela–Ascoli theorem, 337
atomic measure, 282
attitude control, 377

balanced realization, 7, 312, 313, 478
balanced truncation, 310, 314, 318, 321
Banach space, 449, 452, 457
bandwidth, 38, 39, 326, 395
Bezout identity, 80
bilinear transformation, 204, 305, 471
Bode plot, 105, 394
boundary crossing theorem, 211, 223
bounded real lemma, 20, 164, 181, 476
Bryson's rule, 129
Butterworth configuration, 154

Cauchy sequence, 449
Cauchy–Schwartz inequality, 478
causal, 353, 451, 453
characteristic polynomial, 56, 57, 210, 212, 215, 223, 405, 406
cheap control, 142, 145
Chebyshev center, 334, 357
classical control, 5, 19, 29, 42, 93

co-inner system, 88
complementary sensitivity function, 15, 36, 38, 39, 49
condition number, 20, 55, 93, 103, 104, 303, 463
conjugate system, 467
consistency, 326, 344
controllability gramian, 297, 318
ℓ^1 control, 19, 20, 247
   MIMO case, 266
   SISO case, 254
   rational approximation, 285
$\mathcal{H}_2$ control, 127, 133
   disturbance feedforward, 138
   full control, 138
   full information, 136
   output estimation, 139
$\mathcal{H}_\infty$ control, 158
   full control, 172
   full information, 168
   LMI approach, 180
   central controller, 160
   disturbance feedforward, 173
   output estimation, 175
convex hull, 228, 233, 234
coprime factorizations, 80
   double coprime, 82
   left coprime, 80
   right coprime, 80

DC-to-DC converter, 407
   control characteristics, 411
   small signal model, 409
detectability, 294
diameter of information, 331, 335, 357
discrete time, 212, 306, 308, 340, 470, 476, 477
distillation, 2, 104
disturbance, 3, 4, 8, 23, 42
disturbance feedforward problem, 70

disturbance rejection, 127
$dom(Ric)$, 482
dual space, 255, 457
  of ℓ^∞, 458
  of ℓ^p, 457
  of c_o, 458
dual system, 74
duality, 20, 282, 299, 315, 457

edge theorem, 226, 227
eigenvalue, 314, 316, 318, 462
eigenvector, 309, 314
Euler approximating system, 285
experimental phase, 2

feedback linearization, 7, 109
field, 445, 446
finite impulse response, 350
four blocks problem, 197
Fourier transform, 349, 451, 453
Frobenius norm, 311
full control problem, 74
full information problem, 69
functional, 457
  extension of, 458
  linear, 457, 458

gain margin, 8, 23, 29, 30, 56
gain scheduling, 7, 9
generalized $\mathcal{H}_2$ control, 149
generalized inverse, 462

Hahn Banach's theorem, 459, 460
Hamiltonian matrix, 481
Hankel model reduction, 369
Hankel norm, 164, 305, 309, 369
Hankel operator, 162, 163, 166, 305, 308
Hankel singular value, 164, 309
Hermite–Biehler's theorem, 231
Hilbert space, 449, 458
Hurwitz, 231

identification
  ℓ^1, 331, 361, 373
  $\mathcal{H}_\infty$, 331, 342, 361, 369
  algorithm, 326, 329
    frequency point, 343
    interpolatory, 346, 374
    untuned, 338, 362
  classical, 323
  control oriented, 52, 324
  error, 328
    global, 329, 334
  frequency domain, 339

  non-uniformly spaced points, 372
  time domain, 361
image, 301, 467
infimum, 367, 450, 459
inner product, 448, 449, 453
inner system, 88, 130
input normal realization, 313
input sensitivity, 113
input uncertainty, 20, 104
input–output stability, 26, 27, 57, 95, 99
interlacing property, 231
internal model control, 12
internal stability, 26, 27, 64, 95, 97, 213, 242, 385
interpolation constraints
  rank, 266
  zero, 266
intersample behavior, 337
interval plant, 228
invariant set, 268
isometric, 453
isomorphism, 452

Kalman's inequality, 143
kernel, 298, 336, 353, 445, 453
Kharitonov's theorem, 230, 231

linear combination, 446
linear fractional transformation, 19, 57, 95, 384, 470
linear operator, 454
linear quadratic regulator, 5, 129
linear vector space, 446–448
linearization, 7, 33, 380
loop gain, 31
loop shaping, 16, 93, 105, 119
loop shifting, 77
loop transfer matrix, 108, 110
loop transfer recovery, 124, 148
Lyapunov equation, 306, 314
Lyapunov function
  polyhedral, 288
  quadratic, 288

mapping theorem, 225, 233, 234, 406
matrices
  left coprime , 80
  right coprime, 80
matrix square root, 319
maximum modulus theorem, 37, 99
McMillan degree, 306–308, 477
measurement noise, 15, 16, 49, 95, 323, 378, 385, 470
minimal realization, 296, 313, 321

minimum norm problem, 256, 458, 459
minimum phase property of LQR controllers, 145
ℓ^1 model matching, 246
$\mathcal{H}_\infty$ model-matching, 203
model reduction, 21, 293, 295, 304, 310, 373, 470
model uncertainty, 9, 11, 13, 29, 32, 41, 94, 109, 207, 295, 310, 323, 334, 471
multiblock problem, 267
multilinear dependence, 227–229, 406
multiobjective control, 193

Nehari approximation problem, 195, 347
Nehari's theorem, 20, 477
Nevanlinna–Pick Interpolation, 204, 347, 357
nominal performance, 42, 46, 97, 99, 110, 220–222, 243, 385, 473
nominal stability, 44, 97, 99, 110, 125
norm, 300, 310
    euclidean, 311
    induced, 45, 50, 305, 453, 454
        $\mathcal{L}^2 \to \mathcal{L}^2$, 455
        $\mathcal{L}^\infty \to \mathcal{L}^\infty$, 455
    matrix, 57, 300, 311
    operator, 50
ℓ^1 norm
    computation, 476
    definition, 457
    upper bound, 268
$\mathcal{H}_2$ norm, 456
    computation, 475
$\mathcal{H}_\infty$ norm, 46, 307, 310, 318
    computation, 476
    definition, 452
⋆ norm
    minimization via output feedback, 276
    minimization via state feedback, 273
    definition, 273
    LMI analysis, 270
normed space, 447, 459
NP-hard, 215, 227, 237, 323
null space, 297

observability Gramian, 297, 318
observer, 467
observer based controller, 78, 83, 160
offending set, 212
one-block problem, 197, 266
operator, 45, 297, 298, 301, 445, 450, 453, 454, 478
    causal, 247
    strictly causal, 247
optimal Hankel norm approximation, 310

origin reachable set, 268
orthogonal projection, 458
output estimation problem, 74
output feedback problem, 75
output injection, 151
output normal realization, 313
output sensitivity, 95

Padé approximation, 399
parity interlacing property, 91
Parseval's theorem, 45, 453
partition, 211, 236, 315, 321
performance weight, 108, 384, 398
persistent disturbance, 246
phase margin, 23, 29, 30, 56
Pick matrix, 204, 344, 358
plant condition number, 108
pole–zero cancellation, 64
polynomial dependence, 229, 235
polytope, 227–229, 234
power signal, 456
prefilter, 123
principal components, 20, 297, 300
process control, 389, 400
projection operator, 162
projection theorem, 458
proper, 305, 320, 321, 474, 476
    strictly, 469
pseudo-inverse, 298, 299

Q parametrization, 63
quadratic dependence, 406
quadratic stability, 206

radius of information, 341, 343, 356
range, 445, 453
reduced order controllers, 192
regulator, 128
$Ric(H)$, 482
robust identification, 18
robust performance, 19, 23, 50, 51, 100, 219–221, 385
robust stability, 18, 19, 29, 42, 98, 99, 211, 212, 227, 234, 310, 320, 473
Routh–Hurwitz, 229, 236

saddle point condition, 161
sample complexity, 373
sampling, 307, 342, 344, 385
Schmidt pair, 163, 196
sensitivity function, 15, 41, 45, 113, 397
sensor uncertainty, 101
separation structure, 78, 141
simulation phase, 3

simultaneous stabilization, 80, 91
singular problems, 142
singular value, 57, 98, 106–108, 216, 294, 305, 309, 311, 454, 460, 462
singular value decomposition, 99, 311, 353, 462
singular vector, 369, 456
small gain theorem, 249, 253
Smith Predictor, 389, 391, 393
span, 316, 461
spectral factorization, 200
spectral radius, 216, 218
stability
    ℓ^∞, 248
    ℓ^p, 247
    BIBO, 248
stability margin, 32, 41, 42, 210, 214, 224, 227, 341
    multivariable, 214, 236, 400
    of LQR controllers, 144
    of optimal $\mathcal{H}_2$ controllers, 146
stabilizability, 294
stabilization, 34, 35, 43
state feedback, 7, 314, 381, 467, 479
state space realization, 27, 236, 294, 307, 465, 474
stochastic, 8, 223
stochastic methods, 323
strong stabilization, 80, 91
structured singular value, 20, 53, 214–216, 220, 229, 339, 378, 399
subspace, 295, 296, 301, 451, 458
supremum, 96, 333, 450, 460
symmetric set, 332
μ-synthesis, 53, 398

theoretical phase, 5
time delay, 33, 389, 391–393

time varying, 20
Toeplitz operator, 166
trace, 300, 311
transfer matrix, 16
tree structured decomposition, 229

uncertainty
    additive, 224
    block diagonal, 96
    gain, 42, 396
    global dynamic, 32, 208, 216, 221, 238, 310
    LTV, 250
    mixed parametric–structured dynamic, 239
    multiplicative, 40, 50, 320
    NLTI, 253
    parametric, 20, 209, 210, 213, 223, 236, 238, 323, 394
    rank 1, 226
    structured, 53, 54, 207, 210
    unstructured, 96
uncertainty region, 32, 52
unitary, 303, 311, 456, 461, 463, 471
unitary matrix, 297

validation, 338, 339

well posedness, 24, 25, 63, 212, 249, 475
Wiener-Hopf, 124
worst case, 21, 97, 209, 323, 325, 327

X-29 aircraft, 4, 237, 289, 400, 402

Youla parametrization, 63, 80, 193
    general case, 68
    open-loop stable plants, 65